Advanced Topics in Nonsmooth Dynamics

Remco I. Leine · Vincent Acary
Olivier Brüls
Editors

Advanced Topics in Nonsmooth Dynamics

Transactions of the European Network for Nonsmooth Dynamics

Springer

Editors
Remco I. Leine
Institute for Nonlinear Mechanics
University of Stuttgart
Stuttgart
Germany

Vincent Acary
Inria Grenoble—Rhône-Alpes Research Centre
Saint Ismier Cedex
France

Olivier Brüls
Department of Aerospace and Mechanical Engineering
University of Liège
Liège
Belgium

ISBN 978-3-030-09368-6 ISBN 978-3-319-75972-2 (eBook)
https://doi.org/10.1007/978-3-319-75972-2

This Springer imprint is published by the registered company Springer Nature Switzerland AG
The registered company address is: Gewerbestrasse 11, 6330 Cham, Switzerland

Preface

Many open scientific questions and engineering problems in nonlinear dynamics concern dynamical systems with some degree of nonsmoothness or switching behaviour. Switching systems, hybrid systems or, more generally, nonsmooth systems arise in very different disciplines, such as the science of cyber-physical systems, neuroscience and biomathematics and control and electrical circuits theory. Nonsmooth models are abundant in mechanics and related engineering applications. Dry friction and impact laws lead to nonsmooth models in multibody dynamics, whereas switching control laws result in a nonsmooth closed-loop dynamics. Classical theoretical results in nonlinear dynamics, as well as the conventional numerical simulation and optimization algorithms, presuppose a sufficiently smooth behaviour and often fail when applied to nonsmooth systems. Nonsmooth systems, therefore, open up a Pandora's box of unresolved questions and pose a demanding challenge to mechanics and other disciplines, leading to a new research field. The Nonsmooth Dynamics research field studies dynamical systems, for which the state is not required to be a smooth (differentiable or continuous) function of time. Nonsmooth Dynamics finds its roots in nonlinear dynamics with strong connections to stability theory, bifurcation theory and chaos. It uses concepts of nonsmooth analysis, convex analysis, measure theory and nonsmooth optimization (complementarity and variational inequality theory) for the modeling, analysis, simulation, control and design of nonsmooth systems. The study of nonsmooth systems is receiving an increasing amount of attention in the scientific literature, conferences and symposia.

The European Network for Nonsmooth Dynamics (ENNSD, http://ennsd.gforge.inria.fr/) is an initiative that aims to provide a platform for collaborations, symposia, summer schools and workshops on nonsmooth dynamics. The network unites about 40 active specialists in the field of Nonsmooth Dynamics from 9 different countries. It also contributes to the promotion and dissemination of Nonsmooth Dynamics theories and methods throughout the scientific community and to the training of young researchers and newcomers. The symposia of the European Network for Nonsmooth Dynamics took place at ETH in Zürich (2012), INRIA in Grenoble (2013), TUM in Munich (2014), LMGC in Montpellier (2015), University of Liège

(2016), Eindhoven University of Technology (2017) and University of Stuttgart (scheduled for 2018).

The aim of this book is twofold. Firstly, as the title reflects, this edited volume gathers a selection of original research contributions by experts in the field of Nonsmooth Dynamics. Secondly, it plays the role of transactions of the European Network for Nonsmooth Dynamics by documenting some of the scientific knowledge that has been gathered by the network over recent years.

The book covers modeling, analysis, simulation and control of nonsmooth systems. Each chapter is written such that researchers from other research fields can be introduced to the topic.

The editors wish to thank all of the members of ENNSD for their valuable contributions to Nonsmooth Dynamics.

Stuttgart, Germany — Remco I. Leine
Grenoble, France — Vincent Acary
Liège, Belgium — Olivier Brüls
November 2017

Contents

Comparisons of Multiple-Impact Laws For Multibody Systems: Moreau's Law, Binary Impacts, and the LZB Approach

Ngoc Son Nguyen and Bernard Brogliato

Abstract This chapter is dedicated to comparisons of three well-known models that apply to multiple (that is, simultaneous) collisions: Moreau's law, the binary collision law, and the LZB model. First, a brief recall of these three models and the way in which their numerical implementation is done. Then, an analysis based on numerical simulations, in which the LZB outcome is considered as the reference outcome, is presented. It is shown that Moreau's law and the binary collision model possess good prediction capabilities in some few "extreme" cases. The comparisons are made for free chains of aligned grains, and for chains impacting a wall. The elasticity coefficient, coefficients of restitution, mass ratios and contact equivalent stiffnesses are used as varying parameters.

1 Introduction

Multiple impacts are very complex phenomena occurring frequently in multibody systems. Roughly speaking, a *multiple impact* occurs in a multibody system each time the system undergoes several collisions at the same time t_{imp}. In models based on the assumption that the bodies are perfectly rigid at contact, and such that the impacts are instantaneous phenomena, the definition of t_{imp} is clear. When deformations occur, one may consider that an impact is *multiple* whenever the collisions at the m contact/impact points i, which have non-zero durations $[t_{0,i}, t_{f,i}]$ (with $t_{f,i} = +\infty$ for some models –think of an overdamped linear spring-dashpot [1, Sect. 2.1] [2]), overlap, and consequently may influence each other due to dynamic couplings between the various contact points. One subtlety in the definition of a multiple impact is that

N. S. Nguyen (✉)
GeM Institute, University of Nantes, 58 rue Michel Ange, BP 420,
44606 Saint-Nazaire Cedex, France
e-mail: ngocson.nguyen@univ-nantes.fr

B. Brogliato
INRIA Grenoble-Rhône Alpes, Univ. Grenoble-Alpes, Laboratoire
Jean Kuntzman, 655 Avenue de l'Europe, 38334 Saint-Ismier, France
e-mail: bernard.brogliato@inria.fr

R. I. Leine et al. (eds.), *Advanced Topics in Nonsmooth Dynamics*,
https://doi.org/10.1007/978-3-319-75972-2_1

some previously active contacts with zero local relative velocity, may participate in it. This is the case for the two well-known classical systems: chains of aligned balls (like Newton's cradle), in which several balls are in contact before the shock, or the planar rocking block that rotates around one corner. In both cases, one is obliged to take the previously lasting contacts into account, even if the multiple impact is triggered at a single contact. In a Lagrange dynamics framework with generalized coordinates $\mathbf{q}$, impacts are associated with unilateral constraints, which are defined from p gap functions $f_i(\mathbf{q})$ (signed distances) that define an admissible domain Φ for the generalized position, i.e., $\mathbf{q}(t) \in \Phi$ for all $t \geq 0$. Impacts correspond to trajectories hitting the boundary of Φ (denoted bd(Φ)) with a non-zero normal velocity, i.e., $\nabla^T f_i(\mathbf{q}(t))\dot{\mathbf{q}}(t^-) < 0$ if $f_i(\mathbf{q}(t)) = 0$. In most cases, bd(Φ) consists of co-dimension $p' \leq p$ submanifolds $\{\mathbf{q} \in \mathscr{C} \mid f_i(\mathbf{q}) = 0, \text{ for some } 1 \leq i \leq p\}$, of the configuration space $\mathscr{C} \ni \Phi$. When a co-dimension p' boundary submanifold is attained with $p' \geq 2$ (a kind of singularity of bd(Φ) where two smooth hypersurfaces intersect), one speaks of a p'-impact. For instance, the 2-dimensional rocking block with concave base and two corners undergoes a 2-impact during a classical rocking motion. Consider a chain of n aligned spheres, in which one sphere at one end of the chain hits the other $n-1$ ones that are at rest and in contact with no pre-constraint: this is an $n-1$-impact.

Just as for single impacts, several classes of contact/impact models can be used in multiple impacts [3]:

- **(i)** Algebraic models that relate post and pre-impact velocities as $\dot{\mathbf{q}}(t^+) = \mathscr{F}(\dot{\mathbf{q}}(t^-))$ for some function $\mathscr{F}$, which may be explicitly or implicitly defined.
- **(ii)** First-order dynamics following the Darboux-Keller approach [1, Sect. 4.3.5]: positions are assumed constant, and the impact force impulse is used as the new time scale.
- **(iii)** Second-order dynamics that use rheological compliant models with lumped flexibility, like spring-and-dashpot linear (Kelvin-Voigt, Maxwell, Zener) or nonlinear models (Kuwabara-Kono, Simon-Hunt-Crossley, etc.), Discrete Element Method (DEM), or Finite Element Method (FEM).

All models have some advantages and drawbacks. It is not our objective in this chapter to classify or to rank models. Rather, we consider three well-known models that belong to classes **(i)** and **(ii)**, and we compare them in terms of their velocity outcomes, on the benchmark of chains of aligned balls. The results therefore complete those shown in [3, Chap. 6], which is restricted to chains of three aligned balls. Our results also indicate the instances in which Moreau's and the binary laws may provide realistic outcomes. Since multiple impacts in chains of balls are essentially determined by the nonlinear waves that travel through the chain, we pay attention to characterize, when possible, the waves associated with the domains of applicability of these two impact laws.

Remark 1 In this work, we restrict ourselves to frictionless constraints.

Remark 2 Multiple impacts are therefore intrinsically different from infinite sequences of single impacts with an accumulation, like in the bouncing ball system.

However, some approaches for multiple impacts may yield some kind of infinite sequence of impacts, sometimes instantaneously (this may occur, for instance, in the binary collision model, or with the so-called Han-Gilmore algorithm [1, Sect. 6.1.2], which is not always guaranteed to converge in a finite number of steps, or to converge to a unique solution [3, Sect. 3.4]). This is closely related to another feature of multiple impacts, that is, the possible discontinuity of trajectories with respect to the initial data [1, 4].

2 System's Dynamics

In this chapter, we mainly deal with chains of n aligned balls (or more generally, aligned grains not necessarily spherical) with radii $R_i > 0$ whose dynamics is as follows:

$$\begin{cases} \mathbf{M}\ddot{\mathbf{q}}(t) = \boldsymbol{\Lambda}(t) \\ f_i(\mathbf{q}) = q_{i+1} - q_i - (R_{i+1} + R_i) \geq 0, \ 1 \leq i \leq n-1 \\ \mathbf{M} = \mathrm{diag}(m_i), \ 1 \leq i \leq n, \end{cases} \tag{1}$$

where $\mathbf{q} = (q_1, q_2, \ldots, q_n)^T$ is the generalized coordinates of the chain and $\boldsymbol{\Lambda}(t) \in I\!R^n$ is the vector of generalized contact forces between the balls. The gap functions $f_i(\mathbf{q})$ are signed distances between adjacent balls and represent the unilateral constraints in the chain. We have $\boldsymbol{\Lambda} = \nabla \mathbf{f}(\mathbf{q})\boldsymbol{\lambda}$, with $\boldsymbol{\lambda} \in I\!R^{n-1}$ being the vector of Lagrange multipliers associated with the unilateral constraints. We obtain the subsequent equalities that will be useful later:

$$\begin{cases} \nabla^T f_{i+1}(\mathbf{q})\mathbf{M}^{-1}\nabla f_i(\mathbf{q}) = -m_{i+1}^{-1}, \quad \nabla^T f_{i-1}(\mathbf{q})\mathbf{M}^{-1}\nabla f_i(\mathbf{q}) = -m_i^{-1} \\ \nabla^T f_{i-2}(\mathbf{q})\mathbf{M}^{-1}\nabla f_i(\mathbf{q}) = 0, \quad \nabla^T f_i(\mathbf{q})\mathbf{M}^{-1}\nabla f_i(\mathbf{q}) = m_i^{-1} + m_{i+1}^{-1}. \end{cases} \tag{2}$$

In terms of the kinetic angles θ_{ij} between the submanifold (or hypersurfaces) defined by the equalities $f_i(\mathbf{q}) = 0$ and $f_j(\mathbf{q}) = 0$, we obtain (see [1, Eq. (6.66)] for the definition of a kinetic angle), when all masses are equal to $m > 0$:

$$\theta_{i,i+2} = \frac{\pi}{2}, \quad \theta_{i,i+1} = \frac{\pi}{6}. \tag{3}$$

Roughly speaking, and without going into further considerations other than this preliminary geometrical analysis, this means that monodisperse chains of aligned balls may have complex dynamics at impacts because they may not satisfy the conditions that guarantee continuity of trajectories with respect to initial data [4]. As shown in [3, Appendix A], the 3-ball chain is equivalent to a particle in the plane hitting in a corner, whose dynamics may be quite complex [5]. As is well-known, there is another "natural" set of coordinates for the chain, using conservation of linear momentum. Let $z_i = f_i(\mathbf{q})$ for each $1 \le i \le n-1$, and $z_0 = \sum_{i=1}^{n} m_i q_i$. Then, $\ddot{z}_0 = 0$ (by adding the n lines of the dynamics, which just translate Newton's law of action/reaction). We have $\mathbf{z} = \mathbf{N}\mathbf{q} + \mathbf{L}$ for some easily obtained $\mathbf{N} \in I\!R^{n\times n}$ and $\mathbf{L} = (0, R_2 + R_1, \ldots, R_n + R_{n-1})^T$. The $n \times n$ mass matrix becomes in the $\mathbf{z}$ coordinates $\mathbf{N}^{-T}\mathbf{M}\mathbf{N}^{-1} = \begin{pmatrix} 1 & 0 \ldots & 0 \\ 0 & & \\ 0 & \bar{\mathbf{M}} & \\ 0 & & \end{pmatrix}$, with $\bar{\mathbf{M}} = \bar{\mathbf{M}}^T \in I\!R^{(n-1)\times(n-1)}$ positive definite. Let $\bar{\mathbf{z}} = (z_1, \ldots, z_{n-1})^T$; the dynamics in (1) then becomes in a reduced form:

$$\begin{cases} \bar{\mathbf{M}}\ddot{\bar{\mathbf{z}}} = \boldsymbol{\lambda} \\ z_i \ge 0, \quad 1 \le i \le n-1. \end{cases} \tag{4}$$

If all the balls are in contact at the impact time, then $z_i(0) = 0$. Though the dynamics in (4) looks simpler than that in (1), this is not necessarily the case, because $\bar{\mathbf{M}}$ may not be diagonal.

3 The Multiple-Impact Models

In this section, the three models: Moreau's impact law, the binary collision model and the LZB approach are described, and some of their features are analyzed.

3.1 *Moreau's Impact Law*

Moreau's impact law belongs to class (**i**). It is primarily formulated as an extension of Newton's kinematic restitution law, in a Lagrange dynamics framework, and with a *global* coefficient of restitution. Since it can also be expressed in local frames at the contact points, as a linear complementarity problem with the local velocities as unknowns , it is convenient to implement in event-capturing time-stepping schemes. As such, this is the law that is implemented in the software packages SICONOS[1] and LMGC90.[2] It was introduced in [6, 7].

Let us describe it now. We consider a Lagrangian system with generalized coordinates $\mathbf{q} \in I\!R^n$, symmetric positive definite mass matrix $\mathbf{M}(\mathbf{q}) \in I\!R^{n\times n}$, and a set of unilateral constraints $f_i(\mathbf{q}) \ge 0,\ 1 \le i \le m$, defined from the differentiable gap functions $f_i : I\!R^n \to I\!R$, such that $\nabla f_i(\mathbf{q}) \triangleq \left[\frac{\partial f_i}{\partial \mathbf{q}}(\mathbf{q})\right]^T \neq 0$ for all $\mathbf{q}$ such that $f_i(\mathbf{q}) = 0$ (it is assumed that gradients do not vanish on the boundary

[1]http://siconos.gforge.inria.fr/4.1.0/html/index.html.

[2]https://git-xen.lmgc.univ-montp2.fr/lmgc90/lmgc90_user/wikis/home.

of the admissible domain). The non-negative multipliers associated with the unilateral constraints are denoted λ_i, and they are supposed to satisfy complementarity conditions $f_i(\mathbf{q})\lambda_i = 0$. In a compact form, one obtains $\mathbf{0} \leq \boldsymbol{\lambda} \perp \mathbf{f}(\mathbf{q}) \geq \mathbf{0}$, with $\boldsymbol{\lambda} = (\lambda_1, \ldots, \lambda_m)^T, \mathbf{f}(\mathbf{q}) = (f_1(\mathbf{q}), \ldots, f_m(\mathbf{q}))^T$. The right-hand side of the smooth dynamics (outside impacts) is equal to $\Lambda \stackrel{\Delta}{=} \nabla\mathbf{f}(\mathbf{q})\boldsymbol{\lambda}$ with $\mathbf{0} \leq \boldsymbol{\lambda} \perp \mathbf{f}(\mathbf{q}) \geq \mathbf{0}$, which, under some suitable assumptions and using nonsmooth analysis, can be rewritten equivalently as $\boldsymbol{\Lambda}(t) \in -\mathscr{N}_\Phi(\mathbf{q}(t))$, the normal cone being generated by the gradients at the active constraints $f_i(\mathbf{q}) = 0$ (the set of active constraints is denoted as $\mathscr{I}(\mathbf{q})$ in the sequel).

Remark 3 Readers who are not familiar with convex and nonsmooth analysis should simply think of normal and tangent cones as a generalization of normal and tangential subspaces, with normal cones being generated by the gradients of the active constraints on the admissible domain boundary. As we explain next, use of the normal and tangential cones is very useful for understanding particular features of Moreau's impact law, because they provide a clear geometrical picture of the collision process, a point of view that is lost if these tools are not used.

Moreau goes a step further, replacing the normal cone to the admissible domain $\mathscr{N}_\Phi(\mathbf{q})$ with the normal cone to the tangent cone $\mathscr{V}(\mathbf{q}) = \{\mathbf{v} \in I\!R^n | \mathbf{v}^T \nabla f_i(\mathbf{q}) \geq 0$, for all $i \in \mathscr{I}(\mathbf{q})\}$, computed at the right-limit of the velocity, i.e., the following inclusion is proposed: $\boldsymbol{\Lambda}(t) \in -\mathscr{N}_{\mathscr{V}(\mathbf{q}(t))}(\dot{\mathbf{q}}(t^+))$, whose right-hand side we choose to refer to as Moreau's set [1]. We have to assume that $\mathscr{V}(\mathbf{q})$ is non-empty, which may be guaranteed by suitable constraint qualification. We also assume that the pre-impact velocity satisfies $\dot{\mathbf{q}}(t^-) \in -\mathscr{V}(\mathbf{q}(t))$. When no constraints are active, i.e., $\mathscr{I}(\mathbf{q}) = \emptyset$, then one sets $\mathscr{V}(\mathbf{q}) = I\!R^n$. In this case, $\mathscr{N}_{\mathscr{V}(\mathbf{q})}(\cdot) = \{0\}$, as expected (contact forces vanish).

In a more general setting, Moreau's set is computed at $\mathbf{w}(t) \stackrel{\Delta}{=} \frac{\dot{\mathbf{q}}(t^+)+\mathrm{e}\dot{\mathbf{q}}(t^-)}{1+\mathrm{e}}$, where e is a global coefficient of restitution (CoR) (global in the sense that it applies to all the contacts), i.e.: $\boldsymbol{\Lambda}(t) \in -\mathscr{N}_{\mathscr{V}(\mathbf{q}(t))}(\mathbf{w}(t)) \subseteq -\mathscr{N}_\Phi(\mathbf{q}(t))$.[3] One important consequence of using Moreau's set is that, since $\mathscr{V}(\mathbf{q}) \subseteq I\!R^n$ is a convex polyhedral set for velocities (while Φ may be in general non-convex and non-polyhedral), the calculations of the normal cone are doable, as we show next. When an impact occurs at time t, $\boldsymbol{\Lambda}_t = \nabla\mathbf{f}(\mathbf{q}(t))\boldsymbol{\lambda}_t$ is the contact force impulse and the system's dynamics becomes:

$$\mathbf{M}(\mathbf{q}(t))(\dot{\mathbf{q}}(t^+) - \dot{\mathbf{q}}(t^-)) = \nabla\mathbf{f}(\mathbf{q}(t))\boldsymbol{\lambda}_t \in -\mathscr{N}_{\mathscr{V}(\mathbf{q}(t))}(\mathbf{w}(t)). \tag{5}$$

The objective of the above developments may appear obscure to many readers, however, as we show next, they pave the way towards a sound and practical impact law. First of all, one may use a basic result of convex analysis, which

[3]These developments make sense under some well-posedness conditions of the dynamics, which are assumed to hold here. In particular, positions $\mathbf{q}(\cdot)$ are absolutely continuous, velocities $\dot{\mathbf{q}}(\cdot)$ are right continuous of local bounded variations –hence possessing right and left limits everywhere– and accelerations are measures, as well as $\boldsymbol{\lambda}$. See [1, Theorem 5.3] and [8].

states that for a symmetric positive definite matrix $\mathbf{M}$, two vectors $\mathbf{x}$ and $\mathbf{y}$, and a closed non-empty convex set $\mathscr{K}$, $\mathbf{M}(\mathbf{x}-\mathbf{y}) \in -\mathscr{N}_{\mathscr{K}}(\mathbf{x}) \Leftrightarrow \mathbf{x} = \text{proj}_{\mathbf{M}}[\mathscr{K};\mathbf{y}]$, where $\text{proj}_{\mathbf{M}}$ denotes the orthogonal projection in the metric defined by $\mathbf{M}$, i.e.: $\mathbf{x} = \text{argmin}_{\mathbf{z}\in\mathscr{K}} \frac{1}{2}(\mathbf{x}-\mathbf{z})^T\mathbf{M}(\mathbf{x}-\mathbf{z})$. Using this, and after few manipulations, we obtain from (5):

$$\begin{gathered} \mathbf{M}(\mathbf{q}(t))(\dot{\mathbf{q}}(t^+) - \dot{\mathbf{q}}(t^-)) \in -\mathscr{N}_{\mathscr{V}(\mathbf{q}(t))}(\mathbf{w}(t)) \\ \Updownarrow \\ \dot{\mathbf{q}}(t^+) = -\mathrm{e}\dot{\mathbf{q}}(t^-) + (1+\mathrm{e})\text{proj}_{\mathbf{M}(\mathbf{q}(t))}[\mathscr{V}(\mathbf{q}(t)); \dot{\mathbf{q}}(t^-)], \end{gathered} \tag{6}$$

where we used that multiplying both sides of (5) by $\frac{1}{1+\mathrm{e}} > 0$ does not change the right-hand side, which is a cone. Other equivalent formulations exist [1, Eqs. (5.60) (5.61)]. Now, using a corollary of the celebrated Moreau's two cones Lemma [1, Eq. (B.18)], it follows that (6) is equivalent to

$$\dot{\mathbf{q}}(t^+) = \dot{\mathbf{q}}(t^-) - (1+\mathrm{e})\text{proj}_{\mathbf{M}(\mathbf{q}(t))}[\mathscr{N}_{\Phi}(\mathbf{q}(t)); \dot{\mathbf{q}}(t^-)], \tag{7}$$

where, under some constraint qualification (like the so-called Mangasarian-Fromovitz CQ), we can state that $\mathscr{N}_{\Phi}(\mathbf{q})$ is the polar cone to $\mathscr{V}(\mathbf{q})$ (the admissible domain Φ needs not be convex for this). Moreau's law is a global (generalized) law that gives the post-impact velocity in one compact form. The question is then how to compute the projection in a way that is convenient for numerical implementation.

Since the projection is done in the metric defined by $\mathbf{M}(\mathbf{q})$, we can define the (outwards) normal cone as $\mathscr{N}_{\Phi}(\mathbf{q}) = \{\mathbf{w} \in I\!R^n | \mathbf{w} = -\sum_{i\in\mathscr{I}(\mathbf{q})} \lambda_i \mathbf{n}_{\mathbf{q},i},\ \lambda_i \geq 0\}$, with $\mathbf{n}_{\mathbf{q},i} = \frac{\mathbf{M}^{-1}(\mathbf{q})\nabla f_i(\mathbf{q})}{\sqrt{\nabla^T f_i(\mathbf{q})\mathbf{M}^{-1}(\mathbf{q})\nabla f_i(\mathbf{q})}}$, the (inwards) normal vector to the submanifold defined by $f_i(\mathbf{q}) = 0$ in the kinetic metric. It is, however, not trivial to calculate the projection onto a cone in the general case of the kinetic metric. We may start directly from the impact dynamics in (6) to get a more tractable expression. Indeed, Moreau's set can be expressed as $\mathscr{N}_{\mathscr{V}(\mathbf{q}(t))}(\mathbf{w}) = \{\mathbf{z} \in I\!R^n | \mathbf{z} = -\sum_{i\in\mathscr{K}(\mathbf{w})} \lambda_i \nabla g_i(\mathbf{w}),\ \lambda_i \geq 0\}$, with: $g_i(\mathbf{w}) = \mathbf{w}^T\nabla f_i(\mathbf{q})$, $\mathscr{K}(\mathbf{w}) = \{j \in \mathscr{I}(\mathbf{q}) | g_j(\mathbf{w}) = 0\} \subseteq \mathscr{I}(\mathbf{q})$. Thus, $\mathscr{I}(\mathbf{q})$ collects indices of active position constraints, while $\mathscr{K}(\mathbf{w})$ collects indices from active velocity constraints inside position active constraints. We see at once that Moreau's set implies a two-stage process: first look at positions, second look at velocities. In more mathematical language, there is a lexicographical inequality imposed at the contact local kinematics. Notice that we can equivalently rewrite $\mathscr{N}_{\mathscr{V}(\mathbf{q}(t))}(\mathbf{w}) = \{\mathbf{z} \in I\!R^n | \mathbf{z} = -\sum_{i\in\mathscr{I}(\mathbf{q})} \lambda_i \nabla g_i(\mathbf{w}),\quad 0 \leq \lambda_i \perp g_i(\mathbf{w}) \geq 0\}$, and we have $\nabla g_i(\mathbf{w}) = \left[\frac{\partial g_i}{\partial \mathbf{w}}(\mathbf{w})\right]^T = \nabla f_i(\mathbf{q}) = \left[\frac{\partial f_i}{\partial \mathbf{q}}(\mathbf{q})\right]^T$. Then, we obtain:

$$\begin{cases} \mathbf{M}(\mathbf{q}(t))(\dot{\mathbf{q}}(t^+) - \dot{\mathbf{q}}(t^-)) = \sum_{i\in\mathscr{I}(\mathbf{q})} \lambda_i \nabla f_i(\mathbf{q}) \\ \\ 0 \leq \lambda_i \perp g_i(\mathbf{w}) = \nabla^T f_i(\mathbf{q})\mathbf{w} \geq 0. \end{cases} \tag{8}$$

In this approach, the multiplier λ_i has to be interpreted as the contact force impulse at time t, i.e., $\lambda_i = \lambda_{t,i}$. Let $\mathscr{I}(\mathbf{q}) = \{i_1, \ldots, i_l\}$, and denote $\mathbf{f}_{\mathscr{I}(\mathbf{q})}(\mathbf{q}) \stackrel{\Delta}{=} (f_{i_1}(\mathbf{q}), f_{i_2}(\mathbf{q}), \ldots, f_{i_l}(\mathbf{q}))^T$, so that $\nabla \mathbf{f}_{\mathscr{I}(\mathbf{q})}(\mathbf{q}) = (\nabla f_{i_1}(\mathbf{q}), \ldots, \nabla f_{i_l}(\mathbf{q})) \in I\!R^{l \times n}$. In the same way, we denote $\boldsymbol{\lambda}_{t,\mathscr{I}(\mathbf{q})} = (\lambda_{t,i_1}, \ldots, \lambda_{t,i_l})^T$, and $\mathbf{U}_{\mathrm{n},\mathscr{I}(\mathbf{q})} = (U_{\mathrm{n},i_1}, \ldots, U_{\mathrm{n},i_l})^T$, with $\mathbf{U}_{\mathrm{n},i} \stackrel{\Delta}{=} \nabla^T f_i(\mathbf{q})\dot{\mathbf{q}}$ being the normal local velocity at contact i. From (8), and using the expression of $\mathbf{w}(t)$, we obtain:

$$\begin{array}{|l|}\hline \mathbf{U}_{\mathrm{n},\mathscr{I}(\mathbf{q})}(t^+) - \mathbf{U}_{\mathrm{n},\mathscr{I}(\mathbf{q})}(t^-) = \mathbf{D}_{\mathscr{I}(\mathbf{q})}(\mathbf{q})\, \boldsymbol{\lambda}_{t,\mathscr{I}(\mathbf{q})} \\ \mathbf{0} \le \boldsymbol{\lambda}_{t,\mathscr{I}(\mathbf{q})} \perp \mathbf{U}_{\mathrm{n},\mathscr{I}(\mathbf{q})}(t^+) + \mathscr{E}_{\mathrm{nn}}\mathbf{U}_{\mathrm{n},\mathscr{I}(\mathbf{q})}(t^-) \ge \mathbf{0} \\ \mathbf{D}_{\mathscr{I}(\mathbf{q})}(\mathbf{q}) = \nabla^T \mathbf{f}_{\mathscr{I}(\mathbf{q})}(\mathbf{q})\mathbf{M}(\mathbf{q})^{-1}\nabla \mathbf{f}_{\mathscr{I}(\mathbf{q})}(\mathbf{q}), \\ \hline \end{array} \tag{9}$$

with $\mathscr{E}_{\mathrm{nn}} = \mathrm{diag(e)}$. This form of Moreau's impact law for normal local velocities is very interesting, because it takes the form of a Mixed Linear Complementarity Problem (MLCP), which is numerically tractable. See [1, Lemma 5.2, Corollary 5.1] for existence and uniqueness of solutions to this MLCP. It may be seen as a generalized Newton's impact law, however, it is worth noting that it is not a mere application of Newton's law at each active contact. Indeed, there is a complementarity condition and inertial couplings through the Delassus' matrix $\mathbf{D}_{\mathscr{I}(\mathbf{q})}(\mathbf{q}) \in I\!R^{l \times l}$. We see from (9) that Moreau's law is kinetically consistent (non-negative impulse). If $\mathrm{e} \in [0, 1]$, it is also energetically consistent [1, Eq. (5.61)], and it can be shown to be kinematically consistent as well (admissible post-impact velocities) using (6). Indeed, we obtain

$$\dot{\mathbf{q}}(t^+) = \mathrm{proj}_{\mathbf{M}(\mathbf{q}(t))}[\mathscr{V}(\mathbf{q}(t)); \dot{\mathbf{q}}(t^-)] + \mathrm{e}\{-\dot{\mathbf{q}}(t^-) + \mathrm{proj}_{\mathbf{M}(\mathbf{q}(t))}[\mathscr{V}(\mathbf{q}(t)); \dot{\mathbf{q}}(t^-)]\}. \tag{10}$$

The three terms of the right-hand side belong to $\mathscr{V}(\mathbf{q}(t))$, and since $\mathrm{e} \ge 0$, the post-impact velocity also belongs to the tangent cone (a convex cone being closed under addition), and is thus admissible.

Actually, though it is not particularly useful from the calculation point of view, the expression in (6) or in (7) is valuable for visualizing how Moreau's law works from simple geometrical arguments in the plane, as illustrated in Fig. 1. This figure demonstrates that the outcome of Moreau's law is strongly influenced by the (kinetic) angle between the constraints (denoted α in the figure). This is the reason why it can possess good predictability in the case of multiple impacts when waves play a negligible role, but the system's geometry is crucial. For instance, planar rocking blocks follow this intuitive rule: slender blocks have a kinetic angle $\ge \frac{\pi}{2}$ and are likely to rock more easily than flat (or stocky) blocks that have a kinetic angle $\le \frac{\pi}{2}$ [1, Remark 6.10]. This is confirmed in [9, 10], where tangential effects are added to prevent sliding of the block. Contrastingly, in chains of balls, the wave propagation is a crucial mechanical effect that is mainly ruled by contact flexibilities. A purely kinematic impact law that does not contain any information on contact stiffnesses

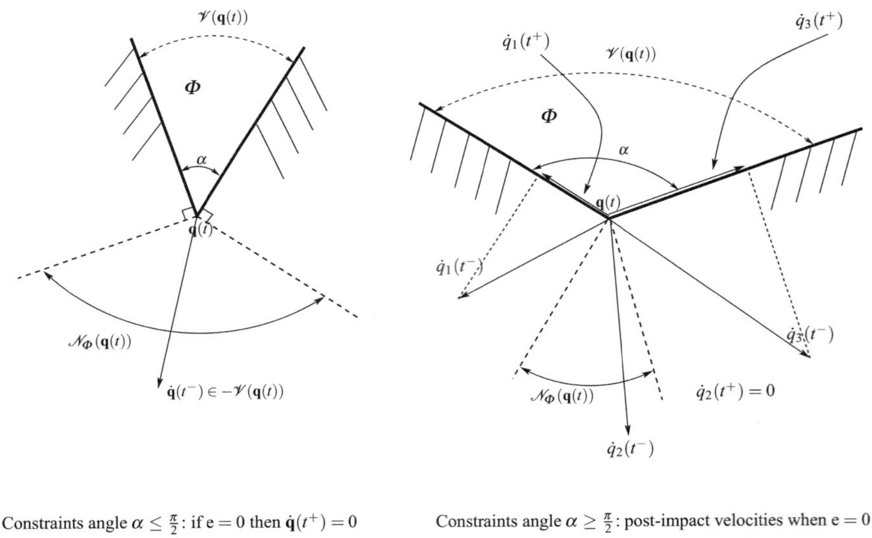

Fig. 1 Moreau's law and constraints angle (planar case)

will, in most cases, fail to predict the outcome correctly. It may, however, in some very particular cases, provide good results, as shown in Sect. 5.

Remark 4 Other kinematic impact laws have been proposed and studied in the literature [11–18], or using a Poisson coefficient and a two-stage linear complementarity problem [19]. It would be worth studying them along the same lines as done in Sect. 5. This is left for future work. Notice, however, that as shown in [20], Poisson-Pfeiffer-Glocker and Moreau's law are equivalent when a unique global CoR is used, though in general, Poisson's hypothesis yields multiple impact laws with a larger post-impact velocity set than Moreau's law [3, Chap. 3]. Finally, Moreau's law may be in some cases formulated as a quadratic problem under non-convex constraints [3, Proposition 3.4], in which the cost function represents the energy dispersion. Most of the above results are taken from [6, 7, 21, 22]; an alternate proof of (9) for Moreau's law can be found in [20, Proposition 5.6]. See also [23] for a geometric analysis of multiple impacts and a characterization of the domain of admissible post-impact velocities.

Let us come back to (9). It implies the following LCP:

$$\mathbf{0} \leq \boldsymbol{\lambda}_{t,\mathscr{I}(\mathbf{q})} \perp \mathbf{D}_{\mathscr{I}(\mathbf{q})}(\mathbf{q})\, \boldsymbol{\lambda}_{t,\mathscr{I}(\mathbf{q})} + (\mathbf{I}_l + \mathscr{E}_{\mathrm{nn}})\mathbf{U}_{\mathrm{n},\mathscr{I}(\mathbf{q})}(t^-) \geq \mathbf{0}. \tag{11}$$

If the active constraints are functionally independent (we have $l \leq n$), then $\mathbf{D}_{\mathscr{I}(\mathbf{q})}(\mathbf{q})$ is positive definite and this LCP always has a unique solution. Let us calculate it for a chain of balls, using (2), where we assume that, during the shock, all the balls are in contact, hence $l = m = n - 1$:

$$\mathbf{D}_{\mathscr{I}(\mathbf{q})} = \begin{pmatrix} (m_1^{-1}+m_2^{-1}) & -m_2^{-1} & 0 & 0 & \dots & & & \dots & 0 \\ -m_2^{-1} & (m_2^{-1}+m_3^{-1}) & -m_3^{-1} & 0 & \dots & & & & 0 \\ 0 & -m_3^{-1} & (m_3^{-1}+m_4^{-1}) & m_4^{-1} & 0 & & & & 0 \\ \vdots & 0 & & & & & & & \vdots \\ \vdots & \vdots & & & & & -m_{n-3}^{-1} & 0 & 0 \\ 0 & \dots & & & & 0 \; -m_{n-3}^{-1} & (m_{n-3}^{-1}+m_{n-2}^{-1}) & -m_{n-2}^{-1} & 0 \\ 0 & \dots & & & & \dots \; 0 & -m_{n-2}^{-1} & (m_{n-2}^{-1}+m_{n-1}^{-1}) & -m_{n-1}^{-1} \\ 0 & 0 & \dots & & & \dots & 0 & -m_{n-1}^{-1} & (m_{n-1}^{-1}+m_n^{-1}) \end{pmatrix} \quad (12)$$

We have the following for an impact occurring at $t = 0$.

Proposition 1 *Consider a chain of n aligned balls in (1). Let $m_i = m > 0$ for all $1 \le i \le n$. Also let $\mathrm{e} = 0$, and the pre-impact conditions are chosen as $\dot{q}_1(0^-) = 1$ m/s and $\dot{q}_i(0^-) = 0$ m/s for $2 \le i \le n$ (hence, $\mathbf{U}_{\mathrm{n},\mathscr{I}(\mathbf{q})}(0^-) = (-1, 0, \dots, 0)^T$). Then, the unique solution of (11) is*

$$\boldsymbol{\lambda}_{0,\mathscr{I}(\mathbf{q})} = \frac{m}{n+1}\begin{pmatrix} n \\ n-1 \\ n-2 \\ \vdots \\ 1 \end{pmatrix}, \textit{ which yields } \mathbf{U}_{\mathrm{n},\mathscr{I}(\mathbf{q})}(0^+) = (0, \dots, 0)^T.$$

Proof In this case,

$$\mathbf{D}_{\mathscr{I}(\mathbf{q})} = \frac{1}{m}\begin{pmatrix} 2 & -1 & 0 & \dots & & & 0 \\ -1 & 2 & -1 & 0 & \dots & & 0 \\ 0 & -1 & 2 & -1 & 0 & & 0 \\ \vdots & & & & & & \vdots \\ 0 & \dots & & & & & 0 \\ 0 & \dots & & 0 & -1 & 2 & -1 \\ 0 & \dots & & & 0 & -1 & 2 \end{pmatrix}, \quad (13)$$

which is positive definite, as Lemma 1 shows. The result follows by inspection, since there is a unique solution to the LCP. ■

Remark 5 (Dependent active coordinates) In case the active constraints are dependent, then $\mathbf{D}_{\mathscr{I}(\mathbf{q})} \succeq \mathbf{0}$, and since it is a symmetric matrix, $\mathbf{D}_{\mathscr{I}(\mathbf{q})}(\boldsymbol{\lambda}^1_{t,\mathscr{I}(\mathbf{q})} - \boldsymbol{\lambda}^2_{t,\mathscr{I}(\mathbf{q})}) = \mathbf{0}$ for any two solutions $\boldsymbol{\lambda}^1_{t,\mathscr{I}(\mathbf{q})}$ and $\boldsymbol{\lambda}^2_{t,\mathscr{I}(\mathbf{q})}$ of the LCP (11). Therefore, $\mathbf{U}_{\mathrm{n},\mathscr{I}(\mathbf{q})}(t^+)$ is uniquely defined from the first line in (9) (see [1, Lemma 5.2, Corollary 5.1] for the same analysis in a slightly more general framework).

Now, we have the next result.

Lemma 1 *The Delassus' matrix in (13) has full rank and is therefore positive definite.*

Proof Consider a matrix as in (13) with dimension $n \times n$, and denote it as $\mathbf{D}_n$. It is not difficult to show that $\det(\mathbf{D}_n) = 2\det(\mathbf{D}_{n-1}) - \det(\mathbf{D}_{n-2})$, for all $n \geq 3$, and letting $\mathbf{D}_1 = 2$. It follows that, provided $\det(\mathbf{D}_{n-1}) = n$ and $\det(\mathbf{D}_{n-2}) = n - 1$, it holds that $\det(\mathbf{D}_n) = n + 1$. One checks that this is true for $n = 3$, since $\det(\mathbf{D}_2) = 3$ and $\det(\mathbf{D}_1) = 2$. Hence, this is true for all $n \geq 3$. Due to the fact that the Delassus' matrix is at least positive semi definite, the result follows. ■

Proposition 1 shows that Moreau's law creates some distance effect with non-zero impulse at all contacts, and that all balls are stuck together after the shock (maximal dispersion of the kinetic energy in accordance with [3, Proposition 3.4]). Notice, however, that $\boldsymbol{\lambda}_{0,\mathscr{I}(\mathbf{q})} > \mathbf{0}$ (component-wise) implies from (11) that $\boldsymbol{\lambda}_{0,\mathscr{I}(\mathbf{q})} = -\mathbf{D}^{-1}_{\mathscr{I}(\mathbf{q})}(\mathbf{I}_l + \mathscr{E}_{\mathrm{nn}})\mathbf{U}_{\mathrm{n},\mathscr{I}(\mathbf{q})}(0^-)$, so that $-\mathscr{E}_{\mathrm{nn}}\mathbf{U}_{\mathrm{n},\mathscr{I}(\mathbf{q})}(0^-) = \mathbf{U}_{\mathrm{n},\mathscr{I}(\mathbf{q})}(0^+)$. In our case, $U_{\mathrm{n},1}(0^-) = -1$ m/s, so this implies that $U_{\mathrm{n},1}(0^+) = \mathrm{e}$ m/s: this is true for $\mathrm{e} = 0$ under the above conditions. Calculations for the 3-ball system show that this is also the case when $\mathrm{e} = 1$ [1, p.271].

Proposition 2 *Consider the chain of n aligned balls in (1) with $m_i = m > 0$. Suppose that $\boldsymbol{\lambda}_{0,\mathscr{I}(\mathbf{q})} > \mathbf{0}$ (each contact undergoes an impact with positive impulse), with pre-impact relative velocity $\mathbf{U}_{\mathrm{n}}(0^-) = (-1, 0, \ldots, 0)^T$ (so that $\mathscr{I}(\mathbf{q}) = \{1, \ldots, n - 1\}$). Then, it holds that $\mathbf{U}_{\mathrm{n}}(0^+) = (\mathrm{e}, 0, \ldots, 0)^T$.*

Let us now state the following result. We still assume that $\dot{q}_1(0^-) = 1$ m/s, and $\dot{q}_i(0^-) = 0$ m/s, $2 \leq i \leq n$.

Proposition 3 *Let $\mathrm{e} = 1$, $m_i = m > 0$, $1 \leq i \leq n$, $\mathbf{U}_{\mathrm{n}}(0^-) = (-1, 0, \ldots, 0)^T$. Assume that $\dot{q}_1(0^+) = \frac{2-n}{n}$ m/s, $\dot{q}_i(0^+) = \frac{2}{n}$ m/s for $2 \leq i \leq n$ (so that $\mathbf{U}_{\mathrm{n}}(0^+) = (1, 0, \ldots, 0)^T$). Then, the kinetic energy is conserved, and $\boldsymbol{\lambda}_{0,\mathscr{I}(\mathbf{q})} = \mathbf{D}^{-1}_{\mathscr{I}(\mathbf{q})}[\mathbf{U}_{\mathrm{n}}(0^+) - \mathbf{U}_{\mathrm{n}}(0^-)] > \mathbf{0}$ (component-wise) is the solution of the LCP in (11).*

Proof Preservation of the kinetic energy follows from a simple calculation. Notice that $\mathbf{U}_{\mathrm{n}}(0^+) - \mathbf{U}_{\mathrm{n}}(0^-) = (2, 0, \ldots, 0)^T$, consequently we only need to know the first column of $\mathbf{D}^{-1}_{\mathscr{I}(\mathbf{q})}$, where $\mathbf{D}_{\mathscr{I}(\mathbf{q})}$ has the structure shown in the proof of Proposition 1. Let us denote matrices with this structure, and of dimension p, as $\mathbf{D}_p$. In fact, it can be shown by induction that the first column of $\mathbf{D}_p^{-1}$ is equal to $\frac{1}{\det(\mathbf{D}_p)}(\det(\mathbf{D}_{p-1}), \det(\mathbf{D}_{p-2}), \ldots, 2, 1)^T$, where $\det(\mathbf{D}_p) = p + 1$. Therefore, since we have $\mathscr{I}(\mathbf{q}) = \{1, \ldots, n - 1\}$, the first column of $\mathbf{D}^{-1}_{\mathscr{I}(\mathbf{q})}$ is equal to

$$\frac{1}{\det(\mathbf{D}_{\mathscr{I}(\mathbf{q})})}(\det(\mathbf{D}_{n-2}), \det(\mathbf{D}_{n-3}), \ldots, 2, 1)^T > 0.$$

Therefore, $\boldsymbol{\lambda}_{0,\mathscr{I}(\mathbf{q})}$ is twice this vector and is positive. We have $\mathbf{D}_{\mathscr{I}(\mathbf{q})}(\mathbf{q})\, \boldsymbol{\lambda}_{0,\mathscr{I}(\mathbf{q})} + (\mathbf{I}_l + \mathscr{E}_{\mathrm{nn}})\mathbf{U}_{\mathrm{n},\mathscr{I}(\mathbf{q})}(0^-) = \mathbf{U}_{\mathrm{n},\mathscr{I}(\mathbf{q})}(0^+) - \mathbf{U}_{\mathrm{n},\mathscr{I}(\mathbf{q})}(0^-) + (1 + \mathrm{e})\mathbf{U}_{\mathrm{n},\mathscr{I}(\mathbf{q})}(0^-) = \mathbf{0}$, which ends the proof, since the impact LCP has a unique solution. ■

It is also checked that the linear momentum of the chain is preserved. Therefore, under the stated assumption, Moreau's impact law is unable to separate the balls 2 to n, while ball 1 "rebounds" on the chain and gives a non-zero velocity to the $n-1$ other balls. It has limited predictability in terms of energy dispersion (see Moreau's line in [3, Fig. 2.6] for the 3-ball system). This is what has motivated researchers to extend it while remaining in a rigid-body approach, and this is what motivates us to analyze which are the cases when it does correctly predict the post-impact velocity in Sect. 5.

Remark 6 Solving the impact LCP in (11) allows one to compute the projection in (7), i.e., the index set $\mathscr{J}(\mathbf{q})$. We could start from the reduced dynamics (4) in which the calculations for the tangent and normal cones are simplified, since the constraints define the first orthant. However, projections are made in the metric defined by $\bar{\mathbf{M}}$, which is no longer a diagonal matrix. In these coordinates, the Delassus' matrix is $\mathbf{D}_{\mathscr{J}(\mathbf{q})} = \bar{\mathbf{M}}$ and $U_{\mathrm{n},i} = \dot{z}_i$. Thus, there is nothing special to gain using (4) instead of (1).

3.2 The Binary Impact Model

Contrary to Moreau's law, which handles all impacts at the same time, the binary impact model handles impacts separately. To do so, the multiple impact problem is assumed to be a succession of binary collisions between rigid particles, so collisions are independent of each other. Each binary collision between two balls can be completely solved by using the conservation law of momentum and Newton's kinematic restitution law:

$$\begin{cases} m_1\dot{q}_1^- + m_2\dot{q}_2^- = m_1\dot{q}_1^+ + m_2\dot{q}_2^+, \\ \dot{q}_2^+ - \dot{q}_1^+ = -e_\mathrm{n}(\dot{q}_2^- - \dot{q}_1^-), \end{cases} \tag{14}$$

where superscripts $(-)$ and $(+)$ indicate the pre- and post-impact velocities, and e_n is the coefficient of normal restitution which takes a value from 0 for purely dissipative collision to 1 for purely elastic collision. Note that each binary collision is assumed to be central: the collision occurs only in the normal direction of the contact, as illustrated in Fig. 2.

The post-impact velocity of each ball is obtained by solving the system of linear equations (14):

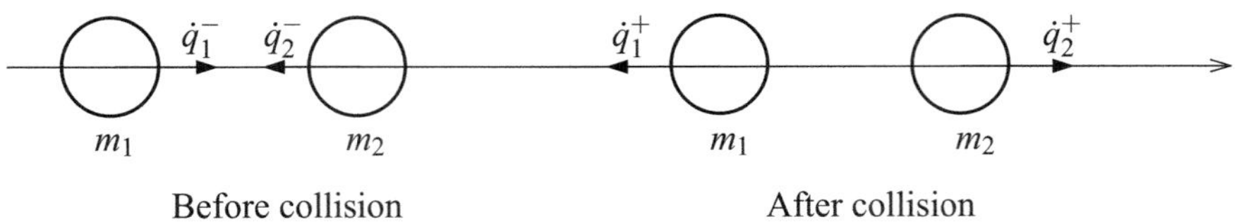

Fig. 2 Two particles before and after a binary collision

$$\begin{cases} \dot{q}_1^+ = \dfrac{m_1 - m_2 e_{\rm n}}{m_1 + m_2}\dot{q}_1^- + \dfrac{(1+e_{\rm n})m_2}{m_1+m_2}\dot{q}_2^-, \\ \dot{q}_2^+ = \dfrac{(1+e_{\rm n})m_1}{m_1+m_2}\dot{q}_1^- + \dfrac{m_2 - e_{\rm n}m_1}{m_1+m_2}\dot{q}_2^-. \end{cases} \tag{15}$$

In the case in which the two balls have the same mass and the first ball comes to collide with the last one at rest, the post-impact velocities are

$$\begin{cases} \dot{q}_1^+ = \dfrac{1-e_{\rm n}}{2}\dot{q}_1^-, \\ \dot{q}_2^+ = \dfrac{(1+e_{\rm n})}{2}\dot{q}_1^-. \end{cases} \tag{16}$$

If the collision is purely elastic ($e_{\rm n} = 1$), the first ball stops and the last one moves forward after collision with a velocity equal to the pre-impact velocity of the first ball. This means that the energy and momentum of the first ball are entirely transferred to the last one.

While the outcome of a binary collision is easily obtained, the definition of the succession of binary collisions is not straightforward. One can try to mimic the wave propagation induced by the shock in a granular media to define the sequence of binary collisions. Let us consider a granular monodisperse chain composed of n elastic identical beads as an example. The beads are numbered $1, 2, \ldots, n$ from the left to the right. When the first ball moves with a velocity of 1 m/s and collides with the other balls which are at rest, a solitary wave is initiated and propagates from the left to the right. According to the wave propagation, the succession of binary collisions can be defined as follows: ball 1 collides with ball 2, then ball 2 collides with ball 3, ..., then ball i collides with ball $i+1$, ..., and at the end, ball $n-1$ collides with ball n. Applying the rule (15) from the first to the last binary collision, we obtain the impact outcome as follows: balls 1 to $n-1$ stop and ball n moves forward with a velocity of 1 m/s. This sequence of binary collisions is also true for a tapered chain in which the bead diameter decreases progressively, and it has been used by several authors to study the momentum and energy propagation in tapered chains [24–27]. It is worth mentioning that for elastic monodisperse chains or tapered chains and for the considered particular initial condition, i.e., the first ball collides with the other balls at rest, the sequence of binary collisions is uniquely defined. However, this is not true in most cases. Let us demonstrate this point by considering a monodisperse chain of 10 dissipative beads. We apply the binary collision rule (15) with the coefficient of restitution $e_{\rm n} = 0.5$ to the sequence of binary collisions defined above. The velocity of each bead after this sequence of binary collisions is shown in Table 1. It can be seen that beads enter into collisions again after the first sequence of binary collisions: there are indeed potential collisions between balls 1 and 2, between balls 2 and 3, and so on. Even for an elastic chain, we can encounter this problem. Let us take an elastic decorated chain (Fig. 3) as an example. For this granular chain, three small balls of mass 0.5 m are placed periodically between four big balls of mass m. Table 2 shows that there are several potential collisions between

Table 1 Bead velocity for a monodisperse chain after a sequence of binary collisions from the left to the right

1	2	3	4	5	6	7	8	9	10
0.25	0.19	0.14	0.11	0.08	0.06	0.04	0.03	0.026	0.08

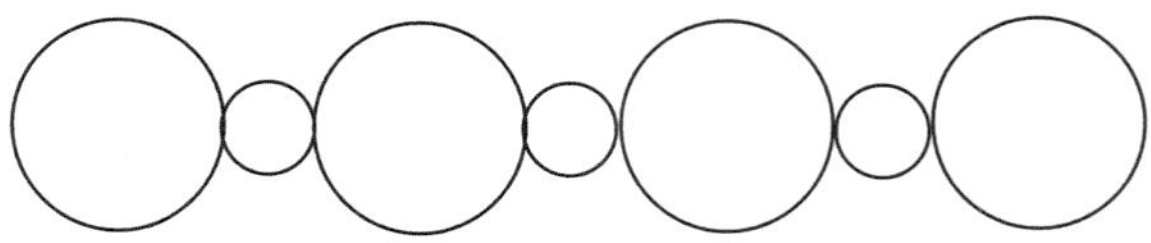

Fig. 3 Illustration of a decorated chain

Table 2 Bead velocity for a decorated chain after a sequence of binary collisions from the left to the right

1	2	3	4	5	6	7
0.3333	−0.4444	0.2963	−0.3951	0.2634	−0.3512	0.7023

balls after the first sequence of binary collisions. A question that arises here regards which order of binary collisions we should consider when there are several binary collisions to be handled. We present here two strategies that can be used for a granular chain in which collisions start at the left end and propagate to the right end.

(1) Binary collisions are always handled from the left to the right. This means that among the set of possible collisions, the collision at the contact with the least value of the index k is handled first.
(2) The order of binary collisions is unimportant, so binary collisions can be randomly handled. This strategy has been adopted in [28, 29].

It is worth mentioning that the selection of binary collisions with the left-to-right order or the random order presented above is not physically justified. Let us apply these two strategies to a disordered chain of 30 elastic balls. For this kind of granular chain, ball masses are randomly distributed. Sequences of binary collisions are randomly selected with the uniform distribution law. Figure 4 shows a comparison between the impact outcomes obtained with the two considered strategies. One can see that the impact outcome depends strongly on the chosen sequence of binary collisions. In addition, different random sequences of binary collisions lead to different impact outcomes, as shown in Fig. 5. This is intimately related to the fact that the trajectories are, in general, discontinuous with respect to initial data, as we pointed out in Sect. 3.1.

Another issue of the binary collision model is that the sequence of binary collisions can tend to infinity before the impact process ends, even for simple cases. For example, Towne and Hadlock [5] have determined analytically that the number of binary collisions for a chain of three balls is infinite if the number z defined in (17) satisfies $z \geq 1$ (see [3] for more discussions):

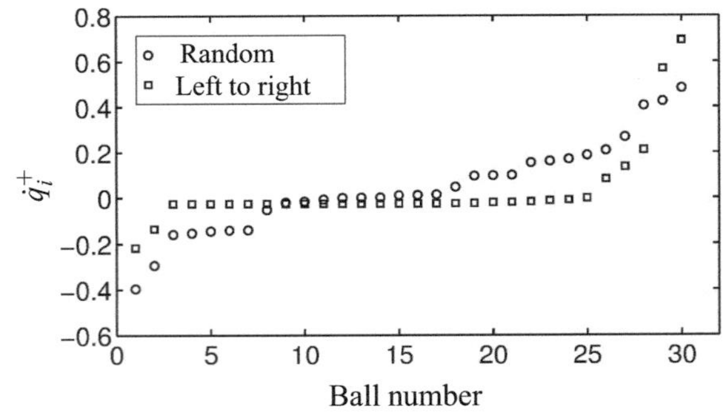

Fig. 4 Ball post-impact velocity versus ball number for a disordered chain obtained with the left-to-right and random sequences of binary collisions

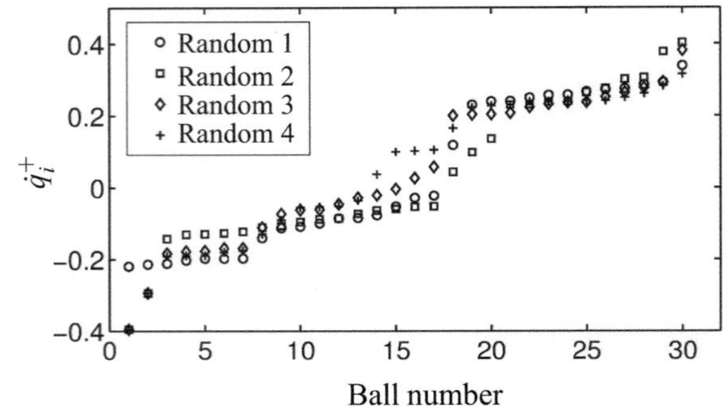

Fig. 5 Ball post-impact velocity versus ball number for a disordered chain obtained with four random sequences of binary collisions

Table 3 Number of binary collisions N_c obtained with the left-to-right (LR) and random (R) orders versus the number of balls n in a disordered chain with $e_{\rm n} = 0.9$

n	10	20	30	40	50	60	70
N_c - LR	53	189	991	18476	4731360	38936068	–
N_c - R	51	153	397	1316	–	–	–

$$z = \frac{1}{2}\left(\sqrt{e_{\rm n}} + \frac{1}{\sqrt{e_{\rm n}}}\right)\frac{1}{\sqrt{\left(1+\frac{m_2}{m_1}\right)\left(1+\frac{m_2}{m_3}\right)}}. \tag{17}$$

The number of binary collisions increases quickly with the number of balls, in particular for dissipative chains ($e_n < 1$), as shown in Table 3 for a disordered chain with $e_{\rm n} = 0.9$. One can see that the binary collision model is not able to determine the impact outcome with 70 balls for the left-to-right order and with 50 balls for the random order, because the number of binary collisions to be handled is too big.

In summary, the binary collision model presents three main drawbacks:

(1) The impact outcome is possibly not unique, which is related to the discontinuity with respect to the initial data;

(2) The impact outcome depends on the chosen order of sequence of binary collisions;
(3) The number of binary collisions to be handled is possibly infinite.

3.3 *The LZB Model*

This way of treating multiple impacts has been introduced in [30–32], and we briefly summarize it in this section. It has been validated through extensive comparisons between experimental and numerical data in [3, 33–37] for chains of balls, rocking blocks, bouncing dimers and other setups. This is a model of the class (**ii**), based on the Darboux-Keller approach [1, Sect. 4.3.5]. As such, it is based on the following fundamental assumptions:

1. Forces other than impact forces are negligible during the collision process.
2. Positions are constant during the collision process.
3. Tangential stiffnesses are infinite.
4. The impact consists of a compression phase followed by an expansion phase.

Then, the impact dynamics consist of a first-order dynamics whose state is the velocity, and the time-scale is replaced by the impact force impulse. Though the Darboux-Keller shock dynamics have a long history for two-body single impacts, it is only recently that its extension to multiple impacts has been proposed with the use of energetic coefficients of restitution (CoRs) [30, 32]. We summarize the LZB dynamics now, when applied to chains of aligned balls. Let us start from (1): $\mathbf{M}\ddot{\mathbf{q}}(t) = \mathbf{W}\boldsymbol{\lambda}(t)$, where $\mathbf{W} \stackrel{\Delta}{=} \nabla\mathbf{f}(\mathbf{q})$ is constant. In this example, $\mathbf{M}$ and $\mathbf{W}$ are constant, so the constant position assumption is useless. During the impact, we will denote the infinitesimal impulse as $d\mathbf{P} \stackrel{\Delta}{=} \boldsymbol{\lambda}dt$, so that the so-called Darboux-Keller dynamics writes $\mathbf{M}d\dot{\mathbf{q}} = \mathbf{W}d\mathbf{P} \Leftrightarrow \mathbf{M}\frac{d\dot{\mathbf{q}}}{d\mathbf{P}} = \mathbf{W}$, after a time rescaling has been performed. The next basic assumption is that at each contact i, one has the force/indentation relation $\lambda_i = K_i(\delta_i)^{\eta_i}$, where K_i is the contact equivalent stiffness and η_i is the elasticity coefficient ($\eta_i = 1$ for linear elasticity, $\eta_i = \frac{3}{2}$ for Hertz' elasticity). More precisely, the LZB model may be designed with a mono-stiffness compression/expansion model, or a bi-stiffness compression/expansion model [30], or even a tri-stiffness model [3, Fig. 4.4]. Let us describe the bi-stiffness model, as shown in Fig. 6. During the compression phase (from the origin to M^i), one has $\lambda_{c,i} = K_i(\delta_i)^{\eta_i}$; during the expansion (or restitution) phase, one has $\lambda_{e,i} = \lambda_{M^i}\left(\frac{\delta_i - \delta_{r^i}}{\delta_{M^i} - \delta_{r^i}}\right)^{\eta_i}$ (see [1, Sect. 4.2.1.2] for a short history about bi-stiffness models). The dashed area corresponds to the dissipated energy during the shock, δ_{M^i} is the maximal indentation, and δ_{r^i} is the residual indentation. The work done by the contact force during the compression phase is $W_{c,i} = \int_0^{\delta_{M^i}} \lambda_i(\delta_i)d\delta_i = \frac{1}{1+\eta_i}K_i(\delta_{M^i})^{\eta_i+1}$, and during the restitution phase $W_{e,i} = \int_{\delta_{M^i}}^{\delta_{r^i}} \lambda_i(\delta_i)d\delta_i = -\frac{1}{1+\eta_i}K_i(\delta_{M^i})^{\eta_i}(\delta_{M^i} - \delta_{r^i})$. Actually, the bi-stiffness model is a piecewise-continuous model, which states that

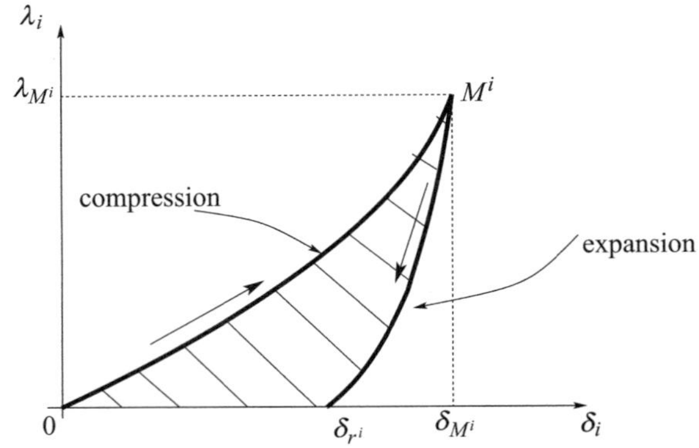

Fig. 6 The bi-stiffness force/indentation model for the LZB model at contact i

$\lambda_i = K_i(\delta_i)^{\eta_i}$ if $\dot{\delta}_i \geq 0$ (compression) and $\lambda_i = K_i^e(\delta_i - \delta_{r^i})^{\eta_i}$ if $\dot{\delta}_i < 0$ (expansion), where $K_i^e = K_i\left(\frac{\delta_{M^i}}{\delta_{M^i}-\delta_{r^i}}\right)^{\eta_i}$. Calculations show that the energetic CoR at contact i satisfies $e_{i,*}^2 = -\frac{W_{e,i}}{W_{c,i}} = 1 - \frac{\delta_{r^i}}{\delta_{M^i}} = \left(\frac{K_i}{K_i^e}\right)^{\frac{1}{\eta_i}}$, hence $\delta_{r^i} = \delta_{M^i}(1 - e_{i,*}^2)$. Perfectly plastic impacts with $e_{i,*} = 0$ imply that $\delta_{r^i} = \delta_{M^i}$, so that the expansion phase has zero duration and the point $(\delta_{M^i}, 0)$ is reached instantaneously from the maximum compression point M^i.

The next step is to calculate the contact force as a function of the potential energy. Starting from $\lambda_i = K_i(\delta_i)^{\eta_i}$, and using $\frac{d\lambda_i}{dt} = \lambda_i \frac{d\lambda_i}{dP_i}$, one finds that

$$\lambda_i(P_i(t)) = \left[(\eta_i + 1)\int_0^{P_i(t)} K_i^{\frac{1}{\eta_i}} \nabla^T f_i \dot{q}\, dP_i\right]^{\frac{\eta_i}{\eta_i+1}}. \tag{18}$$

Further calculations not recalled here allow one to show that, even in case of pre-compression (with $\lambda_i(0) \neq 0$), one has

$$\lambda_i(P_i(t)) = (1 + \eta_i)^{\frac{\eta_i}{\eta_i+1}} K_i^{\frac{1}{\eta_i+1}} (E_i(P_i(t)))^{\frac{\eta_i}{\eta_i+1}}, \tag{19}$$

where $E_i(P_i)$ is the potential energy at contact i, i.e., $E_i(P_i) = E_{0,i} + \int_0^{P_i(t)} \dot{\delta}_i(P_i)\, dP_i$, where $E_{0,i}$ is the potential energy due to pre-compression. Taking pre-compression into account is crucial, because such multiple impacts usually involve *repeated impacts* at the same contact point, which correspond to an impact starting again while the zero indentation has not been reached yet (see Fig. 8). Repeated impacts render the problem more complex. A crucial result is shown in [30, Theorem 3.1]. Let us consider Fig. 7. Then, [30, Theorem 3.1] guarantees that a compression-expansion cycle $\widehat{OR'\delta_{r1}}$ (curves 1 and 3) is equivalent, from the energetic point of view, to a cycle $\widehat{AR\delta_{r2}}$ (curves 2 and 4), where the compression would finish at R' (respectively at R). This allows us to prove the following. When the contact point i

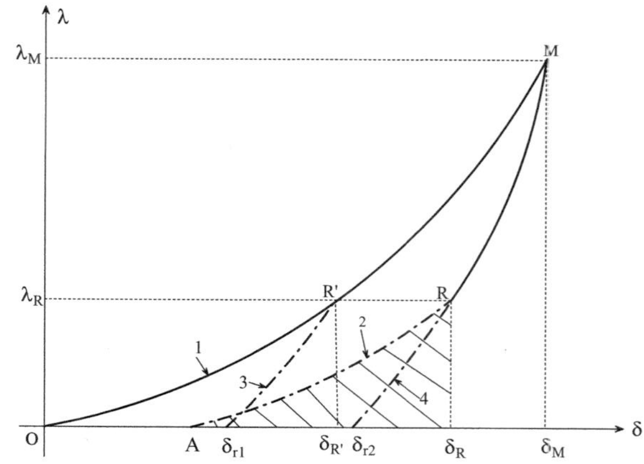

Fig. 7 The potential energy when the contact point is located at the expansion phase

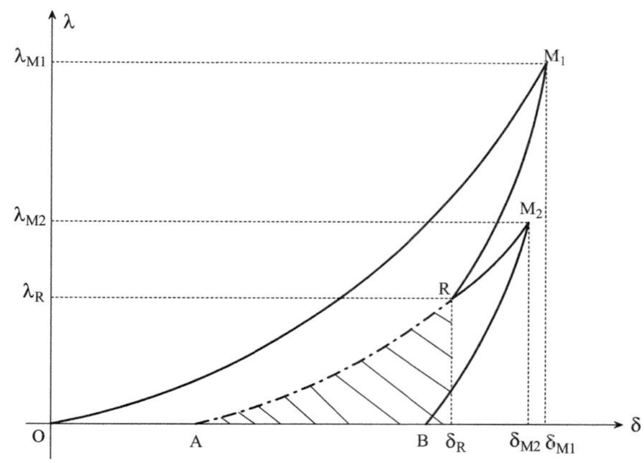

Fig. 8 Repeated impact: the contact point with two compression phases

moves from M^i to R^i along the expansion curve in Fig. 7, the recovered energy is $\int_{\delta_{M,i}}^{\delta_{R,i}} \lambda_i(\delta_i)d\delta_i$, and we obtain at contact i[4]

$$\begin{aligned}\int_{\delta_{M^i}}^{\delta_{R^i}} \lambda_i(\delta_i)d\delta_i &= \int_{\delta_{M^i}}^{\delta_{r2,i}} \lambda_i(\delta_i)d\delta_i - \int_{\delta_{R^i}}^{\delta_{r2,i}} \lambda_i(\delta_i)d\delta_i \\ &= -e_{i,*}^2 \int_{\delta_0}^{\delta_{M^i}} \lambda_i(\delta_i)d\delta_i - \int_{\delta_{R^i}}^{\delta_{r2,i}} \lambda_i(\delta_i)d\delta_i,\end{aligned} \tag{20}$$

where $e_{i,*}$ is the energetic CoR at contact i. According to Stronge [38], the energetic CoR $e_{i,*}$ is defined as $e_{i,*}^2 = -W_i^e/W_i^c$, where W_i^c and W_i^e are the respective works done by the contact force during the compression and expansion phases. The term premultiplied by $-e_{i,*}^2$ is equal to the area enclosed by the curve $\widehat{\delta_{r2}R\delta_R}$ in Fig. 7. Let us assume that the force/indentation relationship remains the same for the second compression/expansion phase (i.e., the elasticity properties do not vary).

[4]In Figs. 7 and 8, the subscript i is not indicated. Thus, R^i is R, and so on.

Using this, and after manipulations, it follows that the potential energy along the repeated impact in Fig. 8 is given as follows, where Q denotes a generic point along the force/indentation curve:

$$E(P(t)) = \begin{cases} E_0 + \int_0^{P(t))} \dot{\delta}(P(s))dP(s) & Q \in \widehat{OM_1} \\ E_{M_1} + \frac{1}{e_*^2} \int_{P_{M_1}}^{P(t)} \dot{\delta}(P(s))dP(s) & Q \in \widehat{M_1 R} \\ E_R + \int_{P_R}^{P(t)} \dot{\delta}(P(s))dP(s) & Q \in \widehat{RM_2} \\ E_{M_2} + \frac{1}{e_*^2} \int_{P_{M_2}}^{P(t)} \dot{\delta}(P(s))dP(s) & Q \in \widehat{M_2 B}, \end{cases} \tag{21}$$

where E_{M_1} is the residual potential energy at point M_1, and so on. As a next step, one can use (19) to derive the *distributing law* between infinitesimal impulses dP_i and dP_j at contact points i and j, respectively:

$$\boxed{\frac{dP_i}{dP_j} = \frac{(1+\eta_i)^{\frac{\eta_i}{1+\eta_i}}}{(1+\eta_j)^{\frac{\eta_j}{1+\eta_j}}} \frac{K_i^{\frac{1}{1+\eta_i}}}{K_j^{\frac{1}{1+\eta_j}}} \frac{(E_i(P_i))^{\frac{\eta_i}{1+\eta_i}}}{(E_j(P_j))^{\frac{\eta_j}{1+\eta_j}}}}. \tag{22}$$

It is noteworthy that if all contacts have the same elasticity coefficient, the distributing law simplifies and shows that the ratio in (22) depends only on the stiffnesses ratio $\left(\frac{K_i}{K_j}\right)^{\frac{1}{1+\eta}}$. It is a well-known fact that the post-impact velocities in chains of aligned balls indeed do not depend on the absolute values of the equivalent contact stiffnesses, but only on their ratio, in the case of linear elasticity (see, for instance, [1, Sect. 6.1.3]). This result generalizes it. In summary, the potential energy can be calculated along (21), while the infinitesimal impulse ratio is given by (22). Now, contrary to the case of a single collision in which one can make a time-scale change, passing from time t to the impact force impulse dP (since the contact/impact forces are always assumed to be non-negative, and positive for times strictly inside the collision interval), one has $dP > 0$, and this time rescaling is valid. In a case of multiple impacts, one has to choose a so-called *primary contact* at which it is guaranteed that the impulse does not become constant, for otherwise, the time rescaling becomes impossible with this impulse. Thus, one chooses the primary impulse as the impulse from contact i at which the potential energy at this contact $E_i(P_i)$ is maximal amongst the various contact points.

We obtain the *multiple impact Darboux-Keller equations*:

1. (contact parameters): $K_j, \eta_j, e_{j,*}, 1 \le j \le n-1$.
2. (dynamical equations):

$$\mathbf{M}d\mathbf{q} = \mathbf{W}d\mathbf{P}, \tag{23}$$

where the impulse increment dP_j at a contact j is related to the impulse increment dP_i at another contact i by the distributing law (22). The impulse increment dP_j can be also related to the time increment dt by the relation

$$dP_j = \lambda_j dt = (1+\eta_j)^{\frac{\eta_j}{\eta_j+1}} K_j^{\frac{1}{\eta_j+1}} (E_j(t))^{\frac{\eta_j}{\eta_j+1}} dt, \tag{24}$$

with the contact force λ_j computed with Eq. (18).

3. (potential energy for the bi-stiffness model):

$$E_j(P_j) = E_{Tra,j} + \frac{1}{Tra} \int_{P_{Tra}(t)}^{P_j(t)} \nabla^T f_j(q)\dot{q}\, dP_j \tag{25}$$

where $Tra = 1$ if $\dot{\delta}_j > 0$ (compression), $Tra = e_{j,*}^2$ if $\dot{\delta}_j < 0$ (expansion), $E_{Tra,j}$ is the accumulated potential energy at the beginning of the integration, and $P_{Tra}(t)$ depends on the impulse value at the beginning of the subphase (see (21)).

4. (impact termination): $E_j(P_j) = 0$ and $\dot{\delta}_j \leq 0$ at all contacts $1 \leq j \leq n-1$.

We have $\dot{\delta}_j = \nabla f_j(q)^T \dot{q} = \dot{q}_{i+1} - \dot{q}_i$. Notice that (25) could be rewritten in its differential form

$$dE_j = \frac{1}{Tra} \nabla f_j(q)^T \dot{q}\, dP_j \Leftrightarrow \frac{dE_j}{dP_i} = \frac{1}{Tra} \dot{\delta}_j(P)\Gamma_{ji}(E_i(P_i), E_j(P_j)), \tag{26}$$

with $\Gamma_{ji} = dP_j/dP_i$ and initial condition $E_j(P_{Tra,j}) = E_{Tra,j}$. The multiple impact Darboux-Keller equations are therefore a set of first-order nonlinear and coupled piecewise smooth differential equations, with states $\dot{\mathbf{q}}$, $\mathbf{\Gamma}$, $\mathbf{E}$, and state-dependent switching conditions at times of maximum compressions ($\dot{\delta}_j = 0$, points M_1, M_2 in Fig. 8) or repeated impacts (point R in Fig. 8).

Remark 7

1. The bi-stiffness model has several drawbacks: it does not model a bounded maximal contact force, it is a rough representation of plasticity (if plastification is the primary source of dissipated energy), and it models dissipation during the expansion phase (while dissipation could also occur during the compression phase). However, it can be improved as described in [3, Sect. 4.2.4].
2. The LZB approach can also be formulated with Coulomb friction at contacts [37].
3. The CoRs $e_{i,*}$ can be estimated off-line from pairwise collisions between two balls.
4. We employ the word "balls", however, the chain may consist of other types of elementary particles than spherical balls, like beads or polyhedral grains.
5. We have written $\dot{\delta}_j(\mathbf{P})$ because, due to dynamical couplings stemming from $\mathbf{M}$ and $\mathbf{W}$ in (23), the local velocity may depend on several contact impulses.
6. As we shall see in Sect. 4.3, it is possible to dispense with the distributing law in (22), which is quite time-consuming during numerical integration (see [3, Chap. 4] for a complete exposition of the event-driven algorithm for the LZB model,

in particular, the algorithm for the primary impulse selection). The distributing law is nevertheless quite interesting, since it highlights the way in which the different contacts interact with each other.

7. We see that the LZB model allows us to include the effects of contact flexibilities (which are crucial in chains of balls impacts) while disregarding position variations. This is done thanks to the distributing law.

4 Numerical Resolution

The numerical algorithms which are used to compute the post-impact velocities, may differ from one impact law to the next. Let us describe now how the above three models of multiple impacts are treated numerically.

4.1 Moreau's Impact Law

As alluded to above, the great advantage of Moreau's law is that it is naturally embedded in the discrete-time version of Moreau's sweeping process for Lagrangian systems, using a suitable event-capturing scheme that stems from Moreau's catching-up algorithm. The numerical aspects of the sweeping process applied to mechanical systems are treated in detail in [1, 7, 21, 22, 39, 40]. Let us briefly introduce the catching-up algorithm. We start from the second order sweeping process:

$$\mathbf{M}(\mathbf{q})d\mathbf{v} + \mathbf{F}(\mathbf{q}, \mathbf{v}, t)dt \in -\mathcal{N}_{\mathcal{V}(\mathbf{q})}(\mathbf{w}), \tag{27}$$

where $\mathbf{v} = \dot{\mathbf{q}}$ almost everywhere and $d\mathbf{v}$ is the so-called differential measure associated with the acceleration (which cannot be a function at impact times, since the velocity has a discontinuity), so that (27) is a measure differential inclusion (MDI). Outside impacts we have $d\mathbf{v} = \ddot{\mathbf{q}}(t)dt$. At an impact time t, the MDI (27) is equivalent to (5), that is $d\mathbf{v} = (\dot{\mathbf{q}}(t^+) - \dot{\mathbf{q}}(t^-))\delta_t$, with δ_t the Dirac measure at t. The basic time-stepping method for (27) is as follows on $[t_k, t_{k+1})$, with constant time-step $h = t_{k+1} - t_k > 0$, $k \geq 0$:

$$\begin{cases} \mathbf{M}(\mathbf{q}_k)(\mathbf{v}_{k+1} - \mathbf{v}_k) + h\ F(\mathbf{q}_k, \mathbf{v}_k, t_k) \in -\mathcal{N}_{\mathcal{V}(\mathbf{q}_k)}\left(\dfrac{\mathbf{v}_{k+1} + \mathrm{e}\mathbf{v}_k}{1+\mathrm{e}}\right) \\ \\ \mathbf{q}_{k+1} = \mathbf{q}_k + h\mathbf{v}_{k+1}. \end{cases} \tag{28}$$

We can proceed as we did in Sect. 3.1 to transform (5). We denote $\mathbf{F}_k \triangleq \mathbf{F}(\mathbf{q}_k, \mathbf{v}_k, t_k)$.

$$
\begin{aligned}
&\frac{\mathbf{v}_{k+1}-\mathbf{v}_k+h\,\mathbf{M}^{-1}(\mathbf{q}_k)\mathbf{F}_k}{1+\mathrm{e}} \in -\mathbf{M}^{-1}(\mathbf{q}_k)\,\mathcal{N}_{\mathcal{V}(\mathbf{q}_k)}\left(\frac{\mathbf{v}_{k+1}+\mathrm{e}\mathbf{v}_k}{1+\mathrm{e}}\right)\\
&\Leftrightarrow \frac{\mathbf{v}_{k+1}+\mathrm{e}\mathbf{v}_k}{1+\mathrm{e}}+\frac{h\,\mathbf{M}^{-1}(\mathbf{q}_k)\mathbf{F}_k}{1+\mathrm{e}}-\frac{\mathrm{e}\mathbf{v}_k}{1+\mathrm{e}}-\frac{\mathbf{v}_k}{1+\mathrm{e}} \in -\mathbf{M}^{-1}(\mathbf{q}_k)\,\mathcal{N}_{\mathcal{V}(\mathbf{q}_k)}\left(\frac{\mathbf{v}_{k+1}+\mathrm{e}\mathbf{v}_k}{1+\mathrm{e}}\right)\\
&\Leftrightarrow \mathbf{v}_{k+1}=-\mathrm{e}\mathbf{v}_k+(1+\mathrm{e})\,\mathrm{proj}_{\mathbf{M}(\mathbf{q}_k)}\left[\mathcal{V}(\mathbf{q}_k);-\frac{h\,\mathbf{M}^{-1}(\mathbf{q}_k)\mathbf{F}_k}{1+\mathrm{e}}+\mathbf{v}_k\right]\\
&\Leftrightarrow \mathbf{v}_{k+1}=-\mathrm{e}\mathbf{v}_k+(1+\mathrm{e})\,\mathrm{argmin}_{\mathbf{z}\in\mathcal{V}(\mathbf{q}_k)}\frac{1}{2}(\mathbf{z}-\bar{\mathbf{v}}_k)^T\mathbf{M}(\mathbf{q}_k)(\mathbf{z}-\bar{\mathbf{v}}_k),
\end{aligned}
\tag{29}
$$

where $\bar{\mathbf{v}}_k = \frac{-h\mathbf{M}^{-1}(\mathbf{q}_k)\mathbf{F}_k}{1+\mathrm{e}} + \mathbf{v}_k$. Notice that if $f_i(\mathbf{q}) > 0$ for all $1 \le i \le n-1$, then $\mathbf{v}_{k+1} = \mathbf{v}_k - h\,\mathbf{M}^{-1}(\mathbf{q}_k)\mathbf{F}_k$. We infer that the next velocity can be computed by solving a quadratic problem under conic varying constraints. A next step is to compute this projection using complementarity. To this aim, we notice first that $\mathcal{N}_{\mathcal{V}(\mathbf{q}_k)}(\mathbf{w}_{k+1}) = \{\mathbf{z} \in I\!R^n \,|\, \mathbf{z} = \sum_{i\in\mathcal{I}(\mathbf{q}_k)} -\lambda_i \nabla f_i(q_k),\ 0 \le \lambda_i \perp \mathbf{w}_{k+1}^T \nabla f_i(\mathbf{q}_k) \ge 0\}$. Thus, we obtain

$$
\begin{aligned}
&\mathbf{M}(\mathbf{q}_k)(\mathbf{v}_{k+1}-\mathbf{v}_k)+h\,\mathbf{F}_k=\nabla\mathbf{f}_{\mathcal{I}(\mathbf{q}_k)}(\mathbf{q}_k)\,\boldsymbol{\lambda}_{\mathcal{I}(\mathbf{q}_k),k+1}\\
&\Leftrightarrow \mathbf{v}_{k+1}-\mathbf{v}_k+h\mathbf{M}^{-1}(\mathbf{q}_k)\mathbf{F}_k=\mathbf{M}^{-1}(\mathbf{q}_k)\nabla\mathbf{f}_{\mathcal{I}(\mathbf{q}_k)}(\mathbf{q}_k)\,\boldsymbol{\lambda}_{\mathcal{I}(\mathbf{q}_k),k+1}\\
&\Leftrightarrow \nabla^T\mathbf{f}_{\mathcal{I}(\mathbf{q}_k)}(\mathbf{q}_k)(\mathbf{v}_{k+1}-\mathbf{v}_k+h\mathbf{M}^{-1}(\mathbf{q}_k)\mathbf{F}_k)=\mathbf{D}_{\mathcal{I}(\mathbf{q}_k)}(\mathbf{q}_k)\,\boldsymbol{\lambda}_{\mathcal{I}(\mathbf{q}_k),k+1},
\end{aligned}
\tag{30}
$$

where $\mathbf{D}_{\mathcal{I}(\mathbf{q}_k)}(\mathbf{q}_k) = \nabla^T\mathbf{f}_{\mathcal{I}(\mathbf{q}_k)}(\mathbf{q}_k)\mathbf{M}^{-1}(\mathbf{q}_k)\nabla\mathbf{f}_{\mathcal{I}(\mathbf{q}_k)}(\mathbf{q}_k)$ is the Delassus'matrix of (position) active constraints at step k. Denoting the local velocities as $\mathbf{U}_{\mathrm{n},\mathcal{I}(\mathbf{q}_k),k}$, we obtain the mixed LCP:

$$
\boxed{
\begin{aligned}
&\mathbf{U}_{\mathrm{n},\mathcal{I}(\mathbf{q}_k),k+1}-\mathbf{U}_{\mathrm{n},\mathcal{I}(\mathbf{q}_k),k}+h\nabla^T\mathbf{f}_{\mathcal{I}(\mathbf{q}_k)}(\mathbf{q}_k)\mathbf{M}^{-1}(\mathbf{q}_k)\mathbf{F}_k=\mathbf{D}_{\mathcal{I}(\mathbf{q}_k)}(\mathbf{q}_k)\,\boldsymbol{\lambda}_{\mathcal{I}(\mathbf{q}_k),k+1}\\
&0\le\boldsymbol{\lambda}_{\mathcal{I}(\mathbf{q}_k),k+1}\perp\mathbf{U}_{\mathrm{n},\mathcal{I}(\mathbf{q}_k),k+1}+\mathrm{e}\,\mathbf{U}_{\mathrm{n},\mathcal{I}(\mathbf{q}_k),k}\ge 0,
\end{aligned}}
\tag{31}
$$

where we used the expression for $\mathbf{w}_{k+1}^T\nabla f_i(\mathbf{q}_k)$ in the complementarity conditions. The similarity between (31) and (9) is obvious. Once the set of active constraints has been computed, one can solve the mixed LCP (31) to compute $\mathbf{U}_{\mathrm{n},\mathcal{I}(\mathbf{q}_k),k+1}$ and $\boldsymbol{\lambda}_{\mathcal{I}(\mathbf{q}_k),k+1}$. Once $\boldsymbol{\lambda}_{\mathcal{I}(\mathbf{q}_k),k+1}$ is known, one can use the first line in (31) to obtain $\mathbf{v}_{k+1}$ and then $\mathbf{q}_{k+1}$. There exist quite efficient algorithms to solve mixed LCPs, some of which are implemented in the INRIA SICONOS software package.

Remark 8 The rationale behind the above is that the elements inside the normal cone $-\mathcal{N}_{\mathcal{V}(\mathbf{q}_k)}\left(\frac{\mathbf{v}_{k+1}+\mathrm{e}\mathbf{v}_k}{1+\mathrm{e}}\right)$ are an approximation of $\nabla\mathbf{f}_{\mathcal{I}(\mathbf{q})}(\mathbf{q})\,\boldsymbol{\lambda}_{t,\mathcal{I}(\mathbf{q})}([t_k, t_{k+1}])$, that is, the measure of the interval $[t_k, t_{k+1}]$ by $\nabla\mathbf{f}_{\mathcal{I}(\mathbf{q})}(\mathbf{q})\,\boldsymbol{\lambda}_{t,\mathcal{I}(\mathbf{q})}$. Thus, even at an impact time, this is a bounded quantity (in fact, the impact magnitude).

In practice, the event-capturing method in (28) can be modified to cope with energy conservation, accuracy, etc [41, 42]. An important feature is that it is shown to converge [8], hence for small time-steps, the numerical solutions must be close to the analytical ones.

4.2 Binary Collision Model

The binary collision model is solved in an iterative manner until no binary collision is found. For a chain of balls where the impact starts at the left end, the balls are numbered $1, 2, \ldots, n$, and the contacts are numbered $1, 2, \ldots, s$ from the left to the right. In this case, we can handle the left-to-right sequence of binary collisions proposed in Sect. 3.2 by using Algorithm 1. This algorithm can also be used to handle a random sequence of binary collisions by randomly selecting a binary collision in set I instead of getting the minimum value in set I.

4.3 LZB Impact Model

The LZB impact model presented in Sect. 3.3 can be integrated with respect to the impulse scale. To do so, the contact at which the potential energy is maximum is chosen as the primary contact for each integration step. The impulse increment dP_j at each contact is related to the one at the primary contact by the distributing law (22). Two singularities may be encountered during the integration. The first singularity may occur at the beginning of the impact process where the potential energy is zero at all contacts. The second one may occur during the impact process when a contact, which has left the impact process previously, enters again into the impact process. When a singularity occurs, the distributing law (22) must be regularized. The interested reader can refer to [3, Sect. 4.2.8] for the regularization techniques and for the integration algorithm. It is worth mentioning that this integration technique requires a significant computational effort to select the primary contact among all contacts at each integration step and to handle the singularities. In addition, when the primary contact changes from one contact to another, the impulse increment dP_j at each contact computed with the distributing law (22) changes brutally, which might slow down the convergence of the algorithm.

The LZB model can also be integrated with respect to the time scale. To do so, the Darboux-Keller equation (23) is first discretized using the Euler explicit method:

$$\dot{\mathbf{q}}^{k+1} = \dot{\mathbf{q}}^k + \mathbf{M}^{-1}\mathbf{W}\Delta\mathbf{P}^k, \tag{32}$$

where k is an integration step ($k = 1, 2, \ldots, N$). The impulse increment dP_j^k at each contact is obtained by integrating (24) with the Euler explicit scheme:

Algorithm 1 Handling binary collisions according to the left-to-right order in a granular chain.

```
Require: q̇_i^-, m_i for all particles i = 1, 2, ..., n
Require: e_j for all contacts j = 1, 2, ..., s
Ensure: q̇_i^+ for all particles i = 1, 2, ..., n
  // Initialize
  for i = 1 → n do
    q̇_i^+ ← q̇_i^-
  end for
  IsTermination ← false
  N ← 0                                   ▷ number of binary collisions handled
  //Iterations
  while IsTermination = false do          ▷ while impact is not yet terminated
    IsTermination ← true
    // Find all binary collisions to be handled
    I ← ∅                                 ▷ set of all binary collisions to be handled
    for j = 1 → s do
      if q̇_{j+1}^+ − q̇_j^+ < 0 then
        Add j to I
        q̇_j^- ← q̇_j^+
        q̇_{j+1}^- ← q̇_{j+1}^+
        IsTermination ← false
      end if
    end for
    // Select a binary collision in set I and handle it
    k ← min(I)                            ▷ get minimum value in I
    q̇_k^+ ← q̇_k^- (m_k − m_{k+1}e_k)/(m_k + m_{k+1}) + q̇_{k+1}^- (1 + e_k)m_{k+1}/(m_k + m_{k+1})
    q̇_{k+1}^+ ← q̇_k^- (1 + e_k)m_k/(m_k + m_{k+1}) + q̇_{k+1}^- (m_{k+1} − e_k m_k)/(m_k + m_{k+1})
    N ← N + 1
  end while
```

$$\Delta P_j^k = \int_{t^k}^{t^{k+1}} \lambda_j(t)dt \approx \lambda_j^k \Delta t = (1+\eta_j)^{\frac{\eta_j}{1+\eta_j}} K_j^{\frac{1}{1+\eta_j}} (E_j^k)^{\frac{\eta_j}{\eta_j+1}} \Delta t. \tag{33}$$

A singularity occurs with (33) when a contact enters into the impact process at an integration step k, i.e., $E_j^k = 0$. In this case, ΔP_j^k can be approximated as

$$\Delta P_j^k = \int_{t^k}^{t^{k+1}} \lambda_j(t)dt \approx \frac{1}{2}\lambda_j^{k+1}\Delta t = \frac{1}{2}K_j(\delta_j^{k+1})^{\eta_j}\Delta t \approx \frac{1}{2}K_j(\dot{\delta}_j^k \Delta t)^{\eta_j}\Delta t. \tag{34}$$

The potential energy is computed by discretizing (25):

$$E_j^{k+1} = E_j^k + \frac{\dot{\delta}_j^k + \dot{\delta}_j^{k+1}}{2}\Delta P_j^k, \text{ if } \dot{\delta}_j^{k+1} \geq 0, \tag{35}$$

$$E_j^{k+1} = E_j^k + \frac{1}{e_{j,*}^2}\frac{\dot{\delta}_j^k + \dot{\delta}_j^{k+1}}{2}\Delta P_j^k, \text{ if } \dot{\delta}_j^{k+1} < 0. \tag{36}$$

The impact process can be considered to be terminated at a step k if

$$E_j^k = 0, \text{ and } \dot{\delta}_j^k \leq 0, \ \forall j = 1, 2, \ldots, s. \tag{37}$$

The interested reader can follow Algorithms 2, 3 and 4 to implement the resolution of the LZB model with respect to time into a programming language.

Algorithm 2 Integration up to the end of the impact process.

Require: $\dot{\mathbf{q}}^0$, $\mathbf{M}$, $\mathbf{W}$, Δt
Require: E_j^0: initial potential energy at all contacts $j = 1, 2, \ldots, s$
Require: K_j, η_j, $e_{j,*}$ for all $j = 1, 2, \ldots, s$
Ensure: $\dot{\mathbf{q}}$, $\mathbf{P}$ at the end of the impact process
  //Initialize
  $\dot{\boldsymbol{\delta}}^0 \leftarrow -\mathbf{W}^T\dot{\mathbf{q}}^0$
  $\mathbf{P}^0 \leftarrow \mathbf{0}$
  //Integration
  $t \leftarrow 0$ ▷ *Time scale*
  $IsTermination \leftarrow false$
  //$IsTermination = true$*: impact is over*
  //$IsTermination = false$*: otherwise*
  $k \leftarrow 0$
  while $IsTermination = false$ **do** ▷ *while the multiple impacts not yet terminated*
    Check status of each contact and the termination condition with Algorithm 3
    Integrate up to the end of the current step with Algorithm 4
    $t \leftarrow t + \Delta t$
    //Advance to the next step
    $k \leftarrow k + 1$
  end while

For a comparison between the two above integration algorithms, we consider a monodisperse chain of 1000 elastic beads, in which the first bead with a velocity of 1 m/s collides with the other beads at rest. The CoR e_* is then equal to 1.0 for all contacts and the Hertz's contact law ($\eta = 3/2$) is used for each contact. The other parameters are: Young's modulus $E = 203$ GPa, Poisson's coefficient $\nu = 0.3$, ball radius $r = 0.01$ m and mass density $\rho = 7780\,\text{kg/m}^3$. For this chain, the post-impact velocities of balls must satisfy the energy conservation. The integration with respect to impulse with a step size $\Delta P = 10^{-6}$ N.s needs about 2.2×10^7 steps and consumes about 380 s of CPU time. The resulting post-impact velocities of balls satisfy the energy conservation with a relative error of about 1.5×10^{-5}. With regard to the integration with respect to time using (34), a step size $\Delta t = 10^{-8}$ s results in about

Algorithm 3 Check status of each contact and the termination condition at the beginning of a step k.

Require: $\dot{\delta}_j^k$, E_j^k for all $j = 1, 2, .., s$
Ensure: $flag_j^k$ for all $j = 1, 2, .., s$
//$flag_j^k = 0$: *contact does not come into collision*
//$flag_j^k = 1$: *contact begins the compression phase*
//$flag_j^k = 2$: *contact is already in the impact process*
Ensure: $IsTermination$
$IsTermination \leftarrow true$
for $j = 1 \rightarrow s$ **do**
    if $E_j^k = 0$ **then**
        if $\dot{\delta}_j^k \leq 0$ **then**
            $flag_j^k \leftarrow 0$
        else ▷ $\dot{\delta}_j^k > 0$
            $flag_j^k \leftarrow 1$
            $IsTermination \leftarrow false$
        end if
    else ▷ $E_j^k > 0$
        $flag_j^k \leftarrow 2$
        $IsTermination \leftarrow false$
    end if
end for

2.9×10^6 steps and about 38 s of CPU time. The resulting post-impact velocities of balls satisfy the energy conservation with a relative error of about 2.0×10^{-7}. The difference between the solutions obtained with the two integration algorithms is about 0.03%. It can be concluded that the integration algorithm with respect to time is about ten times faster than the integration algorithm with respect to impulse for the considered chain. The first one would be more advantageous for systems with higher numbers of particles.

5 Comparisons

In this section, we present a comparison between Moreau's law, the binary collision model and the LZB model. For this comparison, the outcome given by the LZB model is chosen as reference and different parameters such as the elasticity coefficient, the contact stiffness distribution, the coefficient of restitution and the mass distribution are varied. Free chains of aligned balls are first considered in Sect. 5.1, and chains of aligned balls colliding with a wall are then considered in Sect. 5.2.

Algorithm 4 Integration up to the end of each step k.

Require: $\mathbf{M}$, $\mathbf{W}$, Δt, $\dot{\mathbf{q}}^k$
Require: η_j, K_j, $e_{j,*}$, E_j^k, $\dot{\delta}_j^k$, P_j^k for all $j = 1, 2, \dots, s$
Ensure: $\dot{\mathbf{q}}^{k+1}$, $\dot{\delta}_j^{k+1}$, E_j^{k+1}, P_j^{k+1} for all $j = 1, 2, \dots, s$
//Compute the impulse increment at each contact ΔP_j^k
for $j = 1 \rightarrow s$ **do**
  if $flag_j^k = 0$ **then** ▷ *Contact does not come into the collision process*
    $\delta P_j^k \leftarrow 0$
  else if $flag_j^k = 1$ **then** ▷ *Contact begins the collision process*
    $\Delta P_j^k \leftarrow \frac{1}{2} K_j (\dot{\delta}_j^k \Delta t)^{\eta_j} \Delta t$
  else if $flag_j^k = 2$ **then** ▷ *Contact has already been in the collision process*
    $\Delta P_j^k \leftarrow (1+\eta_j)^{\frac{\eta_j}{1+\eta_j}} K_j^{\frac{1}{1+\eta_j}} (E_j^k)^{\frac{\eta_j}{1+\eta_j}} \Delta t$
  end if
end for
//Compute $\dot{\mathbf{q}}^{k+1}$, $\dot{\boldsymbol{\delta}}^{k+1}$
$\dot{\mathbf{q}}^{k+1} \leftarrow \dot{\mathbf{q}}^k + \mathbf{M}^{-1}\mathbf{W}\Delta\mathbf{P}^k$
$\dot{\boldsymbol{\delta}}^{k+1} \leftarrow -\mathbf{W}^T \dot{\mathbf{q}}^{k+1}$
//Compute P_j^{k+1}, E_j^{k+1}, λ_j^{k+1}
for $j = 1 \rightarrow s$ **do**
  $P_j^{k+1} \leftarrow P_j^k + \Delta P_j^k$
  if $\dot{\delta}_j^{k+1} \geq 0$ **then** ▷ *contact located in the compression phase*
    $E_j^{k+1} \leftarrow E_j^k + \frac{\dot{\delta}_j^k + \dot{\delta}_j^{k+1}}{2} \Delta P_j^k$
  else ▷ *contact located in the expansion phase*
    $E_j^{k+1} \leftarrow E_j^k + \frac{1}{e_{j,*}^2} \frac{\dot{\delta}_j^k + \dot{\delta}_j^{k+1}}{2} \Delta P_j^k$
  end if
end for

5.1 Free Chains of Aligned Beads

Let us first consider monodisperse chains of balls in which all the balls have the same mass. The dependence of the impact outcome on the elasticity coefficient, on the contact stiffness distribution and on the coefficient of restitution is analyzed, respectively, in Sects. 5.1.1, 5.1.2 and 5.1.3. The effect of the mass distribution is then analyzed by considering decorated chain (i.e., polydisperse chains) in Sect. 5.1.4.

5.1.1 Varying the Elasticity Coefficient η

To study the effect of the elasticity coefficient η in the LZB model, a monodisperse chain composed of 100 elastic beads ($e_* = 1$) is considered. The stiffness K_i is the same for all contacts, while the elasticity coefficient η is varied. The impact outcomes given by the LZB model for different values of η are shown in Fig. 9 and are compared to the impact outcomes given by Moreau's law and the binary collision

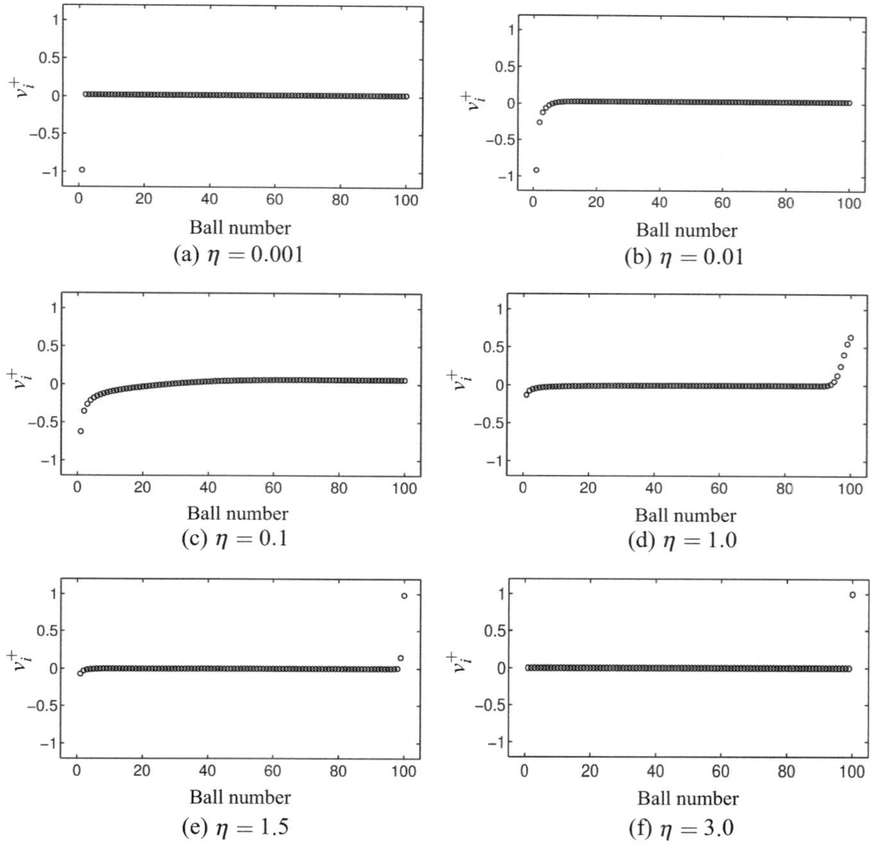

Fig. 9 Post-impact velocities obtained with the LZB model versus ball number for different values of η

law in Fig. 10. It is worth mentioning that the impact outcomes given by Moreau's law and the binary collision model are independent of the elasticity coefficient η. It can be seen that the elasticity coefficient η greatly affects the impact outcome given by the LZB model. For a very small value of η ($\eta = 10^{-3}$, for example), only the first ball bounces back and the remaining balls move forward with almost the same velocity after impact. This is similar to the case where the first ball impacts the other balls which are rigidly bonded. It is interesting to note that this particular impact outcome is given by Moreau's law (Fig. 10). As η increases, fewer balls move forward after impact, as shown in Fig. 11. For a high enough value of η ($\eta = 3$, for example), only the last ball moves forward after impact with a velocity almost equal to the velocity of the first ball before impact, and the other balls are almost at rest, which is the outcome given by the binary collision model (Fig. 10). This is similar to the case where the first ball impacts the other balls which are separated from each other by a gap.

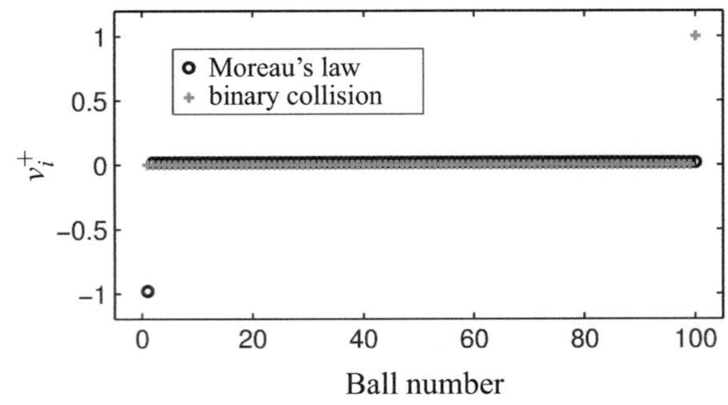

Fig. 10 Post-impact velocities obtained with Moreau's model and the binary collision model versus ball number

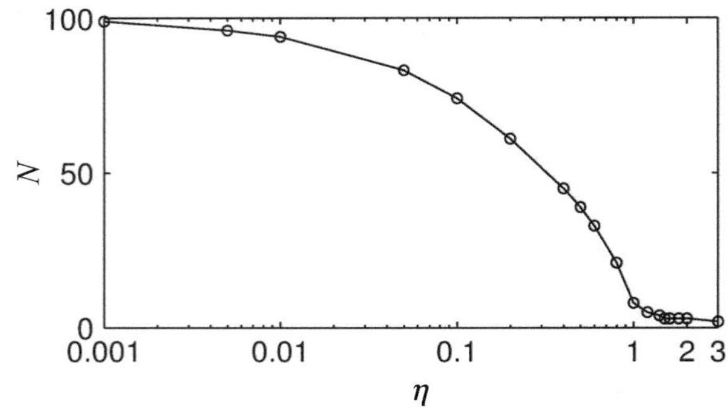

Fig. 11 Number of balls moving forward after impact obtained with the LZB model, versus η

Let us take the outcome given by the LZB model as a reference outcome and then quantify the gap between the impact outcome $\mathbf{v}^+$ obtained with Moreau's law or the binary collision law and the impact outcome $\mathbf{v}^+_{lzb}$ obtained with the LZB model by the following gap measure:

$$\Phi = \frac{\|\mathbf{v}^+ - \mathbf{v}^+_{lzb}\|}{\|\mathbf{v}^+_{lzb}\|} \times 100\%, \tag{38}$$

where $\|.\|$ is the Frobenius norm of a vector. Figure 12 shows the gap measure Φ defined for outcomes given by Moreau's law and the binary collision model versus the elasticity coefficient η. It can be seen that, by varying the elasticity coefficient η from a very small value to a big value, the impact outcome given by the LZB model, initially close to the outcome given by Moreau's law, moves away from the latter, but gets closer to the outcome given by the binary collision model. Except for extreme values of η, the impact outcomes obtained with Moreau's law and the binary collision model are quite far from that given by the LZB model. For spherical homogeneous beads, Hertz's contact model ($\eta = 3/2$) is widely adopted in the literature. In this case, the binary collision model gives an approximation of the impact outcome with an error of about 17% compared to the LZB model, while Moreau's law gives an unrealistic outcome.

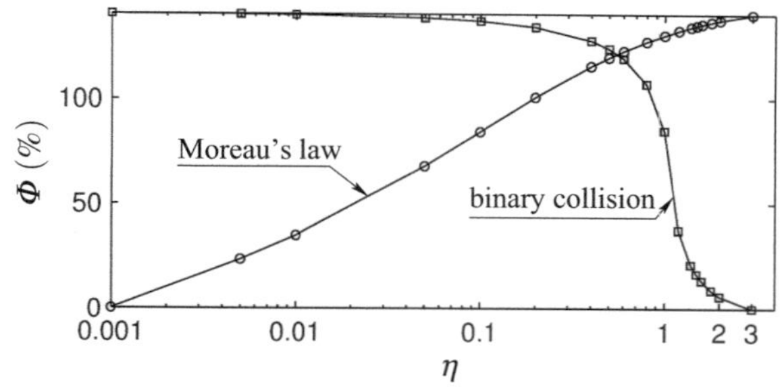

Fig. 12 Gap measure Φ of Moreau's law and the binary collision model versus η used in the LZB model

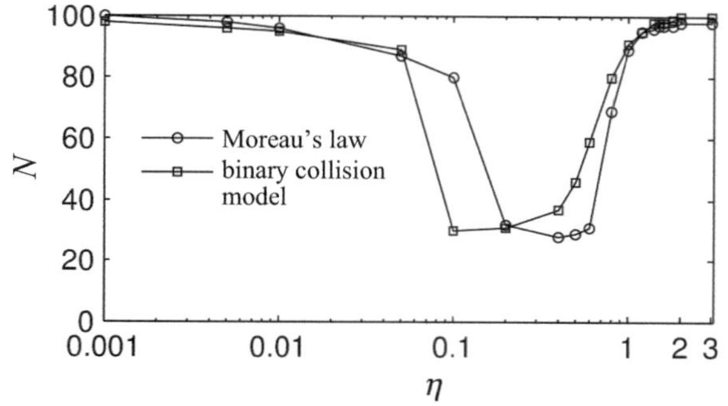

Fig. 13 Number of balls for which the binary collision model and Moreau's model give a good post-impact velocity

It is interesting to note in Fig. 13 that although the outcomes given by the binary collision model and Moreau's law are poor for small and big values of η ($\eta < 0.01$ and $\eta > 1.0$), respectively, they can be considered to be good in terms of the number of balls for which these models give a good post-impact velocity compared to the one given by the LZB model. The post-impact velocity v_i^+ of a ball given by Moreau's law or the binary collision model is considered to be good compared to the result $v_{i,lzb}^+$ obtained with the LZB model if

$$\frac{\mid v_i^+ - v_{i,lzb}^+ \mid}{\|\mathbf{v}_{lzb}^+\|} < \varepsilon, \tag{39}$$

where ε is a precision which is chosen to be equal to 0.05 in this study.

Let us analyze the wave propagation in the considered granular chain when varying the elasticity coefficient η in the LZB model, and the link between the wave propagation and the impact outcome. Figure 14 shows the potential energy E versus time t at the first 20 contacts (from left to right) for different values of η. It can be seen that the wave propagation is greatly affected by the elasticity coefficient η. Three classes can be observed: (i) strongly localized wave at the first contact for very small values of η (Fig. 14a, b), (ii) attenuated and dispersed wave for intermediate values of η (Fig. 14c–e) and (iii) dispersion-free wave for big values of η (Fig. 14f). Herrmann et al. [43] also observed the dispersion-free wave for the elasticity coefficient $\eta = 3.0$. The wave propagation results from the compliance of solid bodies and is

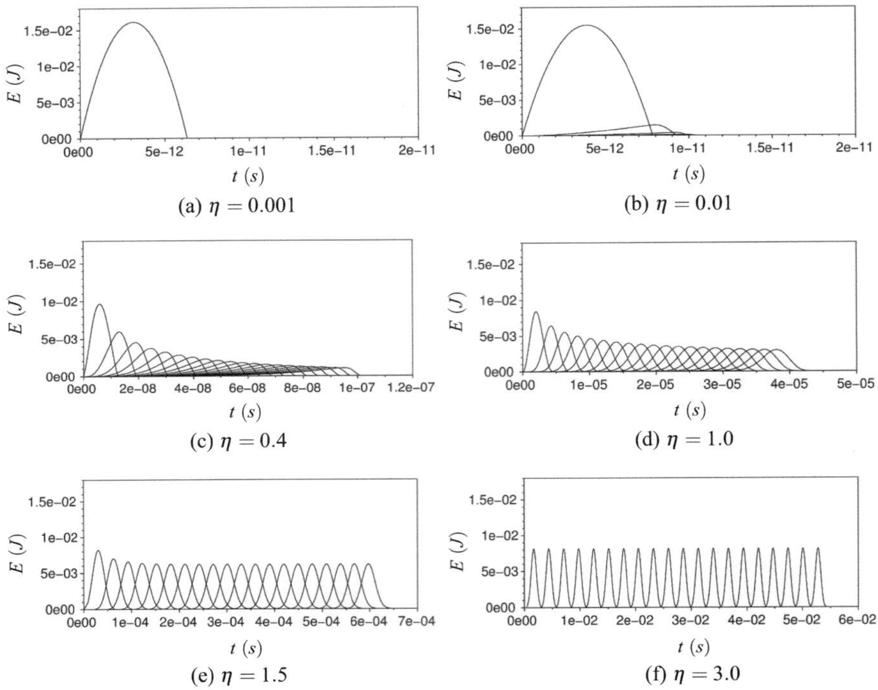

Fig. 14 Potential energy E at the first 20 contacts versus time t for different values of η

an important dynamical effect that should be taken into account in an impact model. The LZB model takes into account this effect by using the contact model shown in Fig. 6. As a result, it is capable of reproducing the wave propagation induced by a shock and subsequently the impact outcome. On the other hand, Moreau's law and the binary collision model completely neglect this compliance effect and make use of two opposite assumptions: the first one assumes that all collisions occur simultaneously, while the second one assumes that collisions occur in a sequential manner. The first assumption can be justified for the wave propagation category (i), so the impact outcome given by the LZB model coincides with the one given by Moreau's law (Figs. 9a, b and 10). The sequential collisions are observed for the wave propagation category (iii); as a consequence, the impact outcome given by the LZB model coincides with the one given by the binary collision model.

The category (ii) corresponds to the wave dispersion for which the shock initiated at the first contact spreads out spatially and the energy induced by impact is shared by many particles. A measure for this dispersion effect in terms of post-impact kinetic energies of balls was introduced in [3]:

$$C_{KE} = \frac{1}{\overline{T}^{+}} \sqrt{\frac{1}{N} \sum_{i=1}^{N} (T_i^{+} - \overline{T}^{+})^2}, \tag{40}$$

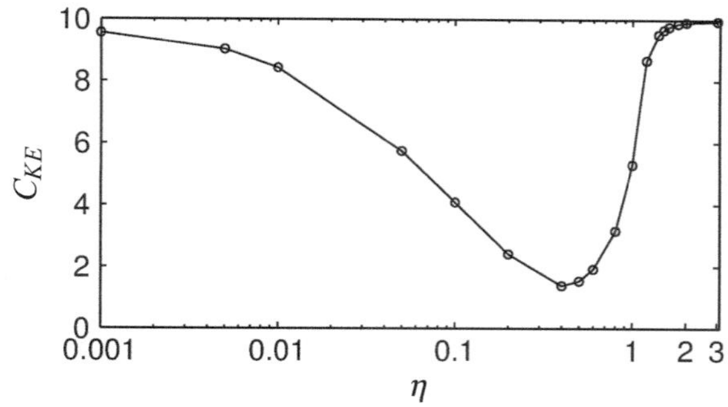

Fig. 15 Dispersion measure C_{KE} versus elasticity coefficient η

where T_i^+ is the post-impact kinetic energy of ball i ($T_i^+ = m_i(\dot{q}_i^+)^2/2$), and $\overline{T}^+$ is the mean post-impact kinetic energy:

$$\overline{T}^+ = \frac{1}{N}\sum_{i=1}^{N} T_i^+. \tag{41}$$

According to (40), the higher the value of C_{KE} is, the lower the dispersion effect is. For a chain of n balls, C_{KE} reaches the maximum value of $\sqrt{n-1}$ for the case where the energy after impact is concentrated in one ball and the other balls are at rest. This chain exhibits zero dispersion effect, also called *dispersion-free* [44, 45]. Figure 15 shows the dispersion measure C_{KE} obtained with the LZB model versus the elasticity coefficient η. The maximum value of C_{KE} for the considered chain of 100 balls is $\sqrt{100-1} \approx 9.95$. It can be seen that the dispersion effect is very weak for very small values of η, and it increases as η increases until $\eta \approx 0.4$, where C_{KE} reaches its minimum value. This means that the dispersion effect is maximum for $\eta \approx 0.4$, for which a strongly dispersed wave propagation can be seen in Fig. 14c. When η increases beyond 0.4, the dispersion effect decreases and almost vanishes for $\eta = 3.0$. It is worth mentioning that the dispersion-free outcome obtained for $\eta = 3.0$ corresponds to the sequential wave propagation shown in Fig. 14f. It was shown in [3] that, for a chain of 3 balls, the dispersion measure C_{KE} increases monotonically with η. However, this monotonic dependency of C_{KE} on η no longer exists for a chain with a high number of balls. Figure 15 also shows that Moreau's law and the binary collision model give good impact outcomes for extreme values of η for which the dispersion effect is very weak.

5.1.2 Varying the Contact Stiffness Distribution

It can be seen in the distributing law (22) that the impact outcome does not depend on the value of the contact stiffness K_i if the latter and the elasticity coefficient η_i are the same for all contacts. In this section, we show how the difference in stiffness between contacts affect the impact outcome and for which cases the outcome of the

LZB model coincides with the ones given by Moreau's law and the binary collision model. For this study, we set the elasticity coefficient $\eta = 3/2$ for all contacts in the monodisperse elastic chain considered in Sect. 5.1.1 and vary the stiffness K_i at each contact according to the following linear law: $K_i = K_{i-1} + \alpha K^*$, with a coefficient α and a reference stiffness K^*. If $\alpha > 0$, the contact stiffness increases progressively from the left to the right of the chain, and the reference stiffness K^* is set to the first contact. Otherwise, the contact stiffness decreases progressively, and the reference stiffness K^* is set to the last contact. It should be noted that the value of the reference stiffness K^* is of no importance.

Figure 16 shows the impact outcome given by the LZB model for different values of α. It can be seen that the impact outcome changes slightly when the contact stiffness is progressively decreased ($\alpha < 0$) and approaches the one given by the binary collision model. Despite a very strong decrease in contact stiffness ($\alpha = -10^4$), we cannot closely reach the latter: the two first balls still bounce back after impact. This

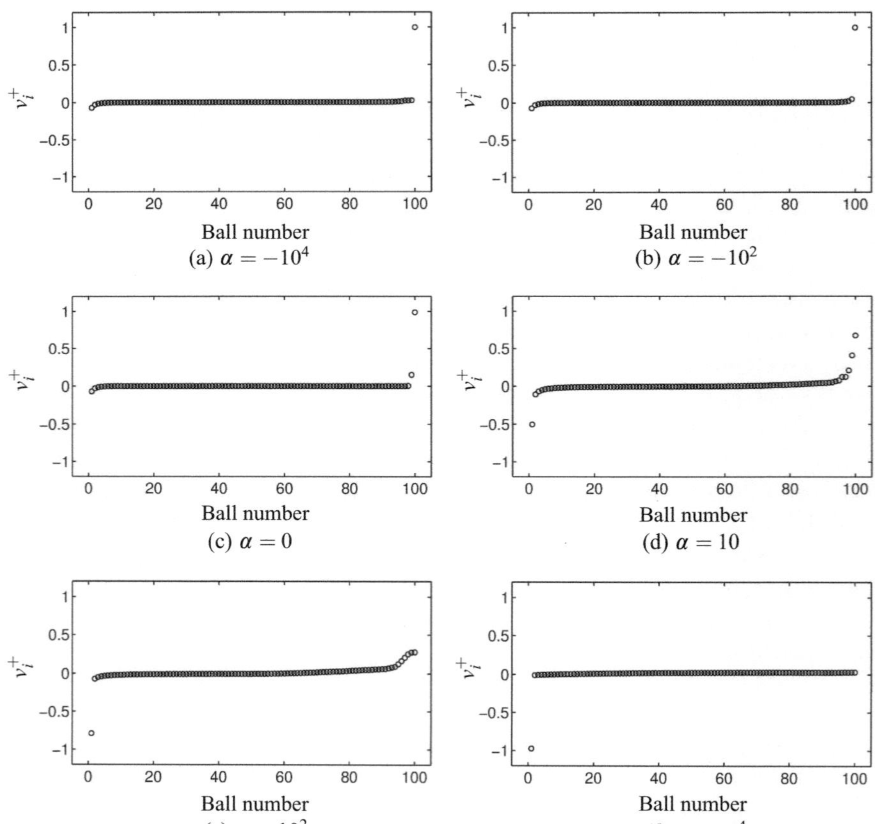

Fig. 16 Post-impact velocities obtained with the LZB model versus ball number for different values of α

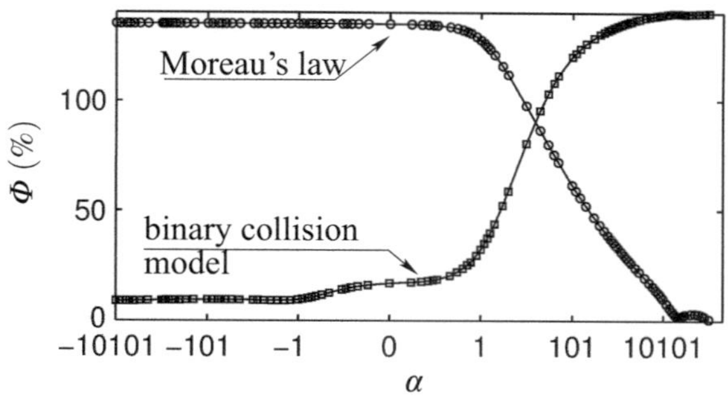

Fig. 17 Gap measure Φ of Moreau's law and the binary collision model versus α

means that the dispersion-free outcome cannot be reached for the considered granular chain if only contact stiffnesses are varied. In fact, Reinsch [45] has developed an analytical analysis for a granular chain with the linear contact model ($\eta = 1$) and has shown that the dispersion-free outcome can only be reached if the mass of each ball and the stiffness of each contact are both varied according to some specific laws. On the other hand, the impact outcome changes greatly when the contact stiffness is progressively increased ($\alpha > 0$) and gets closer to the one given by Moreau's law. The latter one is closely reached for a very strong increase in contact stiffness ($\alpha = 10^4$). Figure 17 shows the gap measure defined in (38) for Moreau's law and the binary collision model versus coefficient α. It is clear that the outcomes given by these two impact laws can be approached by progressively increasing and decreasing the contact stiffness, respectively.

The link between the impact outcome and the wave propagation during impact can be clearly observed in Fig. 18. It can be seen that the solitary wave, which travels in a Hertzian monodisperse chain (Fig. 18c), is not significantly disturbed by a progressive decrease in contact stiffness. On the other hand, a progressive increase in contact stiffness greatly affects the wave propagation in the chain: the wave is more dispersed and more attenuated. This makes the impact outcome significantly different from the one given by the binary collision model. One can also see that for $\alpha = 6$ and 10 (Figs. 18e, f), secondary collisions occur at each contact, making the wave more scattered. With a very strong increase in contact stiffness ($\alpha = 10^4$), the wave is strongly attenuated (Fig. 18h), leading to the impact outcome given by Moreau's law (Fig. 16f). The wave profiles shown in Fig. 18 can explain the non-monotonic dependence of the dispersion measure C_{KE} on α, shown in Fig. 19. The best dispersion effect (the minimum value of C_{KE}) is obtained for $\alpha = 20$. For extreme values of α (a strong decrease or increase in contact stiffness), the dispersion effect is very small, and in these cases Moreau's law and the binary collision model can predict the impact outcome of the chain.

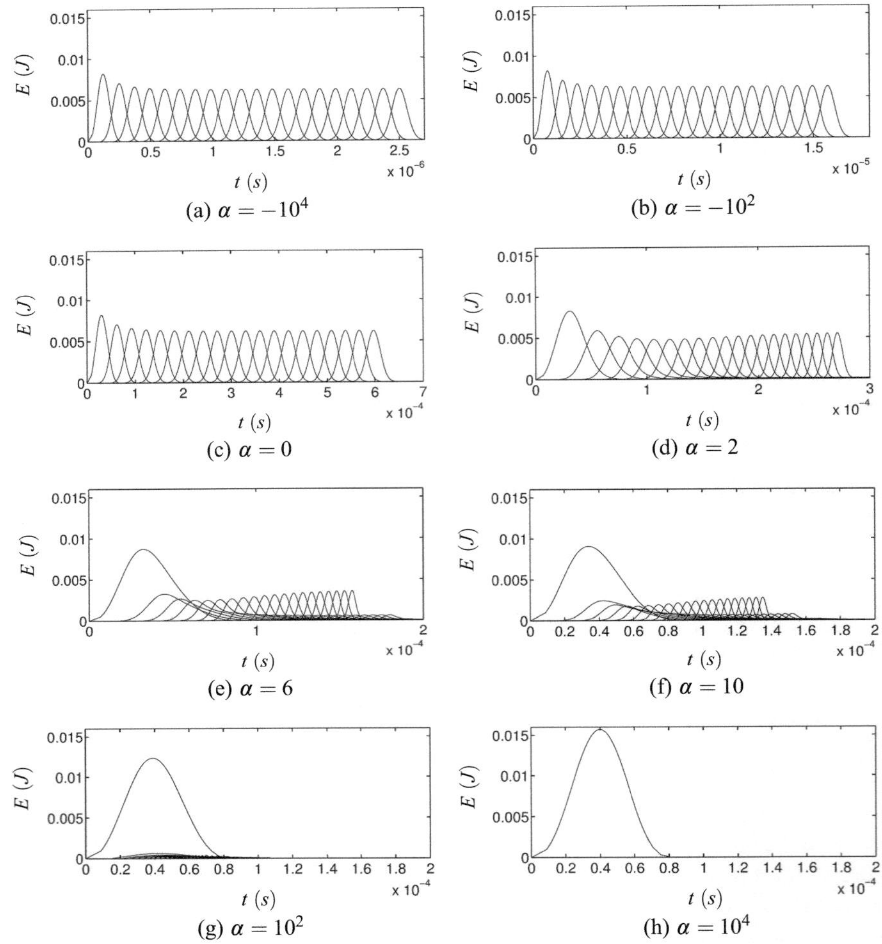

(a) $\alpha = -10^4$ (b) $\alpha = -10^2$

(c) $\alpha = 0$ (d) $\alpha = 2$

(e) $\alpha = 6$ (f) $\alpha = 10$

(g) $\alpha = 10^2$ (h) $\alpha = 10^4$

Fig. 18 Potential energy E at the first 20 contacts versus time t for different values of α

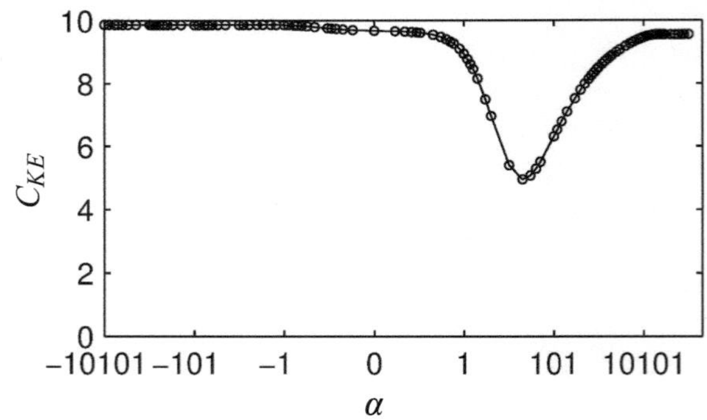

Fig. 19 Dispersion measure C_{KE} versus coefficient α

5.1.3 Varying the Coefficient of Restitution (Dissipation Effect)

So far, we have studied the impact in purely elastic granular chains, i.e., there is no energy dissipation. We will show in the following how these systems behave when contacts between particles are no longer elastic. It is worth mentioning that the perfect elasticity is just an idealized case, and the energy dissipation always exists in the real world. The latter comes from several sources: plasticity or viscosity of the constitutive material, friction at the contact, vibration of the bulk solid, etc. For example, for a collision between two beads constituted of chrome steel, which is a very elastic material, the energetic CoR e_*, defined in Sect. 4.3, is around 0.95 [24]. Let us vary the energetic CoR e_* from 1.0 (purely elastic case) to 0.0 (purely dissipative case) for the monodisperse chain considered in Sect. 5.1.1 with the elasticity coefficient $\eta = 3/2$. Figure 20 shows the impact outcome obtained with the LZB model compared to the one obtained with Moreau's law for different values of the energetic CoR e_*. As mentioned in Sect. 3.2, when using the binary collision model for a dissipative monodisperse chain, there is more than one binary collision to be handled at one time. Two strategies have been proposed for handling these simultaneous collisions. However, the number of binary collisions can be infinite in many cases. Therefore, the binary collision model is not considered for the comparison in this section. The global CoR e used in Moreau's law is equal to the energetic CoR e_* in the LZB model. It can be seen that the CoR e_* greatly affects the impact outcome. Indeed, a decrease of merely 2% in e_* from 1.0 (Fig. 20a) to 0.98 (Fig. 20b) leads to a reduction of 54% in the post-impact velocity of the last ball. Particles tend to be stuck together after impact, i.e., they have almost the same post-impact velocities, as the CoR e_* decreases. We consider that two particles are stuck together if the absolute value of the relative velocity between them is smaller than 0.1% of the pre-impact velocity of the first ball. We define the value of e_* under which particles are stuck together after impact. This value of e_* is 0.86 for the considered chain of 100 balls, and it increases as the number of balls increases (Table 4). According to Moreau's law, the first ball bounces back and the other balls are stuck together after impact for any value of e_*, except for $e_* = 0$ for which all the balls are stuck together. Therefore, the outcome given by Moreau's law is very different from the one given by the LZB model, except for $e_* = 0$ for which these two models give the same outcome. This result is confirmed in Fig. 21, in which the gap measure Φ is plotted against the CoR e_*.

Figure 22 shows a comparison between the two impact models in terms of the kinetic energy ratio KER defined as: $KER = T^+/T^-$ with T^+ and T^- being the kinetic energies before and after impact, respectively. When a granular chain with multiple contacts is subjected to an impact, the induced energy propagates and disperses in the system (Fig. 23), which involves more contacts to participate in the impact process. If the system is dissipative, although each contact dissipates a small amount of energy, the whole system of multiple contacts dissipates a great amount of energy, as shown in Fig. 22. The wave is damped as it propagates through the system. For the considered chain with 100 balls, the energy induced by the shock is almost dissipated when $e_* < 0.9$. With regard to Moreau's law, it underestimates the energy

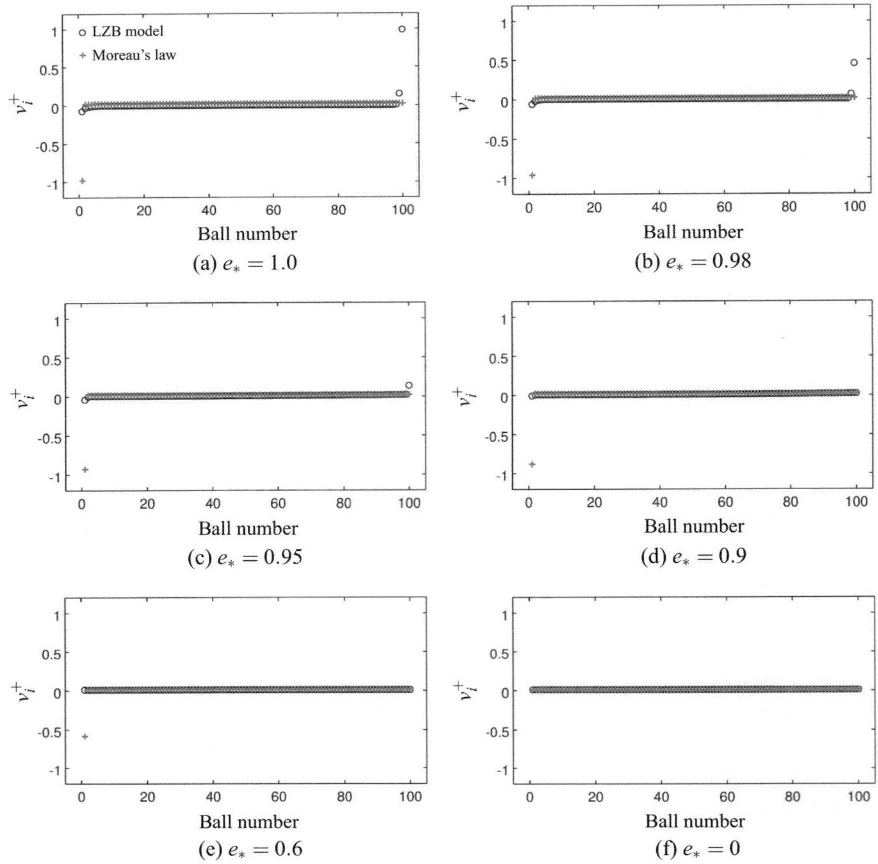

Fig. 20 Post-impact velocities obtained with the LZB model versus ball number for different values of the energetic CoR e_*

dissipation. Indeed, this impact model completely neglects the wave propagation in a system with multiple contacts, so the impact is only localized at the first contact. It should be noted that Moreau's law describes the impact in the considered granular chain as a single impact between the first ball and another solid composed of the other balls. The only case in which this impact law gives the same outcome as the one given by the LZB model is the purely dissipative case ($e_* = 0$), for which the wave is strongly damped and the energy induced by the shock is almost localized at the first contact (Fig. 23f).

Table 4 Value of e_* under which particles are stuck together after impact for different values of the number n of balls

n	2	3	10	20	30	40	50	100	500
e_*	0	0.1	0.5	0.66	0.75	0.79	0.8	0.86	0.88

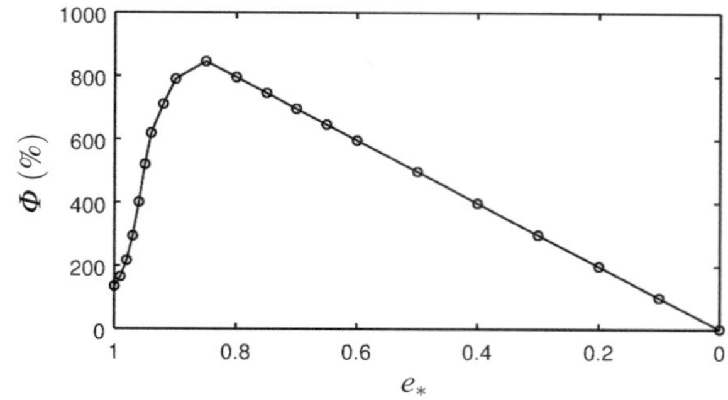

Fig. 21 Gap measure Φ of Moreau's law versus the coefficient of restitution e_*

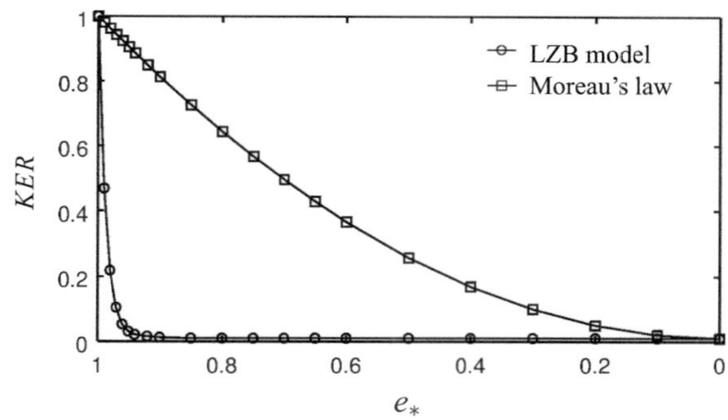

Fig. 22 Kinetic energy ratio KER versus the coefficient of restitution e_*

5.1.4 Decorated Chain

Let us consider a decorated chain to investigate how the distribution of particle masses affects the impact outcome. The considered chain is composed of 101 balls whose masses are distributed as follows: the masses of balls with an odd number (1, 2, 3, ...) are equal to m and the masses of balls with an even number (2, 4, 6, ...) are equal to εm. The mass ratio ε is varied from 0.01 to 100. The Hertz's contact model ($\eta = 3/2$) and the energetic CoR $e_* = 1.0$ are used for simulations performed with the LZB model. Figure 24 shows the impact outcome obtained with the LZB model for different values of the mass ratio ε. One sees that placing small balls between big balls makes the energy more distributed in the chain after impact (Fig. 24b and d), except for very small or very big values of ε. When ε is very small, if we look only at the velocity of the big balls in Fig. 24a, the decorated chain behaves similarly to a monodisperse chain composed of the big balls (Fig. 24c). This means that separating

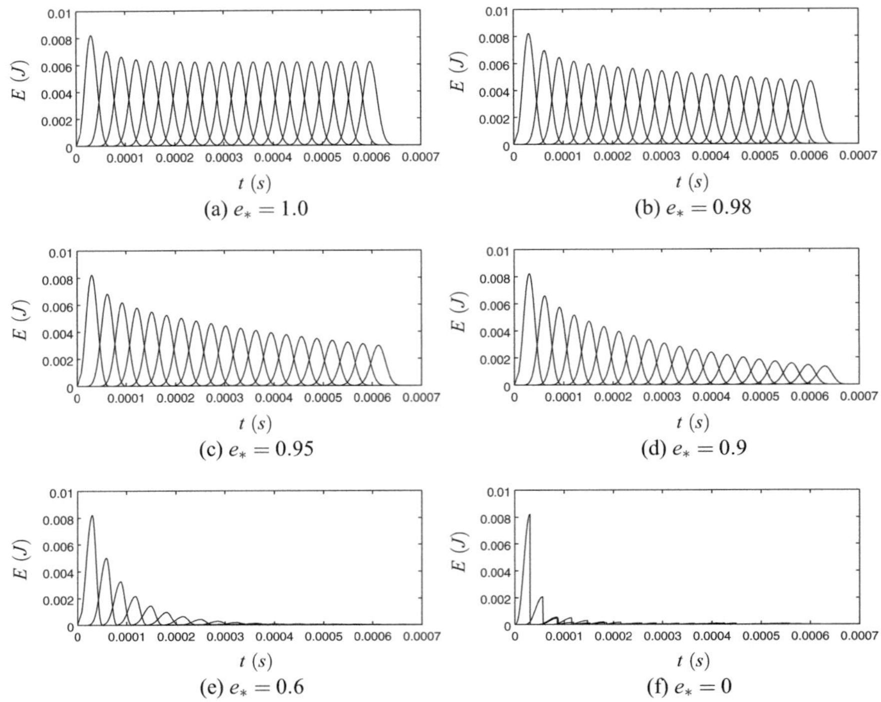

Fig. 23 Potential energy E obtained with the LZB model at the first 20 contacts versus time t for different values of the coefficient of restitution e_*

big balls by very small balls does not significantly change the impact outcome of the big balls. In this case, curves representing the evolution of the potential energy at the two contacts on each small ball almost overlap and we find again the solitary wave which was observed for a monodisperse chain (Fig. 14d). When ε is very big (the first ball is very small compared to the second ball), the first ball bounces back with most of the energy after impact. Concerning Moreau's law, its impact outcome for the considered chain is similar to the one of a single impact between the first ball and the remainder of the chain, independently of the mass distribution. Because the mass of the first ball is very small compared to the remainder of the chain, the first ball bounces back after impact with a velocity almost equal to its velocity before impact. This means that Moreau's law is not capable of predicting the effect of the mass distribution on the outcome of the decorated chain. The only case in which this law gives the same impact outcome as the one given by the LZB model is for a very big value of ε (Fig. 24f). This is due to the fact that the big mass of the second ball compared to the first ball prevents the wave from propagating in the chain, so the collision process is almost localized at the first contact, as shown in Fig. 25b.

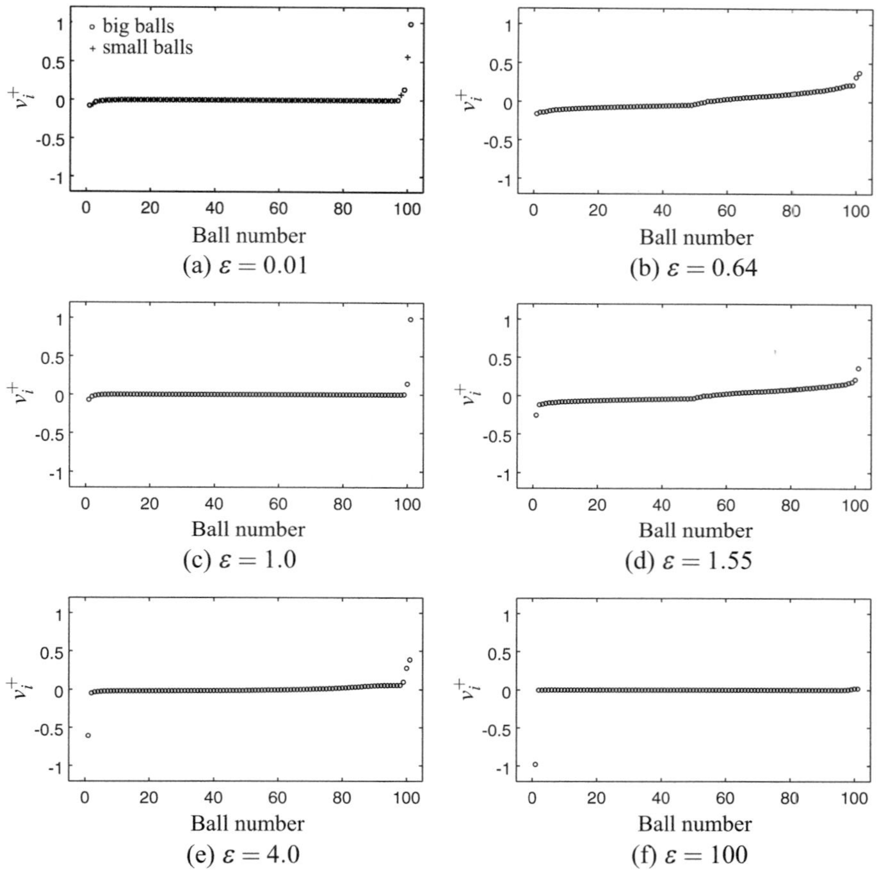

Fig. 24 Post-impact velocities obtained with the LZB model versus ball number for different values of the mass ratio ε

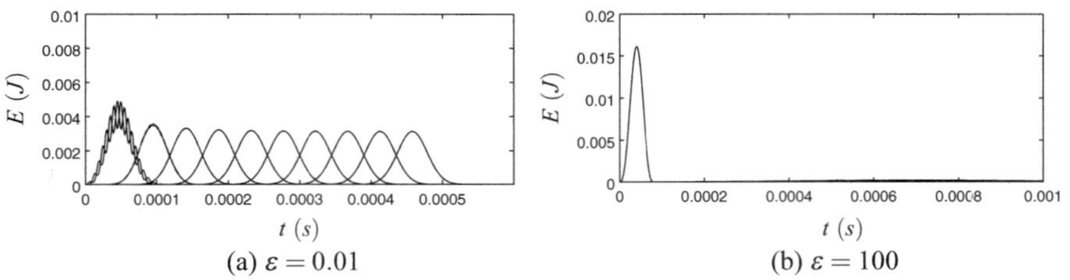

Fig. 25 Potential energy E obtained with the LZB model at the first 20 contacts versus time t for **a** $\varepsilon = 0.01$ and **b** $\varepsilon = 100$

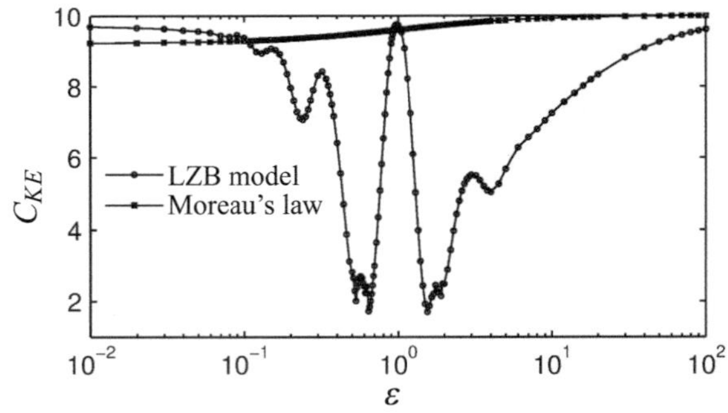

Fig. 26 Dispersion measure C_{KE} versus the mass ratio ε

Remark 9 The binary collision model is not used for this kind of chain, because it leads to undefined impact outcomes after a huge number of binary collisions for several values of the mass ratio ε.

Figure 26 shows the dispersion measure C_{KE} obtained with the LZB model and Moreau's law versus the mass ratio ε. It can be seen that the dispersion of post-impact kinetic energies of balls obtained with Moreau's law is weak and does not change significantly with the mass distribution. Contrastingly, the LZB model predicts a strong effect of the mass distribution on the energy dispersion of the decorated chain. For this kind of chain, the energy induced by the shock is the best dispersed in the chain for $\varepsilon = 0.64$ and 1.55. As stated in [3], the energy dispersion and the force transmission in a granular chain are related to each other. The first value ($\varepsilon = 0.64$) is quite close to the characteristic value $\varepsilon = 0.59$ shown in [46], for which the force transmission in a decorated chain is minimum.

5.1.5 Conclusions

For the tested systems of chains of aligned balls: Moreau's law has good predictive capabilities for small CoR (big dissipation), very small elasticity coefficient, big stiffness increase through the chain, or high mass ratio in decorated chains. In terms of waves, Moreau's law has good prediction capabilities when the wave is localized at the first contact. The binary collision model has good predictive capabilities for large elasticity coefficient or large stiffness decrease through the chain. However, it is very hard to draw conclusions with the binary collision law due to intrinsic issues, like the impossibility of choosing a unique order of collisions (different sequences usually yield different outcomes), and the lack of a criterion that guarantees its convergence (an infinite number of impacts is possible in some cases). For these reasons, this approach should be disregarded most of the time.

Fig. 27 Illustration of a granular chain impacting a wall

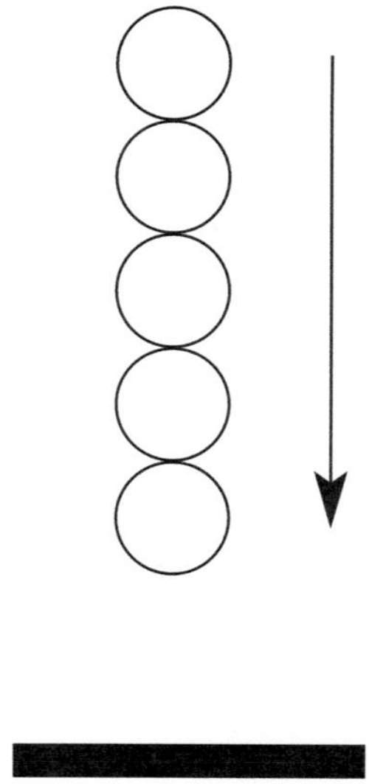

5.2 *Chains Impacting a Rigid Wall*

We have so far considered free granular chains in which the first ball impacts the other stationary balls. In this section, we consider a monodisperse chain of 100 balls in which all balls move with the same velocity and impact a rigid wall, as illustrated in Fig. 27. It is noteworthy that, in contrast to the free chains, in this case, the linear momentum of the 100 balls is not conserved. This kind of impact has been experimentally studied in [47], and a good agreement between the numerical results obtained with the LZB model and the experimental results has been shown in [32]. It was observed that when the chain impacts the wall, the collision process starts at the bottom and then propagates to the top of the chain. The top ball leaves the chain first, is then followed by the next one, and so on. The considered chain is composed of 100 elastic balls and the elasticity coefficient is varied. The balls are numbered from 1 at the top to 100 at the bottom. According to Moreau's law, all the balls are still stuck together and the chain moves upward after the impact with the same velocity as the one before impact. The same impact outcome is obtained with the binary collision model with a sequence of binary collisions from the bottom to the top of the chain to mimic the wave propagation. The post-impact velocities of balls obtained with the LZB model for different values of the elasticity coefficient are shown in Fig. 28. It is shown that when the Hertzian contact model is used ($\eta = 3/2$), balls are detached from each other after impact, except for a few balls in the middle, and the top ball almost doubles its velocity. However, when the linear contact model is used ($\eta = 1$), about 70 balls in the middle come close to being stuck together after impact, this number of balls being about 80 for $\eta = 0.1$. It is expected that for a very small value of η, all the balls are stuck together after impact,

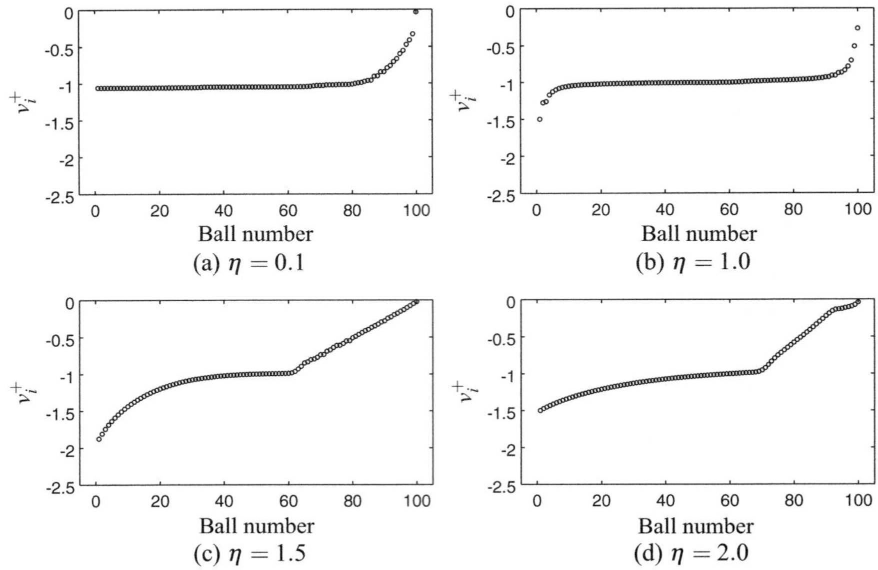

Fig. 28 Post-impact velocities for an elastic monodisperse chain impacting a wall obtained with the LZB model versus the ball number for different values of η

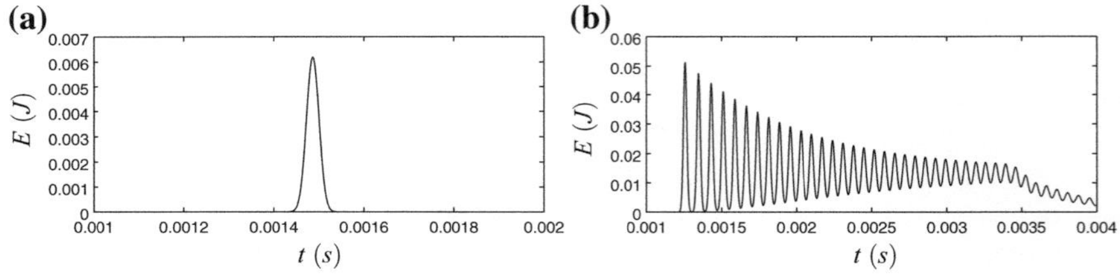

Fig. 29 Potential energy E obtained with the LZB model at the 50th contact versus time t for **a** a free monodisperse chain of 100 balls and **b** for the same chain impacting a wall

which corresponds to the impact outcome given by Moreau's law. However, we were unable to simulate this impact problem for a very small value of η. In fact, when a chain of balls collides with a wall, contacts undergo many repeated collisions, as shown in Fig. 29b. As a consequence, integrating such an impact process is much more difficult than integrating the impact in a free monodisperse chain in which each contact undergoes only one collision (Fig. 29a). One would expect that a value of η higher than 1.5 would make the top ball bounce back with a higher velocity. However, this is not the case for a monodisperse chain impacting a wall, as shown in Fig. 28d where the post-impact velocity of the top ball for $\eta = 2.0$ is lower than for $\eta = 1.5$.

For a granular chain colliding with a wall, a small dissipation at each contact can lead to a large damping effect, since those contacts undergo many repeated collisions,

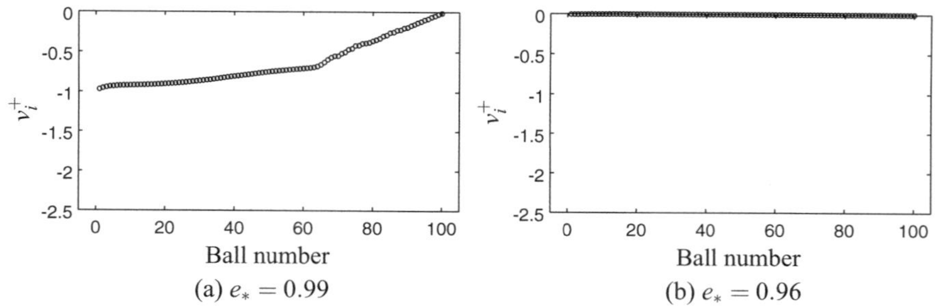

Fig. 30 Post-impact velocities for a monodisperse chain impacting a wall obtained with LZB model versus ball number for **a** $e_* = 0.99$ and **b** $e_* = 0.96$

as mentioned above. As shown in Fig. 30a, a marked change in the impact outcome is observed for $e_* = 0.99$, compared to the elastic case (Fig. 28c), and about 50% of energy is dissipated in this case. When $e_* = 0.96$, the whole energy is dissipated and the chain is stuck to the wall after impact, as shown in Fig. 30b. It is interesting to note that when the considered monodisperse chain impacts a wall, the whole energy is dissipated at a higher value of e_* than when it is free: as shown in Sect. 5.1.3, most of the energy is dissipated for $e_* < 0.9$ for the free chain.

6 Conclusions

In this chapter, we have made comparisons between three classical multiple-impact laws: the binary collision model, Moreau's impact law and the LZB approach. The comparisons are based on chains of aligned beads (free or impacting a wall) in terms of the post-impact velocities and the kinetic energy dispersion, when the coefficients of restitution, the elasticity coefficients, or the contact stiffnesses are varied. The results given by the LZB model are considered as reference solutions. Wave propagation is known to be a crucial effect in such systems. We found that Moreau's law and the binary collision model can predict the impact outcome with accuracy, only in a few "extreme" cases (like a very low or very high elasticity coefficient or mass ratio). Moreau's law gives good outcomes when the wave is localized at the first contact. Its advantage is that it is easy to implement, even in the case of a great number of bodies and contacts. The binary collision law suffers from severe drawbacks, such as the possible infinity of impacts or different outcomes for different sequences of impacts, which make it very delicate to use for reliable computations in most cases. Future studies should focus on two-dimensional granular systems, with Coulomb's friction at contacts.

References

1. Brogliato B (2010) Nonsmooth mechanics. Models, dynamics and control, 3rd edn. Communications and control engineering. Springer International Publication, Switzerland
2. Schwager T, Poschel T (2008) Coefficient of restitution for viscoelastic spheres. Phys Rev E **78**(5): 051304
3. Nguyen NS, Brogliato B (2014) Multiple impacts in dissipative granular chains. Lecture notes in applied and computational mechanics, vol 72. Springer, Berlin Heidelberg
4. Paoli L (2005) Continuous dependence on data for vibro-impact problems. Math Models Methods Appl Sci 35(1):1–41
5. Towne DH, Hadlock CR (1977) One-dimensional collisions and Chebyschev polynomials. Am J Phys 45(3):255–259
6. Moreau JJ (1983) Liaisons unilatérales sans frottement et chocs inélastiques. Comptes-Rendus des Séances de l'Académie des Sciences 296:1473–1476
7. Moreau JJ (1988) Unilateral contact and dry friction in finite freedom dynamics. In: Moreau JJ, Panagiotopoulos PD (eds) Nonsmooth mechanics and applications. CISM Courses and Lectures no 302, International Center for Mechanical Sciences. Springer, pp 1–82
8. Dzonou R, Monteiro Marques MDP, Paoli L (2009) A convergence result for a vibro impact problem with a general inertia operator. Nonlinear Dyn 58(1–2):361–384
9. Giouvanidis AL, Dimitrakopoulos EG (2016) Modeling contact in rocking structures with a nonsmooth dynamics approach. In: ECCOMAS Congress, VII European congress on computational methods in applied sciences and engineering, Crete Island, Greece, 5–10 June 2016
10. Giouvanidis AL, Dimitrakopoulos EG (2017) Nonsmooth dynamics analysis of sticking impacts in rocking structures. Bull Earthquake Eng 15:2273–2304
11. Caselli F, Frémond M (2009) Collision of three balls on a plane. Comput Mech 43:743–754
12. Brogliato B, Zhang H, Liu C (2012) Analysis of a generalized kinematic impact law for multibody-multicontact systems, with application to the planar rocking block and chains of balls. Multibody Syst Dyn 27(3):351–382
13. Brogliato B (2014) Kinetic quasi-velocities in unilaterally constrained Lagragian mechanics with impacts and friction. Multibody Syst Dyn 32(2):175–216
14. Gharib M, Celik A, Hurmuzlu Y (2011) Shock absorption using linear particle chains with multiple impacts. ASME J Appl Mech 78(3):031005
15. Leine RI, van de Wouw N (2008) Stability and convergence of mechanical systems with unilateral constraints. Lecture notes in applied and computational mechanics, vol 36. Springer, Berlin, Heidelberg
16. Najafabadi SAN, Kovecses J, Angeles J (2008) Impacts in multibody systems: modeling and experiments. Multibody Syst Dyn 20(2):163–176
17. Rodriguez A, Bowling A (2015) Study of Newton's cradle using a new discrete approach. Multibody Syst Dyn 33(1):61–92
18. Winandy T, Leine RI (2017) A maximal monotone impact law for the 3-ball Newton's cradle. Multibody Syst Dyn 39:79–94
19. Pfeiffer F, Glocker C (1996) Multibody dynamics with unilateral contacts. Wiley series in nonlinear science (1996)
20. Glocker C (2006) An introduction to impacts. In: CISM courses and lectures no 302, International Center for Mechanical Sciences, Springer, pp 45–101
21. Moreau JJ (1994) Some numerical methods in multibody dynamics: application to granular materials. Eur J Mech A/Solids 13(4):93–114
22. Moreau JJ (1999) Numerical aspects of the sweeping process. Comput Methods Appl Mech Eng 177(3–4):329–349
23. Aeberhard U, Payr M, Glocker C (2006) Theoretical and experimental treatment of perfect multi-contact-collision. In: Proceedings of 3rd Asian conference on multibody dynamics ACMD06, Tokyo, 1–4 Aug 2006
24. Nakagawa M, Agui JH, Wu DT, Extramiana DV (2003) Impulse dispersion in a tapered granular chain. Granular Matter 4(4):167–174

25. Harbola U, Rosas A, Esposito M, Lindenberg K (2009) Pulse propagation in tapered granular chains: an analytic study. Phys Rev E 80(3):031303
26. Machado LP, Rosas A, Lindenberg K (2013) Momentum and energy propagation in tapered granular chains. Granular Matter 15(6):735–746
27. Rosas A, Lindenberg K (2017) Pulse propagation in granular chains: the binary collision approximation. Int J Modern Phys B 31(10):1742016
28. Crassous J, Beladjine D, Valance A (2007) Impact of a projectile on a granular medium described by a collision model. Phys Rev Lett 99(24):248001
29. Valance A, Crassous J (2009) Granular medium impacted by a projectile: experiment and model. Eur Phys J E: Soft Matter Biol Phys 30(1):43–54
30. Liu C, Zhao Z, Brogliato B (2008) Frictionless multiple impacts in multibody systems: Part I. Theoretical framework. Proc R Soc A: Math Phys Eng Sci 464(2100):3193–3211
31. Liu C, Zhao Z, Brogliato B (2008) Energy dissipation and dispersion effects in a granular media. Phys Rev E 78(3):031307
32. Liu C, Zhao Z, Brogliato B (2009) Frictionless multiple impacts in multibody systems: Part II. Numerical algorithm and simulation results. Proc R Soc A: Math Phys Eng Sci 465(2101):1–23
33. Liu C, Zhang H, Zhen Z, Brogliato B (2013) Impact/contact dynamics in a disc-ball system. Proc R Soc A: Math Phys Eng Sci 469:20120741
34. Nguyen NS, Brogliato B (2012) Shock dynamics in granular chains: numerical simulations and comparisons with experimental tests. Granular Matter 14(3):341–362
35. Wang J, Liu C, Zhao Z (2014) Nonsmooth dynamics of a 3D rigid body on a vibrating plate. Multibody Syst Dyn 32(2):217–239
36. Zhang H, Brogliato B, Liu C (2014) Dynamics of planar rocking-blocks with Coulomb friction and unilateral constraints: comparisons between experimental and numerical data. Multibody Syst Dyn 32(1):1–25
37. Zhao Z, Liu C, Brogliato B (2009) Planar dynamics of a rigid body system with frictional impacts. II. Qualitative analysis and numerical simulations. Proc R Soc A: Math Phys Eng Sci 465(2107):2267–2292
38. Stronge WJ (2004) Impact mechanics. Cambridge University Press
39. Acary V, Brogliato B (2008) Numerical methods for nonsmooth dynamical systems. Applications in mechanics and electronics. Lecture notes in applied and computational mechanics, vol 35 Springer, Berlin, Heidelberg
40. Jean M (1999) The non-smooth contact dynamics method. Comput Methods Appl Mech Eng 177(3–4):235–257
41. Acary V (2013) Projected event-capturing time-stepping schemes for nonsmooth mechanical systems with unilateral contact and Coulomb's friction. Comput Methods Appl Mech Eng 256:224–250
42. Acary V (2016) Energy conservation and dissipation properties of time-integration methods for nonsmooth elastodynamics with contact. ZAMM-J Appl Math Mech/Z Angew Math Mechanik 96(5):585–603
43. Herrmann F, Seitz M (1982) How does the ball chain work? Am J Phys 50(11):977–981
44. Herrmann F, Schmälzle P (1981) Simple explanation of a well known collision experiment. Am J Phys 49(8):761–764
45. Reinsch M (1994) Dispersion-free linear chains. Am J Phys 62(3):271–278
46. Jayaprakash KR, Starosvetsky Y, Vakakis AF (2011) New family of solitary waves in granular dimer chains with no precompression. Phys Rev E 83(3):036606
47. Falcon E, Laroche A, Fauve S, Coste C (1998) Collision of a 1-D column of beads with a wall. Eur Phys J B 5:111–131

Variational Analysis of Inequality Impact Laws for Perfect Unilateral Constraints

Tom Winandy, Michael Baumann and Remco I. Leine

Abstract This chapter deals with frictionless instantaneous impacts in rigid multibody dynamics. For autonomous multibody systems which are subjected to perfect unilateral constraints, a geometric description of the impacts on the respective tangent space to the configuration manifold is presented. The mass matrix of a mechanical system endows the configuration manifold with the structure of a Riemannian manifold and provides an isomorphism between the tangent space and the cotangent space at each point of the configuration manifold. Kinematic quantities (virtual displacements, velocities) are elements of the tangent space, while kinetic quantities (forces, impulsive forces) live in the cotangent space, the dual space of the tangent space. Impact laws, as constitutive laws relating primal and dual quantities, are introduced as set-valued mappings between these two spaces. Methods from Convex Analysis permit to study what the implications are if the impact law is maximal monotone. Finally, the generalized Newton's and the generalized Poisson's impact law are considered as illustrative examples.

1 Introduction

The present chapter[1] deals with frictionless instantaneous impacts in rigid multibody dynamics. Our aim is to derive a geometric description of impacts and to reveal the

---

[1] Sections 2–5 were written by the first author. Sections 2–4 gather a wealth of ideas and concepts of various authors, notably Aeberhard [3], Glocker [15, 16, 19], Ballard [5], and Moreau [34, 36] (in reverse chronological order). Sections 5 and 6 are based on the PhD thesis [6] of the second author.

T. Winandy (✉) · M. Baumann · R. I. Leine
Institute for Nonlinear Mechanics, University of Stuttgart, Pfaffenwaldring 9, 70569 Stuttgart, Germany
e-mail: winandy@inm.uni-stuttgart.de

M. Baumann
e-mail: michael.baumann@alumni.ethz.ch

R. I. Leine
e-mail: leine@inm.uni-stuttgart.de

R. I. Leine et al. (eds.), *Advanced Topics in Nonsmooth Dynamics*,
https://doi.org/10.1007/978-3-319-75972-2_2

mathematical structure of the related constitutive laws. We start by considering an autonomous (i.e., time-independent) multibody system without unilateral constraints in the geometric framework of Riemannian manifolds [25, Chap. 13]. It is well-known that the set of all possible configurations of a multibody system can be modelled as a Riemannian configuration manifold whose metric is given by the mass matrix of the system [29]. The dynamics of such multibody systems without unilateral constraints is governed by second-order ordinary differential equations whose solutions are differentiable with respect to time [8].

If the set of possible positions of the multibody system is now restricted by scleronomic (i.e., time-independent) unilateral constraints, a wider class of solutions needs to be considered. Like bilateral constraints, a unilateral constraint comes along with a constraint force that guarantees that the dynamics (i.e., the motion) respects the constraint [15]. Therefore, the velocities may jump instantaneously. The formulation of perfect unilateral constraints in a geometric setting has been presented by Ballard in [5].

In the framework of hard unilateral constraints, the impacts are jumps in the velocities that come along with impulsive forces and that occur at some instants of time. Therefore, the position remains constant over an impact, which means geometrically, that the mathematical description of the impact takes place on the tangent space to the configuration manifold at the respective position where the impact occurs. The perfect unilateral constraints restrict the tangent space (the space of velocities) and its dual, the cotangent space (the space of forces), to conic subsets, as is well-known from non-smooth mechanics [9, 15, 36]. This means that the pre-impact and post-impact velocities, as well as the impulsive contact forces, have to obey certain restrictions, aside from the fact that they are linked by the impact equation. Additionally, the physical requirement can be added that the kinetic energy must not increase over an impact. In general, the so-derived algebraic impact description on the tangent space still allows for multiple post-impact velocities [16, 19]. It needs to be complemented by a constitutive law (the set-valued impact law) to provide a unique post-impact velocity for given initial data. We will investigate the mathematical structure of this constitutive law. The generalized Newton's [17] and Poisson's impact law [18] will serve as two examples of frictionless instantaneous impact laws that are commonly used in multibody dynamics [37].

As clear as the above outline might appear, it involves a substantial amount of mathematics. During their studies in engineering, the authors have followed only a shallow training in mathematics, which seems to be firmly anchored in the curricula at too many engineering faculties. While in the first half of the twentieth century, differential geometry was at the core of analytical mechanics, nowadays, technical mechanics and modern differential geometry have drifted widely apart. For comments on this deplorable development, the reader is referred to [15, Chap. 15]. Because mechanics and mathematics (especially geometry) cannot be separated, Sects. 2–4 have been written with the aim of being accessible for readers with a typical engineering background in mathematics.

Then, we will allow ourselves to draw the reader's attention to the application of convex analysis to mechanics. Convex analysis has been coined by Moreau and Rockafellar. At least for Moreau, convex analysis is tied to mechanics. In his *Fonctionelles Convexes* [34, p. 3], Moreau writes

> L'intérêt de l'auteur pour la convexité est motivé par la théorie des liaisons unilatérales en mécanique [...].

which can be translated as: *The author's interest in convexity originates from the theory of unilateral constraints in mechanics*. Probably due to Moreau's inspiration by mechanics, he formulates his treatise in the context of dual vector spaces, while the books [39, 40] by Rockafellar are based on the consideration of the n-dimensional real vector space $\mathbb{R}^n$ together with its canonical inner product. Therefore, the role played by the duality pairing in the work of Moreau is played by the canonical inner product in the books by Rockafellar.

Because we will need both concepts, we introduce them in Sect. 2. First, the dual space of a finite dimensional real vector space is introduced via the duality pairing between their elements. If a symmetric, positive definite, covariant 2-tensor (see [25, Chap. 12]) is given on a finite dimensional real vector space, then it induces an inner product, as well as an isomorphism between the vector space and its dual. A covariant 2-tensor on a finite dimensional vector space is also called a bilinear form. This section may be skipped by readers familiar with tensor calculus on vector spaces. For a good introduction to linear algebra, including the concept of dual spaces, we refer to [13].

Section 3 introduces the basic differential geometric concepts that are relevant to the description of multibody systems without unilateral constraints. For a concise treatment of differential geometry, we refer to the book by Aubin [4]. A comprehensive treatment with many comments on the historical development can be found in the five volumes by Spivak [41]. Good reference books are [24, 25]. References for geometric mechanics in the framework of a time-independent configuration manifold are [1, 10, 20, 32].

The unilateral constraints are added in Sect. 4. First, a geometric description of the restricted set of positions compatible with the unilateral constraints is introduced on the configuration manifold using gap functions. These restrictions induce cones on the respective tangent and cotangent space of the configuration manifold. These cones are used to define the concept of perfect unilateral constraints, as in [5, 16]. Then, the impact equations are stated in their geometric form. We use a result from [3] and introduce orthogonal projectors on the tangent space that are induced by the unilateral constraints. Section 5 is concerned with the implications of a set-valued impact law that has the property of being maximal monotone. The central concept that we use is the Minty parametrization from the book *Variational Analysis* [40] by Rockafellar and Wets. Therefore, we speak of variational analysis of impact laws [26]. The previously derived projectors now become crucial in order to establish the connection between the description of impacts in contact velocities and the one in generalized velocities. Section 6 deals with two specific frictionless impact laws, the generalized Newton's [17] and the generalized Poisson's impact law [18]. Finally, we use the Sect. 8 to critically review the presented results and to comment on the insight that may be distilled from our considerations.

2 Some Linear Algebra

Let V be a finite dimensional real vector space. An element $\hat{u} \in V$ is called a *vector*. Let $\hat{\omega}$ be a linear real-valued map on V, i.e.,

$$\hat{\omega}\colon V \to \mathbb{R}, \ \hat{u} \mapsto \hat{\omega}(\hat{u}), \tag{1}$$

such that

$$\hat{\omega}(a\hat{u} + b\hat{v}) = a\hat{\omega}(\hat{u}) + b\hat{\omega}(\hat{v}), \tag{2}$$

for any $a, b \in \mathbb{R}$ and $\hat{u}, \hat{v} \in V$. The set V^* of all linear real-valued maps on V is known as the *dual space* of V. It is a real vector space of the same dimension as V. An element $\hat{\omega} \in V^*$ is called a *covector*.

Consider the dual space $V^{**} := (V^*)^*$ of the dual space V^*, i.e., the space of linear real-valued maps on V^*. The identification of V^{**} with V via the canonical isomorphism allows us to write $\hat{u}(\hat{\omega}) = \hat{\omega}(\hat{u})$. This operation between dual vectors is known as *duality pairing* and it is denoted by

$$\langle \hat{\omega}\,,\, \hat{u} \rangle = \langle \hat{u}\,,\, \hat{\omega} \rangle = \hat{\omega} \cdot \hat{u} = \hat{u} \cdot \hat{\omega} = \hat{\omega}(\hat{u}). \tag{3}$$

Let

$$A\cdot\colon V \to W \tag{4}$$

be a linear map between two finite dimensional real vector spaces V and W. It has a unique *transpose* (or dual) *map*

$$A^{\mathrm{T}}\cdot\colon W^* \to V^*, \tag{5}$$

which is defined by the identity

$$\langle A^{\mathrm{T}} \cdot \hat{\omega}\,,\, \hat{u} \rangle = \langle \hat{\omega}\,,\, A \cdot \hat{u} \rangle, \tag{6}$$

which has to hold for any $\hat{u} \in V$ and any $\hat{\omega} \in W^*$. On the left-hand side stands the duality pairing between elements of V^* and V, while on the right-hand side, the angle brackets denote the duality pairing between elements of W^* and W.

The real vector space V can be endowed with an *inner product*, which we denote by

$$(\cdot\,,\,\cdot)\colon V \times V \to \mathbb{R}. \tag{7}$$

An inner product on a real vector space is symmetric, bilinear and positive definite. Symmetric means that $(\hat{u}\,,\,\hat{v}) = (\hat{v}\,,\,\hat{u})$ for all $\hat{u}, \hat{v} \in V$. Bilinearity is the property that

$$\begin{aligned}(a\hat{u} + b\hat{v}\,,\,\hat{w}) &= a(\hat{u}\,,\,\hat{w}) + b(\hat{v}\,,\,\hat{w}),\\ (\hat{u}\,,\,a\hat{v} + b\hat{w}) &= a(\hat{u}\,,\,\hat{v}) + b(\hat{u}\,,\,\hat{w}),\end{aligned} \tag{8}$$

for all $a, b \in \mathbb{R}$ and $\hat{u}, \hat{v}, \hat{w} \in V$. The inner product $(\cdot\,,\,\cdot)$ is positive definite if it satisfies

$$(\hat{u}\,,\,\hat{u}) > 0,\ \forall \hat{u} \in V \setminus \{0\}. \tag{9}$$

Therefore, the inner product on V is nothing other than a symmetric, positive definite, covariant 2-tensor

$$\begin{aligned}\hat{M}\colon V \times V &\to \mathbb{R},\\ (\hat{u}, \hat{v}) &\mapsto \hat{M}(\hat{u}, \hat{v}) = \hat{u} \cdot \hat{M} \cdot \hat{v}.\end{aligned} \tag{10}$$

In the following, we use the notation $(\cdot\,,\,\cdot)_{\hat{M}}$ to designate the inner product, which corresponds to the symmetric, positive definite, covariant 2-tensor $\hat{M}$, i.e.,

$$(\hat{u}\,,\,\hat{v})_{\hat{M}} = \hat{u} \cdot \hat{M} \cdot \hat{v}. \tag{11}$$

If a basis $\{\hat{\mathrm{e}}_1, \ldots, \hat{\mathrm{e}}_n\}$ with $\hat{\mathrm{e}}_i \in V$ is chosen on the n-dimensional real vector space V, then its dual basis is denoted by $\{\hat{\mathrm{e}}^1, \ldots, \hat{\mathrm{e}}^n\}$ with $\hat{\mathrm{e}}^i \in V^*$ and it is defined via the duality pairing

$$\hat{\mathrm{e}}^i(\hat{\mathrm{e}}_j) = \hat{\mathrm{e}}^i \cdot \hat{\mathrm{e}}_j = \delta^i_j = \begin{cases}1, & \text{if } i = j,\\ 0, & \text{if } i \neq j,\end{cases} \tag{12}$$

where δ^i_j is known as the Kronecker delta. A vector $\hat{u} \in V$ is then written as

$$\hat{u} = u^i\,\hat{\mathrm{e}}_i, \tag{13}$$

where we have used index notation together with Einstein's summation convention, i.e., the index i runs from 1 to n and a summation is understood over an index that appears once as a lower and once as an upper index. The n coefficients u^i can be gathered in an $\mathbb{R}^n$-tuple as

$$\mathbf{u} = \left(u^1\ \ldots\ u^n\right)^{\mathrm{T}}. \tag{14}$$

A covector $\hat{\omega} \in V^*$ is expressed analogously as

$$\hat{\omega} = \omega_j \, \hat{\mathrm{e}}^j, \tag{15}$$

and again, the coefficients can be collected in an $\mathbb{R}^n$-tuple as follows:

$$\boldsymbol{\omega} = \left(\omega_1 \, \ldots \, \omega_n\right)^{\mathrm{T}}. \tag{16}$$

In order to keep track of the fact that the $\mathbb{R}^n$-tuple $\boldsymbol{\omega}$ is related to a covector, we will use the notation $\boldsymbol{\omega} \in \mathbb{R}^{n*}$ with an asterisk to distinguish between primal and dual quantities on $\mathbb{R}^n$. The duality pairing can then be written as

$$\langle \hat{\omega} \, , \, \hat{u} \rangle = \hat{\omega} \cdot \hat{u} = \omega_i \, u^i = \boldsymbol{\omega}^{\mathrm{T}} \mathbf{u}. \tag{17}$$

Using the dual basis $\{\hat{\mathrm{e}}^1, \ldots, \hat{\mathrm{e}}^n\}$, the symmetric, positive definite, covariant 2-tensor $\hat{M}$ can be written as

$$\hat{M} = M_{kl} \, \hat{\mathrm{e}}^k \otimes \hat{\mathrm{e}}^l. \tag{18}$$

If we consider the n-by-n matrix

$$\mathbf{M} = \begin{pmatrix} M_{11} & \cdots & M_{1n} \\ \vdots & \ddots & \vdots \\ M_{n1} & \cdots & M_{nn} \end{pmatrix}, \tag{19}$$

then the inner product (11) can be written as

$$(\hat{u} \, , \, \hat{v})_{\hat{M}} = \hat{u} \cdot \hat{M} \cdot \hat{v} = M_{ij} u^i v^j = \mathbf{u}^{\mathrm{T}} \mathbf{M} \mathbf{v}. \tag{20}$$

The symmetric, positive definite, covariant 2-tensor $\hat{M}$, respectively the inner product, provides the two isomorphisms

$$\begin{aligned} \hat{M}\cdot\colon V &\to V^*, \\ \hat{u} &\mapsto \hat{M} \cdot \hat{u} = M_{kl} u^l \, \hat{\mathrm{e}}^k \end{aligned} \tag{21}$$

and

$$\begin{aligned} \hat{M}^{-1}\cdot\colon V^* &\to V, \\ \hat{\omega} &\mapsto \hat{M}^{-1} \cdot \hat{\omega} = M^{ij} \omega_j \, \hat{\mathrm{e}}_i, \end{aligned} \tag{22}$$

where M^{ij} are the elements of the matrix $\mathbf{M}^{-1}$, i.e.,

$$\mathbf{M}^{-1} = \begin{pmatrix} M^{11} & \ldots & M^{1n} \\ \vdots & \ddots & \vdots \\ M^{n1} & \ldots & M^{nn} \end{pmatrix}, \tag{23}$$

with

$$\mathbf{M}\mathbf{M}^{-1} = \mathbf{M}^{-1}\mathbf{M} = \mathbf{I}_n. \tag{24}$$

The tensor $\hat{M}^{-1}$ has the following coordinate expression:

$$\hat{M}^{-1} = M^{ij}\, \hat{\mathrm{e}}_i \otimes \hat{\mathrm{e}}_j. \tag{25}$$

If we consider the special case $V = \mathbb{R}^n$ with the standard basis $\{\hat{\mathrm{e}}_1, \ldots, \hat{\mathrm{e}}_n\} = \{\mathbf{e}_1, \ldots, \mathbf{e}_n\}$, then our notational distinction becomes trivial, because for $\hat{u} \in \mathbb{R}^n$ we have that

$$\hat{u} = u^i\, \hat{\mathrm{e}}_i = u^i\, \mathbf{e}_i = \mathbf{u}, \tag{26}$$

since the basis vectors of the standard basis satisfy

$$\hat{\mathrm{e}}_i = \mathbf{e}_i = (0 \ldots \underset{i}{1} \ldots 0)^T \tag{27}$$

For further details, we refer to [13].

3 Dynamics of Multibody Systems

The position of a multibody system with n degrees-of-freedom is usually denoted by $\mathbf{q} \in \mathbb{R}^n$. The components of the $\mathbb{R}^n$-tuple $\mathbf{q}$, which defines the position of the system, are known as *generalized coordinates*. Geometrically, the *position* (or configuration) of an autonomous system can be seen as a point $\hat{q}$ in an n-dimensional *configuration manifold* Q. For autonomous multibody systems, this set of all possible positions, which is not a vector space in general, is usually [1, 32, 35] modelled as an n-dimensional differentiable manifold, i.e., an n-dimensional topological manifold that is endowed with a differentiable structure. For the definition of a topological and differentiable manifold, we refer to [25]. We only consider autonomous systems. This is not a restriction, since we are interested in the geometric treatment of instantaneous impacts, that occur at fixed instants of time.

A topological manifold is a space that is locally homeomorphic to $\mathbb{R}^n$, i.e., that locally looks like $\mathbb{R}^n$. Such homeomorphisms are known as *charts*. The chart (U_I, ϕ_I) provides local coordinates on the open (w.r.t. the topology on Q) neighbourhood $U_I \subseteq Q$ by

$$\begin{aligned} \phi_I \colon Q \supseteq U_I &\to \mathbb{R}^n, \\ \hat{q} &\mapsto \phi_I(\hat{q}) =: \mathbf{q}. \end{aligned} \tag{28}$$

A global description of Q is given by a collection of charts whose respective domains U_I cover Q. Such a collection of charts is referred to as an *atlas*. The concept of charts is visualized in Fig. 1. The smooth differentiable structure, which we consider

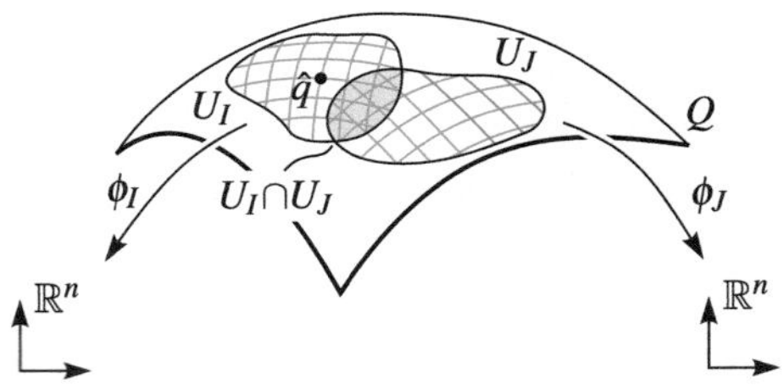

Fig. 1 Differentiable manifold Q with charts (U_I, ϕ_I) and (U_J, ϕ_J)

in this work, is added to the topological manifold by considering an atlas that contains only charts with the property that, for any pair of charts with overlapping domains (i.e., $U_I \cap U_J \neq \emptyset$), the chart transition functions given by

$$\phi_J \circ \phi_I^{-1} : \mathbb{R}^n \supseteq \phi_I(U_I \cap U_J) \to \phi_J(U_I \cap U_J) \subseteq \mathbb{R}^n \tag{29}$$

are diffeomorphisms, i.e., smooth bijections that have a smooth inverse.

By their definition in (28), the generalized coordinates $\mathbf{q}$ provide a local (on the open neighbourhood U_I) description of the configuration manifold as the set of positions $\hat{q}$. To distinguish between a geometrical object and its local representation in a chart, we mark geometrical objects with a hat and use upright bold-faced letters to denote tuples from $\mathbb{R}^n$ (and $\mathbb{R}^{n*}$).

The example of a rigid body that rotates around a fixed point illustrates that the choice of a set of generalized coordinates $\mathbf{q}$ (i.e., the choice of a chart) is not unique and provides, in general, only a local description of Q. The rigid body with a fixed point has three rotational degrees-of-freedom. One choice for $\mathbf{q}$ could be Euler angles. However, due to their singularity, they have to be complemented by Cardan angles in order to obtain a full description of the configuration. This well-known example illustrates that a particular choice of generalized coordinates generally provides only a local description. Alternatively, unit quaternions or an axis angle parameterization could be chosen as generalized coordinates, which shows that their choice is not unique.

We define the *motion* of an autonomous multibody system to be a continuous parametrized curve

$$\hat{q} : \mathbb{R} \supset \mathcal{I} \to Q, \ t \mapsto \hat{q}(t). \tag{30}$$

This curve can be locally represented in the charts of an atlas by considering the chart representations of the restrictions of $\hat{q}(t)$ to the respective neighbourhood U_I, i.e.,

$$\mathbf{q}|_{U_I} : \mathcal{I} \supseteq \mathrm{preIm}_{\hat{q}}(U_I) \to \mathbb{R}^n, \ t \mapsto \phi_I(\hat{q}(t)), \tag{31}$$

where

$$\mathrm{preIm}_{\hat{q}}(U_I) := \{t \in \mathcal{I} \mid \hat{q}(t) \in U_I\} \tag{32}$$

denotes the pre-image of U_I by $\hat{q}$, which is the set of time instants t for which the motion $\hat{q}(t)$ remains in the neighbourhood U_I. The restriction $|_{U_I}$ is usually omitted, and we write $\mathbf{q}(t)$ without referring to a particular chart.

By $T_{\hat{q}}Q$, we denote the tangent space [25] to the configuration manifold Q at the point $\hat{q}$. This n-dimensional real vector space is the habitat of velocities and virtual displacements. Pointwise, the corresponding dual vector space $T^*_{\hat{q}}Q := (T_{\hat{q}}Q)^*$ can be declared as the space of all real-valued linear maps (covectors) on $T_{\hat{q}}Q$, i.e., for $\hat{\omega} \in T^*_{\hat{q}}Q$, it holds that

$$\hat{\omega}\colon T_{\hat{q}}Q \to \mathbb{R},\ \hat{v} \mapsto \hat{\omega}(\hat{v}), \tag{33}$$

with the property that

$$\hat{\omega}(a\hat{v} + b\hat{w}) = a\hat{\omega}(\hat{v}) + b\hat{\omega}(\hat{w}) \tag{34}$$

for all $a, b \in \mathbb{R}$ and $\hat{v}, \hat{w} \in T_{\hat{q}}Q$. Note that the cotangent space has been defined analogously to the dual space V^* in Sect. 2. Therefore, the considerations from Sect. 2 can be pointwisely carried over to the tangent and cotangent spaces of the configuration manifold Q. The duality pairing between a covector $\hat{\omega} \in T^*_{\hat{q}}Q$ and a vector $\hat{v} \in T_{\hat{q}}Q$ is denoted as $\hat{\omega} \cdot \hat{v} = \hat{\omega}(\hat{v})$. In mechanics, the cotangent space $T^*_{\hat{q}}Q$ is considered as the space of forces.

Consider a point $\hat{q} \in Q$ and a chart (U, ϕ) with $\hat{q} \in U$. If we consider a single chart, we omit the index I of the chart in order to get a lighter notation. The chart (U, ϕ) induces a basis

$$\left\{ \frac{\partial}{\partial q^1}\bigg|_{\hat{q}}, \ldots, \frac{\partial}{\partial q^n}\bigg|_{\hat{q}} \right\} \tag{35}$$

of the tangent space $T_{\hat{q}}Q$ such that a $\hat{u}_{\hat{q}} \in T_{\hat{q}}Q$ can be written as

$$\hat{u}_{\hat{q}} = u^i(\hat{q}) \frac{\partial}{\partial q^i}\bigg|_{\hat{q}}. \tag{36}$$

The dual basis to (35) is denoted by

$$\left\{ \mathrm{d}q^1_{\hat{q}}, \ldots, \mathrm{d}q^n_{\hat{q}} \right\} \tag{37}$$

and it holds that

$$\mathrm{d}q^i_{\hat{q}} \left(\frac{\partial}{\partial q^j}\bigg|_{\hat{q}} \right) = \mathrm{d}q^i_{\hat{q}} \cdot \frac{\partial}{\partial q^j}\bigg|_{\hat{q}} = \delta^i_j. \tag{38}$$

Since (37) is a basis of the cotangent space $T^*_{\hat{q}}Q$, an $\hat{\omega}_{\hat{q}} \in T^*_{\hat{q}}Q$ can be written as

$$\hat{\omega}_{\hat{q}} = \omega_i(\hat{q})\, \mathrm{d}q^i_{\hat{q}}. \tag{39}$$

As in Sect. 2, a summation is understood in (36) and (39) over the index i from 1 to n (dimension of the configuration manifold Q) according to the Einstein summation convention. The coefficients $u^i(\hat{q})$ can be written as an $\mathbb{R}^n$-tuple

$$\mathbf{u}(\hat{q}) = \mathbf{u}(\mathbf{q}) = \left(u^1(\hat{q}) \ \dots \ u^n(\hat{q})\right)^{\mathrm{T}}. \tag{40}$$

Often, the point is omitted, and we write $\mathbf{u} = (u^1 \ \dots \ u^n)^{\mathrm{T}}$. If we consider tangent vectors along a curve, then it is common to write

$$\mathbf{u}(t) = \mathbf{u}(\hat{q}(t)) = \mathbf{u}(\mathbf{q}(t)) = \left(u^1(\hat{q}(t)) \ \dots \ u^n(\hat{q}(t))\right)^{\mathrm{T}}. \tag{41}$$

For covectors, an analogue notation applies.

Between the tangent space $T_{\hat{q}}Q$ with the basis (35) induced by a specific chart (U, ϕ) and $\mathbb{R}^n$ with the standard basis $\{\mathbf{e}_1, \dots, \mathbf{e}_n\}$, the following isomorphism holds:

$$\begin{aligned} \mathrm{d}\boldsymbol{\phi}\cdot\colon T_{\hat{q}}Q &\to \mathbb{R}^n, \\ \hat{u}_{\hat{q}} &\mapsto \mathrm{d}\boldsymbol{\phi}\cdot\hat{u}_{\hat{q}} = \mathbf{u}, \end{aligned} \tag{42}$$

where

$$\mathrm{d}\boldsymbol{\phi} = \mathbf{e}_i \otimes \mathrm{d}\phi^i_{\hat{q}} = \mathbf{e}_i \otimes \mathrm{d}q^i_{\hat{q}}. \tag{43}$$

In (43), $\phi^i : U \to \mathbb{R}$ denotes the ith component of the chart ϕ and $\mathrm{d}\phi^i$ denotes its exterior derivative [25]. Finally, $\mathrm{d}\phi^i_{\hat{q}}$ stands for the latter evaluated at the point $\hat{q}$. The inverse map of $\mathrm{d}\boldsymbol{\phi}$ is given by

$$\begin{aligned} \mathrm{d}\boldsymbol{\phi}^{-1}\cdot\colon \mathbb{R}^n &\to T_{\hat{q}}Q, \\ \mathbf{u} &\mapsto \mathrm{d}\boldsymbol{\phi}^{-1}\cdot\mathbf{u} = \hat{u}_{\hat{q}} \end{aligned} \tag{44}$$

and

$$\mathrm{d}\boldsymbol{\phi}^{-1} = \left.\frac{\partial}{\partial q^i}\right|_{\hat{q}} \otimes \mathbf{e}^i. \tag{45}$$

The transpose map of (44), i.e., $\mathrm{d}\boldsymbol{\phi}^{-\mathrm{T}}\cdot\colon T^*_{\hat{q}}Q \to \mathbb{R}^{n*}$ is an isomorphism between $T^*_{\hat{q}}Q$ (the dual of $T_{\hat{q}}Q$) and $\mathbb{R}^{n*}$ (the dual of $\mathbb{R}^n$). The linear maps (42), (44) and their dual maps are shown in Fig. 2.

The symmetric positive definite mass matrix of a finite dimensional mechanical system endows the configuration manifold with a Riemannian metric $\hat{M}$, which is a symmetric covariant 2-tensor field that is positive definite. Pointwise, it holds that

$$\hat{M}(\hat{q})\colon T_{\hat{q}}Q \times T_{\hat{q}}Q \to \mathbb{R}, \tag{46}$$

with $\hat{q} \in Q$. In local coordinates, the Riemannian metric (46) is given by

$$\hat{M}(\hat{q}) = M_{kl}\,\mathrm{d}q^k_{\hat{q}} \otimes \mathrm{d}q^l_{\hat{q}}. \tag{47}$$

Its inverse is denoted by

$$\hat{M}^{-1}(\hat{q})\colon T^*_{\hat{q}}Q \times T^*_{\hat{q}}Q \to \mathbb{R}, \tag{48}$$

and it is locally given by

$$\hat{M}^{-1}(\hat{q}) = M^{ij} \left.\frac{\partial}{\partial q^i}\right|_{\hat{q}} \otimes \left.\frac{\partial}{\partial q^j}\right|_{\hat{q}}. \tag{49}$$

The elements M_{kl} from (47) can be written as an n-by-n matrix $\mathbf{M} = M_{kl}\,\mathbf{e}^k \otimes \mathbf{e}^l$, where $\mathbf{e}^k$ and $\mathbf{e}^l$ denote the dual basis to the standard basis of $\mathbb{R}^n$. The inverse matrix of $\mathbf{M}$ from (49) can be written using the standard basis of $\mathbb{R}^n$ as $\mathbf{M}^{-1} = M^{ij}\,\mathbf{e}_i \otimes \mathbf{e}_j$. This is known as the pushforward with the chart of the Riemannian metric (47) and of its inverse (49) to $\mathbb{R}^n$ and $\mathbb{R}^{n*}$, respectively. Using the isomorphism (42), the diagrams from Fig. 2 can be drawn. The black diagram commutes if

$$\begin{aligned} \mathbf{M}\cdot &= \mathrm{d}\boldsymbol{\phi}^{-\mathrm{T}} \cdot \hat{M}(\hat{q}) \cdot \mathrm{d}\boldsymbol{\phi}^{-1}\cdot \\ &= M_{kl}\,\mathbf{e}^k \otimes \mathbf{e}^l\cdot, \end{aligned} \tag{50}$$

and the gray diagram commutes if

$$\begin{aligned} \mathbf{M}^{-1}\cdot &= \mathrm{d}\boldsymbol{\phi} \cdot \hat{M}^{-1}(\hat{q}) \cdot \mathrm{d}\boldsymbol{\phi}^{\mathrm{T}}\cdot \\ &= M^{ij}\,\mathbf{e}_i \otimes \mathbf{e}_j \cdot\,. \end{aligned} \tag{51}$$

Note that the matrices $\mathbf{M}$ and $\mathbf{M}^{-1}$ are symmetric positive definite, because the tensors $\hat{M}$ and $\hat{M}^{-1}$ are as well. Therefore, the inner product that can be defined on $T_{\hat{q}}Q$ using $\hat{M}$ directly carries over to $\mathbb{R}^n$. Indeed, the Riemannian metric $\hat{M}$ pointwisely induces an inner product (cf. Sect. 2) on each tangent space $T_{\hat{q}}Q$ for any $\hat{q} \in Q$ such that

$$\left(\hat{u}_{\hat{q}}, \hat{v}_{\hat{q}}\right)_{\hat{M}} := \hat{u}_{\hat{q}} \cdot \hat{M}(\hat{q}) \cdot \hat{v}_{\hat{q}} = \mathbf{u} \cdot \mathbf{M} \cdot \mathbf{v} = \mathbf{u}^{\mathrm{T}}\mathbf{M}\mathbf{v}. \tag{52}$$

The associated norm is denoted as

$$\|\hat{u}_{\hat{q}}\|_{\hat{M}} := \sqrt{\left(\hat{u}_{\hat{q}}\,,\, \hat{u}_{\hat{q}}\right)_{\hat{M}}} = \sqrt{\hat{u}_{\hat{q}} \cdot \hat{M}(\hat{q}) \cdot \hat{u}_{\hat{q}}} = \sqrt{\mathbf{u}^{\mathrm{T}}\mathbf{M}\mathbf{u}}. \tag{53}$$

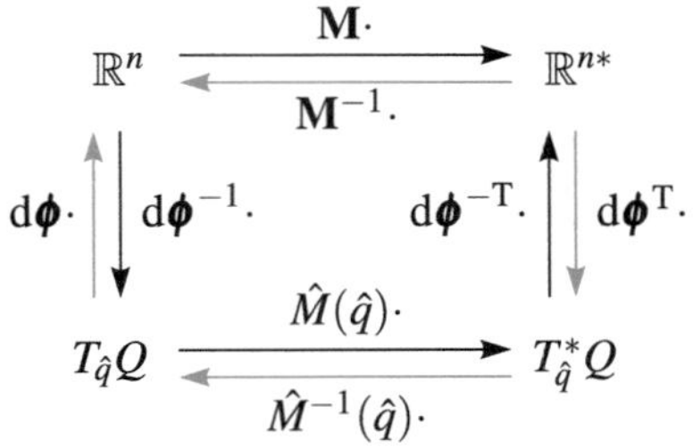

Fig. 2 Isomorphisms induced by the choice of a chart (U, ϕ) and the resulting diagrams, which can be drawn for a Riemannian metric $\hat{M}(\hat{q})$ and its inverse $\hat{M}^{-1}(\hat{q})$

Moreover, the Riemannian metric provides isomorphisms between the tangent and the cotangent space via the pointwise maps

$$\begin{aligned}\hat{M}(\hat{q})\cdot : T_{\hat{q}}Q &\to T^*_{\hat{q}}Q,\\ \hat{v} &\mapsto \hat{M}(\hat{q})\cdot\hat{v} = M_{kl}v^l\,\mathrm{d}q^k_{\hat{q}}\end{aligned} \tag{54}$$

and

$$\begin{aligned}\hat{M}^{-1}(\hat{q})\cdot : T^*_{\hat{q}}Q &\to T_{\hat{q}}Q,\\ \hat{\omega} &\mapsto \hat{M}^{-1}(\hat{q})\cdot\hat{\omega} = M^{ij}\omega_j \left.\frac{\partial}{\partial q^i}\right|_{\hat{q}}.\end{aligned} \tag{55}$$

The analogue holds between $\mathbb{R}^n$ and its dual space $\mathbb{R}^{n*}$

$$\begin{aligned}\mathbf{M}\cdot : \mathbb{R}^n &\to \mathbb{R}^{n*},\\ \mathbf{u} &\mapsto \mathbf{M}\cdot\mathbf{u} = \mathbf{M}\mathbf{u} = M_{kl}u^l\,\mathbf{e}^k\end{aligned} \tag{56}$$

and

$$\begin{aligned}\mathbf{M}^{-1}\cdot : \mathbb{R}^{n*} &\to \mathbb{R}^n,\\ \boldsymbol{\omega} &\mapsto \mathbf{M}^{-1}\cdot\boldsymbol{\omega} = \mathbf{M}^{-1}\boldsymbol{\omega} = M^{ij}\omega_j\,\mathbf{e}_i.\end{aligned} \tag{57}$$

Unfortunately, it would go beyond the scope of the present work to derive the equations of motion of the multibody system in a geometrical context. For this, we refer to [20, 32]. Classically, they can be derived in local coordinates using Lagrange's equations of the second kind or by the projection equation [8], for example. The equations of motion of an autonomous multibody system can be stated in vector notation as

$$\mathbf{M}(\mathbf{q})\ddot{\mathbf{q}} - \mathbf{h}(\mathbf{q},\dot{\mathbf{q}}) = 0, \tag{58}$$

where the mass matrix $\mathbf{M}(\mathbf{q})$ is a symmetric, positive definite matrix, because it is the chart representation of a Riemannian metric $\hat{M}(\hat{q})$.

4 Geometry of Impacts

Now, the positions of the mechanical system are restricted by k scleronomic (i.e., time-independent) constraints to the subset

$$\hat{C} := \left\{\hat{q}\in Q \mid \hat{g}^\beta(\hat{q}) \ge 0,\ \beta = 1,\dots,k\right\} \tag{59}$$

of the configuration manifold Q. The geometric approach that we adopt can be found in [3, 5, 19]. The constraint functions

$$\hat{g}^\beta : Q \to \mathbb{R},\ \hat{q}\mapsto \hat{g}^\beta(\hat{q}) \tag{60}$$

should be continuously differentiable and the points for which $\hat{g}^\beta(\hat{q}) = 0$ holds are assumed to be regular points (cf. [25, Chap. 5]), i.e., $\mathrm{d}\hat{g}^\beta_{\hat{q}} : T_{\hat{q}}Q \to \mathbb{R}$ should be surjective, which means non-zero for a scalar function.

The functions $\hat{g}^\beta$ have the following representation with respect to the chart (U, ϕ)

$$\begin{aligned} g^\beta|_U : \phi(U) &\to \mathbb{R}, \\ \mathbf{q} &\mapsto g^\beta|_U(\mathbf{q}) := \hat{g}^\beta \circ \phi^{-1}(\mathbf{q}), \end{aligned} \tag{61}$$

where the restriction $|_U$ is usually omitted. The functions $\hat{g}^\beta$ (or g^β, respectively) are known as *gap functions* [16]. If $\hat{g}^\beta > 0$ (or $g^\beta > 0$, respectively), then the βth contact is *open*, i.e., the βth constraint is neither *active* $\hat{g}^\beta = 0$ (or $g^\beta = 0$, respectively) nor *violated* $\hat{g}^\beta < 0$ (or $g^\beta < 0$, respectively). The *set of active constraints* $\hat{\mathcal{A}}(\hat{q})$ in $\hat{q}$ is defined as

$$\hat{\mathcal{A}}(\hat{q}) := \left\{\beta \mid \hat{g}^\beta(\hat{q}) = 0\right\} \tag{62}$$

and it has the local expression with respect to the chart (U, ϕ)

$$\mathcal{A}(\mathbf{q}) := \left\{\beta \mid g^\beta|_U(\mathbf{q}) = 0\right\}. \tag{63}$$

Obviously, it holds that $\hat{\mathcal{A}}(\hat{q}) = \mathcal{A}(\mathbf{q})$. The gap functions of the h active constraints can be re-indexed and gathered in a vector

$$\mathbf{g}(\mathbf{q}) = \begin{pmatrix} g^1|_U(\mathbf{q}) \\ \vdots \\ g^h|_U(\mathbf{q}) \end{pmatrix}, \tag{64}$$

such that $\mathbf{g}(\mathbf{q}) = 0$. The dimension h depends on the point $\hat{q}$ and corresponds to the cardinality of the set of active constraints $\hat{\mathcal{A}}(\hat{q})$, i.e., $h = |\hat{\mathcal{A}}(\hat{q})| \leq k$.

The βth contact velocity is defined as

$$\gamma^\beta_{\hat{q}} := \mathrm{d}\hat{g}^\beta_{\hat{q}} \cdot \hat{u}_{\hat{q}}, \tag{65}$$

where $\mathrm{d}\hat{g}^\beta_{\hat{q}}$ denotes the exterior derivative of the function $\hat{g}^\beta$ and is evaluated at the point $\hat{q} \in Q$. With respect to the chart (U, ϕ), the βth contact velocity has the expression

$$\gamma^\beta_{\hat{q}} = \mathrm{d}\hat{g}^\beta_{\hat{q}} \cdot \hat{u}_{\hat{q}} = \left.\frac{\partial g^\beta}{\partial q^i}\right|_{\hat{q}} u^i(\hat{q}) =: \mathbf{w}^{\beta\,\mathrm{T}}(\mathbf{q})\,\mathbf{u}, \tag{66}$$

where u^i can be associated with the ith component of an $\mathbb{R}^n$-tuple $\mathbf{u}$, as in (40), and the generalized direction vector is given by

$$\mathbf{w}^\beta = \left(\left.\tfrac{\partial g^\beta}{\partial q^1}\right|_{\hat{q}} \cdots \left.\tfrac{\partial g^\beta}{\partial q^n}\right|_{\hat{q}} \right)^{\mathrm{T}} \tag{67}$$

because

$$\mathrm{d}\hat{g}_{\hat{q}}^{\beta} = w_m^{\beta}\,\mathrm{d}q_{\hat{q}}^{m}. \tag{68}$$

As we did for the gap functions in (64), the contact velocities of the active constraints can also be gathered in a vector $\boldsymbol{\gamma} \in \mathbb{R}^h$ such that

$$\boldsymbol{\gamma} = \left(\gamma_{\hat{q}}^1 \ \dots \ \gamma_{\hat{q}}^h\right)^{\mathrm{T}} = \mathbf{W}^{\mathrm{T}}(\mathbf{q})\,\mathbf{u}, \tag{69}$$

where

$$\mathbf{W} := \left(\mathbf{w}^1 \ \dots \ \mathbf{w}^h\right) = \left(\frac{\partial \mathbf{g}}{\partial \mathbf{q}}\right)^{\mathrm{T}}. \tag{70}$$

At a point $\hat{q} \in \hat{\mathcal{C}}$, the set $\hat{\mathcal{C}} \subseteq Q$ is locally approximated by the tangent cone

$$\begin{aligned}\mathcal{T}_{\hat{\mathcal{C}}}(\hat{q}) &:= \left\{\hat{u}_{\hat{q}} \in T_{\hat{q}}Q \mid \mathrm{d}\hat{g}_{\hat{q}}^{\beta} \cdot \hat{u}_{\hat{q}} \geq 0,\ \forall \beta \in \hat{\mathcal{A}}(\hat{q})\right\} \\ &= \left\{\hat{u}_{\hat{q}} \in T_{\hat{q}}Q \mid \mathrm{d}\hat{g}_{\hat{q}}^{\alpha} \cdot \hat{u}_{\hat{q}} \geq 0,\ \forall \alpha \in \{1, \dots, h\}\right\},\end{aligned} \tag{71}$$

which is, by definition, a subset of the tangent space, i.e., $\mathcal{T}_{\hat{\mathcal{C}}}(\hat{q}) \subseteq T_{\hat{q}}Q$. The last equality in (71) can be written by re-indexing from β to α. With the tangent cone at the point $\hat{q} \in \hat{\mathcal{C}}$ comes its polar cone, the normal cone

$$\mathcal{N}_{\hat{\mathcal{C}}}(\hat{q}) := \left\{\hat{\omega}_{\hat{q}} \in T_{\hat{q}}^{*}Q \mid \hat{\omega}_{\hat{q}} \cdot \delta\hat{q}_{\hat{q}} \leq 0,\ \forall \delta\hat{q}_{\hat{q}} \in \mathcal{T}_{\hat{\mathcal{C}}}(\hat{q})\right\}, \tag{72}$$

which is a subset of the cotangent space, i.e., $\mathcal{N}_{\hat{\mathcal{C}}}(\hat{q}) \subseteq T_{\hat{q}}^{*}Q$. For points lying in the interior of $\hat{\mathcal{C}}$, it holds that $\hat{\mathcal{A}}(\hat{q}) = \emptyset$. Therefore, $\mathcal{T}_{\hat{\mathcal{C}}}(\hat{q}) = T_{\hat{q}}Q$ and $\mathcal{N}_{\hat{\mathcal{C}}}(\hat{q}) = \{0\} \subset T_{\hat{q}}^{*}Q$, for all $\hat{q} \in \operatorname{int}\hat{\mathcal{C}}$. Given the definitions of the tangent cone (71) and the normal cone (72), it follows from Farkas' lemma (see [39, p. 200]) that the one-forms $\mathrm{d}\hat{g}_{\hat{q}}^{1}, \dots, \mathrm{d}\hat{g}_{\hat{q}}^{k}$ finitely generate the normal cone, i.e.,

$$\mathcal{N}_{\hat{\mathcal{C}}}(\hat{q}) = \left\{-\lambda_{\alpha}\,\mathrm{d}\hat{g}_{\hat{q}}^{\alpha} \in T_{\hat{q}}^{*}Q \mid \lambda_{\alpha} \geq 0,\ \alpha \in \{1, \dots, h\}\right\}. \tag{73}$$

In a local chart (U, ϕ), the set $\hat{\mathcal{C}}$ from (59) has the expression

$$\mathcal{C} = \left\{\mathbf{q} \in \phi(U) \subseteq \mathbb{R}^n \mid g^{\beta}|_U(\mathbf{q}) \geq 0,\ \beta = 1, \dots, k\right\}, \tag{74}$$

where we used (61). Considering (63) and (66), the tangent cone (71) is locally given by

$$\begin{aligned}\mathcal{T}_{\mathcal{C}}(\mathbf{q}) &= \left\{\mathbf{u} \in \mathbb{R}^n \mid \mathbf{w}^{\beta\,\mathrm{T}}\mathbf{u} \geq 0,\ \forall \beta \in \mathcal{A}(\mathbf{q})\right\} \\ &= \left\{\mathbf{u} \in \mathbb{R}^n \mid \mathbf{w}^{\alpha\,\mathrm{T}}\mathbf{u} \geq 0,\ \forall \alpha \in \{1, \dots, h\}\right\} \\ &= \left\{\mathbf{u} \in \mathbb{R}^n \mid \mathbf{W}^{\mathrm{T}}(\mathbf{q})\,\mathbf{u} \geq 0\right\},\end{aligned} \tag{75}$$

where the last equality uses (70).

With the use of (75), the normal cone (72) and (73) can be written locally as

$$
\begin{aligned}
\mathcal{N}_{\mathcal{C}}(\mathbf{q}) &= \left\{\boldsymbol{\omega} \in \mathbb{R}^{n*} \mid \boldsymbol{\omega}^{\mathrm{T}}\delta\mathbf{q} \leq 0,\ \forall \delta\mathbf{q} \in \mathcal{T}_{\mathcal{C}}(\mathbf{q})\right\} \\
&= \left\{-\mathbf{w}^{\alpha}\lambda_{\alpha} \in \mathbb{R}^{n*} \mid \lambda_{\alpha} \geq 0,\ \alpha \in \{1, \ldots, h\}\right\} \\
&= \left\{-\mathbf{W}\boldsymbol{\lambda} \in \mathbb{R}^{n*} \mid \boldsymbol{\lambda} \geq 0\right\},
\end{aligned}
\tag{76}
$$

where the last equality makes use of (70) again.

Now, we want to state the dynamics of the multibody system which is subjected to the k scleronomic unilateral constraints from (59). As long as the position $\hat{q}$ is not on the boundary of $\hat{\mathcal{C}}$, i.e., as long as $\hat{\mathcal{A}}(\hat{q}) = \emptyset$, the dynamics is given by the equations of motion (58). If a motion arrives at the boundary of $\hat{\mathcal{C}}$, we have to add the possibility that jumps in the generalized velocities may occur in order to constrain the motion to the set $\hat{\mathcal{C}}$ for arbitrary initial conditions. Therefore, we extend the local description of the dynamics (58) to

$$
\mathbf{M}(\mathbf{q})\dot{\mathbf{u}} - \mathbf{h}(\mathbf{q}, \mathbf{u}) = \mathbf{f}^{c}, \tag{77}
$$

$$
\mathbf{M}(\mathbf{q})(\mathbf{u}^{+} - \mathbf{u}^{-}) = \mathbf{F}^{c}, \tag{78}
$$

where $\mathbf{u} = \dot{\mathbf{q}}$, whenever it exists, and $\mathbf{u}^{-}$, $\mathbf{u}^{+}$ denote the pre- and the post-impact generalized velocity, respectively. Equation (77) is the equation motion extended with the constraint force $\mathbf{f}^{c}$ on the right-hand side. Equation (78) is referred to as the *impact equation* and relates the velocity jump $\mathbf{u}^{+} - \mathbf{u}^{-}$ to the impulsive constraint force $\mathbf{F}^{c}$. For those time instants $\bar{t}$ for which $\mathbf{u}(\bar{t}) = \mathbf{u}(\mathbf{q}(\bar{t})) = \dot{\mathbf{q}}(\bar{t})$ (impact-free motion), the $\mathbb{R}^{n}$-tuple of the generalized velocities $\mathbf{u}(\bar{t})$ is the chart representation of the velocity vector $\hat{u}_{\hat{q}(\bar{t})} \in T_{\hat{q}(\bar{t})}Q$ of the motion at time $\bar{t}$ with respect to the chart (U, ϕ). At an impact, i.e., at a time instant t^{*} where $\mathbf{u}^{+} \neq \mathbf{u}^{-}$, the pre- and post-impact generalized velocities $\mathbf{u}^{-}$, $\mathbf{u}^{+}$ are the chart representations of the corresponding tangent vectors $\hat{u}^{-}_{\hat{q}(t^{*})}, \hat{u}^{+}_{\hat{q}(t^{*})} \in T_{\hat{q}(t^{*})}Q$ (see Fig. 3).

We model perfect unilateral constraints [5] such that the constraint force $\mathbf{f}^{c}$ and the impulsive constraint force $\mathbf{F}^{c}$ have to obey

$$
-\mathbf{f}^{c} \in \mathcal{N}_{\mathcal{C}}(\mathbf{q}) \qquad \text{and} \qquad -\mathbf{F}^{c} \in \mathcal{N}_{\mathcal{C}}(\mathbf{q}). \tag{79}
$$

The dynamics (77) and (78) together with the constraint (79) can be restated as

$$
\mathbf{M}(\mathbf{q})\dot{\mathbf{u}} - \mathbf{h}(\mathbf{q}, \mathbf{u}) \in -\mathcal{N}_{\mathcal{C}}(\mathbf{q}), \tag{80}
$$

$$
\mathbf{M}(\mathbf{q})(\mathbf{u}^{+} - \mathbf{u}^{-}) \in -\mathcal{N}_{\mathcal{C}}(\mathbf{q}), \tag{81}
$$

with $\dot{\mathbf{q}} = \mathbf{u}$ almost everywhere. Using the local expression (76) of the normal cone, we can write

$$
\mathbf{M}(\mathbf{q})\dot{\mathbf{u}} - \mathbf{h}(\mathbf{q}, \mathbf{u}) = \mathbf{W}\boldsymbol{\lambda}, \tag{82}
$$

$$
\mathbf{M}(\mathbf{q})(\mathbf{u}^{+} - \mathbf{u}^{-}) = \mathbf{W}\boldsymbol{\Lambda}, \tag{83}
$$

with $\dot{\mathbf{q}} = \mathbf{u}$ almost everywhere, $\boldsymbol{\lambda} \geq 0$ and $\boldsymbol{\Lambda} \geq 0$.

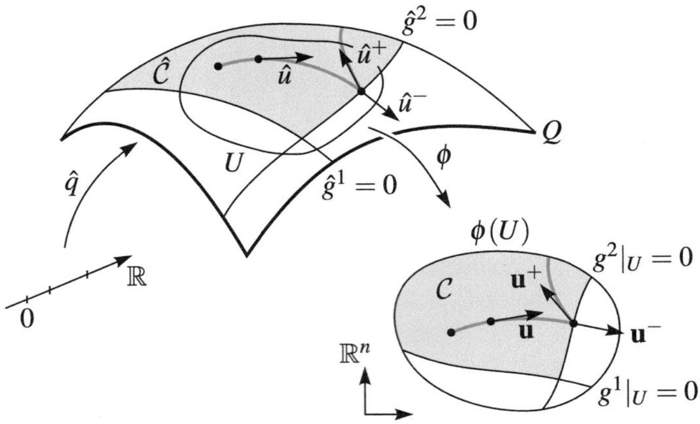

Fig. 3 The generalized velocities $\mathbf{u}$, $\mathbf{u}^-$ and $\mathbf{u}^+$ are the chart representations of the tangent vectors $\hat{u}$, $\hat{u}^-$ and $\hat{u}^+$ to the motion $\hat{q}(t)$

In order to guarantee unique post-impact generalized velocities $\mathbf{u}^+$, the impact equation (81)/(83) has, in general, to be complemented with a constitutive law which we will refer to as the set-valued impact law. In the following, we will investigate the geometry of impacts and under which conditions a set-valued impact law leads to a unique post-impact generalized velocity when combined with the impact equation.

A set-valued impact law together with the impact equation (83) should yield kinematically admissible post-impact contact velocities, i.e.,

$$\gamma^{\alpha+} \geq 0, \quad \text{for all } \alpha \in \{1, \ldots, h\}, \tag{84}$$

which is equivalent to

$$\hat{u}^+ \in \mathcal{T}_{\hat{C}}(\hat{q}), \quad \text{or locally} \quad \mathbf{u}^+ \in \mathcal{T}_C(\mathbf{q}). \tag{85}$$

The restriction (85) is known as *kinematic consistency* [16]. For physically meaningful pre-impact velocities, it holds that

$$\hat{u}^- \in -\mathcal{T}_{\hat{C}}(\hat{q}), \quad \text{or locally} \quad \mathbf{u}^- \in -\mathcal{T}_C(\mathbf{q}). \tag{86}$$

The impact equation (81) is the local expression with respect to the chart (U, ϕ) of the inclusion

$$\hat{M}(\hat{q}) \cdot (\hat{u}^+ - \hat{u}^-) \in -\mathcal{N}_{\hat{C}}(\hat{q}) \tag{87}$$

on the cotangent space. At instantaneous impacts, the positions are constant over the impact, and therefore the impact description reduces to the algebraic expression (87) on $T^*_{\hat{q}}Q$. Using the isomorphism (55), the inclusion (87) can be pushed to the tangent space $T_{\hat{q}}Q$

$$\hat{u}^+ - \hat{u}^- \in -\mathcal{T}_{\hat{\mathcal{C}}}^{\perp}(\hat{q}), \tag{88}$$

where

$$\begin{aligned}\mathcal{T}_{\hat{\mathcal{C}}}^{\perp}(\hat{q}) :&= \left\{\hat{v}_{\hat{q}} \in T_{\hat{q}}Q \mid \hat{v}_{\hat{q}} \cdot \hat{M}(\hat{q}) \cdot \delta\hat{q}_{\hat{q}} \leq 0,\ \forall \delta\hat{q}_{\hat{q}} \in \mathcal{T}_{\hat{\mathcal{C}}}(\hat{q})\right\} \\ &= \hat{M}^{-1}(\hat{q}) \cdot \mathcal{N}_{\hat{\mathcal{C}}}(\hat{q})\end{aligned} \tag{89}$$

is the cone which is orthogonal to the tangent cone $\mathcal{T}_{\hat{\mathcal{C}}}(\hat{q})$ with respect to the inner product on $T_{\hat{q}}Q$ induced by the Riemannian metric $\hat{M}(\hat{q})$. Using the expression (73) of the normal cone, the cone $\mathcal{T}_{\hat{\mathcal{C}}}^{\perp}(\hat{q})$ can be written as

$$\mathcal{T}_{\hat{\mathcal{C}}}^{\perp}(\hat{q}) = \left\{-\lambda_\alpha \nabla \hat{g}_{\hat{q}}^{\alpha} \in T_{\hat{q}}Q \mid \lambda_\alpha \geq 0, \alpha \in \{1, \ldots, h\}\right\}, \tag{90}$$

where we have introduced the gradient on a Riemannian manifold [25, Chap. 13], which is defined as

$$\nabla \hat{g}_{\hat{q}}^{\alpha} := \hat{M}^{-1}(\hat{q}) \cdot \mathrm{d}\hat{g}_{\hat{q}}^{\alpha}. \tag{91}$$

Again, Eqs. (88)–(90) can be written locally as

$$\mathbf{u}^+ - \mathbf{u}^- \in -\mathcal{T}_{\mathcal{C}}^{\perp}(\mathbf{q}), \tag{92}$$

$$\begin{aligned}\mathcal{T}_{\mathcal{C}}^{\perp}(\mathbf{q}) &= \mathbf{M}^{-1}(\mathbf{q})\mathcal{N}_{\mathcal{C}}(\mathbf{q}) \\ &= \left\{\mathbf{v} \in \mathbb{R}^n \mid \mathbf{v}^{\mathrm{T}}\mathbf{M}(\mathbf{q})\delta\mathbf{q} \leq 0,\ \forall \delta\mathbf{q} \in \mathcal{T}_{\mathcal{C}}(\mathbf{q})\right\} \\ &= \left\{-\mathbf{M}^{-1}(\mathbf{q})\mathbf{W}(\mathbf{q})\boldsymbol{\Lambda} \in \mathbb{R}^n \mid \boldsymbol{\Lambda} \geq 0\right\}\end{aligned} \tag{93}$$

and are referred to as *kinetic consistency* [16].

Next, we consider the change of the kinetic energy over an impact. Let T^- and T^+ denote the kinetic energy before and after the impact, respectively. The difference of post-impact and pre-impact kinetic energy is

$$\begin{aligned}T^+ - T^- &= \frac{1}{2}\|\hat{u}_{\hat{q}}^+\|_{\hat{M}}^2 - \frac{1}{2}\|\hat{u}_{\hat{q}}^-\|_{\hat{M}}^2 \\ &= \frac{1}{2}\hat{u}^+ \cdot \hat{M} \cdot \hat{u}^+ - \frac{1}{2}\hat{u}^- \cdot \hat{M} \cdot \hat{u}^- \\ &= \frac{1}{2}\left(\hat{u}^+ + \hat{u}^-\right) \cdot \hat{M} \cdot \left(\hat{u}^+ - \hat{u}^-\right) \\ &= \frac{1}{2}\left(\hat{u}^+ + \hat{u}^-\right) \cdot \left(\Lambda_\alpha \mathrm{d}g^\alpha\right) \\ &= \frac{1}{2}\mathrm{d}g^\alpha \cdot \left(\hat{u}^+ + \hat{u}^-\right)\Lambda_\alpha,\end{aligned} \tag{94}$$

with $\Lambda_\alpha \geq 0$. The fourth equality is based upon (87) and (73). By (65) and with the definitions

$$\bar{\gamma}^\alpha := \frac{1}{2}(\gamma^{\alpha+} + \gamma^{\alpha-}), \tag{95}$$

$$\bar{\boldsymbol{\gamma}} := \left(\bar{\gamma}^1 \ldots \bar{\gamma}^h\right)^{\mathrm{T}}, \tag{96}$$

$$\boldsymbol{\Lambda} := \left(\Lambda_1 \ldots \Lambda_h\right)^{\mathrm{T}}, \tag{97}$$

we can write (94) as

$$\begin{aligned} T^+ - T^- &= \frac{1}{2}\left(\gamma^{\alpha+} + \gamma^{\alpha-}\right)\Lambda_\alpha \\ &= \bar{\gamma}^\alpha \Lambda_\alpha =: \bar{\boldsymbol{\gamma}}^{\mathrm{T}}\boldsymbol{\Lambda}, \end{aligned} \tag{98}$$

with $\boldsymbol{\Lambda} \geq 0$.

From the change of the kinetic energy over an impact (98), $\bar{\boldsymbol{\gamma}}$ and $\boldsymbol{\Lambda}$ can be recognized as dual variables. The assumption that impacts are dissipative, i.e., that the kinetic energy must not increase at impacts, can be written as

$$\hat{u}^+ \in \mathcal{B}_{\|\hat{u}^-\|_{\hat{M}}}(\hat{q}), \quad \text{with } \mathcal{B}_{\|\hat{u}^-\|_{\hat{M}}}(\hat{q}) := \left\{\hat{u}^+ \in T_{\hat{q}}Q \mid \|\hat{u}^+\|_{\hat{M}} \leq \|\hat{u}^-\|_{\hat{M}}\right\} \tag{99}$$

and is referred to as *energetic consistency* [16–18]. To summarize, we restate the three geometric restrictions on the post-impact velocities from (85)–(99):

$$\hat{u}^+ \in \mathcal{T}_{\hat{\mathcal{C}}}(\hat{q}), \qquad \text{(kinematic consistency)} \tag{100}$$

$$\hat{u}^+ \in \hat{u}^- - \mathcal{T}_{\hat{\mathcal{C}}}^{\perp}(\hat{q}), \qquad \text{(kinetic consistency)} \tag{101}$$

$$\hat{u}^+ \in \mathcal{B}_{\|\hat{u}^-\|_{\hat{M}}}(\hat{q}). \qquad \text{(energetic consistency)} \tag{102}$$

Because of the geometric restrictions (100)–(102), the post-impact velocity needs to satisfy the inclusion

$$\hat{u}^+ \in \hat{\mathcal{S}}(\hat{q}, \hat{u}^-) \quad \text{with } \hat{\mathcal{S}}(\hat{q}, \hat{u}^-) := \mathcal{T}_{\hat{\mathcal{C}}}(\hat{q}) \cap \left(\hat{u}^- - \mathcal{T}_{\hat{\mathcal{C}}}^{\perp}(\hat{q})\right) \cap \mathcal{B}_{\|\hat{u}^-\|_{\hat{M}}}(\hat{q}). \tag{103}$$

Figure 4 visualizes the set $\hat{\mathcal{S}}$, which results from the three geometric requirements (100)–(102).

So far, we have seen that the space which is relevant to the description of an impact is the tangent space at the respective boundary point of the set $\hat{\mathcal{C}}$. In the tangent space, we have identified the tangent cone $\mathcal{T}_{\hat{\mathcal{C}}}(\hat{q})$ as the set of admissible post-impact velocities. Moreover, we have seen that its orthogonal cone $\mathcal{T}_{\hat{\mathcal{C}}}^{\perp}(\hat{q})$ is spanned by the gradients of the defining functions $\hat{g}^\alpha$ of the set $\hat{\mathcal{C}}$. These two cones will allow us to push our geometrical considerations further. The decomposition of the tangent space, which will be presented in the next two sections, can be found in the Ph.D. thesis by Aeberhard [3].

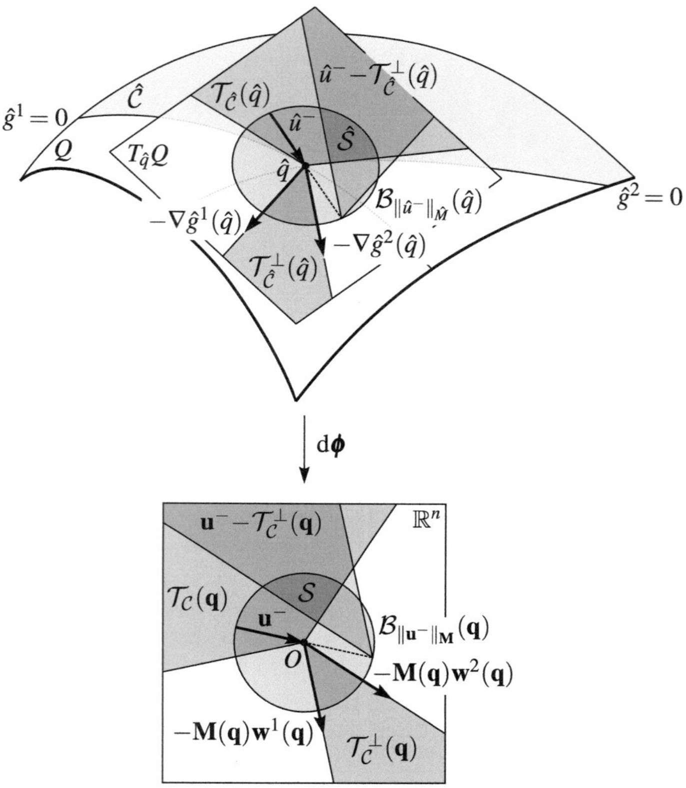

Fig. 4 Set $\hat{\mathcal{S}} \subseteq T_{\hat{q}}Q$ and its local respresentation $\mathcal{S} \subseteq \mathbb{R}^n$ resulting from the requirements of kinematic, kinetic and energetic consistency

4.1 Subspaces of the Tangent Space

A subset $\mathcal{D}$ of $\mathbb{R}^n$ is called *convex* if for any two elements $\mathbf{x}, \mathbf{y} \in \mathcal{D}$ it contains the line between them, i.e., $(1-s)\mathbf{x} + s\mathbf{y} \in \mathcal{D}$ with $0 < s < 1$. A subset $\mathcal{K}$ of $\mathbb{R}^n$ is a *cone* if $0 \in \mathcal{K}$ and $\alpha\mathbf{x} \in \mathcal{K}$ for all $\mathbf{x} \in \mathcal{K}$ and $\alpha > 0$. A cone is not necessarily convex. For convex cones, the following proposition holds. Figure 5 illustrates Propositions 1 and 2 for the example of a three-dimensional vector space V.

Proposition 1 *If $\mathcal{K}$ is a convex cone in $\mathbb{R}^n$, then it holds for the sets $\mathcal{L}_{\mathcal{K}} = \mathcal{K} \cap (-\mathcal{K})$ and $\langle\mathcal{K}\rangle = \mathcal{K} + (-\mathcal{K}) = \{\mathbf{x} + \mathbf{y} \mid \mathbf{x} \in \mathcal{K}, \mathbf{y} \in (-\mathcal{K})\}$ that $\mathcal{L}_{\mathcal{K}} \subseteq \mathcal{K} \subseteq \langle\mathcal{K}\rangle$. Thereby, $\mathcal{L}_{\mathcal{K}}$ is the largest linear subspace in $\mathcal{K}$ and $\langle\mathcal{K}\rangle$ is the linear hull of $\mathcal{K}$, i.e., the smallest linear subspace containing $\mathcal{K}$. The cone $\mathcal{K}$ is itself a linear subspace iff $\mathcal{K} = -\mathcal{K}$.*

The proof is given by Rockafellar and Wets in [40, Proposition 3.8]. The next proposition together with its proof can be found in [3, Corollary 6.1].

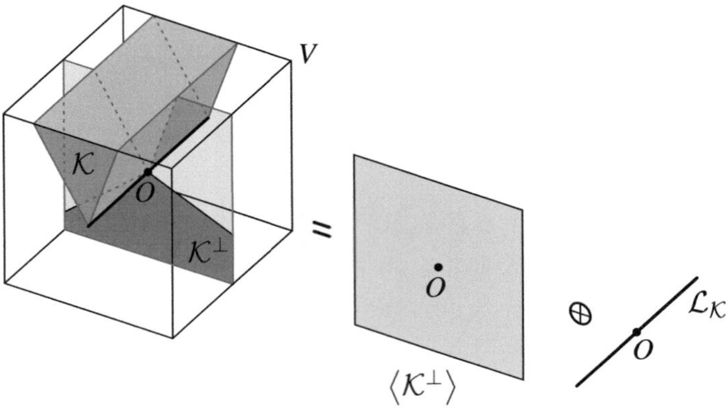

Fig. 5 A closed convex cone $\mathcal{K}$ in a vector space V defines a decomposition of V into the two orthogonal subspaces $\langle\mathcal{K}^\perp\rangle$ and $\mathcal{L}_\mathcal{K}$

Proposition 2 *Let $(V, (\cdot, \cdot))$ be a vector space with inner product $(\cdot, \cdot)$. Let $\mathcal{K} \subseteq V$ be a closed, convex cone and $\mathcal{K}^\perp$ be its orthogonal cone. Let $\mathcal{L}_\mathcal{K} = \mathcal{K} \cap (-\mathcal{K})$ be the largest linear subspace of $\mathcal{K}$ and $\langle\mathcal{K}^\perp\rangle = \mathcal{K}^\perp - \mathcal{K}^\perp = \{\hat{u} - \hat{v} \mid \hat{u}, \hat{v} \in \mathcal{K}^\perp\}$ be the linear hull of $\mathcal{K}^\perp$, i.e., the smallest linear space containing $\mathcal{K}^\perp$. Then, $\mathcal{L}_\mathcal{K}$ and $\langle\mathcal{K}^\perp\rangle$ are orthogonal subspaces of V which span V entirely.*

Proof We start by showing that $\mathcal{L}_\mathcal{K} \subseteq \langle\mathcal{K}^\perp\rangle^\perp$. Consider an arbitrary $\hat{u} \in \mathcal{L}_\mathcal{K}$. For any $\hat{a} \in \mathcal{K}^\perp$, it holds that $(\hat{u}\,,\, \hat{a}) \leq 0$ and $(\hat{u}\,,\, \hat{a}) \geq 0$, so $(\hat{u}, \hat{a}) = 0$. An arbitrary $\hat{s} \in \langle\mathcal{K}^\perp\rangle$ can be written as $\hat{s} = \hat{b} - \hat{c}$ for some $\hat{b}, \hat{c} \in \mathcal{K}^\perp$. It follows from $(\hat{u}\,,\, \hat{b}) = (\hat{u}\,,\, \hat{c}) = 0$ that $(\hat{u}\,,\, \hat{s}) = (\hat{u}\,,\, \hat{b} - \hat{c}) = 0$. Because $\hat{u} \in \mathcal{L}_\mathcal{K}$ has been chosen arbitrarily, it holds that $\mathcal{L}_\mathcal{K} \subseteq \langle\mathcal{K}^\perp\rangle^\perp$.

Now, we show that $\mathcal{L}_\mathcal{K} \supseteq \langle\mathcal{K}^\perp\rangle^\perp$. For an arbitrary $\hat{u} \in \langle\mathcal{K}^\perp\rangle^\perp$, it follows for all $\hat{b}, \hat{c} \in \mathcal{K}^\perp$ that $(\hat{u}\,,\, \hat{b} - \hat{c}) = 0$. If one chooses $\hat{c} = 0$, then for all $\hat{b} \in \mathcal{K}^\perp$ it holds that $(\hat{u}\,,\, \hat{b}) = 0$. This implies that $(\hat{u}\,,\, \hat{b}) \leq 0$ and $(\hat{u}\,,\, \hat{b}) \geq 0$. Since $\hat{b} \in \mathcal{K}^\perp$ is arbitrary, it follows that $\hat{u} \in \mathcal{K}$ and $\hat{u} \in -\mathcal{K}$ and, therefore, $\hat{u} \in \mathcal{L}_\mathcal{K}$. Because $\hat{u} \in \langle\mathcal{K}^\perp\rangle^\perp$ has been chosen arbitrarily, it follows that $\langle\mathcal{K}^\perp\rangle^\perp \subseteq \mathcal{L}_\mathcal{K}$. Hence, we have shown the equality $\langle\mathcal{K}^\perp\rangle^\perp = \mathcal{L}_\mathcal{K}$. Therefore, $\langle\mathcal{K}^\perp\rangle \perp \mathcal{L}_\mathcal{K}$ and $\mathcal{L}_\mathcal{K} \oplus \langle\mathcal{K}^\perp\rangle = V$. □

Considering (71), the largest vector space $\mathcal{L}_{\mathcal{T}_{\hat{C}}(\hat{q})}$ contained in the tangent cone $\mathcal{T}_{\hat{C}}(\hat{q})$ is given by

$$\begin{aligned} \mathcal{L}_{\mathcal{T}_{\hat{C}}(\hat{q})} &= \left(-\mathcal{T}_{\hat{C}}(\hat{q})\right) \cap \mathcal{T}_{\hat{C}}(\hat{q}) \\ &= \left\{\hat{u}_{\hat{q}} \in T_{\hat{q}}Q \mid \mathrm{d}\hat{g}^{\alpha}_{\hat{q}} \cdot \hat{u}_{\hat{q}} = 0,\ \forall \alpha \in \hat{\mathcal{A}}(\hat{q})\right\}. \end{aligned} \tag{104}$$

In the local chart (U, ϕ), we see that

$$\mathcal{L}_{\mathcal{T}_C(\mathbf{q})} = \left\{\mathbf{u} \in \mathbb{R}^n \mid \mathbf{W}^\mathrm{T}(\mathbf{q})\,\mathbf{u} = 0\right\} = \ker\left(\mathbf{W}^\mathrm{T}(\mathbf{q})\right) \tag{105}$$

is given by the kernel of $\mathbf{W}^\mathrm{T}(\mathbf{q})$.

The linear hull of the cone $\mathcal{T}_{\hat{\mathcal{C}}}^{\perp}(\hat{q})$ orthogonal to the tangent cone is given by

$$\left\langle \mathcal{T}_{\hat{\mathcal{C}}}^{\perp}(\hat{q}) \right\rangle = \left\{ -\lambda_\alpha \nabla \hat{g}_{\hat{q}}^{\alpha} \in T_{\hat{q}}Q \mid \lambda_\alpha \in \mathbb{R},\ \alpha \in \{1, \ldots, h\} \right\}, \tag{106}$$

where we have used (90). In the local coordinates (U, ϕ), it can be written as

$$\left\langle \mathcal{T}_{\mathcal{C}}^{\perp}(\mathbf{q}) \right\rangle = \left\{ \mathbf{M}^{-1}(\mathbf{q})\mathbf{W}(\mathbf{q})\boldsymbol{\Lambda} \in \mathbb{R}^n \mid \boldsymbol{\Lambda} \in \mathbb{R}^h \right\}. \tag{107}$$

4.2 Orthogonal Projectors on the Tangent Space

According to Proposition 2, the spaces $\mathcal{L}_{\mathcal{T}_{\hat{\mathcal{C}}}(\hat{q})}$ and $\left\langle \mathcal{T}_{\hat{\mathcal{C}}}^{\perp}(\hat{q}) \right\rangle$ are orthogonal subspaces of the tangent space $T_{\hat{q}}Q$, which span $T_{\hat{q}}Q$ entirely. In the following, we want to derive orthogonal (w.r.t. the inner product on $T_{\hat{q}}Q$) projectors

$$\hat{P}\cdot : T_{\hat{q}}Q \to \mathcal{L}_{\mathcal{T}_{\hat{\mathcal{C}}}(\hat{q})} \tag{108}$$

and

$$\hat{P}^{\perp}\cdot : T_{\hat{q}}Q \to \left\langle \mathcal{T}_{\hat{\mathcal{C}}}^{\perp}(\hat{q}) \right\rangle \tag{109}$$

on the spaces $\mathcal{L}_{\mathcal{T}_{\hat{\mathcal{C}}}(\hat{q})} \subseteq T_{\hat{q}}Q$ and $\left\langle \mathcal{T}_{\hat{\mathcal{C}}}^{\perp}(\hat{q}) \right\rangle \subseteq T_{\hat{q}}Q$, respectively.

Some preliminary remarks are needed before we are able to define the projectors. The covectors $\mathrm{d}\hat{g}_{\hat{q}}^{\alpha}$ of the active constraints that define the tangent cone may be written as a vector-valued one-form

$$\begin{aligned} \mathbf{dg} &= \mathbf{e}_\alpha \otimes \mathrm{d}\hat{g}_{\hat{q}}^{\alpha} = \frac{\partial g^\alpha}{\partial q^i} \mathbf{e}_\alpha \otimes \mathrm{d}q^i \\ &= \left(W^{\mathrm{T}}\right)^{\alpha}{}_{i}\ \mathbf{e}_\alpha \otimes \mathrm{d}q^i = W_i{}^{\alpha}\ \mathbf{e}_\alpha \otimes \mathrm{d}q^i , \end{aligned} \tag{110}$$

where the right expression is the representation in the chart (U, ϕ) and the $\mathbf{e}_\alpha$ with $\alpha \in \{1, \ldots, h\}$ denote the basis vectors of the standard basis of $\mathbb{R}^h$. The $\left(W^{\mathrm{T}}\right)^{\alpha}{}_{i}$ and $W_i{}^{\alpha}$ denote the components of the matrices $\mathbf{W}^{\mathrm{T}}$ and $\mathbf{W}$, respectively. The left index specifies the row, the right one stands for the column. The map

$$\mathbf{dg}\cdot : T_{\hat{q}}Q \to \mathbb{R}^h \tag{111}$$

is a linear map between the two vector spaces $T_{\hat{q}}Q$ and $\mathbb{R}^h$. As a linear map between two vector spaces, it has a dual (or transpose) map

$$\mathbf{dg}^{\mathrm{T}}\cdot : \mathbb{R}^{h*} \to T_{\hat{q}}^{*}Q \tag{112}$$

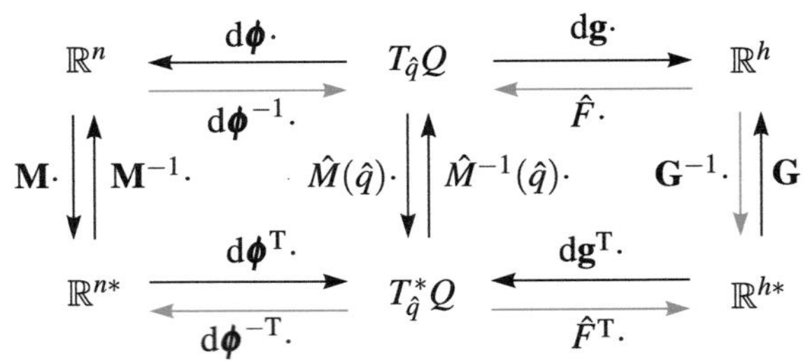

Fig. 6 Relations between primary and dual spaces

between the spaces $T^*_{\hat{q}}Q$ and $\mathbb{R}^{h*}$, which is defined in accordance with (6) by

$$\langle \mathbf{dg}^{\mathrm{T}} \cdot \boldsymbol{\Lambda}, \hat{u}\rangle = \langle \boldsymbol{\Lambda}, \mathbf{dg} \cdot \hat{u}\rangle, \quad \forall \boldsymbol{\Lambda} \in \mathbb{R}^{h*} \text{ and } \hat{u} \in T_{\hat{q}}Q. \tag{113}$$

From the definition, it follows that

$$\begin{aligned} \mathbf{dg}^{\mathrm{T}} &= \mathrm{d}\hat{g}^{\alpha}_{\hat{q}} \otimes \mathbf{e}_{\alpha} = \frac{\partial g^{\alpha}}{\partial q^i}\, \mathrm{d}q^i \otimes \mathbf{e}_{\alpha} \\ &= W_i{}^{\alpha}\, \mathrm{d}q^i \otimes \mathbf{e}_{\alpha} = \left(W^{\mathrm{T}}\right)^{\alpha}{}_i\, \mathrm{d}q^i \otimes \mathbf{e}_{\alpha}. \end{aligned} \tag{114}$$

From the diagram in Fig. 6, we see that, by concatenation of $\mathbf{dg}^{\mathrm{T}}\cdot$, $\hat{M}^{-1}(\hat{q})\cdot$ and $\mathbf{dg}\cdot$, we obtain the map

$$\begin{aligned} \mathbf{G}\cdot &: \mathbb{R}^{h*} \to \mathbb{R}^h, \\ &\boldsymbol{\Lambda} \mapsto \mathbf{G} \cdot \boldsymbol{\Lambda} = \mathbf{dg} \cdot \hat{M}^{-1}(\hat{q}) \cdot \mathbf{dg}^{\mathrm{T}} \cdot \boldsymbol{\Lambda}, \end{aligned} \tag{115}$$

where

$$\boldsymbol{\Lambda} = \Lambda_{\alpha}\, \mathbf{e}^{\alpha} \quad \text{and} \quad \boldsymbol{\Lambda} = \left(\Lambda_1 \ldots \Lambda_h\right)^{\mathrm{T}}. \tag{116}$$

By construction, $\mathbf{G}$ is a symmetric, covariant 2-tensor, because $\hat{M}^{-1}(\hat{q})$ is symmetric. In mechanics, the 2-tensor $\mathbf{G}$ is known as the *Delassus operator*. Moreover, $\mathbf{G}$ is, in general, positive semi-definite. To see this, we consider that

$$\begin{aligned} \boldsymbol{\Lambda} \cdot \mathbf{G} \cdot \boldsymbol{\Lambda} &= \boldsymbol{\Lambda} \cdot \mathbf{dg} \cdot \hat{M}^{-1}(\hat{q}) \cdot \mathbf{dg}^{\mathrm{T}} \cdot \boldsymbol{\Lambda} \\ &= (\mathbf{dg}^{\mathrm{T}} \cdot \boldsymbol{\Lambda}) \cdot \hat{M}^{-1}(\hat{q}) \cdot (\mathbf{dg}^{\mathrm{T}} \cdot \boldsymbol{\Lambda}) \geq 0, \end{aligned} \tag{117}$$

because of the positive definiteness of $\hat{M}(\hat{q})$. For $\mathbf{G}$ being positive definite, it is required that

$$\mathbf{dg}^{\mathrm{T}} \cdot \boldsymbol{\Lambda} = 0 \tag{118}$$

only holds for the trivial solution $\boldsymbol{\Lambda} = 0$. In other words, the coefficient matrix $\mathbf{W}$ needs to have full column rank. The column rank of $\mathbf{W}$ is equivalent to the row rank of $\mathbf{W}^{\mathrm{T}}$, which corresponds to the number of linearly independent $\mathrm{d}\hat{g}^{\alpha}_{\hat{q}}$ in $\mathbf{dg}$.

The maps $\mathbf{dg}\cdot$, $\mathbf{dg}^{\mathrm{T}}\cdot$ and $\mathbf{G}\cdot$ have the property that

$$\ker(\mathbf{dg}^{\mathrm{T}}\cdot) = \ker(\mathbf{G}\cdot), \tag{119}$$

$$\mathrm{im}(\mathbf{dg}\cdot) = \mathrm{im}(\mathbf{G}\cdot). \tag{120}$$

First, we show that $\ker(\mathbf{G}\cdot) \subseteq \ker(\mathbf{dg}^{\mathrm{T}}\cdot)$. For this, we consider a $\boldsymbol{\Lambda} \in \ker(\mathbf{G}\cdot)$, i.e., a $\boldsymbol{\Lambda} \in \mathbb{R}^{h*}$ with the property $\mathbf{G} \cdot \boldsymbol{\Lambda} = 0$. This implies that $\boldsymbol{\Lambda} \cdot \mathbf{G} \cdot \boldsymbol{\Lambda} = 0$, and consequently $\mathbf{dg}^{\mathrm{T}} \cdot \boldsymbol{\Lambda} = 0$ by (117) and the positive definiteness of $\hat{M}$. So, we have shown that $\ker(\mathbf{G}\cdot) \subseteq \ker(\mathbf{dg}^{\mathrm{T}}\cdot)$. The reverse direction $\ker(\mathbf{G}\cdot) \supseteq \ker(\mathbf{dg}^{\mathrm{T}}\cdot)$ follows trivially from the definition of $\mathbf{G}\cdot$.

Concerning the proof of equality (120), the inclusion $\mathrm{im}(\mathbf{G}\cdot) \subseteq \mathrm{im}(\mathbf{dg}\cdot)$ follows from the definition of $\mathbf{G}\cdot$. The rank nullity theorem can then be invoked twice to show equality. For the linear map $\mathbf{dg}^{\mathrm{T}}\cdot\colon \mathbb{R}^{h*} \to T^*_{\hat{q}}Q$, the rank nullity theorem says that $\dim \mathrm{im}(\mathbf{dg}^{\mathrm{T}}\cdot) = h - \dim \ker(\mathbf{dg}^{\mathrm{T}}\cdot)$. According to (119), it follows that $\dim \ker(\mathbf{dg}^{\mathrm{T}}\cdot) = \dim \ker(\mathbf{G}\cdot)$, such that $\dim \mathrm{im}(\mathbf{dg}^{\mathrm{T}}\cdot) = h - \dim \ker(\mathbf{G}\cdot)$. Now, by the rank nullity theorem for $\mathbf{G}\cdot\colon \mathbb{R}^{h*} \to \mathbb{R}^{h}$, it follows that $\dim \mathrm{im}(\mathbf{dg}^{\mathrm{T}}\cdot) = \dim \mathrm{im}(\mathbf{G}\cdot)$. Since the row and column rank of a linear map are identical, it follows that $\dim \mathrm{im}(\mathbf{dg}\cdot) = \dim \mathrm{im}(\mathbf{G}\cdot)$, and therefore $\mathrm{im}(\mathbf{dg}\cdot) = \mathrm{im}(\mathbf{G}\cdot)$.

Next, we consider a couple $\boldsymbol{\gamma} \in \mathbb{R}^{h}$ and $\boldsymbol{\Lambda} \in \mathbb{R}^{h*}$ for which it holds that $\boldsymbol{\gamma} = \mathbf{G} \cdot \boldsymbol{\Lambda}$. The pre-image of $\boldsymbol{\gamma}$ by the possibly singular linear map $\mathbf{G}\cdot$ is then given by all the vectors that are mapped to $\boldsymbol{\gamma}$ by $\mathbf{G}\cdot$, i.e.,

$$\mathrm{preIm}_{\mathbf{G}\cdot}(\boldsymbol{\gamma}) := \{\mathbf{y} \in \mathbb{R}^{h*} \mid \mathbf{G} \cdot \mathbf{y} = \boldsymbol{\gamma}\} = \boldsymbol{\Lambda} + \ker(\mathbf{G}\cdot), \tag{121}$$

since $\mathbf{G} \cdot \ker(\mathbf{G}\cdot) = 0$. Let $\boldsymbol{\Lambda}, \tilde{\boldsymbol{\Lambda}} \in \mathrm{preIm}_{\mathbf{G}\cdot}(\boldsymbol{\gamma})$ be a pair of vectors that are mapped to $\boldsymbol{\gamma}$ under $\mathbf{G}\cdot$, thus it holds that $\mathbf{G} \cdot (\boldsymbol{\Lambda} - \tilde{\boldsymbol{\Lambda}}) = 0$, and therefore $\boldsymbol{\Lambda} - \tilde{\boldsymbol{\Lambda}} \in \ker(\mathbf{G}\cdot)$. Because $\ker(\mathbf{dg}^{\mathrm{T}}\cdot) = \ker(\mathbf{G}\cdot)$, it follows that the set $\mathbf{dg}^{\mathrm{T}} \cdot \mathrm{preIm}_{\mathbf{G}\cdot}(\boldsymbol{\gamma})$ is a singleton (a set with exactly one element) for any $\boldsymbol{\gamma} \in \mathrm{im}(\mathbf{G}\cdot)$. Indeed, for any $\boldsymbol{\gamma} \in \mathrm{im}(\mathbf{G}\cdot)$ and for all $\boldsymbol{\Lambda}, \tilde{\boldsymbol{\Lambda}} \in \mathrm{preIm}_{\mathbf{G}\cdot}(\boldsymbol{\gamma})$, it holds that

$$\mathbf{dg}^{\mathrm{T}} \cdot \boldsymbol{\Lambda} = \mathbf{dg}^{\mathrm{T}} \cdot \tilde{\boldsymbol{\Lambda}}, \tag{122}$$

and therefore

$$\mathbf{dg}^{\mathrm{T}} \cdot \mathrm{preIm}_{\mathbf{G}\cdot}(\boldsymbol{\gamma}) = \{\mathbf{dg}^{\mathrm{T}} \cdot \mathbf{y} \mid \mathbf{y} \in \mathrm{preIm}_{\mathbf{G}\cdot}(\boldsymbol{\gamma})\} \tag{123}$$

is a singleton for any $\boldsymbol{\gamma} \in \mathrm{im}(\mathbf{G}\cdot)$.

Since $\mathrm{im}(\mathbf{dg}\cdot) = \mathrm{im}(\mathbf{G}\cdot)$, the pre-image with $\mathbf{G}\cdot$ of any point $\boldsymbol{\gamma} \in \mathrm{im}(\mathbf{dg}\cdot)$ in the image of $\mathbf{dg}\cdot$ is never empty, i.e., $\mathrm{preIm}_{\mathbf{G}\cdot}(\mathbf{dg} \cdot \hat{u})$ is non-empty for any $\hat{u} \in T_{\hat{q}}Q$. We conclude that

$$\mathbf{dg}^{\mathrm{T}} \cdot \mathrm{preIm}_{\mathbf{G}\cdot}(\cdot)\colon\ \mathrm{im}(\mathbf{dg}\cdot) \to T^*_{\hat{q}}Q \tag{124}$$

is a singleton. This means that the binary relation (124) between the sets $\mathrm{im}(\mathbf{dg}\cdot)$ and $T^*_{\hat{q}}Q$ is a linear map, which we will denote as

$$\mathbf{dg}^{\mathrm{T}} \cdot \mathbf{G}^{-1} \cdot : \ \mathrm{im}(\mathbf{dg} \cdot) \to T^*_{\hat{q}} Q. \tag{125}$$

Note that $\mathbf{dg}^{\mathrm{T}} \cdot \mathbf{G}^{-1} \cdot$ denotes one function in general. In the case when $\mathbf{G}$ is regular, $\mathbf{dg}^{\mathrm{T}} \cdot \mathbf{G}^{-1} \cdot$ is indeed the concatenation of $\mathbf{dg}^{\mathrm{T}} \cdot$ and $\mathbf{G}^{-1} \cdot$.

Using this linear map, we can consider the following two concatenations of maps:

$$\hat{F} \cdot : \ \mathrm{im}(\mathbf{G} \cdot) \to T_{\hat{q}} Q, \quad \text{with} \quad \hat{F} := \hat{M}^{-1}(\hat{q}) \cdot \mathbf{dg}^{\mathrm{T}} \cdot \mathbf{G}^{-1} \tag{126}$$

and

$$\hat{P}^{\perp} \cdot : \ T_{\hat{q}} Q \to T_{\hat{q}} Q, \quad \text{with} \quad \hat{P}^{\perp} := \hat{M}^{-1}(\hat{q}) \cdot \mathbf{dg}^{\mathrm{T}} \cdot \mathbf{G}^{-1} \cdot \mathbf{dg}. \tag{127}$$

We show that (127) defines an orthogonal projector onto the linear hull $\langle \mathcal{T}_C^{\perp}(\hat{q}) \rangle$ of the orthogonal tangent cone $\mathcal{T}_C^{\perp}(\hat{q}) \subseteq T_{\hat{q}} Q$. Straightforward calculation shows that $\hat{P}^{\perp} \cdot$ is idempotent

$$\hat{P}^{\perp} \cdot \hat{P}^{\perp} = \hat{P}^{\perp} \tag{128}$$

and self-adjoint

$$\left(\hat{P}^{\perp} \cdot \hat{u} \, , \, \hat{v} \right)_{\hat{M}} = \left(\hat{u} \, , \, \hat{P}^{\perp} \cdot \hat{v} \right)_{\hat{M}} , \tag{129}$$

for all $\hat{u}, \hat{v} \in T_{\hat{q}} Q$. Therefore, $\hat{P}^{\perp} \cdot$ is an orthogonal projector on $T_{\hat{q}} Q$.

For any $\hat{u} \in T_{\hat{q}} Q$, the set $\mathrm{preIm}_{\mathbf{G} \cdot}(\mathbf{dg} \cdot \hat{u})$ is non-empty, because $\mathrm{im}(\mathbf{dg} \cdot) = \mathrm{im}(\mathbf{G} \cdot)$ according to (120). For an element $\boldsymbol{\Lambda} \in \mathrm{preIm}_{\mathbf{G} \cdot}(\mathbf{dg} \cdot \hat{u})$, it holds that

$$\hat{P}^{\perp} \cdot \hat{u} = \hat{M}^{-1}(\hat{q}) \cdot \mathbf{dg}^{\mathrm{T}} \cdot \boldsymbol{\Lambda}, \tag{130}$$

which can be expressed locally using (49), (114) and $\boldsymbol{\Lambda} = \Lambda_\beta \, \mathbf{e}^\beta$ as

$$\begin{aligned} \hat{P}^{\perp} \cdot \hat{u} &= \hat{M}^{-1}(\hat{q}) \cdot \mathbf{dg}^{\mathrm{T}} \cdot \boldsymbol{\Lambda} \\ &= \left(M^{kl} \frac{\partial}{\partial q^k} \otimes \frac{\partial}{\partial q^l} \right) \cdot \left(\frac{\partial g^\alpha}{\partial q^i} \, \mathrm{d}q^i \otimes \mathbf{e}_\alpha \right) \cdot \left(\Lambda_\beta \, \mathbf{e}^\beta \right) \\ &= M^{kl} \Lambda_\alpha \frac{\partial g^\alpha}{\partial q^l} \frac{\partial}{\partial q^k}. \end{aligned} \tag{131}$$

By comparing (131) and (106), we see that $\hat{P}^{\perp} \cdot \hat{u} \in \langle \mathcal{T}_{\hat{C}}^{\perp}(\hat{q}) \rangle$. Since $\hat{u} \in T_{\hat{q}} Q$ is arbitrary, the statement $\mathrm{im}(\hat{P}^{\perp} \cdot) = \langle \mathcal{T}_{\hat{C}}^{\perp}(\hat{q}) \rangle$ results.

Because the subspace $\mathcal{L}_{\mathcal{T}_{\hat{C}}(\hat{q})}$ is the orthogonal complement to the subspace $\langle \mathcal{T}_C^{\perp}(\hat{q}) \rangle$, the projector $\hat{P} \cdot$, which was announced in (108), can now be defined as

$$\hat{P} \cdot : \ T_{\hat{q}} Q \to T_{\hat{q}} Q, \quad \text{with} \quad \hat{P} := \mathbb{1} - \hat{P}^{\perp} = \mathbb{1} - \hat{M}^{-1}(\hat{q}) \cdot \mathbf{dg}^{\mathrm{T}} \cdot \mathbf{G}^{-1} \cdot \mathbf{dg}, \tag{132}$$

where $\mathbb{1}$ denotes the identity tensor on $T_{\hat{q}}Q$, which can be locally written as

$$\mathbb{1} = \frac{\partial}{\partial q^i} \otimes \mathrm{d}q^i . \tag{133}$$

The projectors $\hat{P}\cdot$ and $\hat{P}^{\perp}\cdot$ satisfy

$$\begin{aligned} &\hat{P}\big|_{\mathcal{L}\mathcal{T}_{\hat{\mathcal{C}}}(\hat{q})} = \mathbb{1}\big|_{\mathcal{L}\mathcal{T}_{\hat{\mathcal{C}}}(\hat{q})}, && \hat{P}^{\perp}\big|_{\mathcal{L}\mathcal{T}_{\hat{\mathcal{C}}}(\hat{q})} = 0, && \\ &\hat{P}\big|_{\langle \mathcal{T}^{\perp}_{\hat{\mathcal{C}}}(\hat{q})\rangle} = 0, && \hat{P}^{\perp}\big|_{\langle \mathcal{T}^{\perp}_{\hat{\mathcal{C}}}(\hat{q})\rangle} = \mathbb{1}\big|_{\langle \mathcal{T}^{\perp}_{\hat{\mathcal{C}}}(\hat{q})\rangle}, && \hat{P} + \hat{P}^{\perp} = \mathbb{1}. \end{aligned} \tag{134}$$

The following properties hold for the linear maps $\mathbf{dg}^{\mathrm{T}}\cdot$, $\mathbf{dg}\cdot$, $\mathbf{G}\cdot$, $\hat{F}\cdot$, $\hat{P}^{\perp}\cdot$ and $\hat{P}\cdot$ introduced so far:

$$\begin{aligned} &\hat{F}\cdot \mathbf{dg} = \hat{P}^{\perp}, && \hat{P}^{\perp}\cdot \hat{F} = \hat{F}, && \mathbf{dg}\big|_{\mathcal{L}\mathcal{T}_{\hat{\mathcal{C}}}(\hat{q})} = 0, \\ &\mathbf{dg}\cdot \hat{F} = \mathbb{1}\big|_{\mathrm{im}(\mathbf{G}\cdot)}, && \hat{P}\cdot \hat{F} = 0, && \hat{F}\cdot \mathbf{G} = \hat{M}^{-1}(\hat{q})\cdot \mathbf{dg}^{\mathrm{T}}. \end{aligned} \tag{135}$$

From the properties (134) and by (88) and (106), it becomes obvious that

$$\begin{aligned} \hat{P}\cdot(\hat{u}^{+} - \hat{u}^{-}) &= 0, \\ \hat{P}^{\perp}\cdot(\hat{u}^{+} - \hat{u}^{-}) &\in -\mathcal{T}^{\perp}_{\hat{\mathcal{C}}}(\hat{q}). \end{aligned} \tag{136}$$

So far, we have introduced geometric objects in the tangent space $T_{\hat{q}}Q$. Often, we want to work in local coordinates, i.e., in $\mathbb{R}^n$. Hence, it is reasonable to ask whether the decomposition of the tangent space given by the orthogonal projectors $\hat{P}$ and $\hat{P}^{\perp}$ can be pushed to $\mathbb{R}^n$ using the isomorphisms from Fig. 6.

Using the standard bases of $\mathbb{R}^n$, $\mathbb{R}^h$ and of their duals $\mathbb{R}^{n*}$, $\mathbb{R}^{h*}$, the matrix $\mathbf{W}$ from (70) and its transpose correspond to the linear map

$$\begin{aligned} \mathbf{W}\cdot : \mathbb{R}^{h*} &\to \mathbb{R}^{n*}, \\ \boldsymbol{\Lambda} &\mapsto \mathbf{F}^{c} = \mathbf{W}\cdot \boldsymbol{\Lambda} := \mathrm{d}\boldsymbol{\phi}^{-\mathrm{T}}\cdot \mathbf{dg}^{\mathrm{T}}\cdot \boldsymbol{\Lambda} \end{aligned} \tag{137}$$

and its transpose

$$\begin{aligned} \mathbf{W}^{\mathrm{T}}\cdot : \mathbb{R}^{n} &\to \mathbb{R}^{h}, \\ \mathbf{u} &\mapsto \boldsymbol{\gamma} = \mathbf{W}^{\mathrm{T}}\cdot \mathbf{u} := \mathbf{dg}\cdot \mathrm{d}\boldsymbol{\phi}^{-1}\cdot \mathbf{u}, \end{aligned} \tag{138}$$

depending on the chart (U, ϕ) with $\hat{q} \in U$. In Eqs. (137) and (138), the chart ϕ appears in the form of the isomorphism (44). Note that the contraction dot $\cdot$ can be omitted when $\mathbf{W}$ is seen as a matrix and $\mathbf{u} \in \mathbb{R}^n$, $\boldsymbol{\Lambda} \in \mathbb{R}^{h*}$ as column vectors.

The orthogonal projectors (108) and (109) are endomorphisms on the tangent space $T_{\hat{q}}Q$. We want to push them from $T_{\hat{q}}Q$ to $\mathbb{R}^n$ using the chart ϕ such that we get corresponding projectors (i.e., endomorphisms) on $\mathbb{R}^n$. We consider the following diagram:

$$\begin{array}{ccc} \mathbb{R}^n & \xrightarrow{\mathbf{P}\cdot,\,\mathbf{P}^{\perp}\cdot} & \mathbb{R}^n \\ \mathrm{d}\phi\cdot \Big\uparrow \Big\downarrow \mathrm{d}\phi^{-1}\cdot & & \mathrm{d}\phi\cdot \Big\uparrow \Big\downarrow \mathrm{d}\phi^{-1}\cdot \\ T_{\hat{q}}Q & \xrightarrow{\hat{P}\cdot,\,\hat{P}^{\perp}\cdot} & T_{\hat{q}}Q \end{array}$$

which commutes if

$$\begin{aligned} \mathbf{P}^{\perp}\cdot &= \mathrm{d}\boldsymbol{\phi}\cdot\hat{P}^{\perp}\cdot\mathrm{d}\boldsymbol{\phi}^{-1}\cdot, && (139)\\ \mathbf{P}\cdot &= \mathrm{d}\boldsymbol{\phi}\cdot\hat{P}\cdot\mathrm{d}\boldsymbol{\phi}^{-1}\cdot\,. && (140) \end{aligned}$$

Using (127), we write (139) as

$$\mathbf{P}^{\perp}\cdot = \mathrm{d}\boldsymbol{\phi}\cdot\left(\hat{M}^{-1}(\hat{q})\cdot\mathbf{dg}^{\mathrm{T}}\cdot\mathbf{G}^{-1}\cdot\mathbf{dg}\right)\cdot\mathrm{d}\boldsymbol{\phi}^{-1}\cdot, \tag{141}$$

where we can insert the identity map $\mathrm{d}\boldsymbol{\phi}^{\mathrm{T}}\cdot\mathrm{d}\boldsymbol{\phi}^{-\mathrm{T}}$ on $\mathbb{R}^{n*}$ such that we can use (51) to obtain

$$\begin{aligned} \mathbf{P}^{\perp}\cdot &= \mathrm{d}\boldsymbol{\phi}\cdot\hat{M}^{-1}(\hat{q})\cdot\mathrm{d}\boldsymbol{\phi}^{\mathrm{T}}\cdot\mathrm{d}\boldsymbol{\phi}^{-\mathrm{T}}\cdot\mathbf{dg}^{\mathrm{T}}\cdot\mathbf{G}^{-1}\cdot\mathbf{dg}\cdot\mathrm{d}\boldsymbol{\phi}^{-1}\cdot\\ &= \mathbf{M}^{-1}\cdot\mathbf{W}\cdot\mathbf{G}^{-1}\cdot\mathbf{W}^{\mathrm{T}}\cdot\,. \end{aligned} \tag{142}$$

Analogously, it can be shown that

$$\mathbf{P}\cdot = \left(\mathbf{I}_n\cdot\right) - \left(\mathbf{M}^{-1}\cdot\mathbf{W}\cdot\mathbf{G}^{-1}\cdot\mathbf{W}^{\mathrm{T}}\cdot\right). \tag{143}$$

It can be checked that $\mathbf{P}\cdot$ and $\mathbf{P}^{\perp}\cdot$ are indeed projectors. Finally, the map $\hat{F}\cdot\colon \mathrm{im}(\mathbf{G}\cdot)\to T_{\hat{q}}Q$ from (126) remains. We define the chart representation of $\hat{F}\cdot$ as

$$\begin{aligned} \mathbf{F}\cdot\colon \mathbb{R}^h \supseteq \mathrm{im}(\mathbf{G}\cdot) &\to \mathbb{R}^n,\\ \boldsymbol{\gamma} &\mapsto \mathbf{F}\cdot\boldsymbol{\gamma} := \mathrm{d}\boldsymbol{\phi}\cdot\hat{F}\cdot\boldsymbol{\gamma}, \end{aligned} \tag{144}$$

i.e., $\mathbf{F} = \mathrm{d}\boldsymbol{\phi}\cdot\hat{F}$. By (126), it follows that

$$\begin{aligned} \mathbf{F}\cdot &= \mathrm{d}\boldsymbol{\phi}\cdot\hat{M}^{-1}(\hat{q})\cdot\mathbf{dg}^{\mathrm{T}}\cdot\mathbf{G}^{-1}\cdot\\ &= \mathrm{d}\boldsymbol{\phi}\cdot\hat{M}^{-1}(\hat{q})\cdot\mathrm{d}\boldsymbol{\phi}^{\mathrm{T}}\cdot\mathrm{d}\boldsymbol{\phi}^{-\mathrm{T}}\cdot\mathbf{dg}^{\mathrm{T}}\cdot\mathbf{G}^{-1}\cdot\\ &= \mathbf{M}^{-1}\cdot\mathbf{W}\cdot\mathbf{G}^{-1}\cdot, \end{aligned} \tag{145}$$

where we used (51) again. The properties (135) remain valid for the linear maps $\mathbf{W}\cdot$, $\mathbf{W}^{\mathrm{T}}\cdot$, $\mathbf{G}\cdot$, $\mathbf{F}\cdot$, $\mathbf{P}^{\perp}\cdot$ and $\mathbf{P}\cdot$ such that

$$\begin{aligned} &\mathbf{F}\cdot\mathbf{W}^{\mathrm{T}} = \mathbf{P}^{\perp}, && \mathbf{P}^{\perp}\cdot\mathbf{F} = \mathbf{F}, && \mathbf{W}^{\mathrm{T}}\big|_{\mathcal{L}_{\mathcal{T}_C(\mathbf{q})}} = 0, \\ &\mathbf{W}^{\mathrm{T}}\cdot\mathbf{F} = \mathbf{I}_h\big|_{\mathrm{im}(\mathbf{G}\cdot)}, && \mathbf{P}\cdot\mathbf{F} = 0, && \mathbf{F}\cdot\mathbf{G} = \mathbf{M}^{-1}\cdot\mathbf{W}. \end{aligned} \tag{146}$$

5 Variational Analysis of Impact Laws

In this section, we give a variational analysis of instantaneous impact laws, as is also done in [6, 26]. Following [6], we restrict the analysis to the special case of a positive definite Delassus operator $\mathbf{G}$.

From now on, we restrict our considerations to the special case in which the level curves $\hat{g}^{\alpha}(\hat{q}) = 0$ are assumed to intersect *transversally*. This means that the $\mathrm{d}\hat{g}^{\alpha}_{\hat{q}} \in T^{*}_{\hat{q}}Q$, which are active in the sense that $\hat{g}^{\alpha}(\hat{q}) = 0$ holds, are linearly independent in $T^{*}_{\hat{q}}Q$. The corresponding matrix $\mathbf{W}$ from (70) then has full column rank, and the corresponding Delassus operator $\mathbf{G}$ from (115) is positive definite such that it has an inverse $\mathbf{G}^{-1}$. Remember that in the previous sections, $\mathbf{G}^{-1}\cdot$ was a shorthand notation to denote the pre-image of $\mathbf{G}\cdot$ for which only the concatenation $\mathbf{dg}^{\mathrm{T}}\cdot\mathbf{G}^{-1}\cdot$ from (125) was a linear map in general. In the following, $\mathbf{G}^{-1}\cdot$ itself is also a linear map, because of $\mathbf{G}$ being positive definite. Moreover, we will work on $\mathbb{R}^n$, $\mathbb{R}^h$ and their duals, rather than on $T_{\hat{q}}Q$, $\mathbb{R}^h$ and their duals.

We start by restating the inclusions (100)–(102) on $\mathbb{R}^n$ as

$$\mathbf{u}^{+} \in \mathcal{T}_C(\mathbf{q}), \qquad \text{(kinematic consistency)} \tag{147}$$

$$\mathbf{u}^{+} \in \mathbf{u}^{-} - \mathcal{T}_C^{\perp}(\mathbf{q}), \qquad \text{(kinetic consistency)} \tag{148}$$

$$\mathbf{u}^{+} \in \mathcal{B}_{\|\mathbf{u}^{-}\|_{\mathbf{M}}}(\mathbf{q}). \qquad \text{(energetic consistency)} \tag{149}$$

In order to obtain a unique post impact velocity $\mathbf{u}^{+}$, the inclusions (147)–(149) have to be complemented by a constitutive law. On the one hand, the description of impacts according to the classical Newton's or Poisson's impact law is done in contact velocities. On the other hand, we have identified the dual variables $\bar{\boldsymbol{\gamma}}$ and $\boldsymbol{\Lambda}$ in (98) such that we assume a binary relation $\mathcal{H}$ as constitutive law, i.e.,

$$(\bar{\boldsymbol{\gamma}}, -\boldsymbol{\Lambda}) \in \operatorname{grph}\mathcal{H}, \tag{150}$$

with $\operatorname{grph}\mathcal{H} \subseteq \mathbb{R}^h \times \mathbb{R}^{h*}$. Following Rockafellar [40], a binary relation $\mathcal{H}$ can be interpreted as set-valued mapping $\mathcal{H}\colon \mathbb{R}^h \rightrightarrows \mathbb{R}^{h*}$ and the binary relation (150) can be denoted as

$$-\boldsymbol{\Lambda} \in \mathcal{H}(\bar{\boldsymbol{\gamma}}). \tag{151}$$

In the following, we define the property of *maximal monotonicity*, which the set-valued impact law (151) may have.

5.1 Convex Analysis

In convex analysis, the domain of a scalar, real-valued function f on a convex set $\mathcal{D} \subset \mathbb{R}^n$ (cf. Sect. 4.1) is usually extended to $\mathbb{R}^n$ by defining $f(\mathbf{x}) = \infty$ for all $\mathbf{x} \notin \mathcal{D}$. Therefore, the functions of interest are $f : \mathbb{R}^n \to]-\infty, \infty]$. The function f is *convex* if for arbitrary $\mathbf{x}_1, \mathbf{x}_2 \in \mathbb{R}^n$ it holds that $f((1-s)\mathbf{x}_1 + s\mathbf{x}_2) \leq (1-s)f(\mathbf{x}_1) + sf(\mathbf{x}_2)$ with $0 < s < 1$. The function is *proper* if $f(\mathbf{x}) < \infty$ for at least one $\mathbf{x} \in \mathbb{R}^n$. The function f is *lower semi-continuous* if $\lim_{\mathbf{x}\to\bar{\mathbf{x}}} f(\mathbf{x}) \geq f(\bar{\mathbf{x}})$ for all $\mathbf{x} \in \mathbb{R}^n$.

In the following, we will introduce some concepts from convex analysis. As we discussed in Sect. 1, we want to introduce the objects in the context of dual vector spaces, as proposed by Moreau in [34]. At some points, we needed to adapt the definitions taken from [39, 40] by replacing the canonical inner product $(\cdot\,,\,\cdot)$ on $\mathbb{R}^n$ with the duality pairing $\langle\cdot\,,\,\cdot\rangle$.

Definition 1 The *subdifferential* of a proper, lower semi-continuous, convex function $f : \mathbb{R}^n \to]-\infty, \infty]$ at $\mathbf{x}_0 \in \mathbb{R}^n$, is defined as the set

$$\partial f(\mathbf{x}_0) = \left\{\mathbf{y} \in \mathbb{R}^{n*} \mid f(\mathbf{x}) \geq f(\mathbf{x}_0) + \left\langle\mathbf{y}\,,\, \mathbf{x} - \mathbf{x}_0\right\rangle,\ \forall \mathbf{x} \in \mathbb{R}^n\right\}.$$

In general, the subdifferential is a set-valued mapping, i.e., $\partial f : \mathbb{R}^n \rightrightarrows \mathbb{R}^{n*}$. Elements of the domain may be mapped to subsets of the image.

If we are given a set-valued mapping $\mathcal{H}\colon \mathbb{R}^n \rightrightarrows \mathbb{R}^{n*}$, then it is an interesting question for which conditions it can be written as the subdifferential of a proper, lower semi-continuous, convex function $f : \mathbb{R}^n \to]-\infty, \infty]$. We start by defining the *graph* of the set-valued mapping $\mathcal{H}$ as

$$\operatorname{grph}\mathcal{H} = \left\{(\mathbf{x}, \mathbf{y}) \in \mathbb{R}^n \times \mathbb{R}^{n*} \mid \mathbf{y} \in \mathcal{H}(\mathbf{x})\right\}. \tag{152}$$

Definition 2 A mapping $\mathcal{H}\colon \mathbb{R}^n \rightrightarrows \mathbb{R}^{n*}$ is called *monotone* if it has the property that

$$\left\langle\mathbf{y}_2 - \mathbf{y}_1\,,\, \mathbf{x}_2 - \mathbf{x}_1\right\rangle \geq 0 \tag{153}$$

whenever $\mathbf{y}_1 \in \mathcal{H}(\mathbf{x}_1)$, $\mathbf{y}_2 \in \mathcal{H}(\mathbf{x}_2)$ and *strictly monotone* if this inequality is strict when $\mathbf{x}_1 \neq \mathbf{x}_2$. The monotone mapping $\mathcal{H}\colon \mathbb{R}^n \rightrightarrows \mathbb{R}^{n*}$ is *maximal monotone* if no enlargement of its graph is possible in $\mathbb{R}^n \times \mathbb{R}^{n*}$ without destroying monotonicity.

From the definition of the subdifferential, it is obvious that subdifferentials of proper, lower semi-continuous, convex functions are monotone. It can be shown that they are maximal monotone (see Theorem 12.17 in [40]). Maximal monotonicity is therefore a necessary condition that a set-valued mapping $\mathcal{H}\colon \mathbb{R}^n \rightrightarrows \mathbb{R}^{n*}$ needs to fulfil in order for it to be written as the subdifferential of a proper, lower semi-continuous, convex function $f : \mathbb{R}^n \to]-\infty, \infty]$. The property of *maximal cyclical monotonicity* provides us with a necessary and sufficient criterion.

Definition 3 A mapping $\mathcal{H}: \mathbb{R}^n \rightrightarrows \mathbb{R}^{n*}$ is *cyclically monotone* if for any choice of points $\mathbf{x}_0, \mathbf{x}_1, \ldots, \mathbf{x}_m$ (for arbitrary $m \geq 1$) and elements $\mathbf{y}_i \in \mathcal{H}(\mathbf{x}_i)$, it holds that

$$\langle \mathbf{y}_0 , \mathbf{x}_1 - \mathbf{x}_0 \rangle + \langle \mathbf{y}_1 , \mathbf{x}_2 - \mathbf{x}_1 \rangle + \ldots + \langle \mathbf{y}_m , \mathbf{x}_0 - \mathbf{x}_m \rangle \leq 0. \tag{154}$$

It is *maximal cyclically monotone* if it is cyclically monotone and its graph cannot be enlarged without destroying this property.

For $m = 1$, the condition (154) for cyclical monotonicity corresponds to the condition (153) for monotonicity. It can be shown that every *maximal* cyclically monotone mapping is *maximal* monotone.

Theorem 1 *A mapping $\mathcal{H}: \mathbb{R}^n \rightrightarrows \mathbb{R}^{n*}$ has the form $\mathcal{H} = \partial f$ for some proper, lower semi-continuous, convex function $f: \mathbb{R}^n \to]-\infty, \infty]$ iff $\mathcal{H}$ is maximal cyclically monotone. Then, f is determined by $\mathcal{H}$ uniquely up to an additive constant.*

For the proof, we refer to [40, Theorem 12.25].

The above definitions can be stated for mappings from V to itself. The inner product on V then takes the place of the duality pairing. We confine our considerations to the definition of maximal monotonicity in this context.

Definition 4 Let $\mathbb{R}^n$ be a vector space with inner product $(\cdot\,, \cdot)$. A mapping $\mathcal{T}: \mathbb{R}^n \rightrightarrows \mathbb{R}^n$ is called *monotone* if it has the property that

$$(\mathbf{y}_2 - \mathbf{y}_1 , \mathbf{x}_2 - \mathbf{x}_1) \geq 0$$

whenever $\mathbf{y}_1 \in \mathcal{T}(\mathbf{x}_1)$, $\mathbf{y}_2 \in \mathcal{T}(\mathbf{x}_2)$ and *strictly monotone* if this inequality is strict when $\mathbf{x}_1 \neq \mathbf{x}_2$. The monotone mapping $\mathcal{T}: \mathbb{R}^n \rightrightarrows \mathbb{R}^n$ is *maximal monotone* if no enlargement of its graph is possible in $\mathbb{R}^n \times \mathbb{R}^n$ without destroying monotonicity.

In terms of the norm $\|\cdot\|$, which is induced by an inner product $(\cdot\,, \cdot)$ as

$$\|\mathbf{x}\| = \sqrt{(\mathbf{x}\,, \mathbf{x})}, \tag{155}$$

the property of nonexpansivity can be defined as follows.

Definition 5 Let $\mathbb{R}^n$ be a vector space with inner product $(\cdot\,, \cdot)$. A mapping $\mathcal{S}: \mathbb{R}^n \rightrightarrows \mathbb{R}^n$ is called *nonexpansive* if

$$\|\mathbf{w}_1 - \mathbf{w}_0\| \leq \|\mathbf{z}_1 - \mathbf{z}_0\| \quad \text{whenever} \quad \mathbf{w}_1 \in \mathcal{S}(\mathbf{z}_1),\ \mathbf{w}_0 \in \mathcal{S}(\mathbf{z}_0)$$

and *contractive* if the inequality is strict when $\mathbf{z}_1 \neq \mathbf{z}_0$.

Rockafellar and Wets [40] state the following theorem about the so-called Minty parametrization, which connects maximal monotone and nonexpansive mappings.

Theorem 2 *Let $\mathcal{T}: \mathbb{R}^n \rightrightarrows \mathbb{R}^n$ be maximal monotone. Then, the mappings*

$$\mathcal{P} = (\mathbf{I} + \mathcal{T})^{-1}, \qquad \mathcal{Q} = \left(\mathbf{I} + \mathcal{T}^{-1}\right)^{-1}$$

are single-valued, in fact maximal monotone and nonexpansive, and the mapping $\mathbf{z} \mapsto (\mathcal{Q}(\mathbf{z}), \mathcal{P}(\mathbf{z}))$ *is one-to-one from* $\mathbb{R}^n$ *to* grph $\mathcal{T}$. *This provides a parametrization of* grph $\mathcal{T}$ *that is Lipschitz continuous in both directions:*

$$(\mathcal{P}(\mathbf{z}), \mathcal{Q}(\mathbf{z})) = (\mathbf{x}, \mathbf{y}) \quad \Leftrightarrow \quad \mathbf{z} = \mathbf{x} + \mathbf{y},\ (\mathbf{x}, \mathbf{y}) \in \operatorname{grph} \mathcal{T}.$$

Let $\mathcal{D} \subseteq \mathbb{R}^n$ be a closed, non-empty, convex set. The *indicator function of the set* $\mathcal{D}$ is defined as

$$\Psi_{\mathcal{D}}(\mathbf{x}) := \begin{cases} 0 & \text{if } \mathbf{x} \in \mathcal{D}, \\ \infty & \text{if } \mathbf{x} \notin \mathcal{D}. \end{cases} \tag{156}$$

We can take the subdifferential of the indicator function $\Psi_{\mathcal{D}}$ because it is a lower semi-continuous, proper, convex function, and we obtain

$$\begin{aligned} \partial\Psi_{\mathcal{D}}(\mathbf{x}) &= \left\{\mathbf{y} \in \mathbb{R}^{n*} \mid \Psi_{\mathcal{D}}(\mathbf{x}') \geq \Psi_{\mathcal{D}}(\mathbf{x}) + \langle \mathbf{y}\,,\ \mathbf{x}' - \mathbf{x}\rangle,\ \forall \mathbf{x}' \in \mathbb{R}^n\right\} \\ &= \left\{\mathbf{y} \in \mathbb{R}^{n*} \mid 0 \geq \langle \mathbf{y}\,,\ \mathbf{x}' - \mathbf{x}\rangle,\ \forall \mathbf{x}' \in \mathcal{D}\right\}, \end{aligned} \tag{157}$$

which is the normal cone to the set $\mathcal{D}$

$$\mathcal{N}_{\mathcal{D}}(\mathbf{x}) = \left\{\mathbf{y} \in \mathbb{R}^{n*} \mid \langle \mathbf{y}\,,\ \mathbf{x}' - \mathbf{x}\rangle \leq 0,\ \forall \mathbf{x}' \in \mathcal{D}\right\}. \tag{158}$$

Considering that the tangent cone to the set $\mathcal{D}$ at the point $\mathbf{x}$ is defined as

$$\mathcal{T}_{\mathcal{D}}(\mathbf{x}) = \mathrm{cl}\left\{\mathbf{v} \in \mathbb{R}^n \mid \mathbf{v} = \mathrm{s}(\mathbf{x}' - \mathbf{x}),\ \forall \mathbf{x}' \in \mathcal{D}, \mathrm{s} \geq 0\right\}, \tag{159}$$

the normal cone (158) can be written in the same form as (72), i.e.,

$$\mathcal{N}_{\mathcal{D}}(\mathbf{x}) = \left\{\mathbf{y} \in \mathbb{R}^{n*} \mid \langle \mathbf{y}\,,\ \mathbf{v}\rangle \leq 0,\ \forall \mathbf{v} \in \mathcal{T}_{\mathcal{D}}(\mathbf{x})\right\}. \tag{160}$$

If the vector space $\mathbb{R}^n$ is equipped with an inner product $(\cdot\,,\ \cdot)_{\mathbf{R}}$ such that, according to (11),

$$(\mathbf{u}\,,\ \mathbf{v})_{\mathbf{R}} = \mathbf{u}^{\mathrm{T}}\mathbf{R}\mathbf{v} \tag{161}$$

for any $\mathbf{u}, \mathbf{v} \in \mathbb{R}^n$, then the proximal point function $\mathrm{prox}_{\mathcal{D}}^{\mathbf{R}} : \mathbb{R}^n \to \mathcal{D}$ of a point $\mathbf{z} \in \mathbb{R}^n$ to a nonempty closed convex set $\mathcal{D} \subseteq \mathbb{R}^n$ can be defined as

$$\mathrm{prox}_{\mathcal{D}}^{\mathbf{R}}(\mathbf{z}) := \mathrm{argmin}_{\mathbf{x}' \in \mathcal{D}}\ \|\mathbf{z} - \mathbf{x}'\|_{\mathbf{R}}. \tag{162}$$

The proximal point function (162) provides the closest point of $\mathbf{z}$ in the set $\mathcal{D}$ with respect to the norm corresponding to the inner product $(\cdot\,,\ \cdot)_{\mathbf{R}}$. The $\mathbf{R}$ in the notation of the proximal point function serves as a mnemonic for the inner product space.

The proximal point function of a set $\mathcal{D}$ is related to its normal cone as follows:

$$\begin{aligned} \mathbf{x} = \mathrm{prox}_{\mathcal{D}}^{\mathbf{R}}(\mathbf{z}) &\Leftrightarrow \mathbf{x} = \mathrm{argmin}_{\mathbf{x}' \in \mathcal{D}} \|\mathbf{z} - \mathbf{x}'\|_{\mathbf{R}} \\ &\Leftrightarrow \mathbf{x} = \mathrm{argmin}_{\mathbf{x}' \in \mathbb{R}^n} \frac{1}{2}\|\mathbf{z} - \mathbf{x}'\|_{\mathbf{R}}^2 + \Psi_{\mathcal{D}}(\mathbf{x}') \\ &\Leftrightarrow \mathbf{0} \in -\mathbf{R}(\mathbf{z} - \mathbf{x}) + \mathcal{N}_{\mathcal{D}}(\mathbf{x}) \\ &\Leftrightarrow \mathbf{z} \in \mathbf{x} + \mathbf{R}^{-1}\mathcal{N}_{\mathcal{D}}(\mathbf{x}). \end{aligned} \tag{163}$$

The distance function $\mathrm{dist}_{\mathcal{D}}^{\mathbf{R}}(\mathbf{z}) := \|\mathbf{z} - \mathrm{prox}_{\mathcal{D}}^{\mathbf{R}}(\mathbf{z})\|_{\mathbf{R}}$ gives the distance in the $\mathbf{R}$-norm of a point $\mathbf{z} \in \mathbb{R}^n$ to the proximal point in the set $\mathcal{D}$. It holds that

$$\partial \frac{1}{2}\left(\mathrm{dist}_{\mathcal{D}}^{\mathbf{R}}(\mathbf{z})\right)^2 = \mathbf{R}\left(\mathbf{z} - \mathrm{prox}_{\mathcal{D}}^{\mathbf{R}}(\mathbf{z})\right), \tag{164}$$

which is proven in [27, Proposition 2.33] for $\mathbf{R} = \mathbf{I}$ and can be directly extended to a more general inner product $(\cdot\,,\ \cdot)_{\mathbf{R}}$.

For a given pair of orthogonal closed convex cones $\mathcal{K}, \mathcal{K}^{\perp} \subseteq \mathbb{R}^n$, any vector $\mathbf{u} \in \mathbb{R}^n$ can be represented uniquely in the form

$$\mathbf{u} = \mathbf{v} + \mathbf{v}^{\perp}, \quad \mathbf{v} \in \mathcal{K}, \quad \mathbf{v}^{\perp} \in \mathcal{K}^{\perp}, \quad \mathbf{v} \perp \mathbf{v}^{\perp}, \tag{165}$$

where $\perp$ denotes orthogonality with respect to the inner product, i.e., $(\mathbf{v}\,,\ \mathbf{v}^{\perp})_{\mathbf{R}} = 0$. The decomposition $\mathbf{u} = \mathbf{v} + \mathbf{v}^{\perp}$ may be written using the proximal point function as

$$\mathbf{u} = \mathrm{prox}_{\mathcal{K}}^{\mathbf{R}}(\mathbf{u}) + \mathrm{prox}_{\mathcal{K}^{\perp}}^{\mathbf{R}}(\mathbf{u}), \tag{166}$$

according to [33].

Next, we consider the following diagram:

$$\begin{array}{ccc} \mathbb{R}^n, (\cdot\,, \cdot)_{\mathbf{R}} & \xrightarrow{\mathrm{prox}_{\mathbf{A}\mathcal{D}}^{\mathbf{R}}(\mathbf{x})} & \mathbf{A}\mathcal{D} \subseteq \mathbb{R}^n \\ \mathbf{A}\uparrow & & \mathbf{A}\uparrow \\ \mathbb{R}^m, (\cdot\,, \cdot)_{\mathbf{A}^{\mathrm{T}}\mathbf{R}\mathbf{A}} & \xrightarrow{\mathrm{prox}_{\mathcal{D}}^{\mathbf{A}^{\mathrm{T}}\mathbf{R}\mathbf{A}}(\mathbf{x})} & \mathcal{D} \subseteq \mathbb{R}^m \end{array}$$

where $m \leq n$ and $\mathbf{A} \in \mathbb{R}^{n \times m*}$ is a matrix with full column rank. Note that the inner products $(\cdot\,,\ \cdot)_{\mathbf{R}}$ on $\mathbb{R}^n$ and $(\cdot\,,\ \cdot)_{\mathbf{A}^{\mathrm{T}}\mathbf{R}\mathbf{A}}$ on $\mathbb{R}^m$ are compatible in the sense that

$$(\mathbf{A}\mathbf{u}\,,\ \mathbf{A}\mathbf{v})_{\mathbf{R}} = (\mathbf{u}\,,\ \mathbf{v})_{\mathbf{A}^{\mathrm{T}}\mathbf{R}\mathbf{A}} \tag{167}$$

for all $\mathbf{u}, \mathbf{v} \in \mathbb{R}^m$. The above diagram commutes if

$$\mathbf{A}\,\mathrm{prox}_{\mathcal{D}}^{\mathbf{A}^{\mathrm{T}}\mathbf{R}\mathbf{A}}(\mathbf{x}) = \mathrm{prox}_{\mathbf{A}\mathcal{D}}^{\mathbf{R}}(\mathbf{A}\mathbf{x}), \tag{168}$$

which is the transformation rule from [6, 31]. Lastly, we state the identities

$$\mathrm{prox}_{\mathcal{T}_C}^{\mathbf{M}}(\mathbf{P}\mathbf{x}+\mathbf{y}) = \mathbf{P}\mathbf{x} + \mathrm{prox}_{\mathcal{T}_C}^{\mathbf{M}}(\mathbf{y}), \tag{169}$$

$$\mathrm{prox}_{\mathcal{T}_C^{\perp}}^{\mathbf{M}}(\mathbf{x}+\mathbf{y}) = \mathrm{prox}_{\mathcal{T}_C^{\perp}}^{\mathbf{M}}(\mathbf{P}^{\perp}\mathbf{x}+\mathbf{y}), \tag{170}$$

from [6] without proof.

5.2 Impact Map in the Generalized Velocities

If we assume that the constitutive law $-\boldsymbol{\Lambda} \in \mathcal{H}(\bar{\boldsymbol{\gamma}})$ is maximal monotone, then we can derive a single-valued map

$$\begin{aligned} Z\colon \mathbb{R}^n &\to \mathbb{R}^n, \\ \mathbf{u}^- &\mapsto \mathbf{u}^+ = Z(\mathbf{u}^-). \end{aligned} \tag{171}$$

We consider the Minty parametrization of the maximal monotone set-valued map on $\mathbb{R}^n$

$$\begin{aligned} \mathcal{T}_u\colon \mathbb{R}^n &\rightrightarrows \mathbb{R}^n, \\ \bar{\mathbf{u}} &\mapsto \mathcal{T}_u(\bar{\mathbf{u}}) = \frac{1}{2}\mathbf{M}^{-1}\cdot\mathbf{W}\cdot\mathcal{H}\circ\mathbf{W}^{\mathrm{T}}\cdot\bar{\mathbf{u}}, \end{aligned} \tag{172}$$

as shown in Fig. 7. The variables $\mathbf{x}$ and $\mathbf{y}$ from Theorem 2 are then given by $\bar{\mathbf{u}}$ and $-\frac{1}{2}\mathbf{M}^{-1}\mathbf{W}\boldsymbol{\Lambda}$, respectively. Considering the impact equation (83)

$$\mathbf{u}^+ - \mathbf{u}^- = \mathbf{M}^{-1}\mathbf{W}\boldsymbol{\Lambda} \tag{173}$$

and defining

$$\bar{\mathbf{u}} = \frac{1}{2}(\mathbf{u}^+ + \mathbf{u}^-), \tag{174}$$

it follows that

$$\mathbf{z} = \mathbf{x} + \mathbf{y} = \bar{\mathbf{u}} + \left(-\frac{1}{2}\mathbf{M}^{-1}\mathbf{W}\boldsymbol{\Lambda}\right) = \mathbf{u}^- \tag{175}$$

as desired. According to Theorem 2, the mapping

$$\bar{\mathbf{u}} = \mathcal{P}(\mathbf{u}^-) = (\mathbf{I}_n + \mathcal{T}_u)^{-1}(\mathbf{u}^-) = \left(\mathbf{I}_n + \frac{1}{2}\mathbf{M}^{-1}\cdot\mathbf{W}\cdot\mathcal{H}\circ\mathbf{W}^{\mathrm{T}}\right)^{-1}(\mathbf{u}^-) \tag{176}$$

is single-valued. Using (174), we can state the desired impact map in generalized velocities

$$\mathbf{u}^+ = 2\left(\mathbf{I}_n + \frac{1}{2}\mathbf{M}^{-1}\cdot\mathbf{W}\cdot\mathcal{H}\circ\mathbf{W}^{\mathrm{T}}\right)^{-1}(\mathbf{u}^-) - \mathbf{u}^- \tag{177}$$

such that

$$Z = 2\left(\mathbf{I}_n + \frac{1}{2}\mathbf{M}^{-1}\cdot\mathbf{W}\cdot\mathcal{H}\circ\mathbf{W}^{\mathrm{T}}\right)^{-1} - \mathbf{I}_n. \tag{178}$$

5.3 Impact Map in the Contact Velocities

As in the previous section, we now want to obtain a single-valued map

$$\begin{aligned} S\colon \mathbb{R}^h &\to \mathbb{R}^h,\\ \boldsymbol{\gamma}^- &\mapsto \boldsymbol{\gamma}^+ = S(\boldsymbol{\gamma}^-), \end{aligned} \tag{179}$$

under the assumption that the constitutive law $-\boldsymbol{\Lambda} \in \mathcal{H}(\bar{\boldsymbol{\gamma}})$ is maximal monotone. To do so, we consider the set-valued map

$$\begin{aligned} \mathcal{T}_\gamma : \mathbb{R}^h &\rightrightarrows \mathbb{R}^h,\\ \bar{\boldsymbol{\gamma}} &\mapsto \mathcal{T}_\gamma(\bar{\boldsymbol{\gamma}}) = \frac{1}{2}\mathbf{G}\cdot\mathcal{H}(\bar{\boldsymbol{\gamma}}), \end{aligned} \tag{180}$$

together with its Minty parametrization (see Fig. 7). The variables $\mathbf{x}$ and $\mathbf{y}$ from Theorem 2 are then given by $\bar{\boldsymbol{\gamma}}$ and $-\frac{1}{2}\mathbf{G}\boldsymbol{\Lambda}$, respectively.

First, we consider that the Delassus operator from (115) can be written using (51), (137) and (138) as

$$\begin{aligned} \mathbf{G} &= \mathbf{dg}\cdot\hat{M}^{-1}(\hat{q})\cdot\mathbf{dg}^{\mathrm{T}}\\ &= \mathbf{dg}\cdot\mathbf{d}\boldsymbol{\phi}^{-1}\cdot\mathbf{d}\boldsymbol{\phi}\cdot\hat{M}^{-1}(\hat{q})\cdot\mathbf{d}\boldsymbol{\phi}^{\mathrm{T}}\cdot\mathbf{d}\boldsymbol{\phi}^{-\mathrm{T}}\cdot\mathbf{dg}^{\mathrm{T}}\\ &= \mathbf{W}^{\mathrm{T}}\mathbf{M}^{-1}\mathbf{W}. \end{aligned} \tag{181}$$

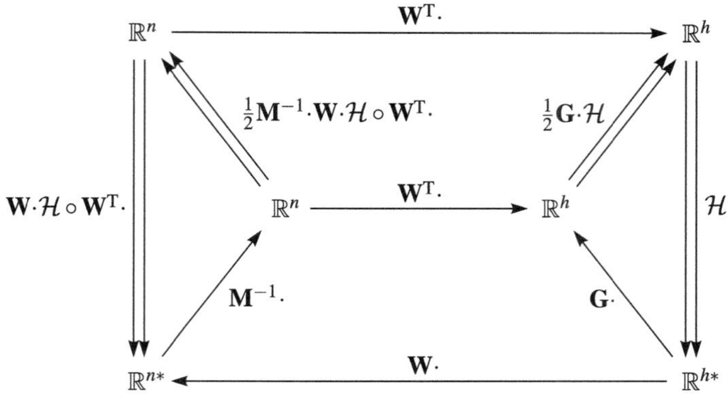

Fig. 7 Set-valued maps $\frac{1}{2}\mathbf{G}\cdot\mathcal{H}$ on the inner product space $\left(\mathbb{R}^h, (\cdot,\cdot)_{\mathbf{G}^{-1}}\right)$ and $\frac{1}{2}\mathbf{M}^{-1}\cdot\mathbf{W}\cdot\mathcal{H}\circ\mathbf{W}^{\mathrm{T}}$ on $(\mathbb{R}^n, (\cdot,\cdot)_{\mathbf{M}})$ associated with a set-valued map $\mathcal{H}\colon \mathbb{R}^h \rightrightarrows \mathbb{R}^{h*}$

Then, we multiply (173) from the left by $\mathbf{W}^{\mathrm{T}}$ and obtain

$$\gamma^+ - \gamma^- = \mathbf{G}\Lambda. \tag{182}$$

With (96) and (182), it follows that

$$\mathbf{z} = \mathbf{x} + \mathbf{y} = \bar{\gamma} + \left(-\frac{1}{2}\mathbf{G}\Lambda\right) = \gamma^-. \tag{183}$$

According to Theorem 2, the mapping

$$\bar{\gamma} = \mathcal{P}(\gamma^-) = \left(\mathbf{I}_h + \mathcal{T}_\gamma\right)^{-1}(\gamma^-) = \left(\mathbf{I}_h + \frac{1}{2}\mathbf{G}\cdot\mathcal{H}\right)^{-1}(\gamma^-) \tag{184}$$

is single-valued. Using (96), we can state the desired impact map in contact velocities

$$\gamma^+ = 2\left(\mathbf{I}_h + \frac{1}{2}\mathbf{G}\cdot\mathcal{H}\right)^{-1}(\gamma^-) - \gamma^- \tag{185}$$

such that

$$S = 2\left(\mathbf{I}_h + \frac{1}{2}\mathbf{G}\cdot\mathcal{H}\right)^{-1} - \mathbf{I}_h. \tag{186}$$

5.4 *Interrelations Between the Representations of an Impact Law*

Figure 8 summarizes the interrelations between the maps $\mathcal{H}$, S and Z. In Sect. 5.3, we have derived the mapping S when we are given $\mathcal{H}$. In the opposite direction, we can solve (186) for $\mathcal{H}$ such that

$$\mathcal{H} = 2\mathbf{G}^{-1}\cdot\left(\left(\frac{S + \mathbf{I}_h}{2}\right)^{-1} - \mathbf{I}_h\right). \tag{187}$$

Similarly, we can solve (178) for $\mathcal{H}$. In a first step, we can write

$$\mathbf{W}\cdot\mathcal{H}\circ\mathbf{W}^{\mathrm{T}} = 2\mathbf{M}\cdot\left(\left(\frac{Z + \mathbf{I}_n}{2}\right)^{-1} - \mathbf{I}_n\right). \tag{188}$$

Then, $\mathbf{F}$ can be applied from the right to yield

$$\mathbf{W}\cdot\mathcal{H} = 2\mathbf{M}\cdot\left(\left(\frac{Z + \mathbf{I}_n}{2}\right)^{-1} - \mathbf{I}_n\right)\circ\mathbf{F} \tag{189}$$

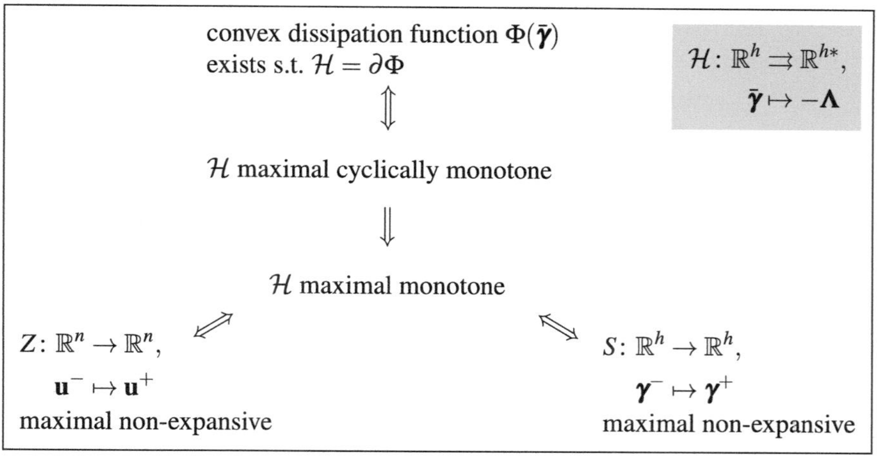

Fig. 8 Interrelations of the monotonicity properties of an impact constitutive law $\mathcal{H}$ and the corresponding impact mappings Z and S

because of the properties (146). Next, we apply $\mathbf{F}^{\mathrm{T}}\colon \mathbb{R}^{h*} \to \mathbb{R}^{n*}$, the transpose map of $\mathbf{F}$, from the left to obtain

$$\mathcal{H} = 2\mathbf{F}^{\mathrm{T}} \cdot \mathbf{M} \cdot \left(\left(\frac{Z + \mathbf{I}_n}{2} \right)^{-1} - \mathbf{I}_n \right) \circ \mathbf{F}, \tag{190}$$

which is the desired relation.

Our considerations from Sect. 4.2 can be used to derive a direct relation between S and Z. Using the orthogonal projectors on $\mathbb{R}^n$ from (139) and (140), we have that

$$\begin{aligned} \mathbf{u}^+ &= \mathbf{P} \cdot \mathbf{u}^+ + \mathbf{P}^{\perp} \cdot \mathbf{u}^+ \\ &= \mathbf{P} \cdot \mathbf{u}^- + \mathbf{F} \cdot \mathbf{W}^{\mathrm{T}} \cdot \mathbf{u}^+ \\ &= \mathbf{P} \cdot \mathbf{u}^- + \mathbf{F} \cdot \boldsymbol{\gamma}^+ \\ &= \mathbf{P} \cdot \mathbf{u}^- + \mathbf{F} \cdot S(\boldsymbol{\gamma}^-) \\ &= \mathbf{P} \cdot \mathbf{u}^- + \mathbf{F} \cdot S(\mathbf{W}^{\mathrm{T}} \cdot \mathbf{u}^-). \end{aligned} \tag{191}$$

At the second equality, we have used (136), (142), (145), (146) and (173). At the third and fifth equality we have used that $\boldsymbol{\gamma}^{\pm} = \mathbf{W}^{\mathrm{T}} \cdot \mathbf{u}^{\pm}$. Finally, we have

$$\begin{aligned} Z &= \mathbf{P} + \mathbf{F} \cdot S \circ \mathbf{W}^{\mathrm{T}} \\ &= \mathbf{P} + \mathbf{F} \cdot \left(2 \left(\mathbf{I}_h + \frac{1}{2} \mathbf{G} \cdot \mathcal{H} \right)^{-1} \circ \mathbf{W}^{\mathrm{T}} - \mathbf{W}^{\mathrm{T}} \right), \end{aligned} \tag{192}$$

where we inserted the expression (186) for S. The expression (192) can be solved for S by applying $\mathbf{W}^{\mathrm{T}}\cdot$ from the left as

$$\begin{aligned}\mathbf{W}^{\mathrm{T}} \cdot Z &= \mathbf{W}^{\mathrm{T}} \cdot \mathbf{P} + \mathbf{W}^{\mathrm{T}} \cdot \mathbf{F} \cdot S \circ \mathbf{W}^{\mathrm{T}} \\ &= S \circ \mathbf{W}^{\mathrm{T}},\end{aligned} \tag{193}$$

because it follows from (143) that $\mathbf{W}^{\mathrm{T}} \cdot \mathbf{P} = 0$ and by the properties (146) $\mathbf{W}^{\mathrm{T}} \cdot \mathbf{F} \cdot S \circ \mathbf{W}^{\mathrm{T}} = S \circ \mathbf{W}^{\mathrm{T}}$. After applying $\mathbf{F}$ from the right, we can invoke the properties (146) again to conclude that

$$S = \mathbf{W}^{\mathrm{T}} \cdot Z \circ \mathbf{F}. \tag{194}$$

6 Specific Impact Laws

In this section, which is an abridged version of [6, Chap. 3.3.4], we will apply the variational analysis of impact laws to two specific impact laws, those being Newton's impact law and Poisson's impact law, which are typically used in rigid multibody dynamics [38]. Both of these impact laws have originally been stated for collisions with a single unilateral constraint. Here, we will consider generalizations of these impact laws to multi-contact collisions that are referred to as the generalized Newton's impact law and the generalized Poisson's impact law, respectively.

6.1 Generalized Newton's Impact Law

The classical Newton's impact law considers a collision with a single active unilateral constraint $g^1 = 0$ with a pre-impact velocity $\gamma^{1-} < 0$, or $\gamma^{1-} \leq 0$ if zero-velocity collisions are also considered. Newton's impact law simply inverts the pre-impact contact velocity and scales it with what is called a Newtonian restitution coefficient ε^1, i.e.,

$$\gamma^{1+} = -\varepsilon^1 \gamma^{1-}, \tag{195}$$

thereby giving a kinematically admissible post-impact contact velocity $\gamma^{1+} \geq 0$, as the coefficient of restitution is chosen in the interval $\varepsilon^1 \in [0,\ 1]$. The case $\varepsilon^1 = 1$ corresponds to a completely elastic impact (energy conservation), whereas $\varepsilon^1 = 0$ corresponds to a completely inelastic impact (maximal energy dissipation). The impact is accompanied by a contact impulse Λ_1. The assumption of an adhesion-free contact poses the restriction $\Lambda_1 \geq 0$ on the contact impulse, which is naturally fulfilled for single-contact collisions. In a multi-contact collision, one cannot simply apply Newton's restitution rule to all active constraints $\alpha = 1, \ldots, h$. Simple application of Newton's restitution rule may result in a negative contact impulse for some constraints. The classical Newton's impact law has therefore to be generalized for multi-contact collisions by explicitly allowing for superfluous constraints. A unilateral constraint $g^\alpha \geq 0$ is called *superfluous* if it does not participate in the impact (i.e., $\Lambda_\alpha = 0$) although the unilateral constraint is active (i.e., $g^\alpha = 0$). For $\gamma^{\alpha-} < 0$, the occurrence of such a superfluous constraint only happens for multi-constraint collisions. Following [14, 36], we generalize the classical Newton's impact law to account for superfluous constraints by allowing post-impact contact velocities larger than prescribed by Newton's restitution rule, i.e., $\gamma^{\alpha+} \geq -\varepsilon^\alpha \gamma^{\alpha-}$. Summarizing, two cases can occur at an active unilateral constraint α:

1. The unilateral constraint is actively participating in the impact process, i.e., $\Lambda_\alpha > 0$ and $\gamma^{\alpha+} = -\varepsilon^\alpha \gamma^{\alpha-}$.
2. The unilateral constraint is superfluous, i.e., $\Lambda_\alpha = 0$ and $\gamma^{\alpha+} \geq -\varepsilon^\alpha \gamma^{\alpha-}$.

These two cases are combined in an inequality complementarity impact law on velocity–impulse level with $\xi^\alpha := \gamma^{\alpha+} + \varepsilon^\alpha \gamma^{\alpha-}$ as

$$-\Lambda_\alpha \in \mathcal{N}_{\mathbb{R}_0^+}(\xi^\alpha). \tag{196}$$

The generalized Newton's impact law (196) can be written in vector notation using the restitution coefficient matrix $\mathbf{E} = \operatorname{diag}\left(\{\varepsilon^1, \ldots, \varepsilon^h\}\right) \in \mathbb{R}^{h\times h}$ as

$$-\boldsymbol{\Lambda} \in \mathcal{N}_{\mathbb{R}_0^{h+}}(\boldsymbol{\xi}), \tag{197}$$

where $\boldsymbol{\xi} = \boldsymbol{\gamma}^+ + \mathbf{E}\boldsymbol{\gamma}^-$. Alternatively, we may write the generalized Newton's impact law as the inequality complementarity

$$\mathbb{R}_0^{h*+} \ni \boldsymbol{\Lambda} \perp \boldsymbol{\xi} \in \mathbb{R}_0^{h+}, \tag{198}$$

where $\boldsymbol{\Lambda} \perp \boldsymbol{\xi}$ denotes $\langle \boldsymbol{\Lambda}, \boldsymbol{\xi} \rangle = 0$. The symmetry in the inequality complementarity (198) reveals that we can invert the normal cone inclusion (196) into

$$-\boldsymbol{\xi} \in \mathcal{N}_{\mathbb{R}_0^{h*+}}(\boldsymbol{\Lambda}). \tag{199}$$

We will consider the simpler case of a global restitution coefficient ε, meaning that all restitution coefficients are identical, $\varepsilon^\alpha = \varepsilon$ for all $\alpha = 1, \ldots, h$, and correspondingly $\mathbf{E} = \varepsilon\mathbf{I}$. The following theorem is proven in [6].

Theorem 3 *Consider the impact equation* (83) *together with the generalized Newton's impact law* (197) *with a global restitution coefficient* $\varepsilon \in [0, 1]$, *that is,* $\boldsymbol{\xi} = \boldsymbol{\gamma}^+ + \varepsilon\boldsymbol{\gamma}^-$. *Then, the set-valued impact law* $\mathcal{H}$ *and the impact mappings* S *and* Z *are given by*

$$-\boldsymbol{\Lambda} \in \mathcal{H}(\bar{\boldsymbol{\gamma}}) = \begin{cases} -\frac{2(1+\varepsilon)}{1-\varepsilon} \operatorname{prox}^{\mathbf{G}}_{\mathbb{R}_0^{h*+}}(-\mathbf{G}^{-1}\bar{\boldsymbol{\gamma}}) & 0 \leq \varepsilon < 1, \\ \mathcal{N}_{\mathbb{R}_0^{h+}}(\bar{\boldsymbol{\gamma}}) & \varepsilon = 1, \end{cases} \tag{200}$$

$$\boldsymbol{\gamma}^+ = S(\boldsymbol{\gamma}^-) = \mathrm{prox}^{\mathbf{G}^{-1}}_{\mathbb{R}_0^{h+}}(\boldsymbol{\gamma}^-) - \varepsilon\ \mathrm{prox}^{\mathbf{G}^{-1}}_{(\mathbb{R}_0^{h+})^\perp}(\boldsymbol{\gamma}^-), \tag{201}$$

$$\mathbf{u}^+ = Z(\mathbf{u}^-) = \mathrm{prox}^{\mathbf{M}}_{\mathcal{T}_C}\left(\mathbf{u}^-\right) - \varepsilon\ \mathrm{prox}^{\mathbf{M}}_{\mathcal{T}_C^\perp}\left(\mathbf{u}^-\right). \tag{202}$$

Furthermore, the set-valued impact law $-\boldsymbol{\Lambda} \in \mathcal{H}(\bar{\boldsymbol{\gamma}}) = \partial\Phi(\bar{\boldsymbol{\gamma}})$ *is cyclically maximal monotone with the corresponding dissipation function*

$$\Phi(\bar{\boldsymbol{\gamma}}) = \begin{cases} \frac{1+\varepsilon}{1-\varepsilon}\left(\mathrm{dist}^{\mathbf{G}^{-1}}_{\mathbb{R}_0^{h+}}(\bar{\boldsymbol{\gamma}})\right)^2 & 0 \le \varepsilon < 1, \\ \Psi_{\mathbb{R}_0^{h+}}(\bar{\boldsymbol{\gamma}}) & \varepsilon = 1. \end{cases} \tag{203}$$

Proof In the first part of the proof, we derive the impact mappings S and Z. For this purpose, the generalized Newton's impact law (197) is written using the local impact equation (182) together with $\boldsymbol{\xi} = \boldsymbol{\gamma}^+ + \varepsilon\boldsymbol{\gamma}^-$ as

$$-(\boldsymbol{\gamma}^+ - \boldsymbol{\gamma}^-) \in \mathbf{G}\mathcal{N}_{\mathbb{R}_0^{h+}}(\boldsymbol{\gamma}^+ + \varepsilon\boldsymbol{\gamma}^-). \tag{204}$$

The normal cone inclusion transforms by (163) to the explicit equation

$$\boldsymbol{\gamma}^+ = -\varepsilon\boldsymbol{\gamma}^- + \mathrm{prox}^{\mathbf{G}^{-1}}_{\mathbb{R}_0^{h+}}\left((1+\varepsilon)\boldsymbol{\gamma}^-\right). \tag{205}$$

Using the positive homogeneity of $\mathrm{prox}^{\mathbf{G}^{-1}}_{\mathbb{R}_0^{h+}}(\cdot)$, we arrive at

$$\boldsymbol{\gamma}^+ = -\varepsilon\boldsymbol{\gamma}^- + (1+\varepsilon)\,\mathrm{prox}^{\mathbf{G}^{-1}}_{\mathbb{R}_0^{h+}}(\boldsymbol{\gamma}^-), \tag{206}$$

which is already a formulation for S. Using the orthogonal cone decomposition (166), we may put the mapping S in its final form

$$\boldsymbol{\gamma}^+ = \mathrm{prox}^{\mathbf{G}^{-1}}_{\mathbb{R}_0^{h+}}(\boldsymbol{\gamma}^-) - \varepsilon\,\mathrm{prox}^{\mathbf{G}^{-1}}_{(\mathbb{R}_0^{h+})^\perp}(\boldsymbol{\gamma}^-). \tag{207}$$

The mapping Z is obtained from the mapping S using (192) as

$$\begin{aligned} \mathbf{u}^+ &= \left(\mathbf{P} + \mathbf{F}S \circ \mathbf{W}^{\mathrm{T}}\right)(\mathbf{u}^-) \\ &= \left(\mathbf{I} + \mathbf{F}(S - \mathbf{I}) \circ \mathbf{W}^{\mathrm{T}}\right)(\mathbf{u}^-) \\ &= \mathbf{u}^- - (1+\varepsilon)\mathbf{F}\cdot\mathrm{prox}^{\mathbf{G}^{-1}}_{(\mathbb{R}_0^{h+})^\perp}\left(\mathbf{W}^{\mathrm{T}}\cdot\mathbf{u}^-\right), \end{aligned} \tag{208}$$

in which $\mathbf{P} = \mathbf{I} - \mathbf{P}^\perp = \mathbf{I} - \mathbf{F}\cdot\mathbf{W}^{\mathrm{T}}$ has been used. Because $\mathbf{G}^{-1} = \mathbf{F}^{\mathrm{T}}\cdot\mathbf{M}\cdot\mathbf{F}$, the proximal point function on $(\mathbb{R}^h, (\cdot\,,\,\cdot)_{\mathbf{G}^{-1}})$ can be replaced by the one on $(\mathbb{R}^n, (\cdot\,,\,\cdot)_{\mathbf{M}})$ using the transformation rule (168)

$$\begin{aligned} \mathbf{u}^+ &= \mathbf{u}^- - (1+\varepsilon)\mathbf{F}\cdot\mathrm{prox}^{\mathbf{F}^{\mathrm{T}}\mathbf{M}\mathbf{F}}_{(\mathbb{R}_0^{h+})^\perp}\left(\mathbf{W}^{\mathrm{T}}\cdot\mathbf{u}^-\right) \\ &= \mathbf{u}^- - (1+\varepsilon)\,\mathrm{prox}^{\mathbf{M}}_{\mathbf{F}(\mathbb{R}_0^{h+})^\perp}\left(\mathbf{F}\cdot\mathbf{W}^{\mathrm{T}}\cdot\mathbf{u}^-\right). \end{aligned} \tag{209}$$

The formulation (202) of the mapping Z is obtained by again using the orthogonal cone decomposition together with $\mathcal{T}_C^\perp = \mathbf{F}(\mathbb{R}_0^{h+})^\perp$ and the identity (170), i.e.,

$$\begin{aligned}\mathbf{u}^+ &= \left(\mathbf{u}^- - \operatorname{prox}^{\mathbf{M}}_{\mathcal{T}_C^\perp}\left(\mathbf{u}^-\right)\right) - \varepsilon \operatorname{prox}^{\mathbf{M}}_{\mathcal{T}_C^\perp}\left(\mathbf{u}^-\right) \\ &= \operatorname{prox}^{\mathbf{M}}_{\mathcal{T}_C}\left(\mathbf{u}^-\right) - \varepsilon \operatorname{prox}^{\mathbf{M}}_{\mathcal{T}_C^\perp}\left(\mathbf{u}^-\right).\end{aligned} \tag{210}$$

We continue by deriving the set-valued impact law $\mathcal{H}$ in (200). Using the mean contact velocity and (182), we express $\boldsymbol{\xi}$ as

$$\begin{aligned}\boldsymbol{\xi} &= \boldsymbol{\gamma}^+ + \varepsilon\boldsymbol{\gamma}^- \\ &= \frac{1+\varepsilon}{2}(\boldsymbol{\gamma}^+ + \boldsymbol{\gamma}^-) + \frac{1-\varepsilon}{2}(\boldsymbol{\gamma}^+ - \boldsymbol{\gamma}^-) \\ &= (1+\varepsilon)\bar{\boldsymbol{\gamma}} + \frac{1-\varepsilon}{2}\mathbf{G}\boldsymbol{\Lambda}.\end{aligned} \tag{211}$$

The generalized Newton's impact law $-\boldsymbol{\Lambda} \in \mathcal{N}_{\mathbb{R}_0^{h+}}(\boldsymbol{\xi})$ for $\varepsilon = 1$ is equivalent to $-\boldsymbol{\Lambda} \in \mathcal{N}_{\mathbb{R}_0^{h+}}(\bar{\boldsymbol{\gamma}})$, where the cone property of the normal cone $\mathcal{N}_{\mathbb{R}_0^{h+}}$ has been used. For $\varepsilon \in [0,\ 1)$, we proceed by inverting the normal cone into (199) and employ the cone condition to obtain

$$-\frac{2(1+\varepsilon)}{1-\varepsilon}\bar{\boldsymbol{\gamma}} - \mathbf{G}\boldsymbol{\Lambda} \in \mathcal{N}_{\mathbb{R}_0^{h*+}}(\boldsymbol{\Lambda}). \tag{212}$$

The equivalence (163) allows us to write

$$-\boldsymbol{\Lambda} = -\frac{2(1+\varepsilon)}{1-\varepsilon}\operatorname{prox}^{\mathbf{G}}_{\mathbb{R}_0^{h*+}}(-\mathbf{G}^{-1}\bar{\boldsymbol{\gamma}}), \tag{213}$$

in which the positive homogeneity of $\operatorname{prox}^{\mathbf{G}}_{\mathbb{R}_0^{h*+}}(\cdot)$ has been used.

It remains to be proven that the function Φ given in (203) is the dissipation function of the set-valued impact law $\mathcal{H}$. The case $\varepsilon = 1$ is trivial. For the case $\varepsilon \in [0,\ 1)$, we use (164) to obtain the subdifferential of the dissipation function as

$$\begin{aligned}\partial\Phi(\bar{\boldsymbol{\gamma}}) &= \frac{2(1+\varepsilon)}{1-\varepsilon}\partial\frac{1}{2}\left(\operatorname{dist}^{\mathbf{G}^{-1}}_{\mathbb{R}_0^{h+}}(\bar{\boldsymbol{\gamma}})\right)^2 \\ &= \frac{2(1+\varepsilon)}{1-\varepsilon}\mathbf{G}^{-1}\left(\bar{\boldsymbol{\gamma}} - \operatorname{prox}^{\mathbf{G}^{-1}}_{\mathbb{R}_0^{h+}}(\bar{\boldsymbol{\gamma}})\right) \\ &= \frac{2(1+\varepsilon)}{1-\varepsilon}\mathbf{G}^{-1}\operatorname{prox}^{\mathbf{G}^{-1}}_{(\mathbb{R}_0^{h+})^\perp}(\bar{\boldsymbol{\gamma}}) \\ &= -\frac{2(1+\varepsilon)}{1-\varepsilon}\operatorname{prox}^{\mathbf{G}}_{\mathbb{R}_0^{h*+}}\left(-\mathbf{G}^{-1}\bar{\boldsymbol{\gamma}}\right),\end{aligned} \tag{214}$$

where the proximal point transformation (168) together with $-\mathbf{G}^{-1}(\mathbb{R}_0^{h+})^\perp = \mathbb{R}_0^{h*+}$ has been used. Hence, the generalized Newton's impact law with a global restitution coefficient has a set-valued impact law $\mathcal{H}$, which is cyclically maximal monotone. The corresponding convex dissipation function Φ is positively homogeneous of degree 2. □

Theorem 3 considers the case of a global restitution coefficient, i.e., $\mathbf{E} = \varepsilon\mathbf{I}$. In [6], the impact mappings S, Z and the set-valued impact law $\mathcal{H}$ are also derived for the more general case of a diagonal matrix $\mathbf{E} = \operatorname{diag}\left(\{\varepsilon^1, \ldots, \varepsilon^h\}\right)$. However, the set-valued impact law $\mathcal{H}$ looses its property of *cyclic* maximal monotonicity and a dissipation function Φ for the impact law no longer exists. Furthermore, if the restitution coefficients ε^α differ much such that $\mathbf{G} - \mathbf{EGE} \geq 0$ no longer holds, then the property of monotonicity itself is also lost. For a further example of an instantaneous impact law in which the corresponding set-valued impact law is maximal but not cyclically maximal monotone, we refer to [43].

6.2 Generalized Poisson's Impact Law

The generalized Poisson's impact law [14, 18] distinguishes between a compression and an expansion phase. In the compression phase, the impulsive forces reduce the normal relative velocities until standstill, thereby maximizing the reduction of the kinetic energy that is accessible by the constraint forces. The compression phase corresponds to a completely inelastic impact, and can thus be represented by the generalized Newton's impact law with $\varepsilon^\alpha = 0$ as

$$\mathbf{M}(\mathbf{q})(\mathbf{u}^C - \mathbf{u}^-) = \mathbf{W}(\mathbf{q})\boldsymbol{\Lambda}^C, \quad -\boldsymbol{\Lambda}^C \in \mathcal{N}_{\mathbb{R}_0^{h+}}(\boldsymbol{\gamma}^C). \tag{215}$$

The compression constraint impulses $\boldsymbol{\Lambda}^C$ and the constraint velocities after the compression phase $\boldsymbol{\gamma}^C = \mathbf{W}^\mathrm{T}(\mathbf{q})\mathbf{u}^C$ are therefore nonnegative by components.

The deformation energy gained during the compression is partly released during the expansion phase and reconverted into kinetic energy. The dissipative behavior is expressed by the Poisson's restitution coefficients $\varepsilon^\alpha \in [0, 1]$, which relate the expansion impulse to the compression impulse. The expansion phase is described by an inequality complementarity as

$$\mathbf{M}(\mathbf{q})(\mathbf{u}^+ - \mathbf{u}^C) = \mathbf{W}(\mathbf{q})\boldsymbol{\Lambda}^E, \quad -(\boldsymbol{\Lambda}^E - \mathbf{E}\boldsymbol{\Lambda}^C) \in \mathcal{N}_{\mathbb{R}_0^{h+}}(\boldsymbol{\gamma}^+). \tag{216}$$

The impact equation (83) with the total constraint impulses $\boldsymbol{\Lambda} = \boldsymbol{\Lambda}^C + \boldsymbol{\Lambda}^E$ is obtained by addition of the impact equations (215) and (216). The generalized Poisson's impact law is able to describe certain restitution effects of multi-constraint collisions, which are not possible to describe with the generalized Newton's impact law. The differences between the generalized Newton's impact law and the generalized Poisson's impact law are explained in detail in [14, 17, 18].

In the following theorem, which is an adapted version of [6], we give the mappings S and Z for the generalized Poisson's impact law and also prove maximal monotonicity of $\mathcal{H}$ without giving an explicit formulation for $\mathcal{H}$ itself.

Theorem 4 *Consider the impact equation* (83) *together with the generalized Poisson's impact law* (215)–(216) *with a global coefficient of restitution* $\varepsilon \in [0, 1]$, *that is,* $\mathbf{E} = \varepsilon \mathbf{I}$. *Then, the impact mappings S and Z are given by*

$$\boldsymbol{\gamma}^+ = S(\boldsymbol{\gamma}^-) = \operatorname{prox}^{\mathbf{G}^{-1}}_{\mathbb{R}_0^{h+}} \left(\operatorname{prox}^{\mathbf{G}^{-1}}_{\mathbb{R}_0^{h+}}(\boldsymbol{\gamma}^-) - \varepsilon \operatorname{prox}^{\mathbf{G}^{-1}}_{(\mathbb{R}_0^{h+})^\perp}(\boldsymbol{\gamma}^-) \right), \tag{217}$$

$$\mathbf{u}^+ = Z(\mathbf{u}^-) = \operatorname{prox}^{\mathbf{M}}_{\mathcal{T}_C} \left(\operatorname{prox}^{\mathbf{M}}_{\mathcal{T}_C}(\mathbf{u}^-) - \varepsilon \operatorname{prox}^{\mathbf{M}}_{\mathcal{T}_C^\perp}(\mathbf{u}^-) \right). \tag{218}$$

Furthermore, the set-valued impact law $-\boldsymbol{\Lambda} \in \mathcal{H}(\bar{\boldsymbol{\gamma}})$ *is maximal monotone.*

Proof In the first part of the proof, we derive the impact mappings S and Z. The impact law of the compression phase of the generalized Poisson's impact law (215) is written using $\boldsymbol{\gamma}^C = \boldsymbol{\gamma}^- + \mathbf{G}\boldsymbol{\Lambda}^C$ as

$$-\boldsymbol{\Lambda}^C \in \mathcal{N}_{\mathbb{R}_0^{h+}}(\boldsymbol{\gamma}^- + \mathbf{G}\boldsymbol{\Lambda}^C). \tag{219}$$

The compression impulses $\boldsymbol{\Lambda}^C$ as a function of the pre-impact contact velocities $\boldsymbol{\gamma}^-$ are obtained by inverting the normal cone and transforming (219) using (163) and (170) as

$$-\boldsymbol{\gamma}^- \in \mathcal{N}_{\mathbb{R}_0^{h*+}}(\boldsymbol{\Lambda}^C) + \mathbf{G}\boldsymbol{\Lambda}^C, \tag{220}$$

$$\boldsymbol{\Lambda}^C = \operatorname{prox}^{\mathbf{G}}_{\mathbb{R}_0^{h*+}}(-\mathbf{G}^{-1}\boldsymbol{\gamma}^-), \tag{221}$$

$$\boldsymbol{\Lambda}^C = -\mathbf{G}^{-1} \operatorname{prox}^{\mathbf{G}^{-1}}_{(\mathbb{R}_0^{h+})^\perp}(\boldsymbol{\gamma}^-). \tag{222}$$

The impact law of the expansion phase (216) is written using $\boldsymbol{\Lambda}^E = \boldsymbol{\Lambda} - \boldsymbol{\Lambda}^C = \mathbf{G}^{-1}(\boldsymbol{\gamma}^+ - \boldsymbol{\gamma}^-) - \boldsymbol{\Lambda}^C$ as

$$\boldsymbol{\gamma}^- + (1+\varepsilon)\mathbf{G}\boldsymbol{\Lambda}^C \in \mathbf{G}\mathcal{N}_{\mathbb{R}_0^{h+}}(\boldsymbol{\gamma}^+) + \boldsymbol{\gamma}^+, \tag{223}$$

which transforms by (163) to the explicit form

$$\boldsymbol{\gamma}^+ = \operatorname{prox}^{\mathbf{G}^{-1}}_{\mathbb{R}_0^{h+}} \left(\boldsymbol{\gamma}^- + (1+\varepsilon)\mathbf{G}\boldsymbol{\Lambda}^C \right). \tag{224}$$

Replacing $\boldsymbol{\Lambda}^C$ using (222) and (166) yields the mapping S in (217)

$$\boldsymbol{\gamma}^+ = \operatorname{prox}^{\mathbf{G}^{-1}}_{\mathbb{R}_0^{h+}} \left(\boldsymbol{\gamma}^- - (1+\varepsilon) \operatorname{prox}^{\mathbf{G}^{-1}}_{(\mathbb{R}_0^{h+})^\perp}(\boldsymbol{\gamma}^-) \right) \tag{225}$$

$$= \operatorname{prox}^{\mathbf{G}^{-1}}_{\mathbb{R}_0^{h+}} \left(\operatorname{prox}^{\mathbf{G}^{-1}}_{\mathbb{R}_0^{h+}}(\boldsymbol{\gamma}^-) - \varepsilon \operatorname{prox}^{\mathbf{G}^{-1}}_{(\mathbb{R}_0^{h+})^\perp}(\boldsymbol{\gamma}^-) \right). \tag{226}$$

The mapping Z in (218) is obtained by substituting (225) in (192) as

$$\begin{aligned}\mathbf{u}^+ &= Z(\mathbf{u}^-) = (\mathbf{P} + \mathbf{F}S \circ \mathbf{W}^\mathrm{T})(\mathbf{u}^-)\\ &= \mathbf{P}\mathbf{u}^- + \mathbf{F}\,\mathrm{prox}^{\mathbf{G}^{-1}}_{\mathbb{R}_0^{h+}}\left(\mathbf{W}^\mathrm{T}\mathbf{u}^- - (1+\varepsilon)\,\mathrm{prox}^{\mathbf{G}^{-1}}_{(\mathbb{R}_0^{h+})^\perp}\left(\mathbf{W}^\mathrm{T}\mathbf{u}^-\right)\right)\\ &= \mathbf{P}\mathbf{u}^- + \mathrm{prox}^{\mathbf{M}}_{\mathbf{F}\mathbb{R}_0^{h+}}\left(\mathbf{F}\mathbf{W}^\mathrm{T}\mathbf{u}^- - (1+\varepsilon)\mathbf{F}\,\mathrm{prox}^{\mathbf{G}^{-1}}_{(\mathbb{R}_0^{h+})^\perp}\left(\mathbf{W}^\mathrm{T}\mathbf{u}^-\right)\right)\\ &= \mathbf{P}\mathbf{u}^- + \mathrm{prox}^{\mathbf{M}}_{\mathbf{F}\mathbb{R}_0^{h+}}\left(\mathbf{F}\mathbf{W}^\mathrm{T}\mathbf{u}^- - (1+\varepsilon)\,\mathrm{prox}^{\mathbf{M}}_{\mathbf{F}(\mathbb{R}_0^{h+})^\perp}\left(\mathbf{F}\mathbf{W}^\mathrm{T}\mathbf{u}^-\right)\right)\\ &= \mathrm{prox}^{\mathbf{M}}_{\mathcal{T}_C}\left(\mathbf{P}\mathbf{u}^- + \mathbf{P}^\perp\mathbf{u}^- - (1+\varepsilon)\,\mathrm{prox}^{\mathbf{M}}_{\mathcal{T}_C^\perp}\left(\mathbf{P}^\perp\mathbf{u}^-\right)\right)\\ &= \mathrm{prox}^{\mathbf{M}}_{\mathcal{T}_C}\left(\mathbf{u}^- - (1+\varepsilon)\,\mathrm{prox}^{\mathbf{M}}_{\mathcal{T}_C^\perp}\left(\mathbf{u}^-\right)\right)\\ &= \mathrm{prox}^{\mathbf{M}}_{\mathcal{T}_C}\left(\mathrm{prox}^{\mathbf{M}}_{\mathcal{T}_C}\left(\mathbf{u}^-\right) - \varepsilon\,\mathrm{prox}^{\mathbf{M}}_{\mathcal{T}_C^\perp}\left(\mathbf{u}^-\right)\right),\end{aligned} \tag{227}$$

where the transformation rule (168), the identities (169) and (170) and the same steps as in the derivation of Z for Newton's impact law have been used.

The non-expansivity of the impact mapping S can directly be shown by writing (225) as

$$\begin{aligned}\boldsymbol{\gamma}^+ &= S(\boldsymbol{\gamma}^-)\\ &= \mathrm{prox}^{\mathbf{G}^{-1}}_{\mathbb{R}_0^{h+}}\left(\frac{1+\varepsilon}{2}\boldsymbol{\gamma}^- + \frac{1-\varepsilon}{2}\left(\mathrm{prox}^{\mathbf{G}^{-1}}_{\mathbb{R}_0^{h+}}(\boldsymbol{\gamma}^-) - \mathrm{prox}^{\mathbf{G}^{-1}}_{(\mathbb{R}_0^{h+})^\perp}(\boldsymbol{\gamma}^-)\right)\right).\end{aligned} \tag{228}$$

The mappings I, $\mathrm{prox}^{\mathbf{G}^{-1}}_{\mathbb{R}_0^{h+}}$ and $\mathrm{prox}^{\mathbf{G}^{-1}}_{\mathbb{R}_0^{h+}} - \mathrm{prox}^{\mathbf{G}^{-1}}_{(\mathbb{R}_0^{h+})^\perp}$ are non-expansive, as shown in [6]. Furthermore, if any two functions $\mathbf{f}, \mathbf{g}$ are non-expansive, then the functions $\mathbf{f} \circ \mathbf{g}$ and $\lambda_0\mathbf{f} + \lambda_1\mathbf{g}$ with $|\lambda_0| + |\lambda_1| \le 1$ are non-expansive as well. The non-expansivity of S implies that Z is also non-expansive and that $\mathcal{H}$ is monotone. Maximal monotonicity of $\mathcal{H}$ follows from the fact that S is maximal non-expansive, as its domain is $\mathbb{R}^h$. □

In [6], the derivation for S and Z of the more general case with different restitution coefficients is given. Moreover, an example is given which shows that the set-valued impact law $\mathcal{H}$ of the generalized Poisson's impact law is not cyclically maximal monotone in general.

7 Discussion on Variational Analysis of Impact Laws

We attempt to give a critical discussion of the results on the variational analysis of impact laws that was presented in Sect. 5 and the specific impact laws in Sect. 6:

- Closed form expressions (201) and (202), respectively (217) and (218), for the impact mappings S and Z have been derived for the generalized Newton's and Poisson's impact law. It has to be remarked that the numerical evaluation of the proximal point functions in such global mappings become cumbersome when multiple unilateral constraints are considered. Instead, a numerical scheme will usually solve a combinatorial problem to find the post-impact velocities using the impact laws in the local form (199), e.g., by setting up a linear complementarity problem or by using Alart and Curnier's solution method [2]. Furthermore, the relations (186) and (178) that express the mappings S and Z in terms of the set-valued impact law $\mathcal{H}$ are not constructive in the sense that they do not allow for the systematic derivation of the expressions for S and Z in terms of proximal point functions for a given set-valued impact law $\mathcal{H}$, as they are derived in Sect. 6. However, in the research fields Hybrid Systems [21, 23, 28, 30, 42] and Robotic Locomotion (see the survey paper [22]), one uses reset maps to describe instantaneous jumps in the state of the system, either for simulation or control purposes. A description of mechanical systems with unilateral constraints within the theory of Hybrid Systems requires an explicit expression for Z. The closed form expressions for the impact mappings S and Z therefore help to unite related research fields.

- In this work, as well as in earlier work of the authors, it has been shown that under some conditions on the restitution coefficients, the impact maps S and Z of the generalized Newton's and Poisson's impact law are maximal non-expansive. Furthermore, it has been shown that S and Z are related to $\mathcal{H}$ through a Minty parametrization. The maximal non-expansivity of S and Z is therefore equivalent to the maximal monotonicity of $\mathcal{H}$. This property is used to derive stability and synchronization properties of unilaterally constrained systems [27] and attempts are made to derive state-observers for such systems based on this property [7]. The validity of the assumption of maximal monotonicity of $\mathcal{H}$ in application problems remains an open question, being closely related to the chosen discretization level. A fine discretization will result in small values of the restitution coefficients and also lead to a decoupling in the Delassus matrix $\mathbf{G}$. Both these effects render the impact map $\mathcal{H}$ maximal monotone.
- The concept of a dissipation function as pseudo-potential for the impact law has been presented. The introduction of this concept strengthens the relationship with convex optimization theory and puts impact theory within the general framework of set-valued force laws, as expounded upon in the book by Glocker [15]. However, such a dissipation function only exists for very specific impact laws, notably the generalized Newton's impact law with global restitution coefficient. The practical relevance of dissipation functions to characterize impact laws is therefore extremely limited. Furthermore, the implication of cyclic maximal monotonicity of $\mathcal{H}$ for the dynamics of systems with unilateral constraints is still unexplored.

Clearly, the above discussion shows that our work does not immediately contribute to, for instance, improved simulation methods, lead to improved impact laws or solve any other practical issue related to systems with contact. We merely conclude that a variational analysis of impact laws does bring a mathematical structure to impact theory and also reveals properties that can be favorable from a dynamic analysis point of view.

8 Concluding Remarks

In Sects. 2–4, we have presented the geometric setting for the description of impacts in rigid multibody dynamics. One may observe that these sections are mainly based on linear algebra and basic differential geometry. The strength of the geometric description of impacts lies in the identification of the involved primal and dual quantities, i.e., the geometric description allows one to formulate a mechanical theory that distinguishes between kinematic and kinetic quantities. It is well-known that this geometric structure is applicable to other domains in mechanics, e.g., in continuum mechanics [11, 12] and analytical mechanics [1, 10, 20]. Therefore, the value of these sections may be found in the generality of the underlying concepts.

In Sects. 5 and 6, we have given a variational analysis of impact laws and applied it to two specific impact laws. These sections apply convex analysis, in particular, the Minty parametrization, to impact theory. A critical discussion has been given in Sect. 7.

Acknowledgements This research is supported by the Fonds National de la Recherche, Luxembourg (Proj. Ref. 8864427).

References

1. Abraham R, Marsden JE (1987) Foundations of mechanics, 2nd edn. Addison-Wesley, Boston
2. Acary V, Brogliato B (2008) Numerical methods for nonsmooth dynamical systems. Applications in mechanics and electronics, vol 35. Lecture notes in applied and computational mechanics. Springer, Berlin
3. Aeberhard U (2008) Geometrische Behandlung idealer Stösse. Ph.D. Thesis, ETH Zurich
4. Aubin T (2001) A course in differential geometry, vol 27. Graduate studies in mathematics. American Mathematical Society, Providence
5. Ballard P (2000) The dynamics of discrete mechanical systems with perfect unilateral constraints. Arch Ration Mech Anal 154:199–274
6. Baumann M (2017) Synchronization of nonsmooth mechanical systems with impulsive motion. Ph.D Thesis, ETH Zurich
7. Baumann M, Leine RI (2016) A synchronization-based state observer for impact oscillators using only collision time information. Int J Robust Nonlinear Control 26(12):2542–2563
8. Bremer H (2008) Elastic multibody dynamics: a direct ritz approach, vol 35. Intelligent systems, control, and automation: science and engineering. Springer, Berlin
9. Brogliato B (1999) Nonsmooth mechanics: models dynamics and control. Springer, London
10. de León M, Rodrigues PR (1989) Methods of differential geometry in analytical mechanics, vol 158. Elsevier
11. Epstein M (2010) The geometrical language of continuum mechanics. Cambridge University Press
12. Eugster SR (2015) Geometric continuum mechanics and induced beam theories, vol 75. Lecture notes in applied and computational mechanics. Springer, Berlin
13. Fischer G (2010) Lineare algebra, 17th edn. Grundkurs Mathematik: Studium. Vieweg+Teubner Verlag, Berlin
14. Glocker Ch (2001) On frictionless impact models in rigid-body systems. Philos Trans R Soc Lond A 359:2385–2404

15. Glocker Ch (2001) Set-valued force laws: dynamics of non-smooth systems, vol 1. Lecture notes in applied mechanics. Springer, Berlin
16. Glocker Ch (2006) An introduction to impacts. In: Haslinger J, Stavroulakis G (eds) Nonsmooth mechanics of solids. CISM courses and lectures, vol 485. Springer, Wien, New York, pp 45–101
17. Glocker Ch (2013) Energetic consistency conditions for standard impacts. Part I: Newton-type inequality impact laws and Kane's example. Multibody Syst Dyn 29(1):77–117
18. Glocker Ch (2013) Energetic consistency conditions for standard impacts. Part II: Poisson-type inequality impact laws. Multibody Syst Dyn 32(4):1–65
19. Glocker Ch, Aeberhard U (2006) The geometry of Newton's cradle. In: Nonsmooth mechanics and analysis. Springer, pp 185–194
20. Godbillon C (1969) Géométrie Différentielle et Mécanique Analytique. Hermann, Collection Méthodes
21. Goebel R, Sanfelice R, Teel AR (2012) Hybrid dynamical systems: modeling, stability, and robustness. Princeton University Press
22. Grizzle JW, Chevallereau C, Sinnet RW, Ames AD (2014) Models, feedback control, and open problems of 3D bipedal robotic walking. Automatica 50:1955–1988
23. Haddad W, Chellaboina V, Nersesov S (2006) Impulsive and hybrid dynamical systems: stability, dissipativity, and control. Princeton series in applied mathematics. Princeton University Press, Princeton
24. Lee JM (2009) Manifolds and differential geometry, vol 107. American Mathematical Society, Providence
25. Lee JM (2012) Introduction to smooth manifolds, vol 218, 2nd edn. Graduate texts in mathematics. Springer, New York
26. Leine RI, Baumann M (2014) Variational analysis of inequality impact laws. In: Proceedings of the 8th EUROMECH nonlinear dynamics conference (ENOC 2014). Vienna, Austria
27. Leine RI, van de Wouw N (2008) Stability and convergence of mechanical systems with unilateral constraints, vol 36. Lecture notes in applied and computational mechanics. Springer, Berlin
28. Liberzon D (2003) Switching in systems and control. Birkhäuser, Boston
29. Maisser P (1991) A differential-geometric approach to the multi-body system dynamics. Zeitschrift für Angewandte Mathematik und Physik 71:T116–T119
30. Matveev AS, Savkin AV (2000) Qualitative theory of hybrid dynamical systems. Control engineering series. Birkhäuser, Boston
31. Möller M (2012) Consistent integrators for non-smooth dynamical systems. Ph.D. Thesis, ETH Zurich, Switzerland
32. Morandi G, Ferrario C, Lo Vecchio G, Marmo G, Rubano C (1990) The inverse problem in the calculus of variations and the geometry of the tangent bundle. Phys Rep 188(3–4):147–284
33. Moreau JJ (1962) Décomposition orthogonale d'un espace hilbertien selon deux cones mutuellement polaires. Comptes Rendus de l'Académie des Sciences 255:238–240
34. Moreau JJ (1966) Fonctionnelles Convexes, Séminaire sur les Équations aux Dérivées Partielles, Collège de France, 1966, et Edizioni del Dipartimento di Ingeneria Civile dell'Università di Roma Tor Vergata. Roma, Séminaire Jean Leray
35. Moreau JJ (1987) Une formulation de la dynamique classique. Comptes rendus de l'Académie des sciences. Série II, Mécanique, Physique, Chimie, Sciences de l'univers. Sciences de la Terre 304(5):191–194
36. Moreau JJ (1988) Unilateral contact and dry friction in finite freedom dynamics. In: Moreau JJ, Panagiotopoulos PD (eds) Non-smooth mechanics and applications. CISM courses and lectures. Springer, Wien, pp 1–82
37. Pfeiffer F (2007) The TUM walking machines. Philos Trans R Soc A: Math Phys Eng Sci 365(1850):109–131
38. Pfeiffer F, Glocker Ch (1996) Multibody dynamics with unilateral contacts. Wiley, New York
39. Rockafellar RT (1970) Convex analysis. Princeton mathematical series. Princeton University Press, New Jersey
40. Rockafellar RT, Wets R-B (2009) Variational analysis. Springer, Berlin

41. Spivak M (1999) A comprehensive introduction to differential geometry, 3 edn., vol 1–5. Publish or Perish, Houston, Texas
42. van der Schaft AJ, Schumacher JM (2000) An introduction to hybrid dynamical systems, vol 251. Lecture notes in control and information sciences. Springer, London
43. Winandy T, Leine RI (2017) A maximal monotone impact law for the 3-ball Newton's cradle. Multibody Syst Dyn 39(1–2):79–94

Periodic Motions of Coupled Impact Oscillators

Guillaume James, Vincent Acary and Franck Pérignon

Abstract We study the existence and stability of time-periodic oscillations in a chain of coupled impact oscillators, for rigid impacts without energy dissipation. We formulate the search for periodic solutions as a boundary value problem incorporating unilateral constraints. This problem is solved analytically in the vicinity of the uncoupled limit and numerically for larger coupling constants. Different solution branches corresponding to nonlinear localized modes (breathers) and normal modes are computed.

1 Introduction

Understanding the dynamics of nonlinear lattices (i.e., large networks of coupled nonlinear oscillators) is a problem of fundamental importance in mechanics, condensed matter physics and biology. One of the major issues concerns the mathematical analysis and numerical computation of special classes of nonlinear time-periodic oscillation that organize the dynamics in many situations. In particular, spatially periodic waves (standing waves or periodic traveling waves) and spatially localized waves (breathers) are the object of intensive research [16, 41]. In this context, many

G. James (✉) · V. Acary · F. Pérignon
Université Grenoble Alpes, CNRS, Inria, Grenoble INP (Institute of Engineering, Université Grenoble Alpes), LJK, 38000 Grenoble, France
e-mail: guillaume.james@inria.fr

V. Acary
e-mail: vincent.acary@inria.fr

F. Pérignon
e-mail: franck.perignon@univ-grenoble-alpes.fr

R. I. Leine et al. (eds.), *Advanced Topics in Nonsmooth Dynamics*,
https://doi.org/10.1007/978-3-319-75972-2_3

theoretical and numerical works have focused on smooth nonlinear systems, whereas relatively few mathematical existence results are available for waves in nonsmooth infinite lattices [17, 18, 28, 39]. Developing theoretical and numerical tools for the analysis of nonlinear waves in nonsmooth systems is extremely important for applications, in particular, in the context of impact mechanics in which unilateral contacts and friction come into play [1, 5, 6, 15, 23]. Spatially discrete lattice models are frequently encountered in this context, in particular, for the modeling of waves in multibody mechanical systems (e.g., granular media) or in finite element models of continuum systems. A classical example illustrating the latter case concerns thin oscillating mechanical structures (a string under tension or a clamped beam) contacting rigid obstacles [5, 6, 23]. Such a structure can be described by a one-dimensional finite-element model involving a large number of degrees of freedom [2, 37]. The contact force between the string/beam and a rigid obstacle is either measure-valued (for rebounds with velocity jumps at contact times) or set-valued (if a wrapping of the string on the obstacle occurs) see, e.g., [13].

Although nonlinear modes of oscillation have been observed in experiments on impacting mechanical systems (see, e.g., [3, 6]), relatively little is known from a mathematical point of view about their existence and stability. Existence theorems for periodic and almost-periodic oscillations have been obtained in particular cases, for a continuum string model with point-mass or plane obstacle [9–11, 14, 20, 36] (see also [12] for a review). In addition, several analytical approaches have been used to obtain time-periodic solutions formally for different types of piecewise-linear dynamical systems with rigid impacts. One can mention Fourier and Green function methods [4–6, 17–19, 24–26, 33, 39], modal decomposition [29, 40] and sawtooth time transformations [34]. Most of the results obtained for discrete systems concern impacts localized on a *single particle*, and different types of wave have been constructed. In [29, 34, 40], nonsmooth normal modes have been obtained for general classes of conservative multiple degrees-of-freedom systems (the analysis in [34] is performed for a single or two impacting particles). Spatially-localized oscillations (breathers) with a single impacting node have also been studied for different classes of infinite or finite system. Breather existence and stability has been analyzed for oscillator chains with linear nearest-neighbor coupling and a symmetric local vibroimpact potential (including, in some cases, a linear component), both for conservative systems [18] and forced systems with dissipative impacts [17, 33, 39].

One of the main difficulties with the above techniques is the need to check analytically that the formal solutions to the piecewise-linear systems are consistent, i.e., that they satisfy the inequality constraints corresponding to non-penetration of the obstacles. This has been achieved in a number of works in the case of breathers [17, 18, 39] and for nonsmooth modes close to grazing linear normal modes [29]. In [19], the analysis from [33] has been extended to several impacting particles, but the verification of the inequality constraints is still an open problem in that case.

In this work, we study the existence and stability of time-periodic oscillations in an infinite chain of linearly coupled impact oscillators reminiscent of a model analyzed in [19, 33], for rigid impacts without energy dissipation. We show the existence of exact solutions (i.e., check the non-penetration conditions) for an arbitrary number of

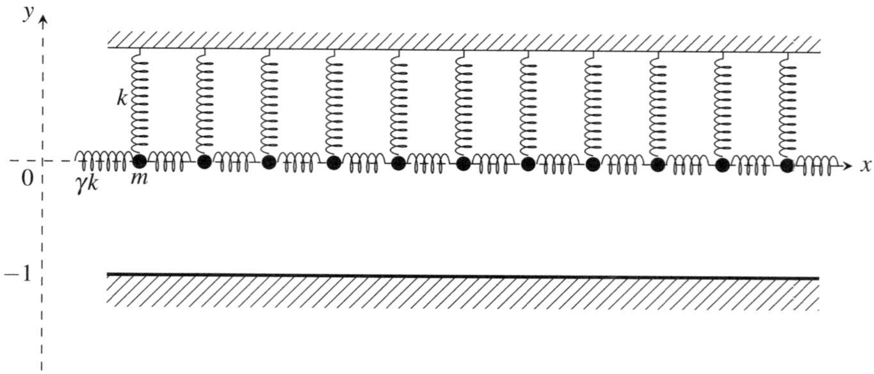

Fig. 1 A chain of identical impact oscillators with linear nearest-neighbor coupling. The chain is allowed to oscillate above a straight obstacle. After suitable rescaling, the obstacle position is fixed to $y = -1$, and the masses m of particles and local stiffness k are set to unity

impacting particles when the coupling between oscillators is small, and we compute solution branches numerically for larger couplings. The system under consideration is depicted in Fig. 1. Particle positions are denoted as $y(t) = (y_n(t))_{n \in \mathbb{Z}}$ and satisfy the following complementarity system:

$$\ddot{y}_n + y_n - \gamma\,(\Delta y)_n = \lambda_n, \quad n \in \mathbb{Z}, \tag{1}$$

$$0 \leq \lambda \perp (y + \mathbb{1}) \geq 0, \tag{2}$$

$$\text{if } \dot{y}_n(t^-) < 0 \text{ and } y_n(t) = -1 \text{ then } \dot{y}_n(t^+) = -\dot{y}_n(t^-), \tag{3}$$

where $(\Delta y)_n = y_{n+1} - 2\,y_n + y_{n-1}$ defines a discrete Laplacian operator, $\mathbb{1}$ denotes the constant sequence with all terms equal to unity and $\gamma \geq 0$ is a parameter. Nondissipative impacts occur for $y_n(t) = -1$ and give rise to impulsive reaction forces $\lambda_n(t)$. This configuration differs from the case of a symmetric local vibroimpact potential considered in [19, 33], which introduces an additional barrier above the chain.

Our analytical results are presented in Sect. 2. We start by describing in Sect. 2.1 some simple examples of nonsmooth modes of oscillations (in-phase, out-of-phase, and some symmetry-breaking bifurcations from these modes). In Sect. 2.2, we reformulate the search for periodic solutions of (1)–(3) as a boundary value problem incorporating unilateral constraints. This formulation, together with an appropriate notion of nondegenerate modes introduced in Sect. 2.3, allows us to construct nonsmooth modes of oscillations (spatially localized or extended) at small coupling (see Theorems 1 and 2). This approach is an adaptation of the idea of an "anticontinuum" limit [16, 30, 38] to the nonsmooth setting. Section 2.4 deals with the linear stability of time-periodic solutions to (1)–(3). We provide a formula for the monodromy matrix that determines spectral stability in the presence of simple impacts, following the lines of [32]. In Sect. 3, the above results are used for the numerical computation of time-periodic solutions. Solution branches are continued for fixed values of T, varying the linear stiffness γ (and starting from the limit $\gamma = 0$) or by fixing γ and varying T. In this way, we compute some families of breathers and extended modes and study their linear stability. Dynamical instabilities are illustrated by integrating (1)–(3) numerically. These computations are performed with the Siconos software for nonsmooth dynamical systems [1, 22].

2 Analytical Study of Nonsmooth Modes

2.1 *Definitions and Basic Examples*

We look for T-periodic solutions to (1)–(3) that are even in time, and assume each particle undergoes at most one impact during each period of oscillation. Consequently, for a given particle, impacts either occur at half-period multiples or do not occur at all. We denote by $I_k \subset \mathbb{Z}$ with $k = 1$ or 2 the index sets of particles impacting at $t = (2m + k)\,T/2$ for all $m \in \mathbb{Z}$ (i.e., $y_n((2m + k)\,T/2) = -1$), and by $I_0 := \mathbb{Z} \setminus (I_1 \cup I_2)$ the index set corresponding to non-impacting particles (i.e., $y_n(t) > -1$ for all t). We thus have $\lambda_n = 0$ for all $n \in I_0$ and

$$\lambda_n = 2\,\dot{y}_n(\frac{kT^+}{2}) \sum_{m\in\mathbb{Z}} \delta_{(m+\frac{k}{2})T} \quad \text{for all } n \in I_k. \tag{4}$$

The triplet (I_0, I_1, I_2) will be denoted as the *pattern* of the periodic solution. A *nonsmooth mode* corresponds to a continuous one-parameter family of periodic solutions (typically parameterized by T) sharing a given pattern with $I_0 \neq \mathbb{Z}$ (i.e., impacts occur).

We provide below some simple examples of nonsmooth modes. The simplest case corresponds to the in-phase mode with $I_1 = \mathbb{Z}$ (or equivalently, $I_2 = \mathbb{Z}$ up to a phase shift). This solution exists for $T \in (\pi, 2\pi)$ and reads as

$$y_n(t) = -\frac{\cos t}{\cos{(T/2)}} \text{ for } |t| \leq T/2, \tag{5}$$

where (5) is extended by periodicity outside the interval $(-T/2, T/2)$. The impact velocity in particular, reads as $\dot{y}_1((T/2)^+) = -\dot{y}_1((T/2)^-) = -\tan{(T/2)}$. The amplitude of oscillations diverges when $T \to \pi$ and becomes unity for $T = 2\pi$. In that case, the impact becomes grazing (i.e., occurs at zero velocity), and one recovers the *linear* in-phase mode $y_n(t) = \cos t$, which a solution to (1) with $\lambda = 0$. Notice that, for $T \neq 2\,k\,\pi$ outside the interval $(\pi, 2\pi)$, expression (5) does not provide a solution to (1)–(3), because the constraint $y_n \geq -1$ is violated.

Another example concerns nonsmooth modes with spatial period two, i.e., which satisfy $y_{n+2}(t) = y_n(t)$. Nonsmooth modes in two degrees-of-freedom impacting systems have been studied in a number of works (see, e.g., [31, 42] for a case of symmetric constraints and [23] for more references). In what follows, we discuss the case when I_1 and I_2 consist of the sets of odd and even integers, respectively. Moreover, we assume that all impact velocities are identical and nonzero. In order to compute such modes, we introduce the relative displacement $r = y_2 - y_1$, the center of mass $q = (y_1 + y_2)/2$ and the impact velocity $v = \dot{y}_2(0^+) = \dot{y}_1((T/2)^+) \neq 0$. From Eqs. (1) and (4) taken at $n = 1, 2$, and considering the spatial period two of the mode, one obtains

$$\ddot{r} + \Omega^2 r = 2v \sum_{m\in\mathbb{Z}} (-1)^m \, \delta_{m\frac{T}{2}}, \tag{6}$$

where $\Omega = \sqrt{1+4\gamma}$. Note that Ω is the frequency of the linear out-of-phase mode $y_n(t) = (-1)^n \cos(\Omega t)$, which is a solution to (1) with $\lambda = 0$. If the non-resonance condition $(2m+1)(2\pi/T) \neq \Omega$ holds true for all integers m, there exists an even T-periodic solution to (6) defined by

$$r(t) = \frac{v}{\Omega} \frac{\sin(\Omega(t - \frac{T}{4}))}{\cos(\Omega T/4)} \quad \text{for } t \in [0, T/2], \tag{7}$$

where the integration constants have been determined from the conditions $v = \dot{r}(0^+) = \dot{r}((T/2)^-)$. In addition, the T-periodic solution is unique if $m(2\pi/T) \neq \Omega$ for all integers m. From expression (7), and using the fact that r is T-periodic and even, one can see that $r(\frac{T}{4} + t) = -r(\frac{T}{4} - t)$ for all $t \in \mathbb{R}$.

Similarly, the center of mass satisfies

$$\ddot{q} + q = v \sum_{m\in\mathbb{Z}} \delta_{m\frac{T}{2}}. \tag{8}$$

Let us assume that the non-resonance condition $T \neq 4m\pi$ holds true for all integers m. In that case, Eq. (8) admits an even $T/2$-periodic solution. Indeed, since $v/2 = \dot{q}(0^+) = -\dot{q}((T/2)^-)$, we find

$$q(t) = \frac{v}{2} \frac{\cos(t - \frac{T}{4})}{\sin(T/4)} \quad \text{for } t \in [0, T/2], \tag{9}$$

and q is defined as the $T/2$-periodic extension of (9). The symmetry $q(\frac{T}{4} + t) = q(\frac{T}{4} - t)$ for all $t \in \mathbb{R}$ and the fact that q is $T/2$-periodic imply that q is even. In addition, (8) does not possess additional T-periodic solutions if the non-resonance condition $T \neq 2m\pi$ holds true for all integers m.

Particle displacements are obtained from the identities

$$y_1 = q - \frac{r}{2}, \quad y_2 = q + \frac{r}{2}.$$

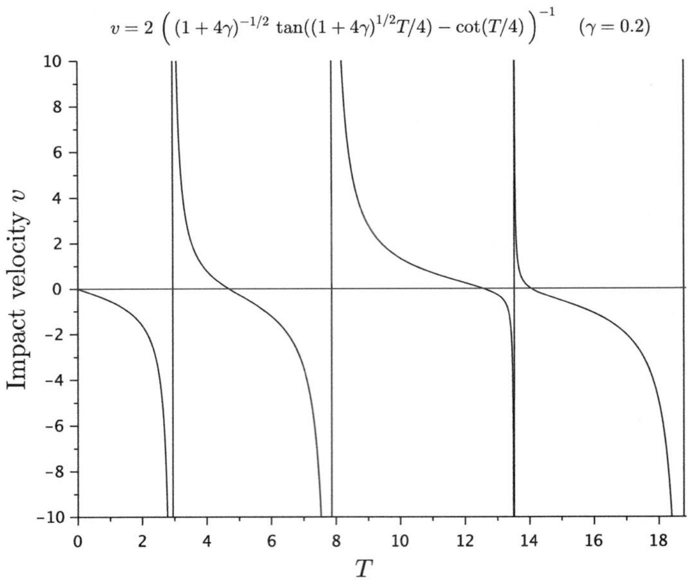

Fig. 2 Impact velocity as a function of the period

One can check that $\dot{y}_1(0^+) = 0$, hence $\dot{y}_1(0^-) = 0$ and y_1 is smooth everywhere except at the impact times $t = (2k+1)T/2$ with $k \in \mathbb{Z}$. Moreover, it follows from the symmetries of r that $y_2(t) = y_1(t + T/2) = y_1(t - T/2)$.

We use the constraint $y_2(0) = -1$ to determine v from T, which yields

$$v = 2\left(\frac{1}{\Omega}\tan(\Omega\, T/4) - \cot(T/4)\right)^{-1} \tag{10}$$

and implies that $y_1(T/2) = -1$. The expression in (10) is depicted in Fig. 2. In the uncoupled case $\gamma = 0$, expression (10) simplifies to $v = -\tan(T/2)$, and one recovers the case $n = 1$ of (5). Moreover, in the limit cases $T \to (2k+1)\,2\pi/\Omega$ ($k \in \mathbb{N}_0$) and $T \to 4m\pi$ ($m \in \mathbb{N}$), one obtains $v \to 0$, i.e., a grazing impact. When $T \to (2k+1)\,2\pi/\Omega$ and $\Omega \neq (2k+1)/(2m)$ for all $m \in \mathbb{N}$, the above solution converges towards the linear out-of-phase mode $y_n(t) = (-1)^{n+1}\cos(\Omega\, t)$, while $T \to 4m\pi$ and $\Omega \neq (2k+1)/(2m)$ for all $k \in \mathbb{N}_0$ leads to a convergence towards the linear in-phase mode $y_n(t) = -\cos t$.

In order to obtain solutions to (1)–(3), there remains to check the values of parameters γ, T for which the constraint $y_1 \geq -1$ is satisfied. Let us examine this problem when the coupling constant γ is fixed and T is varied. A necessary condition is $v \geq 0$, which is achieved for values of $T > 0$ within an infinite and unbounded sequence of disjoint intervals depending on γ. The lower bounds of these intervals are the roots of v^{-1}, and the upper bounds take the form $T = (2k+1)\,2\pi/\Omega$ with $k \in \mathbb{N}_0$ or $T = 4m\pi$ with $m \in \mathbb{N}$ (values leading to $v = 0$). In particular, the first interval takes the form $(T_0(\gamma), 2\pi/\Omega]$, where $T_0(\gamma)$ is implicitely defined through

$$\frac{1}{\Omega}\tan(\Omega\, T_0/4) = \cot(T_0/4), \quad T_0 \in (0, 2\pi/\Omega). \tag{11}$$

Note that $\lim_{\gamma\to+\infty} T_0(\gamma) = 0$ (since $T_0 < 2\pi\,(1+4\gamma)^{-1/2}$), $\lim_{\gamma\to 0} T_0(\gamma) = \pi$ (the case $\Omega = 1$ of (11)), and T_0 is a decreasing function of γ (since the left side of (11) increases with Ω or γ), hence $T_0(\gamma) < \pi$ for $\gamma > 0$. The upper bound $T = 2\pi/\Omega$ yields $v = 0$ (grazing impact), as previously outlined, whereas in the case $T \to T_0(\gamma)^+$, one obtains $v \to +\infty$.

Now, let us check the constraint $y_1(t) \geq -1$ in the case $T \in (T_0(\gamma), 2\pi/\Omega)$. One can restrict the discussion to $t \in [0, T/2]$ without loss of generality (since y_1 is even and T-periodic). In that case, we deduce from the above computations that

$$\dot{y}_1(t) = -\frac{v}{2}\left(\frac{\sin(t-\frac{T}{4})}{\sin(T/4)} + \frac{\cos(\Omega\,(t-\frac{T}{4}))}{\cos(\Omega\, T/4)}\right).$$

Consequently, the conditions $T < 2\pi/\Omega < 2\pi$ and $v > 0$ (which follows from $T \in (T_0(\gamma), 2\pi/\Omega)$) imply that y_1 decreases on $[T/4, T/2]$, hence $y_1(t) > -1 = y_1(T/2)$ for all $t \in [T/4, T/2)$. In addition, expressions (7) and (9) show that $r \leq 0$ and $q > 0$ on $[0, T/4]$, hence $y_1 > 0$ on $[0, T/4]$. This shows that $y_1(t) > -1$ for all $t \in [0, T/2)$.

As a result, we have obtained a family of even and time-periodic solutions to (1)–(3), parameterized by their period $T \in (T_0(\gamma), 2\pi/\Omega)$. These solutions have spatial period two and possess the symmetry $y_{n+1}(t) = y_n(t + T/2)$. When $T \to T_0(\gamma)^+$, the impact velocity v and amplitude of oscillations $y_1(0)$ diverge. When $T \to 2\pi/\Omega$, the mode converges towards the linear out-of-phase mode. This family of solutions will be denoted as the *nonsmooth out-of-phase mode*. They are illustrated for several values of T in Fig. 3.

There exist other nonsmooth modes with spatial period 2 and $I_0 = \varnothing$, $I_2 = 2\mathbb{Z}$ not discussed above, for example, a branch of solutions emerging above $T = 4\pi/\Omega$. For $T = 4\pi/\Omega$, odd particles undergo a grazing impact at $t = 0$ (we conjecture the existence of a nonsmooth mode with two impacts per period and $T < 4\pi/\Omega$). When T increases above $4\pi/\Omega$, no impacts occur at $t = 0$ for odd particles and the branch of solutions can evolve in different ways depending on γ. If $\gamma < 5/16$ (so that $4\pi < 6\pi/\Omega$), the mode converges towards the linear in-phase mode when $T \to 4\pi^-$ (this corresponds to a period-doubling bifurcation of the in-phase mode), a limit in which odd particles again display a grazing impact at $t = 0$. If $\gamma > 5/16$ (the case $6\pi/\Omega < 4\pi$), convergence towards the linear out-of-phase mode takes place when $T \to (6\pi/\Omega)^-$ (period-tripling bifurcation of the out-of-phase mode). In this limit, odd particles undergo a grazing impact at $t = \pi/\Omega$. Illustrations of period doubling bifurcations are displayed in Fig. 4 and those period tripling bifurcations in Fig. 5.

Fig. 3 Nonsmooth out-of-phase modes for several values of T

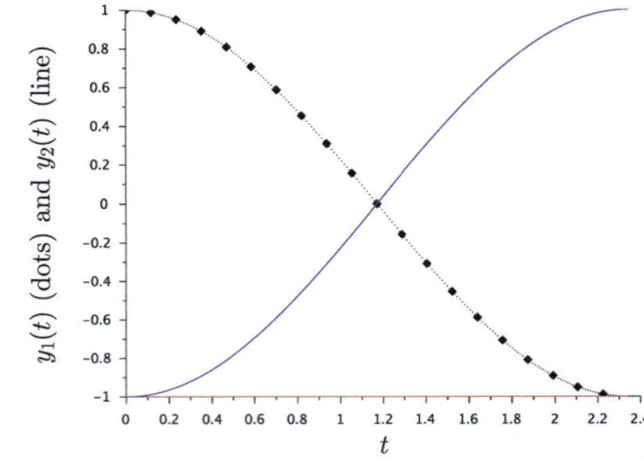

(a) Particle oscillations for $\gamma = 0.2, \quad T = 2\pi(1+4\gamma)^{-1/2} \approx 4.68$

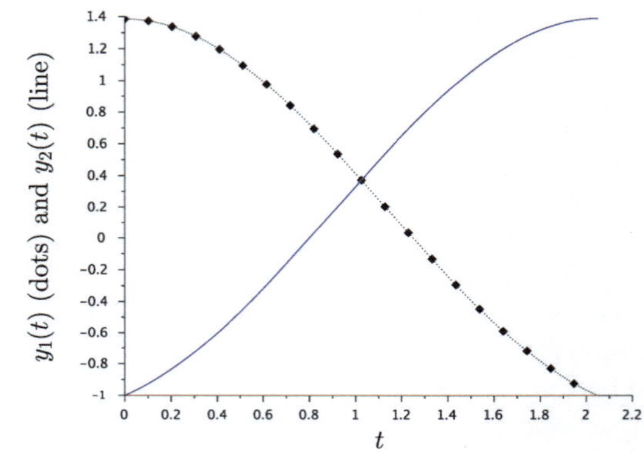

(b) Particle oscillations for $\gamma = 0.2, \quad T = 4.1$

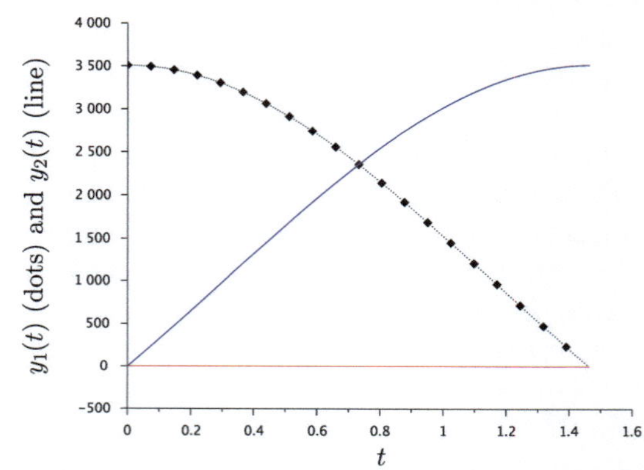

(c) Particle oscillations for $\gamma = 0.2, \quad T = 2.926$

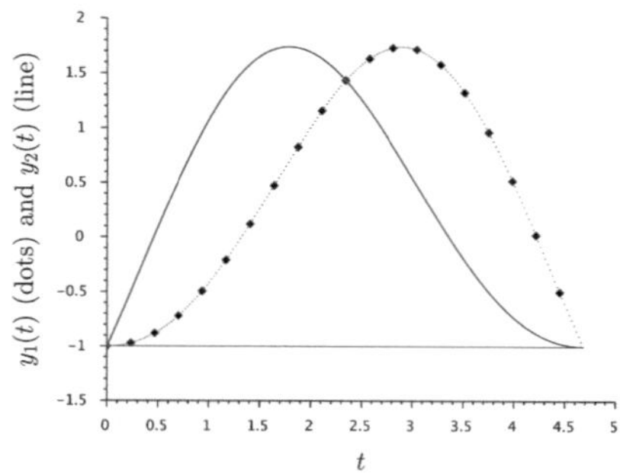

(a) Particle oscillations for $\gamma = 0.2, T = 4\pi(1 + 4\gamma)^{-1/2}$

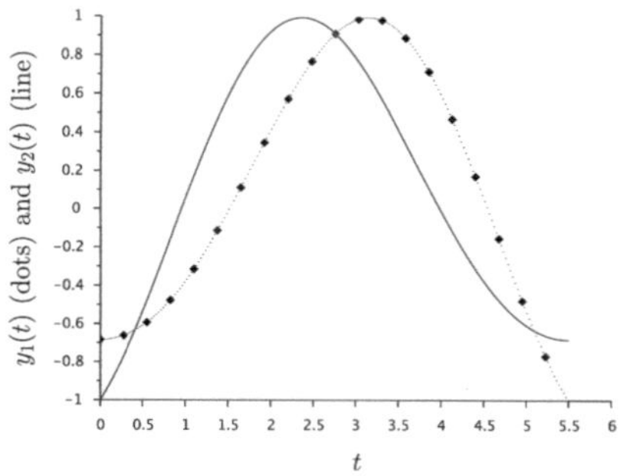

(b) Particle oscillations for $\gamma = 0.2, T = 11$

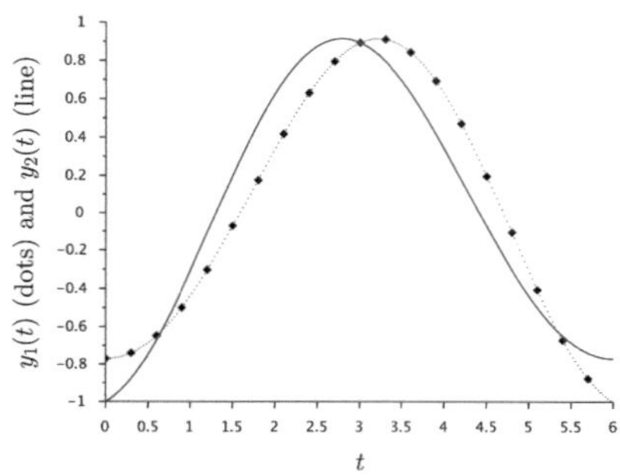

(c) Particle oscillations for $\gamma = 0.2, T = 12$

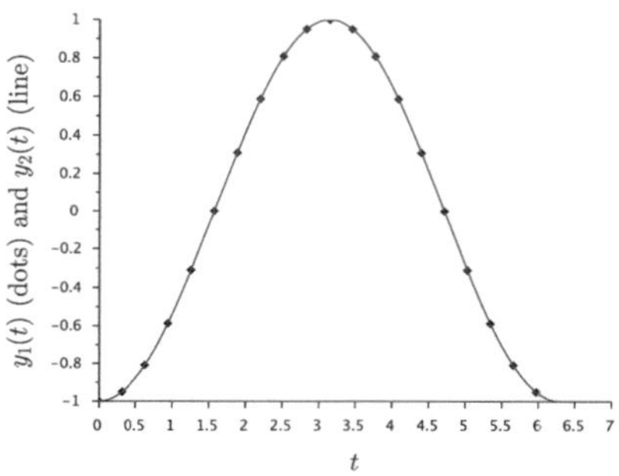

(d) Particle oscillations for $\gamma = 0.2, T = 4\pi$

Fig. 4 Period doubling bifurcation

2.2 Boundary Value Problem

In the sequel, E denotes either the Banach space $\ell_\infty(\mathbb{Z})$ of real bounded sequences on $\mathbb{Z}$, the Hilbert space $\ell_2(\mathbb{Z})$ of square-summable sequences, or the Hilbert space $\mathscr{P}^p$ of p-periodic sequences (isomorphic to the Euclidean space $\mathbb{R}^p$) for a fixed integer p. The case $E = \ell_2(\mathbb{Z})$ will be relevant for the study of localized modes, and the periodic case will be considered for numerical computations. We consider a chain of impact ocillators with positions described by a vector $y(t) \in E$ solution to the complementarity system (1)–(3). We look for T-periodic solutions even in time, with a prescribed pattern (I_0, I_1, I_2) (as defined in Sect. 2.1) such that $I_0 \neq \mathbb{Z}$.

The splitting $\mathbb{Z} = I_0 \cup I_1 \cup I_2$ allows one to identify E with $E^{(0)} \times E^{(1)} \times E^{(2)}$, where $E^{(k)}$ is a space of sequences indexed by $n \in I_k$, equiped with the same norm as E ($\| \|_2$ or $\| \|_\infty$). For all $y \in E$, we shall use the notation $y = (y^{(0)}, y^{(1)}, y^{(2)})$ with $y^{(k)} = (y_n)_{n \in I_k} \in E^{(k)}$. Any solution to the linear differential equation (12) satisfies $\dot{y}(t) \in E$, therefore we shall denote $\dot{y} = (\dot{y}^{(0)}, \dot{y}^{(1)}, \dot{y}^{(2)})$ with $\dot{y}^{(k)} \in E^{(k)}$. The above problem can be reformulated as a boundary value problem on a half-period interval $(0, T/2)$,

$$\ddot{y}_n + y_n - \gamma\,(\Delta y)_n = 0, \quad n \in \mathbb{Z}, \quad t \in (0, T/2), \tag{12}$$

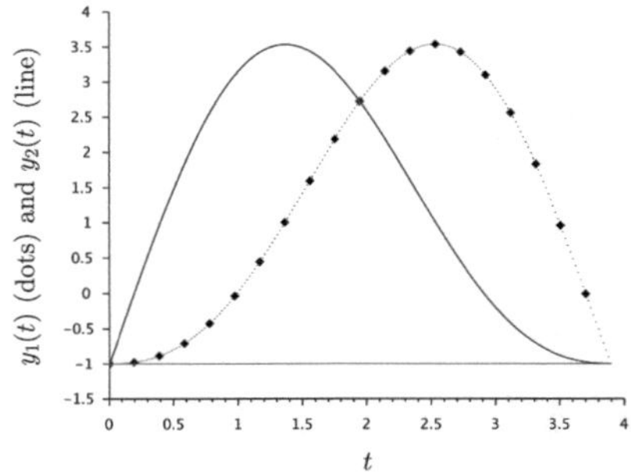

(a) Particle oscillations for $\gamma = 0.4, T = 4\pi(1+4\gamma)^{-1/2}$

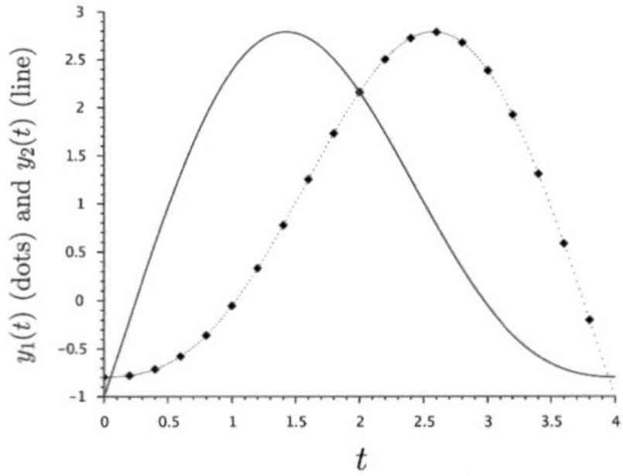

(b) Particle oscillations for $\gamma = 0.4, T = 8$

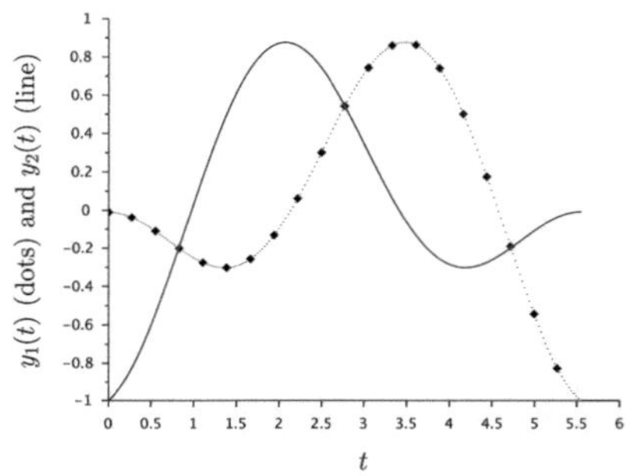

(c) Particle oscillations for $\gamma = 0.4, T = 11.1$

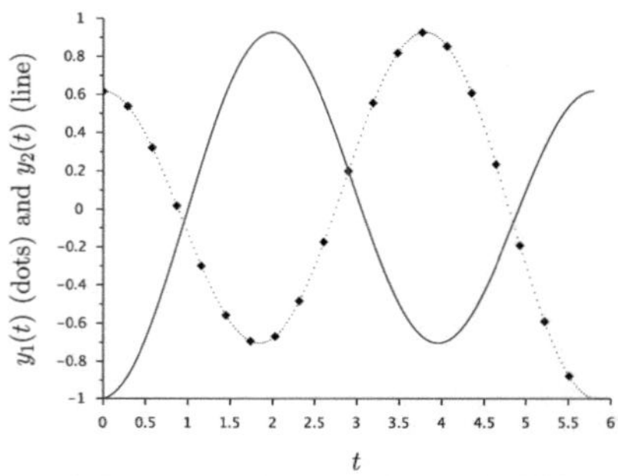

(d) Particle oscillations for $\gamma = 0.4, T = 11.6$

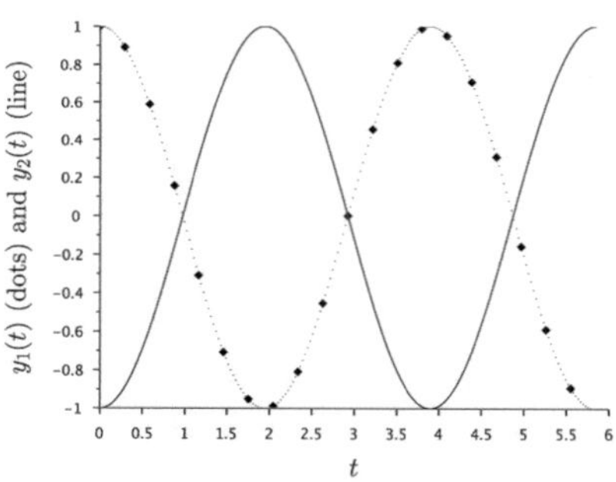

(e) Particle oscillations for $\gamma = 0.4, T = 6\pi(1+4\gamma)^{-1/2}$

Fig. 5 Period tripling bifurcation

with boundary conditions

$$\begin{aligned} \dot{y}^{(i)}(0) &= 0 \text{ for } i \in I_0 \cup I_1 \,, \; y^{(2)}(0) = -\mathbb{1}, \\ \dot{y}^{(i)}(T/2) &= 0 \text{ for } i \in I_0 \cup I_2 \,, \; y^{(1)}(T/2) = -\mathbb{1}, \end{aligned} \tag{13}$$

and constraint

$$y(t) + \mathbb{1} > 0, \quad t \in (0, T/2). \tag{14}$$

Indeed, it is immediately apparent that any even T-periodic solution to (1)–(3) with pattern (I_0, I_1, I_2) satisfies (12)–(14). Moreover, every solution to (12)–(14) can be extended to an even T-periodic function y, which, in turn, defines a solution to (1)–(3). Indeed, since $\dot{y}$ is odd, we have $\dot{y}(0^-) = -\dot{y}(0^+)$, and thus $\dot{y}((k\,T)^-) = -\dot{y}((k\,T)^+)$ for all $k \in \mathbb{Z}$ because $\dot{y}$ is T-periodic. In the same way, since $\dot{y}$ is odd and T-periodic, we have $\dot{y}((T/2)^-) = -\dot{y}((-T/2)^+) = -\dot{y}((T/2)^+)$, and thus we have, by periodicity, $\dot{y}(((2k+1)T/2)^-) = -\dot{y}(((2k+1)T/2)^+)$ for all $k \in \mathbb{Z}$.

In what follows, we reformulate the boundary value problem (12)–(13) as a linear system for $\xi = (y^{(0)}(0), y^{(1)}(0), \dot{y}^{(2)}(0)) \in E^{(0)} \times E^{(1)} \times E^{(2)}$, i.e., as an affine equation in E. For this purpose, we define the projection $P : E \times E \to E$ through

$$P\,(y, \dot{y}) = (\dot{y}^{(0)}, y^{(1)}, \dot{y}^{(2)})$$

and an embedding $N : E \to E \times E$ by

$$N\,(y^{(0)}, y^{(1)}, \dot{y}^{(2)}) = (u, v), \quad u = (y^{(0)}, y^{(1)}, 0), \; v = (0, 0, \dot{y}^{(2)}) \text{ in } E^{(0)} \times E^{(1)} \times E^{(2)}.$$

Introducing $Y = (y, \dot{y})^T \in E \times E$, the linear differential equation (12) takes the form

$$\dot{Y} = J\,Y + \gamma\,L\,Y, \tag{15}$$

where

$$J = \begin{pmatrix} 0 & \mathbb{I} \\ -\mathbb{I} & 0 \end{pmatrix}, \quad L = \begin{pmatrix} 0 & 0 \\ \Delta & 0 \end{pmatrix}$$

and $\mathbb{I}$ is the identity map in E. Let us denote by $S_\gamma(t) = e^{(J+\gamma L)t} \in \mathscr{L}(E \times E)$ the flow of (15).

The boundary condition at $t = 0$ defined in (13) takes the form $Y(0) = N\,\xi - B$, where $B = (\mathbb{1}_{I_2}, 0)^T \in E \times E$ and $\mathbb{1}_{I_2}$ denotes the indicator function of I_2. Moreover, the boundary condition at $t = T/2$ in (13) reads as $P\,Y(T/2) = -\mathbb{1}_{I_1}$. Consequently, the boundary value problem (12)–(13) is equivalent to

$$M_{\gamma,T}\,\xi = \eta, \tag{16}$$

where $M_{\gamma,T} = P\,S_\gamma(T/2)\,N \in \mathscr{L}(E)$ and $\eta = P\,S_\gamma(T/2)\,B - \mathbb{1}_{I_1}$.

In the case $E = \mathscr{P}^p$ (periodic boundary conditions with period p), E is isomorphic to $\mathbb{R}^p$ and (16) takes the form of a p-dimensional linear system. The solution $\xi \in E$ can be identified with a vector $x \in \mathbb{R}^p$ defined by

$$x_i = y_i \ \text{ if } \ i \in I_0 \cup I_1, \quad x_i = \dot{y}_i \ \text{ if } \ i \in I_2.$$

The matrix $P \in M_{p,2p}(\mathbb{R})$ reads as

$$P_{j,j} = 1 \ \text{ if } \ j \in I_1, \quad P_{j,j+p} = 1 \ \text{ if } \ j \in I_0 \cup I_2, \quad P_{i,j} = 0 \ \text{ elsewhere.}$$

The matrix $N \in M_{2p,p}(\mathbb{R})$ is defined by

$$N_{i,i} = 1 \ \text{ if } \ i \in I_0 \cup I_1, \quad N_{i+p,i} = 1 \ \text{ if } \ i \in I_2, \quad N_{i,j} = 0 \ \text{ elsewhere.}$$

2.3 Nondegenerate Modes and Continuation at Small Coupling

Consider an even T-periodic solution to (1)–(3) with pattern (I_0, I_1, I_2) (recall that under these assumptions, each particle undergoes at most one impact per period). The reduced initial condition $\xi = (y^{(0)}(0), y^{(1)}(0), \dot{y}^{(2)}(0)) \in E^{(0)} \times E^{(1)} \times E^{(2)}$ defines a solution to the linear problem (16). This leads us to introduce the following notion of a *nondegenerate* periodic solution.

Definition 1 An even T-periodic solution to (1)–(3) with pattern (I_0, I_1, I_2) is nondegenerate if the map $M_{\gamma,T}$ is invertible and

$$\dot{y}_n((T/2)^-) < 0 \, \forall n \in I_1, \quad \dot{y}_n(0^+) > 0 \, \forall n \in I_2. \tag{17}$$

Let us consider any nondegenerate periodic solution to (1)–(3). Since $M_{\gamma,T}$ depends analytically on γ, T, the corresponding solution to (16) *locally* admits a unique continuation with respect to (γ, T) denoted by $\xi_{\gamma,T}$, which is analytic in (γ, T) in some open set [43]. It follows that

$$Y_{\gamma,T}(t) = (y_{\gamma,T}(t), \dot{y}_{\gamma,T}(t))^T = S_\gamma(t)\,(N\,\xi_{\gamma,T} - B) \tag{18}$$

is a solution to (12) satisfying (13).

In order to check the constraint (14), we define $u_{\gamma,T}(t) = y_{\gamma,T}(\frac{T}{2}\,t) + \mathbb{1}$ and introduce the Banach space

$$X = \{ u \in C^1([0,1], E), \quad u_n(1) = 0 \, \forall n \in I_1, \quad u_n(0) = 0 \, \forall n \in I_2 \},$$

equiped with the C^1-norm. We consider the *open* set

$$\begin{aligned}\Omega = \{ u \in X,\ &\forall n \in I_0,\ u_n > 0 \text{ on } [0,1],\\ &\forall n \in I_1,\ u_n > 0 \text{ on } [0,1),\ \dot{u}_n(1^-) < 0,\\ &\forall n \in I_2,\ u_n > 0 \text{ on } (0,1],\ \dot{u}_n(0^+) > 0 \}.\end{aligned}$$

Thanks to assumption (17), the nondegenerate periodic solution belongs to Ω. Since the map $(\gamma, T) \mapsto u_{\gamma,T}$ is continuous in X, the local continuation with respect to (γ, T) of the nondegenerate solution stays locally in Ω, and thus the constraint (14) is satisfied by $y_{\gamma,T}$ when (γ, T) lies in some open set $\mathscr{U}$. Consequently, we have obtained a family of solutions to the boundary value problem (12)–(14) parameterized by (γ, T), which provides in turn a family of solutions to (1)–(3). As a result, we have shown the following.

Theorem 1 *Any nondegenerate even periodic solution to (1)–(3) with a given pattern persists for values of the coupling constant* γ *and period* T *lying in an open set* $\mathscr{U}$. *Moreover, these solutions take the form* $y(t) = y_{\gamma,T}(t)$ *for all* $t \in [0, T/2]$, *where the map* $(t, \gamma, T) \mapsto y_{\gamma,T}(t)$ *is analytic in* $\mathbb{R} \times \mathscr{U}$ *and defined in (18).*

In particular, the above result shows that any nondegenerate periodic solution is part of a continuous branch of periodic solutions parameterized by T and forming a nonsmooth mode. The continuation may stop when a new grazing impact takes place for $n \in I_0$ or if an impact occurring for $n \in I_1$ or I_2 becomes grazing. In such cases, the branch of periodic solutions might be continued with a different pattern or by allowing several impacts per period or sticking contacts, but these extensions are outside of the scope of the present study.

Another case when the above continuation theorem does not apply corresponds to the noninvertibility of $M_{\gamma,T}$. This situation may lead to a divergence of the solution (i.e., divergence of $\|(y^{(0)}(0), y^{(1)}(0), \dot{y}^{(2)}(0))\|$) or to a bifurcation of periodic solutions.

The solution to (12)–(13) is non-unique, or equivalently, $M_{\gamma,T}$ admits a nontrivial kernel if, and only if, the homogeneous boundary value problem given by (12) and

$$\begin{aligned}\dot{y}^{(i)}(0) = 0 \text{ for } i \in I_0 \cup I_1,\ y^{(2)}(0) = 0,\\ \dot{y}^{(i)}(T/2) = 0 \text{ for } i \in I_0 \cup I_2,\ y^{(1)}(T/2) = 0,\end{aligned} \tag{19}$$

admits nontrivial solutions $y(t) \in E$. Let us fix $E = \ell_\infty(\mathbb{Z})$ and discuss some *resonant* cases when this phenomenon occurs. The linear equation (12) admits normal mode solutions (or "phonons")

$$y_n(t) = a \cos(\Omega_q t + \varphi) \cos(q n + \psi), \tag{20}$$

whose frequencies $\Omega_q = (1 + 4\gamma \sin^2(q/2))^{1/2}$ span the phonon band $[1, \Omega]$, the highest frequency $\Omega = \sqrt{1 + 4\gamma}$ corresponding to the out-of-phase mode with $q = \pi$. For nonsmooth modes having certain patterns, simple nontrivial solutions to (12)–(19) can be found in the form (20) if some multiple of π/T belongs to the phonon band.

For example, if $I_1 = \mathbb{Z}$ or $I_2 = \mathbb{Z}$ (this is the case for the in-phase mode) and if one has a resonance $(2m+1)\,\pi/T = \Omega_q$ for some integer m and $q \in [0, \pi]$, then (20) provides nontrivial solutions to (12)–(19), and thus $M_{\gamma,T}$ is non-invertible. This occurs, e.g., for $T = \pi$ $(m = 0,\, q = 0)$, where the amplitude of the in-phase mode becomes infinite.

Moreover, if one considers a localized pattern $I_0 = \mathbb{Z} \setminus \{n_0\}$ for some integer n_0, then the resonance $m\,(2\pi/T) = \Omega_q$ $(m \in \mathbb{N})$ leads to nontrivial solutions to (12)–(19) (obtained by choosing $\psi = \frac{\pi}{2} - q\, n_0$ in (20)), and thus $M_{\gamma,T}$ is non-invertible.

In the case $E = \mathscr{P}^p$ (p-periodic sequences), the phonon band becomes discrete (wavenumbers take the form $q = k\, 2\pi/p$ with $k \in \mathbb{Z}$), but the above resonance conditions remain valid when $I_1 = \mathbb{Z}$ or $I_2 = \mathbb{Z}$, or if $I_0 = \mathbb{Z} \setminus \{n_0 + p\,\mathbb{Z}\}$.

As an application of Theorem 1, we now prove the existence of nonsmooth modes having any type of pattern, close to the uncoupled (or "anticontinuum") limit $\gamma = 0$. In Theorem 2 below, the mode pattern $I = (I_0, I_1, I_2)$ must be compatible with the choice of E. For $E = \mathscr{P}^p$, the sets I_k are assumed invariant modulo p, and for $E = \ell_2(\mathbb{Z})$, the sets I_1 and I_2 have to be finite (no impacts occur at infinity when oscillations are spatially localized). In the case $E = \ell_\infty(\mathbb{Z})$, there are no restrictions on the mode pattern.

Theorem 2 *Fix a mode pattern $I = (I_0, I_1, I_2)$ compatible with E. There exists an open set $\mathscr{V} \subset \mathbb{R}^2$ including the segment $\{0\} \times (\pi, 2\pi)$ such that for all $(\gamma, T) \in \mathscr{V}$, system (1)–(3) admits a unique even periodic solution with pattern I, which is defined by (18).*

Proof It suffices to check that for $\gamma = 0$ and all $T \in (\pi, 2\pi)$, system (1)–(3) admits a unique nondegenerate periodic solution with pattern I. Then, the result follows by direct application of Theorem 1.

Let us denote by $y_n^{\text{ip}}(t)$ the in-phase mode defined by (5) with period $T \in (\pi, 2\pi)$. For $\gamma = 0$, system (1)–(3) consists of uncoupled impact oscillators. Consequently, the unique T-periodic solution with pattern I is given by $y_n = y_n^{\text{ip}}$ for all $n \in I_1$, $y_n(t) = y_n^{\text{ip}}(t + T/2)$ for all $n \in I_2$, and $y_n = 0$ for all $n \in I_0$ (for $\gamma = 0$, all non-impacting nontrivial solutions are 2π-periodic, and we have assumed that $T < 2\pi$). It follows that the condition (17) of non-grazing impacts is satistied for $T \in (\pi, 2\pi)$. In order to show that the T-periodic solution obtained for $\gamma = 0$ is nondegenerate, there remains to check that the linear map $M_{0,T}$ of (16) is invertible. We have, for all $\xi = (\xi^{(0)}, \xi^{(1)}, \xi^{(2)}) \in E^{(0)} \times E^{(1)} \times E^{(2)}$,

$$M_{0,T}\, \xi = P\, e^{J\,T/2} \begin{pmatrix} u \\ v \end{pmatrix}, \tag{21}$$

where $u, v \in E = E^{(0)} \times E^{(1)} \times E^{(2)}$ are defined as follows:

$$u = (\xi^{(0)}, \xi^{(1)}, 0), \quad v = (0, 0, \xi^{(2)}).$$

Moreover, we have in the block form

$$e^{Jt} = \begin{pmatrix} \cos t & \sin t \\ -\sin t & \cos t \end{pmatrix} \in \mathscr{L}(E \times E),$$

hence (21) yields

$$M_{0,T}\,\xi = P\,(y, \dot{y}),$$

where $y, \dot{y} \in E = E^{(0)} \times E^{(1)} \times E^{(2)}$ are defined by

$$y = (\cos(T/2)\,\xi^{(0)}, \cos(T/2)\,\xi^{(1)}, \sin(T/2)\,\xi^{(2)}),$$

$$\dot{y} = (-\sin(T/2)\,\xi^{(0)}, -\sin(T/2)\,\xi^{(1)}, \cos(T/2)\,\xi^{(2)}).$$

Consequently, $M_{0,T} \in \mathscr{L}(E^{(0)} \times E^{(1)} \times E^{(2)})$ takes the following diagonal form:

$$M_{0,T}\,\xi = (-\sin(T/2)\,\xi^{(0)}, \cos(T/2)\,\xi^{(1)}, \cos(T/2)\,\xi^{(2)}).$$

It follows that $M_{0,T}$ is invertible because the coefficients $\cos(T/2)$ and $\sin(T/2)$ do not vanish for $T \in (\pi, 2\pi)$. □

It is interesting to compare the local continuation result of Theorem 2 and the explicit computations of the nonsmooth in-phase and out-of-phase modes performed in Sect. 2. The in-phase mode actually exists for all $\gamma \in \mathbb{R}$ and $T \in (\pi, 2\pi)$. Moreover, the out-of-phase mode exists for all $\gamma \geq 0$ (and even for γ slightly negative) and $T \in (T_0(\gamma), 2\pi(1+4\gamma)^{-1/2})$.

2.4 Stability

In this section, the linear stability of periodic solutions is analyzed through the eigenvalues of an associated monodromy matrix. Since the trajectory of the state of the system is nonsmooth at impact times, some precautions must be taken into account to compute the monodromy matrix. The computation of the monodromy follows the line of the work in [32].

In this section, we will consider the finite-dimensional case $E = \mathscr{P}^p$. For a given initial condition $Y_0 = (y(t_0), \dot{y}(t_0))^T \in \mathbb{R}^{2p}$, the conservative system (1)–(3) admits a unique solution (without accumulation of impacts) that is analytic in time between impacts [7, 8, 35]. Let us define the trajectory of the flow of (1)–(3) for the initial conditions (t_0, Y_0) as

$$\begin{aligned} \phi : \mathbb{R} \times \mathbb{R} \times \mathbb{R}^{2p} &\to \mathbb{R}^{2p} \\ (t, t_0, Y_0) &\mapsto \phi(t, t_0, Y_0). \end{aligned} \tag{22}$$

The flow ϕ satisfies $\phi(t_0, t_0, Y_0) = Y_0$. The trajectory of the system for the initial condition (t_0, Y_0) is $Y(t) = \phi(t, t_0, Y_0)$. In the sequel, we consider a time t and an initial time t_0 at which no impact occurs. The computation of the monodromy amounts to performing the differentiation of the flow ϕ at time t for the initial time t_0 with respect to the initial condition Y_0, that is,

$$M(t) = \frac{d\phi(t, t_0, Y_0)}{dY_0}. \tag{23}$$

This matrix can be approximated by finite differences. As noted in [32], the application of a finite-difference scheme may result in a poor approximation of the monodromy matrix. Since, in our application, the flow can be defined as a concatenation of piecewise smooth flows between impact times, we present here a closed-form formula for the monodromy matrix based on the computation of a saltation matrix that takes into account how the impact times evolve with the initial conditions. This closed-form formula is based on the assumption that the impacts are *simple impacts* in the sense that only one particle impacts at a given time. Moreover, we consider non-grazing impacts, i.e., impact at nonzero velocities.

The case of a simple impact at time $t_\star > t_0$:

Let us assume that we have a unique and simple impact in the interval (t_0, t) at time $t_\star(Y_0)$. The notation outlines its dependency on the initial condition. At the impact time $t_\star(Y_0)$, the trajectory is reset using the elastic Newton impact law, which can be written as follows:

$$Y(t_\star^+(Y_0)) = R_{t_\star} Y(t_\star^-(Y_0)), \tag{24}$$

where $R_{t_\star} \in \mathbb{R}^{2p\times 2p}$ is the reset matrix. Let us denote by $i_{t_\star}$ the index of the impacting particle at $t_\star(Y_0)$, i.e.,

$$y_{i_{t_\star}}(t_\star(Y_0)) = -1. \tag{25}$$

The reset matrix can be written as

$$R_{t_\star} = \begin{bmatrix} I & 0 \\ 0 & E \end{bmatrix}, \tag{26}$$

where the matrix $E \in \mathbb{R}^{p\times p}$ is given by its components as

$$E_{ij} = \begin{cases} 0, & \text{if } i \neq j, \\ 1, & \text{if } i = j \neq i_{t_\star}, \\ -1, & \text{if } i = j = i_{t_\star}. \end{cases} \tag{27}$$

The state of the system at time t can be written as

$$\begin{aligned} Y(t) &= \phi(t, t_0, Y_0) = \phi(t, t_\star^+(Y_0), Y(t_\star^+(Y_0))) \\ &= \phi(t, t_\star^+(Y_0), R_{t_\star} Y(t_\star^-(Y_0))) = \phi(t, t_\star^+(Y_0), R_{t_\star}\phi(t_\star^-(Y_0), t_0, Y_0)). \end{aligned} \tag{28}$$

The differentiation of the previous expression amounts to differentiating, with respect to Y_0, a composition of smooth functions

$$\begin{aligned}\frac{d\phi(t,t_0,Y_0)}{dY_0} &= D_2\phi(t,t_\star^+(Y_0),R_{t_\star}\phi(t_\star^-(Y_0),t_0,Y_0))\frac{dt_\star(Y_0)}{dY_0}\\ &+D_3\phi(t,t_\star^+(Y_0),R_{t_\star}\phi(t_\star^-(Y_0),t_0,Y_0))R_{t_\star}\frac{d\phi(t_\star^-(Y_0),t_0,Y_0)}{dY_0}\end{aligned} \tag{29}$$

with

$$\frac{d\phi(t_\star^-(Y_0),t_0,Y_0)}{dY_0} = D_1\phi(t_\star^-(Y_0),t_0,Y_0)\frac{dt_\star(Y_0)}{dY_0} + D_3\phi(t_\star^-(Y_0),t_0,Y_0). \tag{30}$$

The notation $D_k\phi$ denotes the partial derivatives of ϕ with respect to its k-th argument. If the smooth flow is known between impacts, the only difficult part that remains to compute is the derivative of the time of impact $t_\star$ with respect to Y_0. Let us split the flow ϕ such that

$$Y(t) = \phi(t,t_0,Y_0) = \begin{bmatrix}\phi_y(t,t_0,Y_0)\\ \phi_{\dot y}(t,t_0,Y_0)\end{bmatrix} = \begin{bmatrix}y(t)\\ \dot y(t)\end{bmatrix}. \tag{31}$$

We have assumed that only one particle of index $i_{t_\star}$ is impacting at $t_\star(Y_0)$. The constraint (25) can be written as

$$\phi_{y,i_{t_\star}}(t_\star,t_0,Y_0) = -1. \tag{32}$$

Since $\partial_t\phi_{y,i_{t_\star}}(t_\star^-,t_0,Y_0) = \dot y_{i_{t_\star}}(t_\star^-(Y_0)) < 0$ (non-grazing impact) and the flow is smooth (analytic) between impacts, the implicit function theorem guarantees that the impact persists upon small variations of Y_0, with an impact time $t_\star$ being a smooth (analytic) function of Y_0. Moreover, defining a projection matrix $P_i \in \mathbb{R}^{1\times 2p}$ such that

$$D_3\phi_{y,i}(t_\star^-(Y_0),t_0,Y_0) = P_i D_3\phi(t_\star^-(Y_0),t_0,Y_0), \tag{33}$$

we have

$$\frac{dt_{t_\star}(Y_0)}{dY_0} = -\frac{1}{\dot y_{i_{t_\star}}(t_\star^-(Y_0))}P_{i_{t_\star}}D_3\phi(t_\star^-(Y_0),t_0,Y_0). \tag{34}$$

In order to simplify the expression of the monodromy matrix given by (29) and (30), we observe that

$$D_2\phi(t,t_\star^+,Y(t_\star^+(Y_0))) = -D_3\phi(t,t_\star^+(Y_0),Y(t_\star^+(Y_0)))\dot Y(t_\star^+(Y_0)). \tag{35}$$

Indeed, since $\phi(t,\tilde t,\phi(\tilde t,t_\star^+,Y_\star)) = \phi(t,t_\star^+,Y_\star)$ is independent of $\tilde t$, the identity $\partial_{\tilde t}\phi(t,\tilde t,\phi(\tilde t,t_\star^+,Y_\star)) = 0$ evaluated at $\tilde t = t_\star^+$ and $Y_\star = Y(t_\star^+(Y_0))$ yields identity (35). Using (29), (30) and (35), the monodromy matrix simplifies to

$$\frac{d\phi(t,t_0,Y_0)}{dY_0} = D_3\phi(t,t_\star^+,Y(t_\star^+(Y_0))) \left[[R_{t_\star}\dot{Y}(t_\star^-(Y_0)) - \dot{Y}(t_\star^+(Y_0))]\frac{dt_\star(Y_0)}{dY_0} + R_{t_\star}D_3\phi(t_\star^-(Y_0),t_0,Y_0)\right]. \tag{36}$$

Finally, using the relation (34), the monodromy matrix is expressed as follows:

$$\frac{d\phi(t,t_0,Y_0)}{dY_0} = D_3\phi(t,t_\star^+(Y_0),Y(t_\star^+(Y_0)))S_{t_\star}D_3\phi(t_\star^-(Y_0),t_0,Y_0), \quad t > t_\star(Y_0), \tag{37}$$

where the so-called saltation matrix $S_{t_\star}$ is defined by

$$S_{t_\star} = -\frac{1}{\dot{y}_{i_{t_\star}}(t_\star^-(Y_0))}[R_{t_\star}\dot{Y}(t_\star^-(Y_0)) - \dot{Y}(t_\star^+(Y_0))]P_{i_{t_\star}} + R_{t_\star}. \tag{38}$$

Note that the monodromy matrix is obtained as the product of the Jacobian matrices of the flow with respect to the initial condition in each smooth phase separated by the saltation matrix.

The Case of Two Simple Impacts at Times $t_{\star,2} > t_{\star,1} > t_0$:

For the two simple impacts at time $t_{\star,2} > t_{\star,1} > t_0$, the computation of the monodromy matrix follows the same line. It is also a product of the Jacobian matrices of the flow with respect to the initial condition in each smooth phase separated by the saltation matrix:

$$\frac{d\phi(t,t_0,Y_0)}{dY_0} = D_3\phi(t,t_{\star,2}^+(Y_0),Y(t_{\star,2}^+(Y_0)))S_{t_{\star,2}} D_3\phi(t,t_{\star,1}^+(Y_0),Y(t_{\star,1}^+(Y_0)))S_{t_{\star,1}}D_3\phi(t_{\star,1}^-(Y_0),t_0,Y_0), \quad t > t_{\star,2}(Y_0). \tag{39}$$

Computation of the Monodromy for the Piecewise Linear System :

In our case of a piecewise-linear dynamics, the flow of the system between two impacts is given by

$$\phi(t,t_0,Y_0) = \exp(D(t-t_0))\,Y_0, \; t_0 \le t \le t_{\star,1}(Y_0), \tag{40}$$

$$\phi(t,t_{\star,1}^+(Y_0),Y(t_{\star,1}^+(Y_0))) = \exp(D(t-t_{\star,1}(Y_0)))\,Y(t_{\star,1}^+(Y_0)), \; t_{\star,1}(Y_0) \le t \le t_{\star,2}(Y_0) \tag{41}$$

$$\phi(t,t_{\star,2}^+(Y_0),Y(t_{\star,2}^+(Y_0))) = \exp(D(t-t_{\star,2}(Y_0)))\,Y(t_{\star,2}^+(Y_0)), \; t \ge t_{\star,2}(Y_0), \tag{42}$$

with $D = J + \gamma L$. As indicated above in the derivation of the monodromy matrix, the piecewise linear flow is smooth (analytic). If we consider the explicit formula of the linear flow (40)–(42) between impacting times at $t_{\star,1} = T/2$ and $t_{\star,2} = T$, we get, for the monodromy matrix,

$$\frac{d\phi(t,t_0,Y_0)}{dY_0} = \exp(D(t-T))\, S_T\, \exp(D(T/2))\, S_{T/2}\, \exp(D(T/2-t_0)), \quad t > T, \tag{43}$$

where $t_0 < T/2$. In Sect. 3, we shall fix $t_0 = T/4$ and $t = t_0 + T = 5T/4$ to compute the monodromy matrix of a T-periodic solution with impact times multiple of $T/2$. This leads to

$$\frac{d\phi(5T/4, T/4, Y_0)}{dY_0} = \exp(DT/4)\, S_T\, \exp(D(T/2))\, S_{T/2}\, \exp(DT/4). \tag{44}$$

The periodic solution will be unstable if this monodromy matrix admits an eigenvalue with modulus greater than unity, and spectrally stable if all eigenvalues lie on the unit circle (due to time-reversal symmetry, the Floquet spectrum has the invariance $\sigma \to \sigma^{-1}$). The spectrum of the above monodromy is the same as for $S_T\, \exp(D(T/2))\, S_{T/2}\, \exp(DT/2)$.

3 Numerical Computation of Nonsmooth Modes

We solve problem (12)–(13) numerically for a chain of p oscillators with periodic boundary conditions. Unless explicitly stated otherwise, we fix $p = 100$. Although the system (12)–(13) is a standard linear system, we use a general shooting method, i.e., determine a vector $\xi = (y^{(0)}(0), y^{(1)}(0), \dot{y}^{(2)}(0)) \in \mathbb{R}^p$ such that the three boundary conditions of (13) at $t = 0$ and $t = T/2$ are satisfied through Newton iterations. For each Newton iteration, this requires solving a linear system for ξ obtained through time-integration of the linear ODE (12). This time integration is equivalent to computing the exponential matrix of the linear flow numerically. When the coupling parameter is chosen far from the degeneracy case of the BVP matrix, the shooting technique converges in one iteration. When we are in the neighborhood of the degenerate cases, the number of Newton iterations may increase, indicating an ill-conditioned linear system of the BVP. Thanks to the general shooting technique, the case of nonlinear local or interaction potentials could be similarly addressed. The constraint (14) is checked *a posteriori*. To this end, we integrate (1)–(3) numerically using an event-driven scheme for nonsmooth dynamical systems implemented in the Siconos software [22]. For the shooting technique and validation of the constraints, the linear ODE is integrated thanks to ODEPACK [21] embedded in the Siconos software.

Usually, the solution branches are first continued for fixed values of T, varying the coupling parameter γ. For all fixed value $T \in (\pi, 2\pi)$, a choice of impacting particles and phases (determined by I_1, I_2) selects a unique solution for $\gamma = 0$, which can be continued up to some maximal value of the coupling parameter γ. We shall see in the sequel that some continuations are also done with respect to the period.

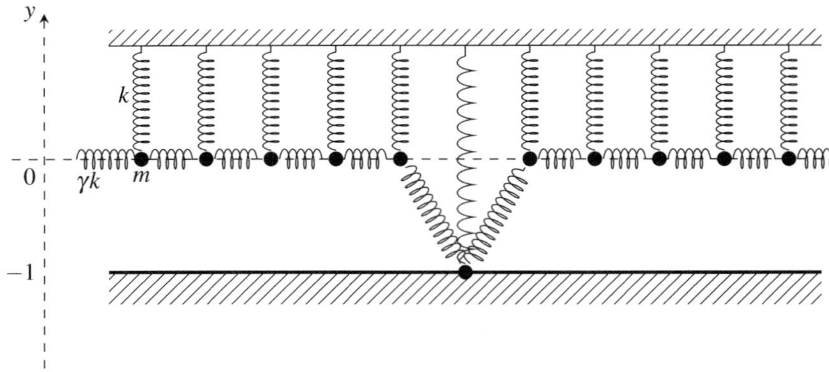

Fig. 6 Mode pattern for the site-centered breather

3.1 Site-Centered Breathers

In this section, we illustrate the site-centered breather for the mode pattern $I_2 = \{50\}$, $I_1 = \emptyset$ depicted in Fig. 6. The period is $T = \frac{3\pi}{2}$. The periodic solution has been successfully computed for $\gamma \in [0, \gamma_c]$ with

$$\gamma_c = \frac{1}{4}\left(\left(\frac{2\pi}{T}\right)^2 - 1\right), \tag{45}$$

the critical value of γ for which we expect to reach the out-of-phase mode. For $T = \frac{3\pi}{2}$, we have $\gamma_c \approx 0.1944$. In Fig. 7, the initial positions and velocities are displayed for the particle indices between 40 and 60 and for 4 different values of γ. We observe that, for small values of the coupling parameter γ, the breather is localized on a few particles. With the increasing values of γ, the support of the solution is increasing to reach the out-of-phase linear grazing mode for $\gamma = \gamma_c$. Let us note that the velocity of the central particle 50 is decreasing to the grazing solution for all the particles.

In Fig. 8, the eigenvalues of the monodromy matrix are displayed. In Fig. 8a, we remark that the eigenvalues have a modulus equal to 1 up to a critical value γ_s between 0.129 and 0.142 for which a pair of eigenvalues is leaving the unit circle. In Fig. 8b, c and d, all the eigenvalues are plotted in the complex plane for three different values of $\gamma \in \{0, 0.064, 0.181\}$. For $\gamma = 0$, a pair of eigenvalues are equal to $+1$ and all the other conjugate eigenvalues pairs are equal to i or $-i$. For $\gamma < \gamma_s$, the conjugate eigenvalue pairs, equal to i and $-i$ for $\gamma = 0$, start to slide on the unit circle toward the pair of eigenvalues that remains at $+1$. For $\gamma = \gamma_s$, a collision occurs at $+1$. Finally, for $\gamma > \gamma_s$, a pair of real inverse eigenvalues leaves the unit circle to slide on the real line while a pair of eigenvalues remains at $+1$. In that case, the stability of the periodic solution is lost. For $\gamma = 0.181$, one of the eigenvalues of modulus around 5.71 is not displayed. To illustrate this loss of stability, we report, in Fig. 9, several time integrations of the system with constraints and impacts for different values of

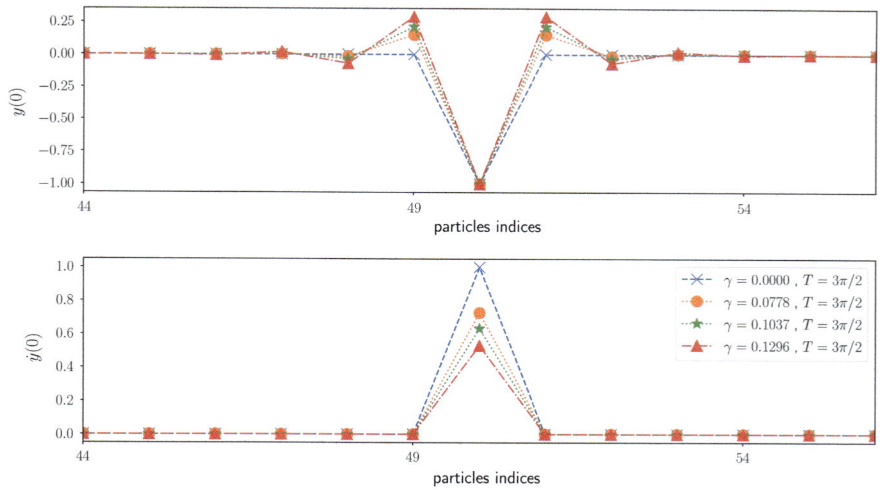

Fig. 7 Site-centered breather with pattern $I_1 = \emptyset$, $I_2 = \{50\}$

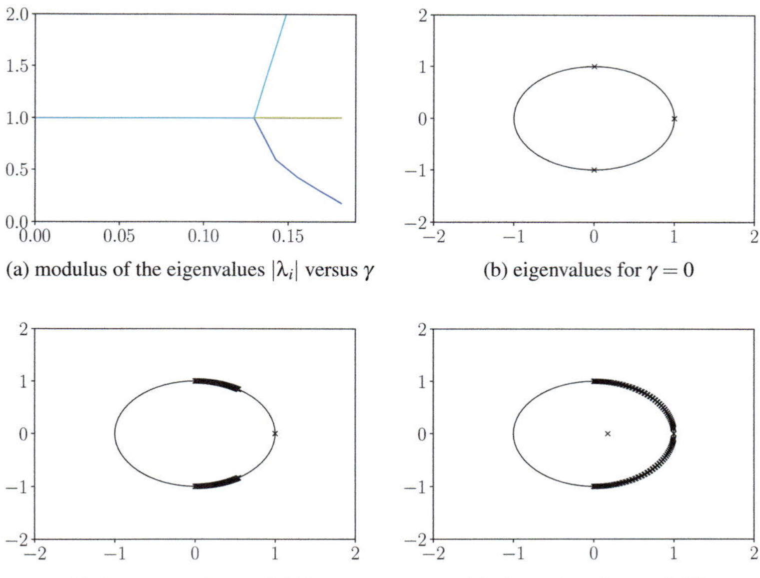

(a) modulus of the eigenvalues $|\lambda_i|$ versus γ

(b) eigenvalues for $\gamma = 0$

(c) eigenvalues for $\gamma = 0.064$

(d) eigenvalues for $\gamma = 0.181$

Fig. 8 Eigenvalues of the monodromy matrix for the site-centered breather with pattern $I_1 = \emptyset$, $I_2 = \{50\}$

γ over the time interval $[0, 25T]$. Although the system is numerically integrated with high accuracy Runge-Kutta schemes in ODEPACK with very tight tolerances (10^{-14}), the periodic solutions for $\gamma = 0.181$ are destabilized.

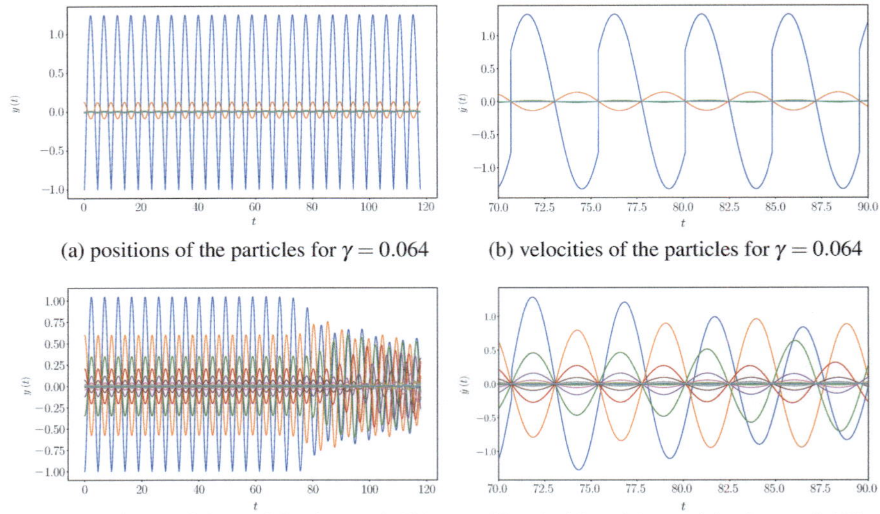

(a) positions of the particles for $\gamma = 0.064$ (b) velocities of the particles for $\gamma = 0.064$

(c) positions of the particles for $\gamma = 0.181$ (d) velocities of the particles for $\gamma = 0.181$

Fig. 9 Time integration of the periodic solutions for the site-centered breather with pattern $I_1 = \emptyset$, $I_2 = \{50\}$

We also perform a continuation of the solution with respect to the period. We start for a value of (γ, T) equal to $(0.15, 3\pi/2)$ and we decrease the period following a solution with a fixed pattern. The numerical solutions are displayed in Fig. 10a. We can observe that a family of site-centered breathers is found with an increasing amplitude of the initial state. For the uncoupled case ($\gamma = 0.0$), we know that the amplitude of the solution goes to infinity when $T \to \pi$. The same phenomenon is observed for a given coupling parameter $\gamma = 0.15$. In Fig. 10b, we plot the maximum amplitude of the position $\|y(0)\|_\infty$ and the velocity $\|\dot{y}(0)\|_\infty$ as a function of T. An asymptotic value of the period clearly appears for which the amplitude of the solution blows up. In this specific case, the asymptotic value of the period is about $0.58(3\pi/2) \approx 2.78$. Let us note that this value is below π.

To conclude this section, an exploration of the viability of the site-centered breathers has been performed for $(\gamma, T) \in [0, 1.1] \times [2, 2\pi]$ and $p = 30$ particles. We select a mesh grid in the plane (γ, T) and solve the boundary value problem for each pair (γ, T). The results are reported in Fig. 11. The light areas correspond to a numerical computation of a periodic solution to (12)–(13) with the satisfaction of the constraint (14) and the pattern $I_1 = \emptyset$, $I_2 = \{15\}$. The red dashed curve is given by the out-of-phase grazing linear mode whose period is related to γ by

$$T(\gamma) = 2\pi\,(1 + 4\gamma)^{-1/2}. \tag{46}$$

As expected with the previous computations, we observe that there exists a large light area bounded above by the relation (46) and corresponding to site-centered

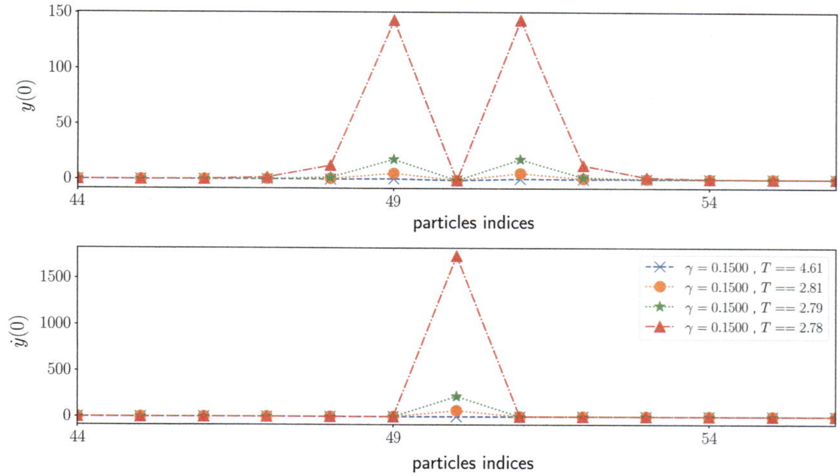

(a) positions and velocities of the particles

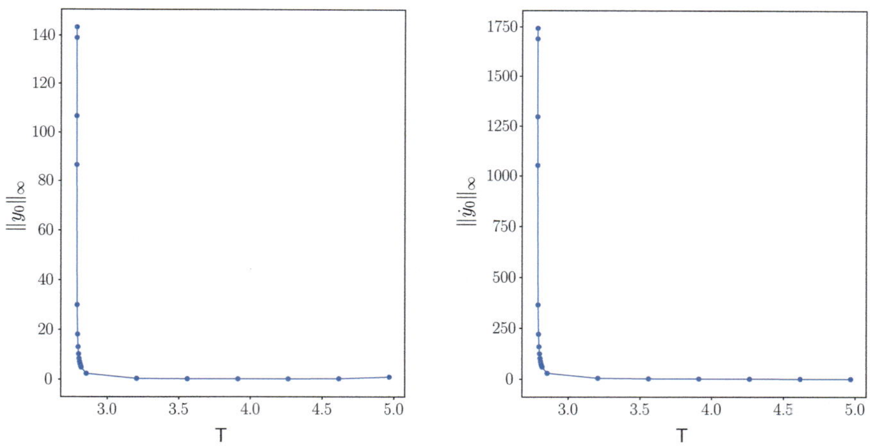

(b) maximum amplitude of the position $\|y(0)\|_\infty$ and the velocity $\|\dot{y}(0)\|_\infty$ as a function of T

Fig. 10 Continuation with a decreasing period of the site-centered breather with pattern $I_1 = \emptyset$, $I_2 = \{50\}$ for $\gamma = 0.15$

breathers. This area is also bounded below by another curve that corresponds to modes whose amplitudes go to infinity, as we have already discussed for a particular value of $\gamma = 0.15$ in Fig. 10. Quite interestingly, other light areas are present above the red curve. To explain these areas, we plot the graphs of the periods with respect to γ for larger wavenumber q given by

$$T_n(\gamma) = 2\pi \left(1 + 4\gamma \sin^2(q/2)\right)^{-1/2}, \text{ with } q = n2\pi/p, \quad n = 1, \ldots, 15. \tag{47}$$

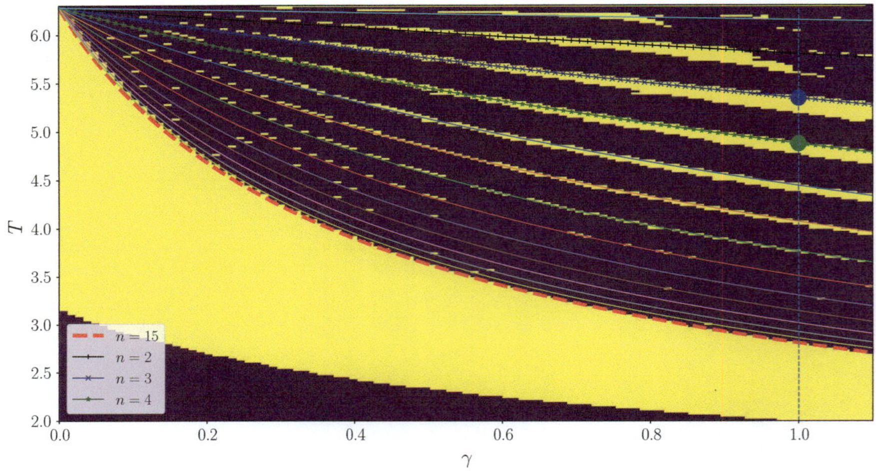

Fig. 11 Continuation of periodic solutions with pattern $I_1 = \emptyset$, $I_2 = \{15\}$ (light areas) for $(\gamma, T) \in [0, 1.1] \times [2, 2\pi]$. Graphs of $T_n(\gamma) = 2\pi\,(1 + 4\gamma \sin^2(q/2))^{-1/2}$, with $q = n2\pi/p$, for $n = 1, \ldots, 15$ and $p = 30$

We can observe the existence of modulated waves near the linear grazing solutions. In order to illustrate the solutions obtained in these areas, we plot, in Fig. 12, the results of two continuations over the period for $\gamma = 1$, $T_3 \approx 5.34$ and $T_4 \approx 4.87$ (large dots in Fig. 11). We can observe that these solutions are not exactly normal nonsmooth modes that emerge from the linear grazing modes, but rather spatial modulations of nonsmooth normal modes. For the computation of what could be called a *nonsmooth normal mode*, we refer to Sect. 3.4. There, other solutions are computed (with long-wavelength near T_1) with preservation of the normal mode pattern at the start of continuation.

3.2 Bond-Centered Breathers

In this section, some bond-centered breathers are computed with two different patterns.

Bond-Centered Breathers With Pattern $I_1 = \{49\}$, $I_2 = \{50\}$

Let us start with the out-of-phase pattern $I_1 = \{49\}$, $I_2 = \{50\}$, illustrated in Fig. 13. We again choose a period equal to $\frac{3\pi}{2}$, and the periodic solution has successfully been computed in the range $[0, \gamma_c]$, with γ_c given by (45). The initial conditions of the periodic solutions are displayed in Fig. 14 for the particle indices in [40, 60]. Again, we can observe that the breather is localized over a few particles for small values of the coupling parameter. Once again, the solution reaches the out-of-phase

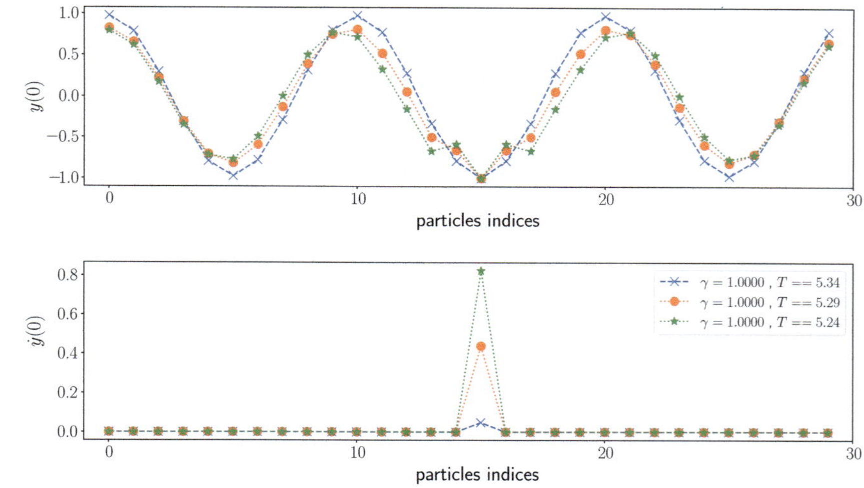

(a) Continuation for the value of the period around $T_3 = 2\pi\,(1+4\gamma\sin^2(3\pi/30))^{-1/2} \approx 5.34$

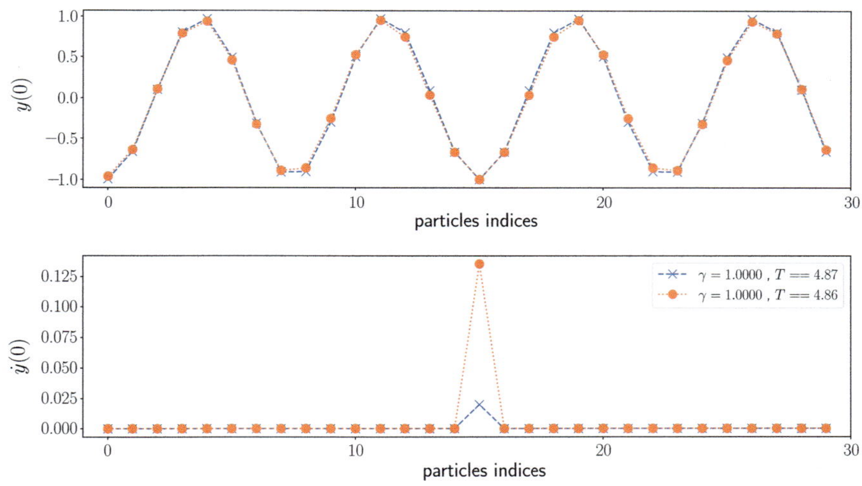

(b) Continuation for the value of the period around $T_4 = 2\pi\,(1+4\gamma\sin^2(4\pi/30))^{-1/2} \approx 4.87$

Fig. 12 Continuation of spatially-modulated nonsmooth normal modes with pattern $I_1 = \emptyset$, $I_2 = \{15\}$ for $\gamma = 1$

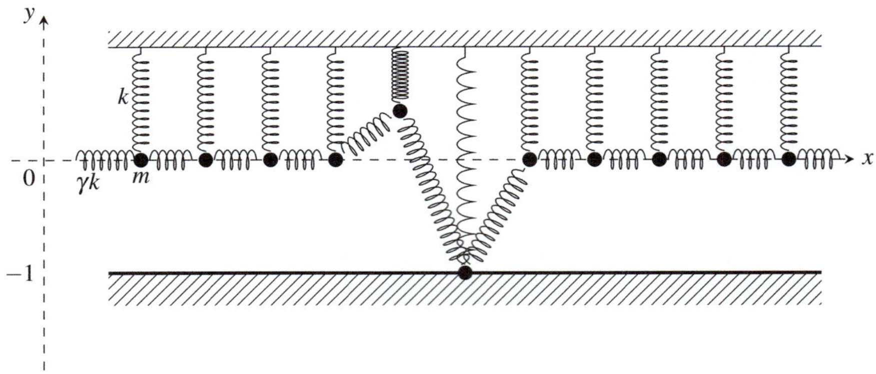

Fig. 13 Mode pattern for the bond-centered breather

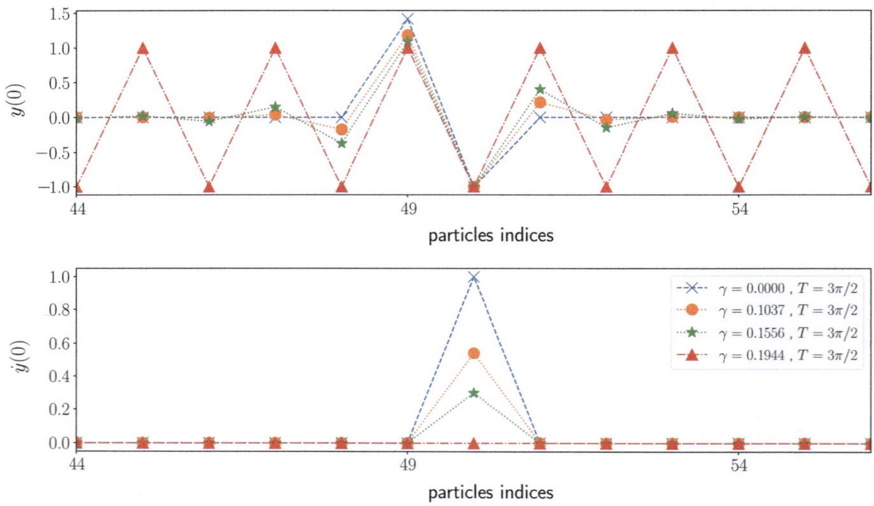

Fig. 14 Bond-centered breather with pattern $I_1 = \{49\}$, $I_2 = \{50\}$

linear grazing mode for $\gamma = \gamma_c$ while the velocity of the central particle decreases at time 0.

In Fig. 15, we depict the eigenvalues of the monodromy matrix. In Fig. 15b, for $\gamma = 0$, we have two pairs of eigenvalues in $+1$. All the other pairs of conjugate eigenvalues are equal to i or $-i$. We observe, in Fig. 15a and b, that for $\gamma > 0$, a pair of real inverse eigenvalues slides from $+1$ on the real line as γ increases, while the other pair remains equal to $+1$. The others pairs of conjugate eigenvalues slide on the unit circle toward the pair of real eigenvalues in $+1$. A collision occurs again at $+1$ for $\gamma = \gamma_s \in [0.142, 0.155]$. Then, a second pair of inverse real eigenvalues slides on the real line. For $\gamma > 0$, the stability of the periodic solutions is lost. We

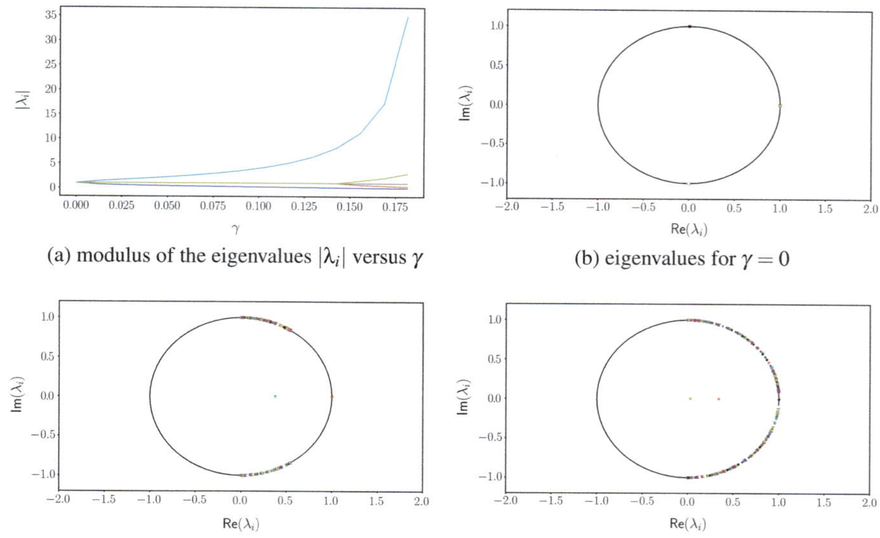

(a) modulus of the eigenvalues $|\lambda_i|$ versus γ

(b) eigenvalues for $\gamma = 0$

(c) eigenvalues for $\gamma = 0.064$

(d) eigenvalues for $\gamma = 0.181$

Fig. 15 Eigenvalues of the monodromy matrix for the bond-centered breather with pattern $I_1 = \{49\}$, $I_2 = \{50\}$

attempt to illustrate this phenomena with numerical time integration of the periodic solutions over a long time interval $[0, 35T]$ in Fig. 16.

Bond-Centered Breathers With Pattern $I_1 = \emptyset$, $I_2 = \{49, 50\}$

For the pattern $I_1 = \emptyset$, $I_2 = \{49, 50\}$, the solution for the initial conditions is depicted for the whole chain in Fig. 17a and for the particles with indices in $[40, 60]$ in Fig. 17b. The period is again $\frac{3\pi}{2}$, and we succesfully perform a continuation of the solution over $[0, \gamma_c]$ with γ_c given by (45). The main difference with the previous breathers concerns the solution when $\gamma \to \gamma_c$. In this latter case, it seems that we do not converge towards a grazing linear mode. This has to be confirmed with a more accurate study of the critical value of γ.

In Fig. 18, we depict the eigenvalues of the monodromy matrix computed by finite differences. In this case, the closed form formula of the monodromy (44) no longer applies, since we have multiple impacts. Although the approximation of the eigenvalues may contain some numerical errors, we observe a more complicated behavior of the evolution with respect to γ of the eigenvalues. For $\gamma = 0$, two pairs of real eigenvalues are equal to $+1$ and the others are conjugated pairs of eigenvalues equal to i and $-i$. For increasing values of γ, one of the pairs of real eigenvalues starts to slide on the unit circle, respectively towards i and $-i$, while the other pairs of conjugate eigenvalues slide on the unit circle from i and $-i$ towards $+1$. A first collision occurs on the unit circle for $\gamma \in [0.051, 0.064]$ and two pairs of eigenvalues leave the unit circle. Several other collisions of different types occur when we increase the value of γ up to γ_c.

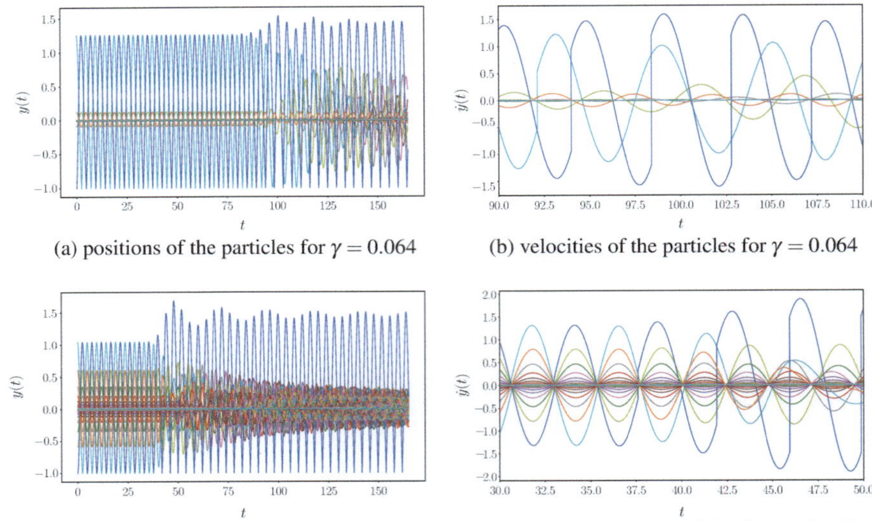

(a) positions of the particles for $\gamma = 0.064$

(b) velocities of the particles for $\gamma = 0.064$

(c) positions of the particles for $\gamma = 0.181$

(d) velocities of the particles for $\gamma = 0.181$

Fig. 16 Time integration of the periodic solutions for the bond-centered breather with pattern $I_1 = \{49\}$, $I_2 = \{50\}$

3.3 Multiple Impacting Particles

In this section, we illustrate wave patterns with multiple impacts, where the pattern is either spatially periodic or localized on several particles (multi-site breathers).

Out-of-Phase Mode with Spatial Period Two

We start with the nonsmooth mode of spatial period two described in Sect. 2. The pattern is given by $I_1 = \{2k + 1\}_{k=0,\ldots,49}$, $I_2 = \{2k\}_{k=0,\ldots,49}$, which corresponds to the sets of odd and even integers, respectively. In Fig. 19, the initial conditions for the periodic solutions are given for $T = \frac{3\pi}{2}$. For this example, we are able to continue the solution over the range $[0, \gamma_c]$ up to reaching the out-of-phase linear grazing mode. In Fig. 20, the eigenvalues of the monodromy matrix computed by finite differences are depicted. For $\gamma = 0$, all the eigenvalues are equal to $+1$. For $\gamma > 0$, the pairs of inverse real eigenvalues slide on the real line. The periodic solutions are therefore unstable for $\gamma > 0$. This is illustrated in Fig. 21, where long time integration simulations have been performed over the time interval $[0, 35T]$.

Periodic Wave with Spatial Period Six

Another example of nonsmooth spatially periodic standing wave is displayed in Fig. 22. The spatial period is six and the time period is again $\frac{3\pi}{2}$. The mode profiles are depicted for several values of γ in $[0, \gamma_c]$.

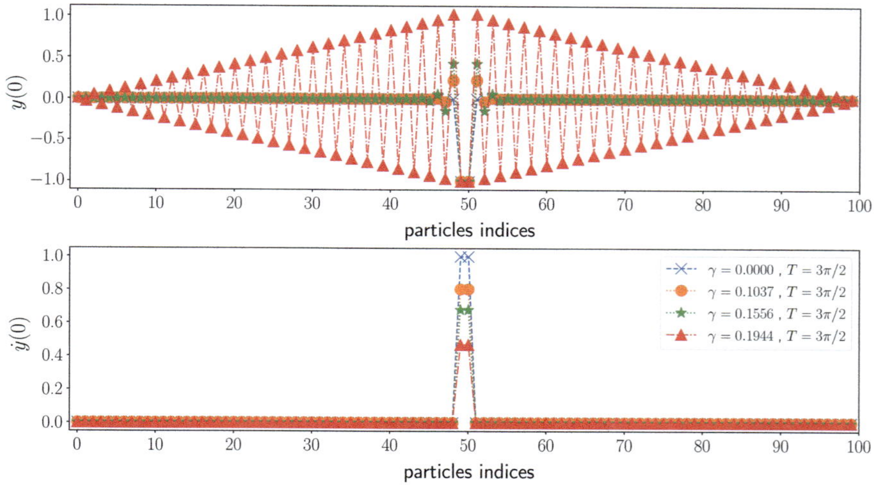

(a) initial positions and velocities of the particles

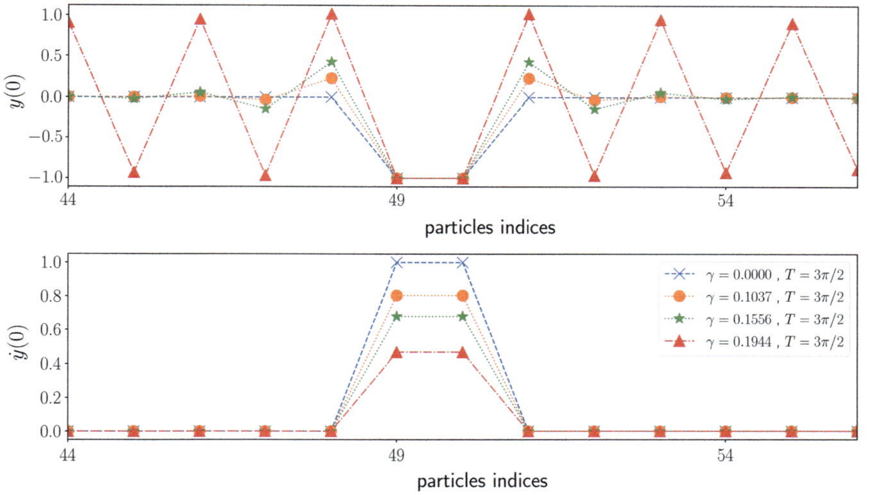

(b) initial positions and velocities of the particles with indices in $[40, 60]$

Fig. 17 Bond-centered breather with pattern $I_1 = \emptyset$, $I_2 = \{49, 50\}$

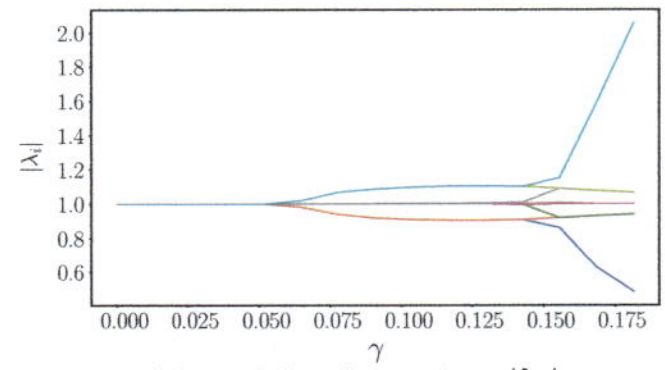

(a) modulus of the eigenvalues $|\lambda_i|$ versus γ

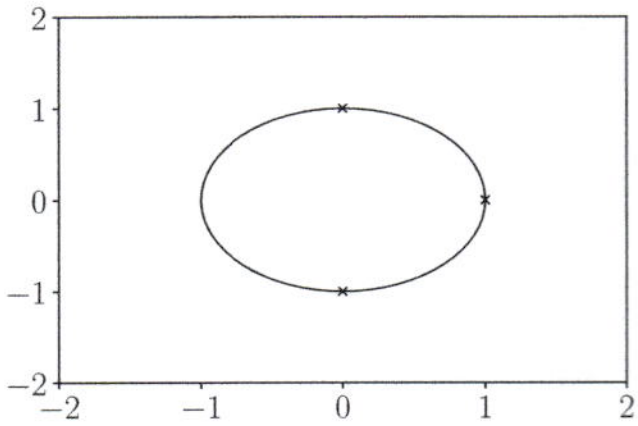

(b) eigenvalues for $\gamma = 0$

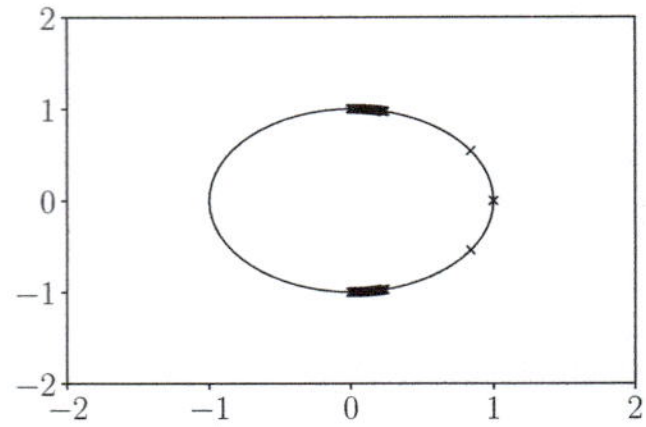

(c) eigenvalues for $\gamma = 0.025$

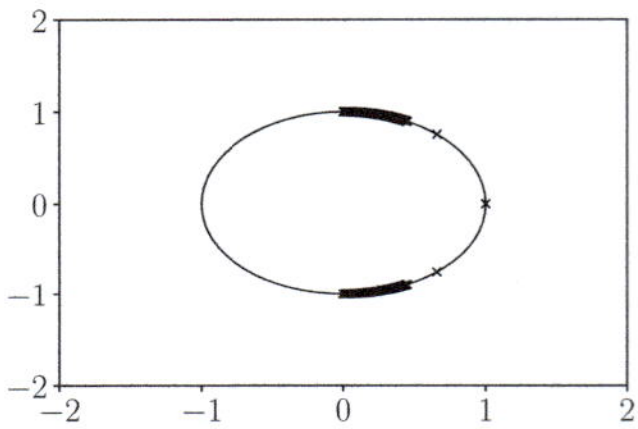

(d) eigenvalues for $\gamma = 0.051$

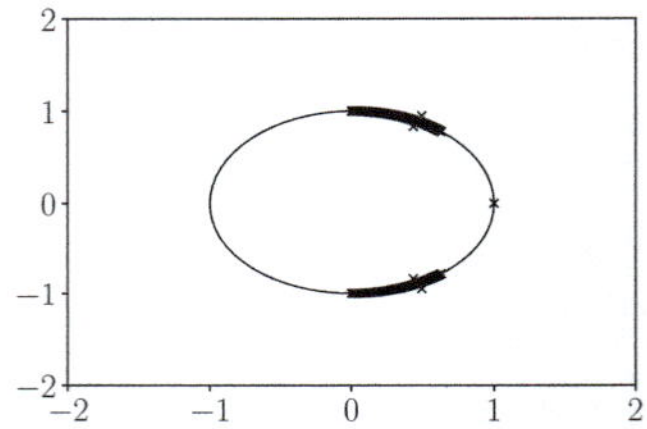

(e) eigenvalues for $\gamma = 0.077$

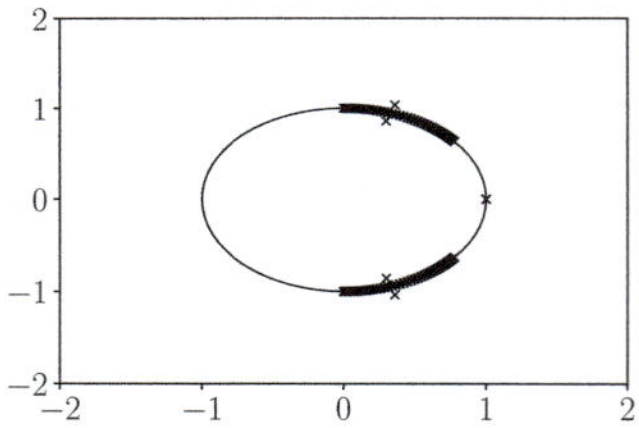

(f) eigenvalues for $\gamma = 0.103$

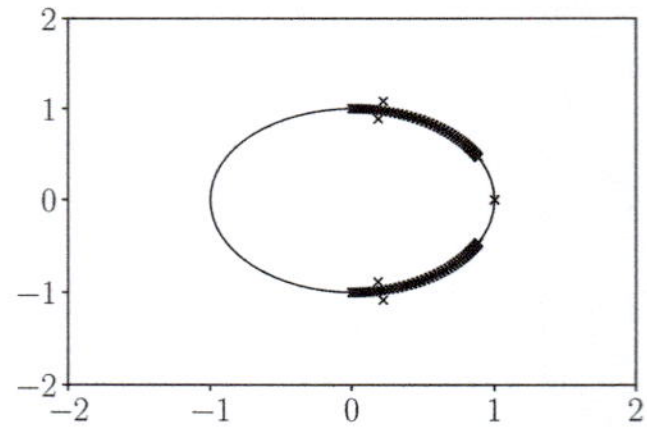

(g) eigenvalues for $\gamma = 0.129$

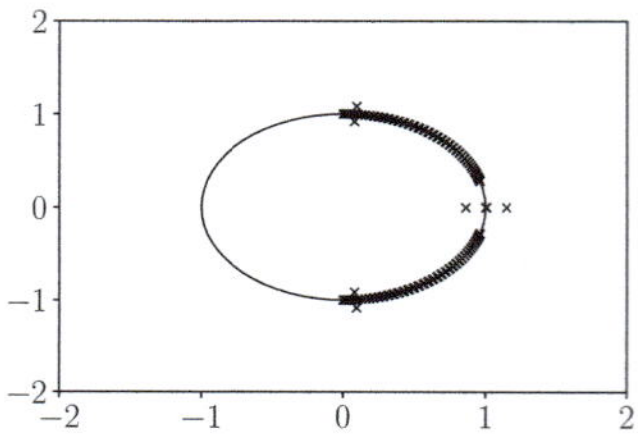

(h) eigenvalues for $\gamma = 0.155$

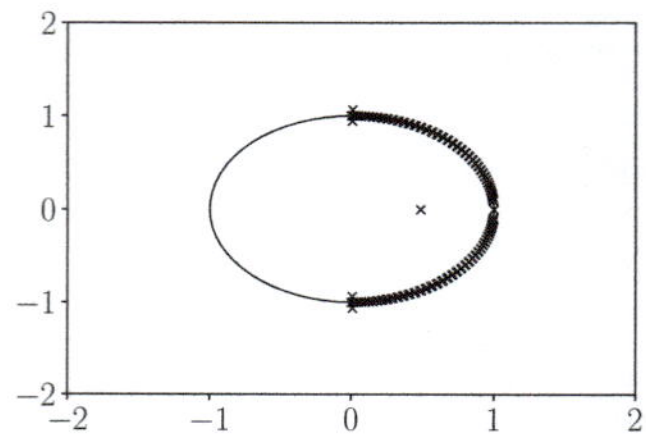

(i) eigenvalues for $\gamma = 0.181$

Fig. 18 Eigenvalues of the monodromy matrix computed by finite differences for the bond-centered breather with pattern $I_1 = \emptyset$, $I_2 = \{49, 50\}$

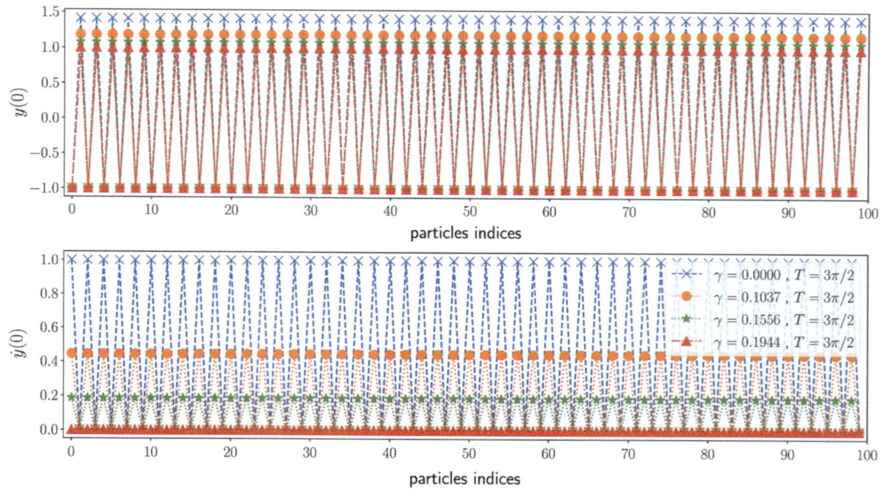

Fig. 19 Out-of-phase mode with pattern $I_1 = \{2k+1\}_{k=0,\ldots,49}$, $I_2 = \{2k\}_{k=0,\ldots,49}$

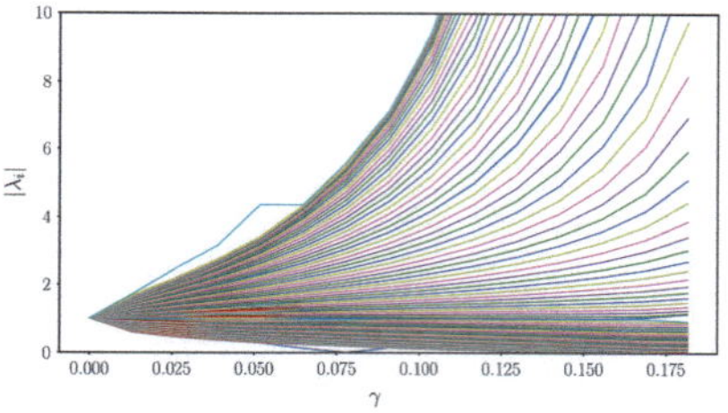

(a) modulus of the eigenvalues $|\lambda_i|$ versus γ

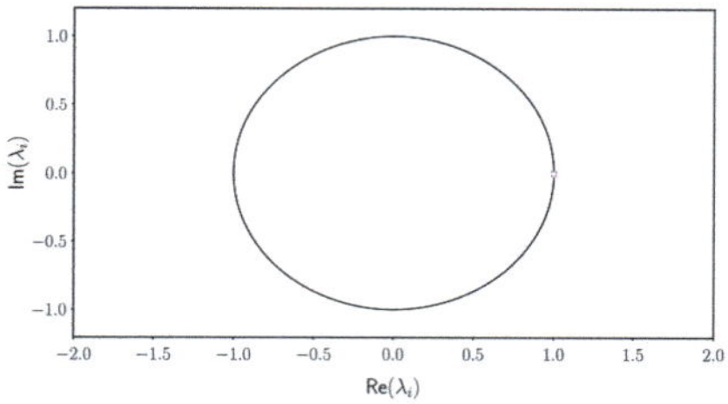

(b) eigenvalues for $\gamma = 0$

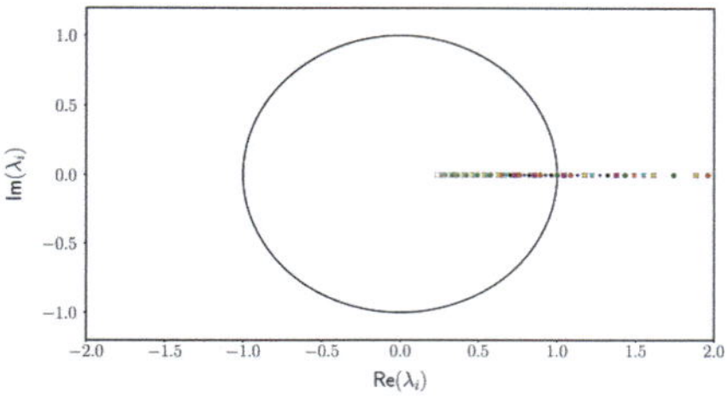

(c) eigenvalues for $\gamma = 0.064$

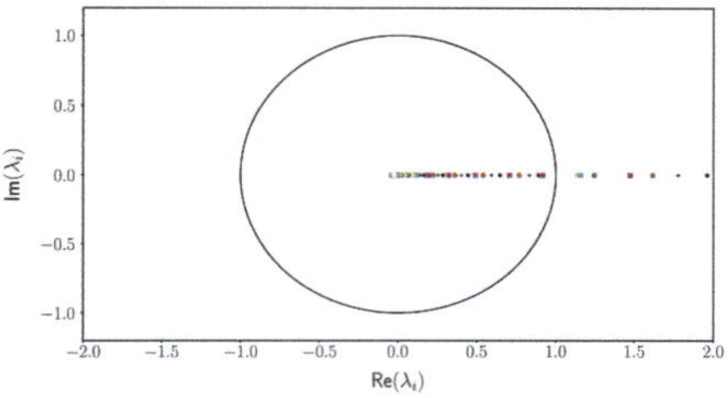

(d) eigenvalues for $\gamma = 0.181$

Fig. 20 Eigenvalues of the monodromy matrix computed by finite differences for the out-of-phase mode with pattern $I_1 = \{2k+1\}_{k=0,\ldots,49}$, $I_2 = \{2k\}_{k=0,\ldots,49}$

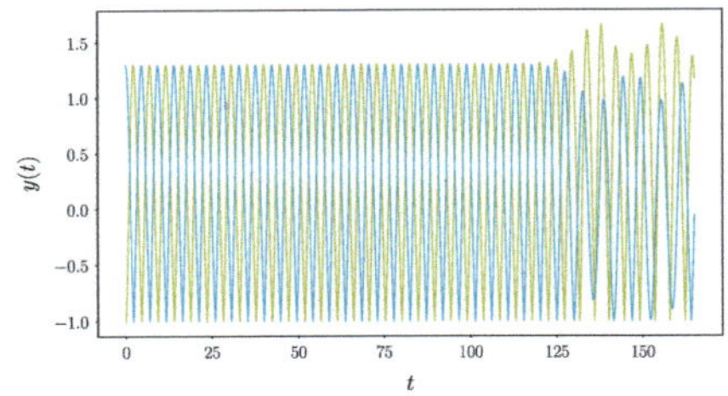

(a) positions of the particles for $\gamma = 0.051$

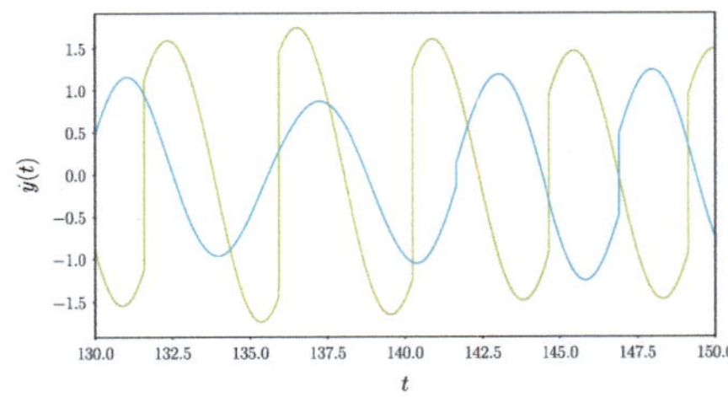

(b) velocities of the particles for $\gamma = 0.051$

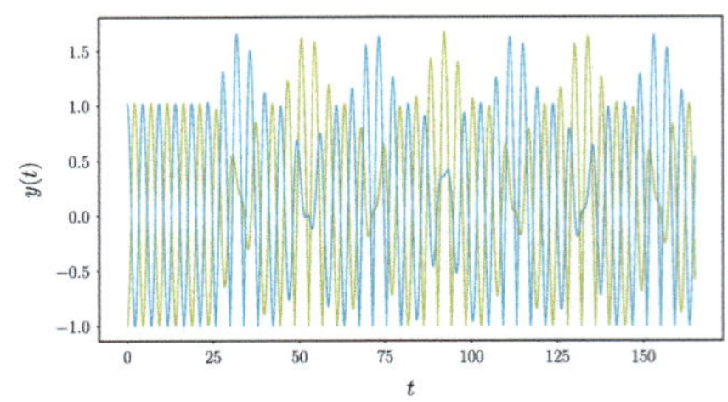

(c) positions of the particles for $\gamma = 0.181$

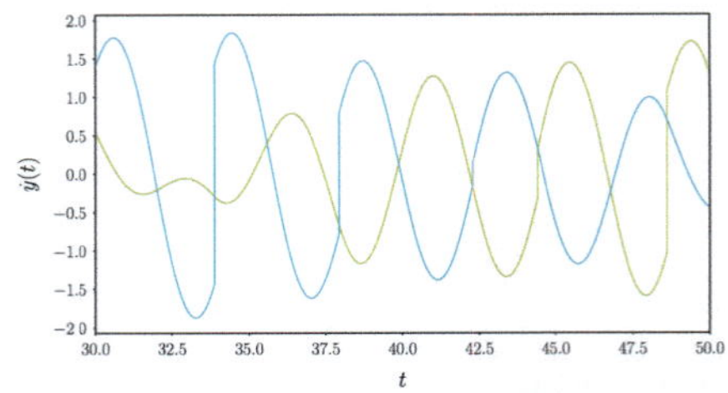

(d) velocities of the particles for $\gamma = 0.181$

Fig. 21 Time integration of the periodic solutions for the out-of-phase mode with pattern $I_1 = \{2k+1\}_{k=0,\ldots,49}$, $I_2 = \{2k\}_{k=0,\ldots,49}$

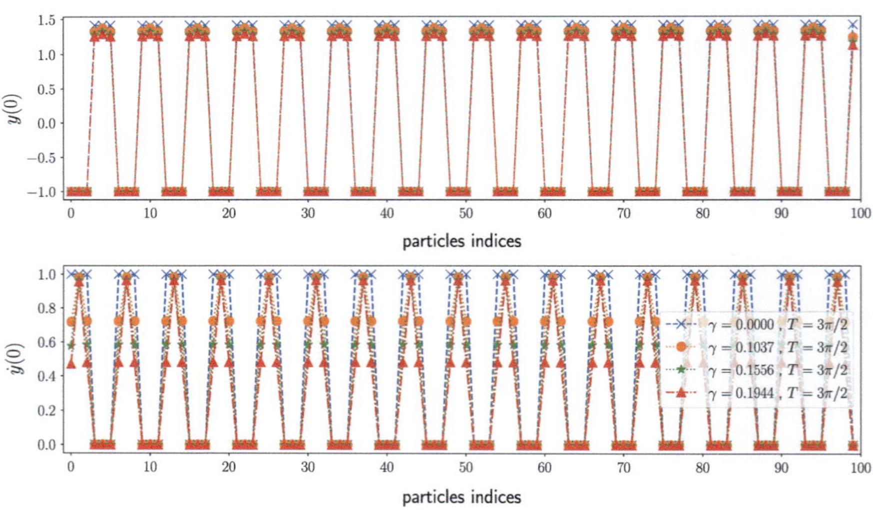

Fig. 22 Periodic wave with pattern of spatial period 6 : $I_1 = \{6k+3, 6k+4, 6k+5\}_{k=0,3,\ldots}$, $I_2 = \{6k, 6k+1, 6k+2\}_{k=0,3,\ldots}$

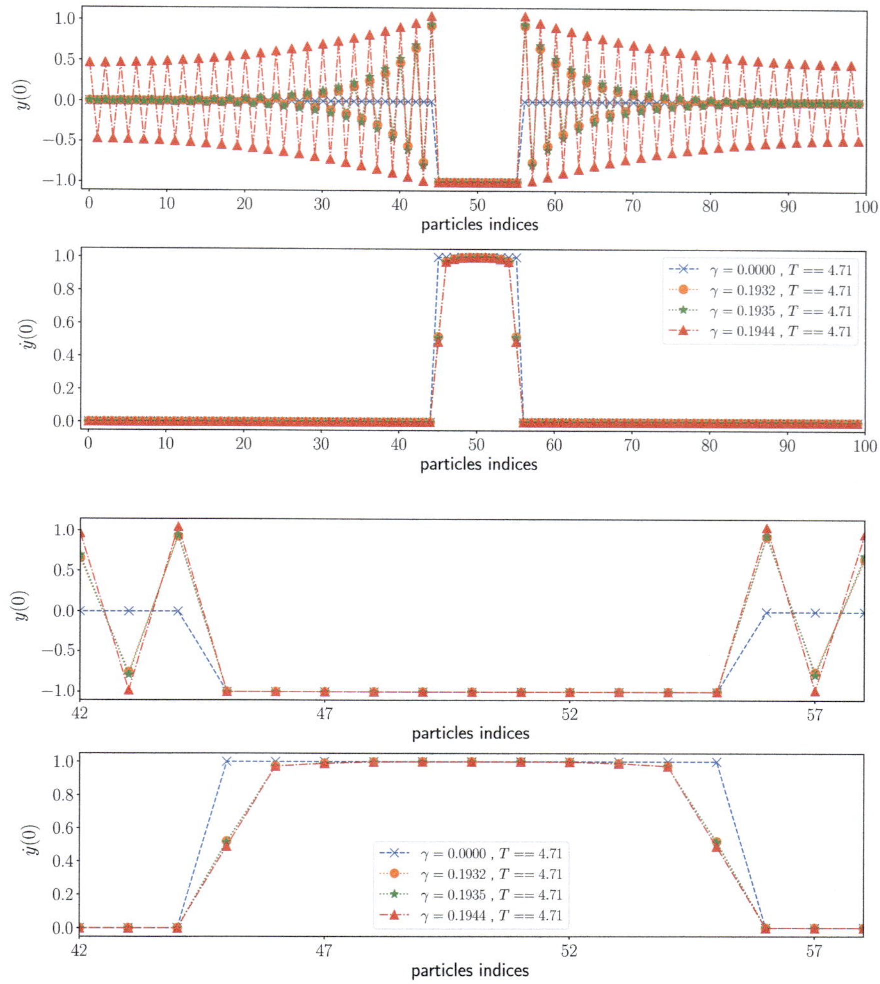

Fig. 23 Multi-site breather with pattern $I_1 = \emptyset$, $I_2 = \{45, \ldots, 55\}$

Multi-site Breather Localized on 10 Particles

In Fig. 23, a multi-site breather with pattern $I_1 = \emptyset$, $I_2 = \{45, \ldots, 55\}$ is displayed for $T = \frac{3\pi}{2}$. For $\gamma \to \gamma_c$, the computation of the solutions is more difficult. The largest value of γ for which a solution is displayed is $0.1944096 < \gamma_c$. We can observe that the particles in I_0 are still not grazing.

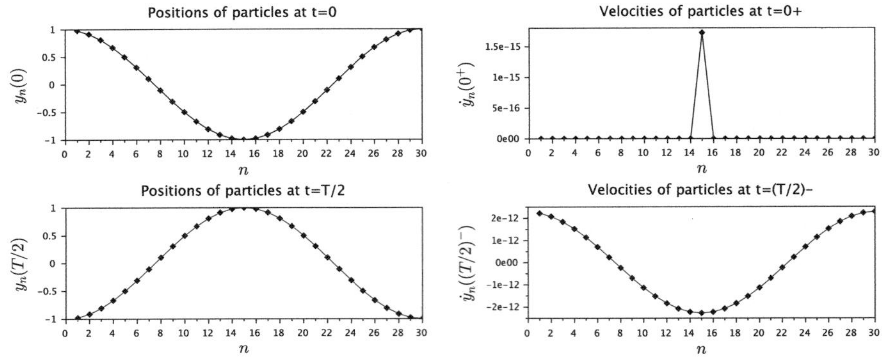

(a) Positions for $T = T_1$ (b) Velocities for $T = T_1$

Fig. 24 Main linear grazing mode for $\gamma = 1$ and $T_1 = 2\pi\,(1 + 4\gamma \sin^2(\pi/30))^{-1/2}$

3.4 Long-Wavelength Modes

We also compute spatially extended long-wavelength modes close to the main linear mode with wavenumber $q = 2\pi/p$, that is depicted in Fig. 24. The period of the linear mode for a given wavenumber q is

$$T_1 = 2\pi\,(1 + 4\gamma \sin^2(q/2))^{-1/2}. \tag{48}$$

Our computations are performed for $\gamma = 1$ and $p = 30$ particles and we get $T_1 \approx 6.150$.

A First Branch of Solutions

We are able to follow a first continuous branch of solutions depicted in Fig. 25 with periods $T \in [\alpha_7\, T_1, \alpha_1\, T_1]$, and $\alpha_1 = 0.99056$ and $\alpha_7 = 0.5035988$. The mode amplitude diverges when $T \to \alpha_7\, T_1^+$, and two particles at $n = 15, 30$ (the antinodes, i.e., the particles that reach maximal height) undergo grazing impacts when $T \to \alpha_1\, T_1^-$. The number of impacting particles decreases from 30 to 10 when T is increased. More precisely, for T in intervals of the form $[\alpha_j\, T_1, \alpha_{j-1}\, T_1]$, we find $4j + 2$ impacting particles with pattern $I_1 = \{\, 1, 2, \ldots, j, p - j, p - j + 1, \ldots, p \,\}$, $I_2 = \{\, 15 - j, \ldots, 15 + j \,\}$. We find $\alpha_6 \approx 0.5798$, $\alpha_5 \approx 0.7641$, $\alpha_4 \approx 0.92$, $\alpha_3 \approx 0.9618$, $\alpha_2 \approx 0.9771$.

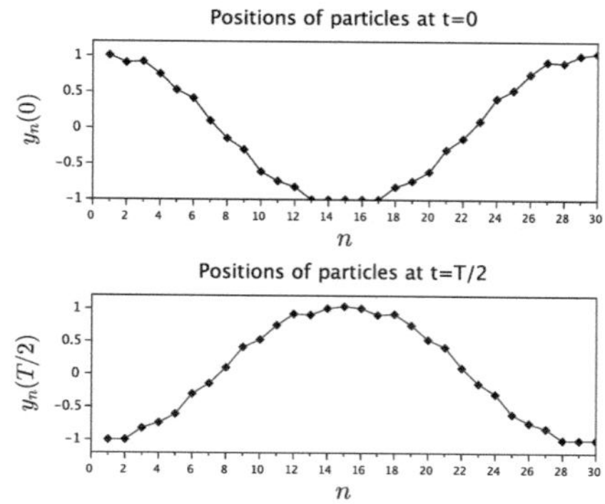

(a) Positions for $T = \alpha_1 T_1$

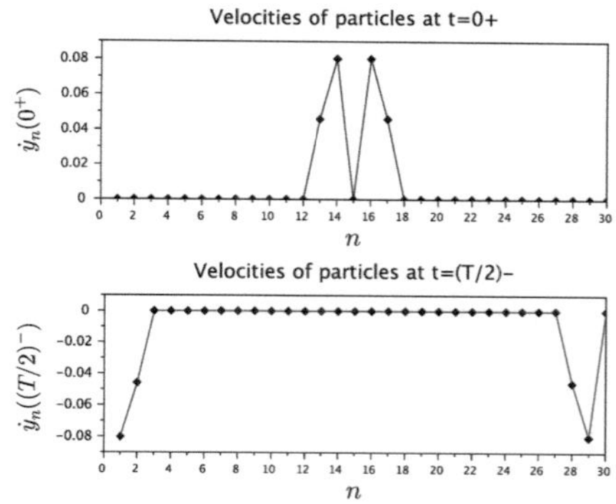

(b) Velocities for $T = \alpha_1 T_1$

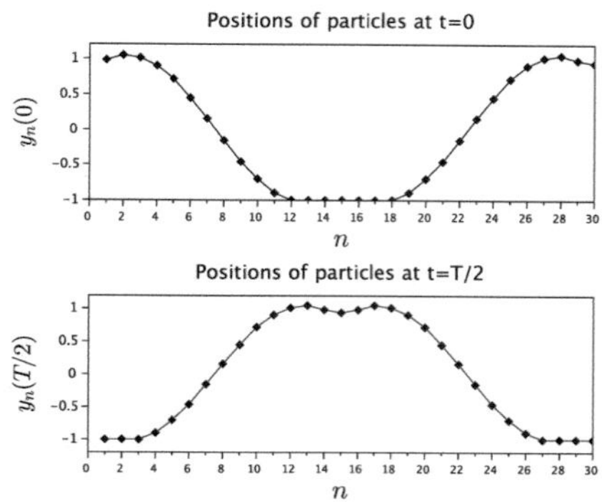

(c) Positions for $T = \alpha_2 T_1$

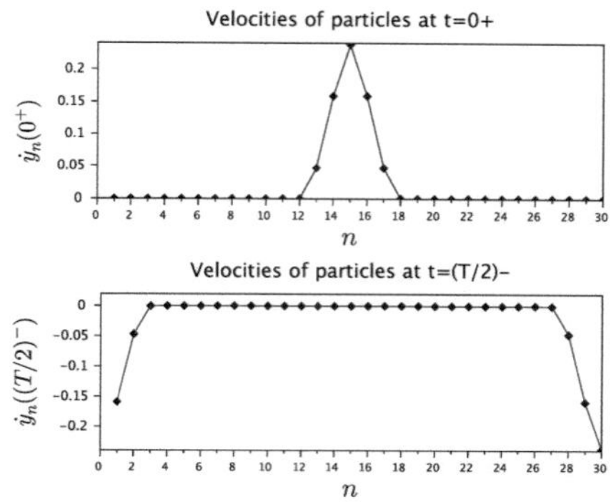

(d) Velocities for $T = \alpha_2 T_1$

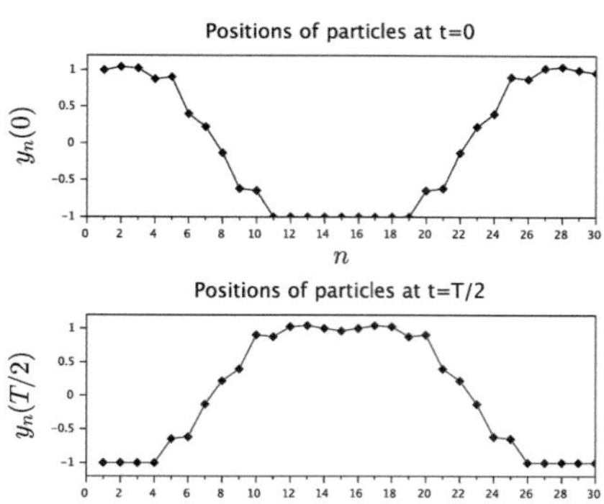

(e) Positions for $T = \alpha_3 T_1$

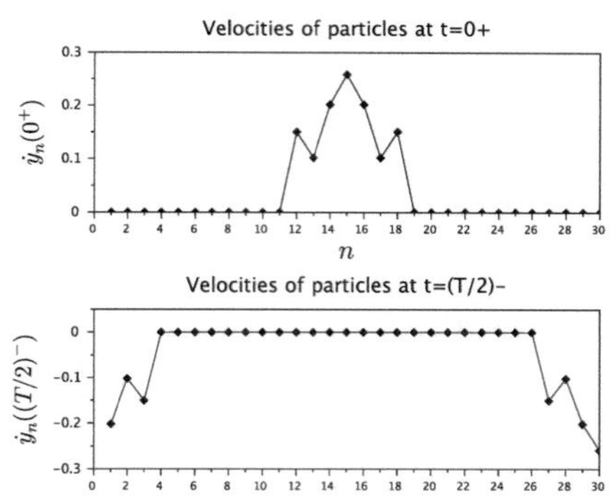

(f) Velocities for $T = \alpha_3 T_1$

Fig. 25 A first branch of long-wavelength normal modes for $\gamma = 1$ and $T_1 = 2\pi (1 + 4\gamma \sin^2(\pi/30))^{-1/2}$

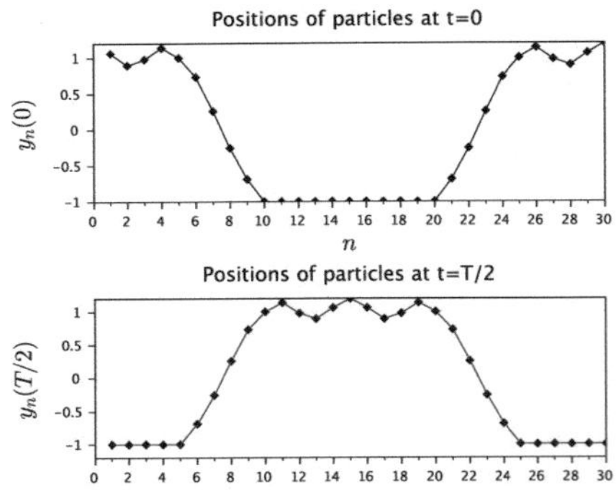

(g) Positions for $T = \alpha_4 T_1$

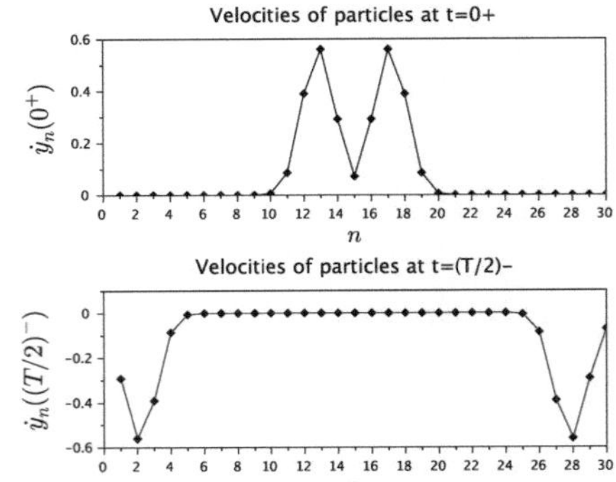

(h) Velocities for $T = \alpha_4 T_1$

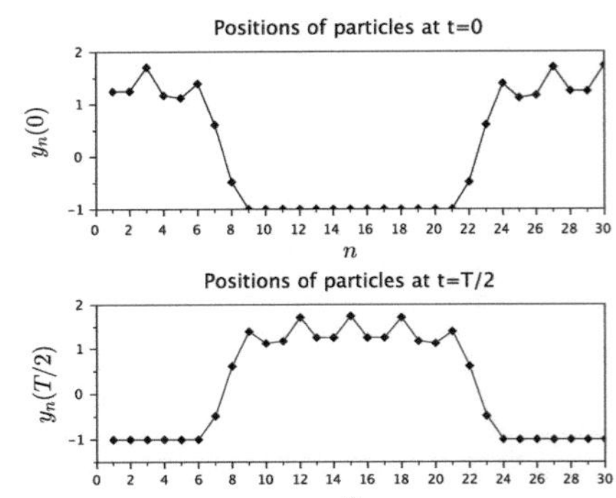

(i) Positions for $T = \alpha_5 T_1$

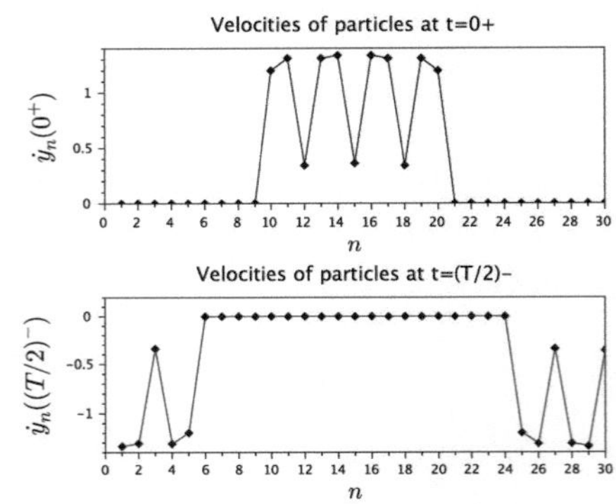

(j) Velocities for $T = \alpha_5 T_1$

Fig. 25 (continued)

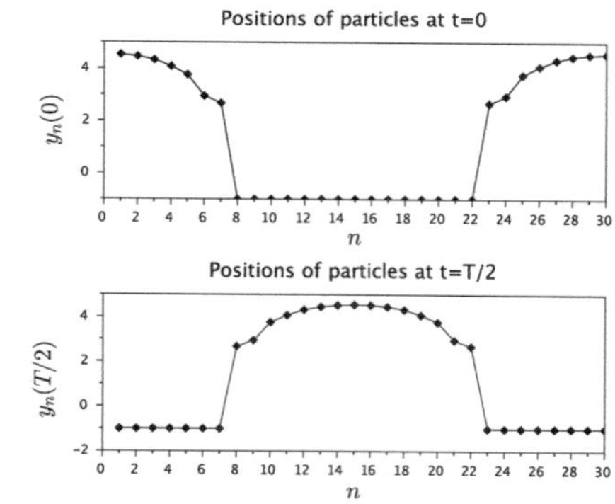

(k) Positions for $T = \alpha_6 T_1$

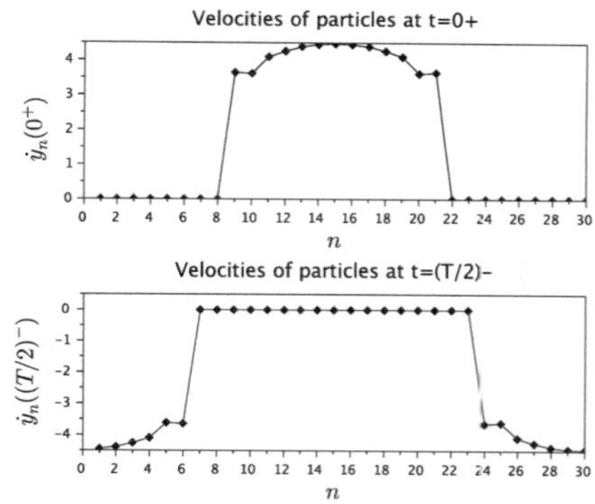

(l) Velocities for $T = \alpha_6 T_1$

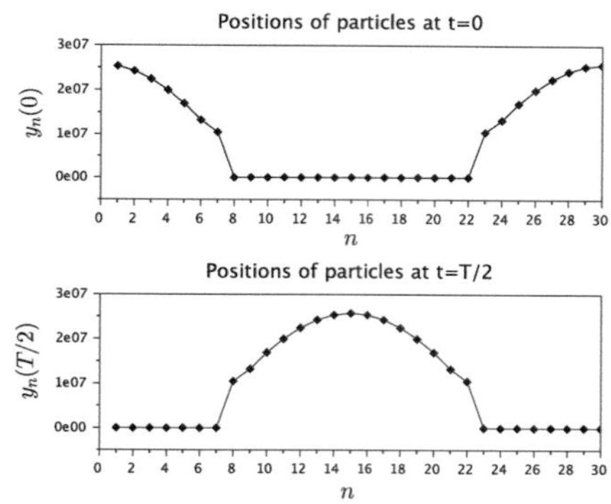

(m) Positions for $T = \alpha_7 T_1$

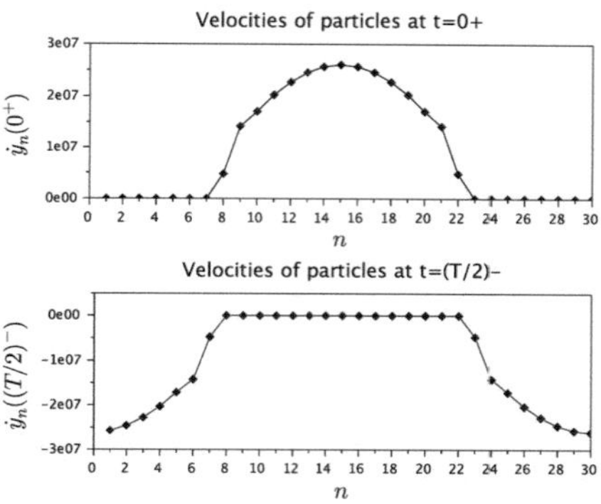

(n) Velocities for $T = \alpha_7 T_1$

Fig. 25 (continued)

A Second Branch of Solutions

We find another branch of solutions whose period $T \in [0.81 \cdot T_1, T_1)$ can approach T_1 arbitrary closely. These solutions emerge from the linear grazing mode when $T \to T_1$. Let us set $T = \alpha\, T_1$ and describe the mode pattern depending on α. We only describe I_2, given that $I_1 = I_2 + 15 (\mathrm{mod}\, 30)$. We have $I_2 = \{15\}$ for $\alpha \in [0.991, 1)$, $I_2 = \{14, 15, 16\}$ for $\alpha \in [0.9825921, 0.99]$, $I_2 = \{12, 14, 15, 16, 18\}$ for $\alpha \in [0.965, 0.9825924]$, $I_2 = \{11, 12, 14, 15, 16, 18, 19\}$ for $\alpha \in [0.85, 0.964]$, $I_2 = \{9, 11, 12, 14, 15, 16, 18, 19, 21\}$ for $\alpha \in [0.836, 0.849]$, and for $\alpha \in [0.81, 0.835]$, we find $I_2 = \{9, 11, 12, 13, 14, 15, 16, 17, 18, 19, 21\}$. Mode profiles are shown in Fig. 26.

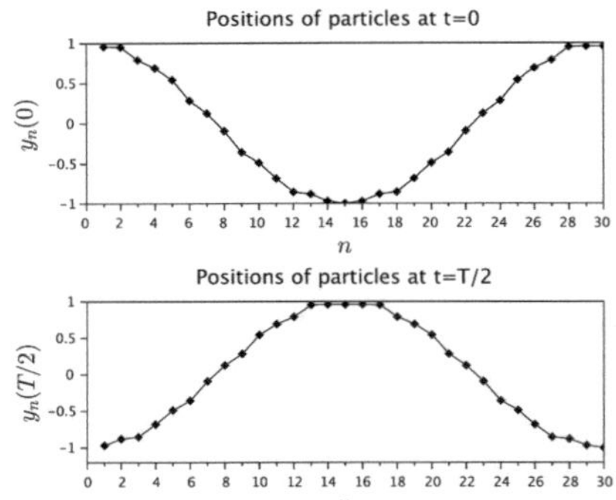

(a) Positions for $T = 0.997T_1$

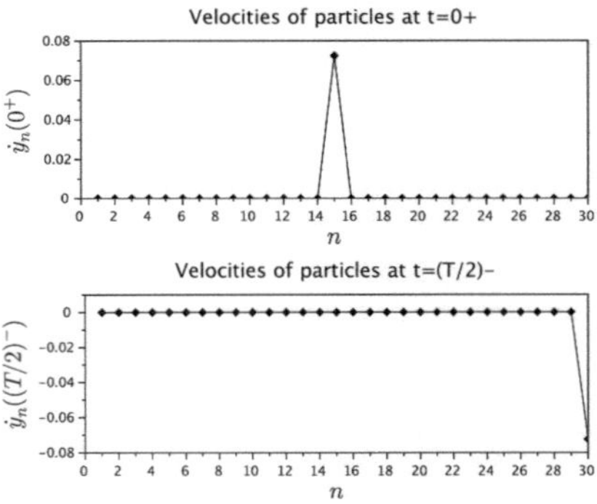

(b) Velocities for $T = 0.997T_1$

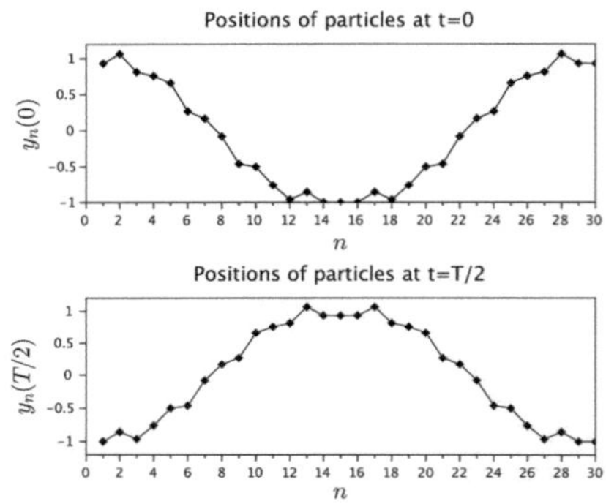

(c) Positions for $T = 0.99T_1$

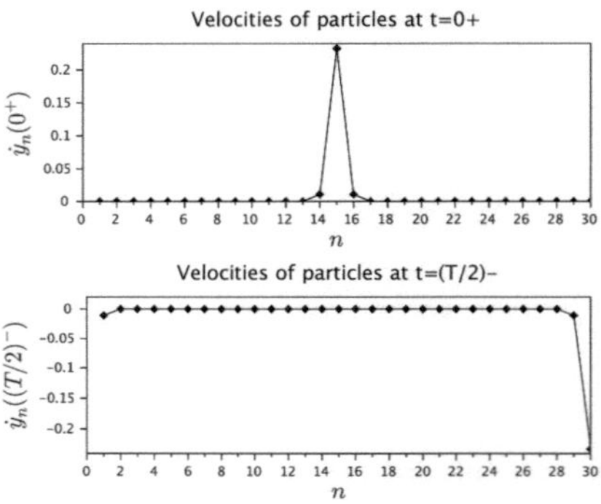

(d) Velocities for $T = 0.99T_1$

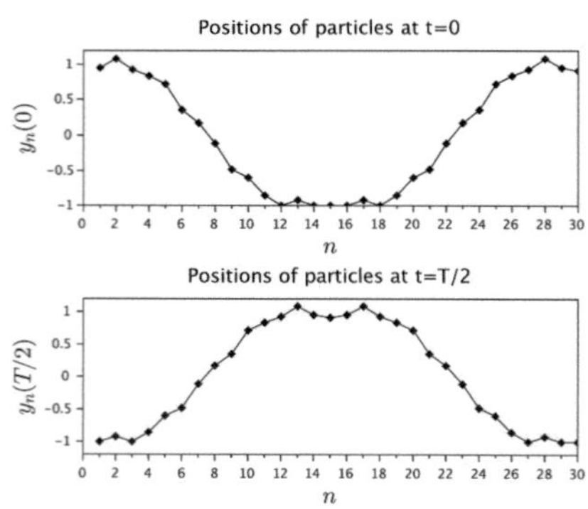

(e) Positions for $T = 0.9825924T_1$

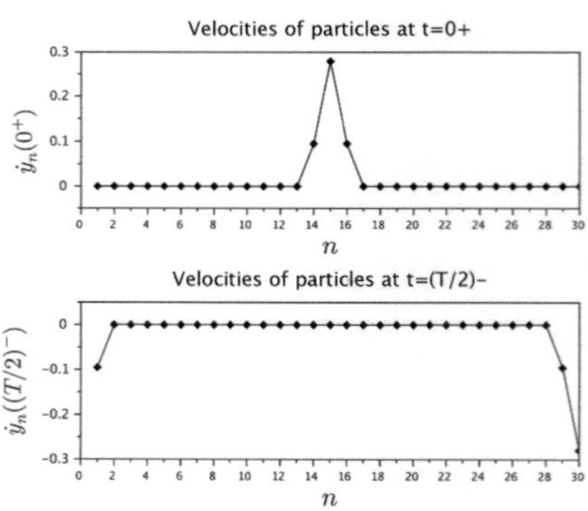

(f) Velocities for $T = 0.9825924T_1$

Fig. 26 A second branch of long-wavelength normal modes for $\gamma = 1$ and $T_1 = 2\pi\,(1 + 4\gamma \sin^2(\pi/30))^{-1/2}$

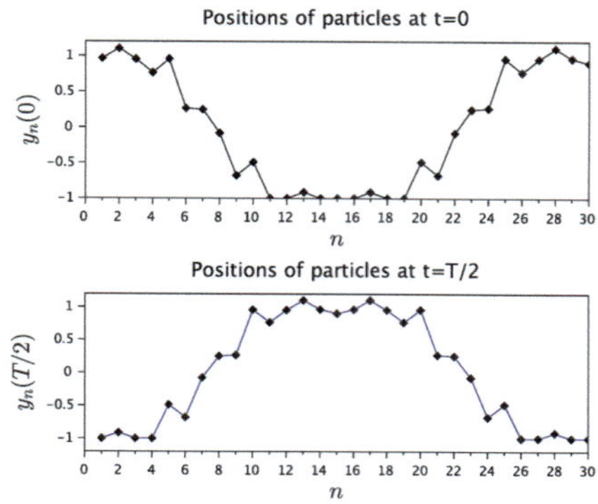

(g) Positions for $T = 0.964T_1$

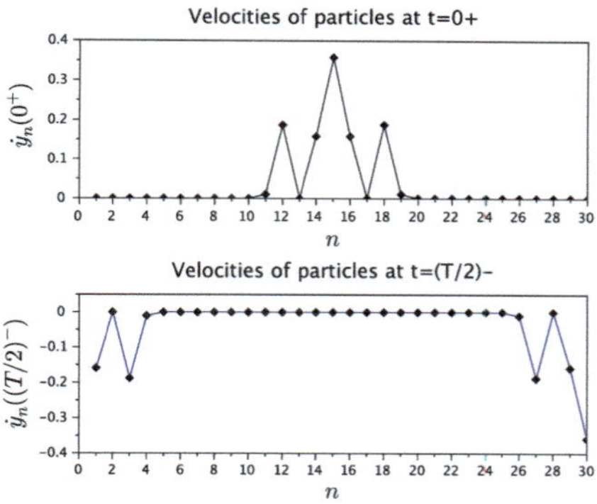

(h) Velocities for $T = 0.964T_1$

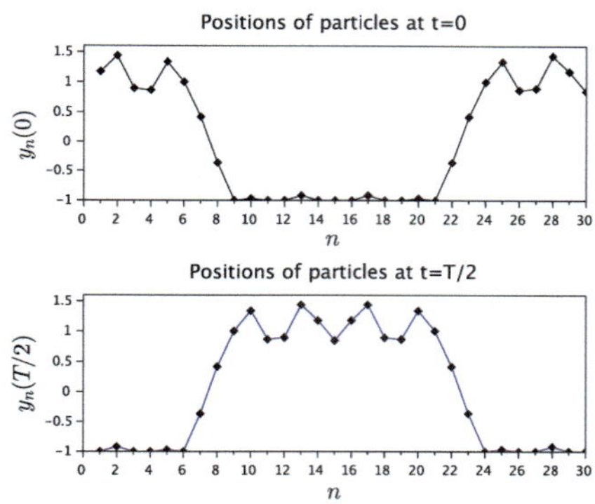

(i) Positions for $T = 0.849T_1$

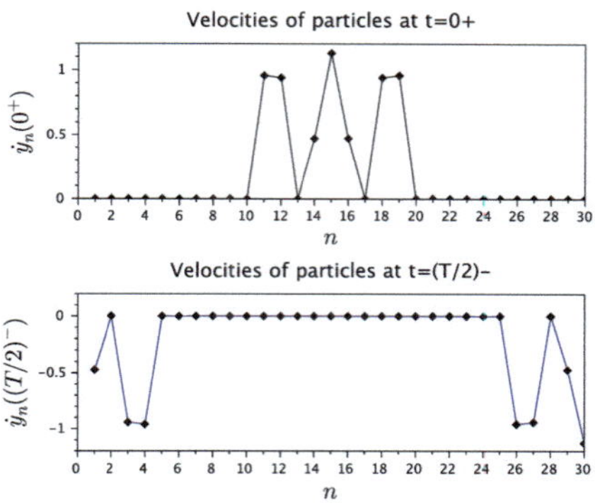

(j) Velocities for $T = 0.849T_1$

Fig. 26 (continued)

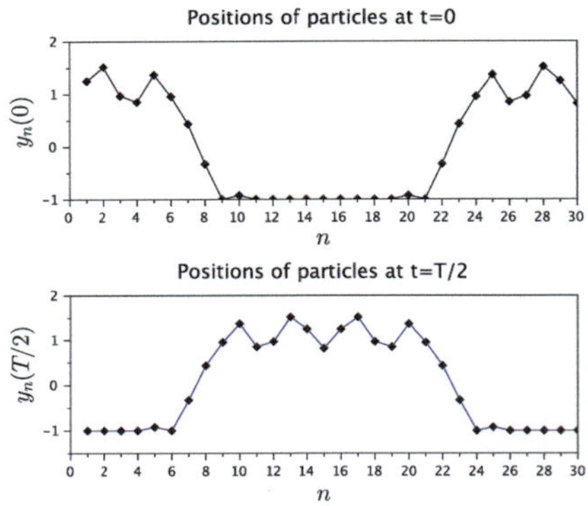

(k) Positions for $T = 0.835T_1$

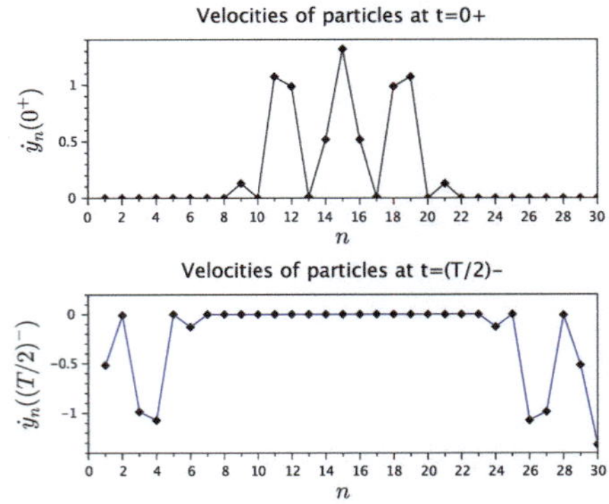

(l) Velocities for $T = 0.835T_1$

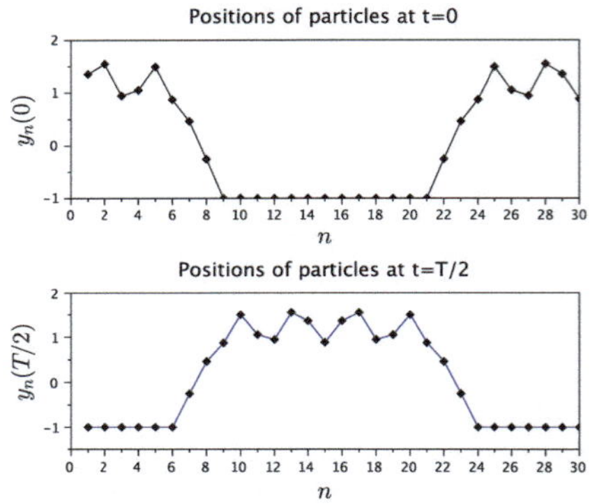

(m) Positions for $T = 0.81T_1$

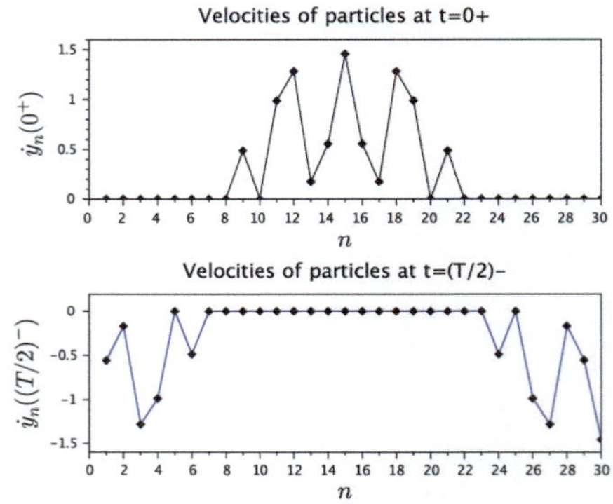

(n) Velocities for $T = 0.81T_1$

Fig. 26 (continued)

4 Discussion

In this work, we have studied the existence and stability of nonsmooth modes (either spatially localized or extended) in a chain of coupled impact oscillators, for rigid impacts without energy dissipation. We have obtained analytical solutions with an arbitrary number of impacting particles at small coupling, and have computed such solutions numerically for larger coupling constants. Different solution branches corresponding to stable or unstable breathers, multibreathers and nonsmooth normal modes have been found.

The computation of periodic solutions based on the above approach is much more effective than numerical continuation of periodic solutions based on stiff compliant models. In the latter case, impacts are described by smooth nonlinear Hertzian type potentials leading to stiff ODE and costly numerical continuation.

Several extensions of this work could be considered. It would be interesting to perform the continuation of periodic solutions while allowing switches in the mode patterns. In addition, the study of more complex types of nonsmooth mode would be of great interest. In particular, one could allow particles to realize several impacts per period [40] or display sticking phases after a grazing contact [27]. The inclusion of dissipative impacts and forcing and the application of the method to more complex finite-element models of continuous impacting systems constitute additional challenging directions.

Acknowledgements The authors are grateful to Oleg Gendelman and Itay Grinberg for all of the stimulating discussions.

References

1. Acary V, Brogliato B (2008) In: Numerical methods for nonsmooth dynamical systems. Applications in mechanics and electronics. Lecture notes in applied and computational mechanics, vol 35. Springer, Berlin
2. Ahn J, Stewart DE (2006) Existence of solutions for a class of impact problems without viscosity. SIAM J Math Anal 38:37–63
3. Astashev VK, Krupenin VL (2001) Experimental investigation of vibrations of strings interacting with point obstacles. Doklady Phys 46:522–525
4. Astashev VK, Krupenin VL (2007) Longitudinal vibrations of a thin rod interacting with an immobile limiter. J Mach Manuf Reliab 36:535–541
5. Babitsky VI (1998) Theory of vibro-impact systems and applications. Foundations of engineering mechanics. Springer, Berlin
6. Babitsky VI, Krupenin VL (2001) Vibration of strongly nonlinear discontinuous systems. Foundations of engineering mechanics. Springer
7. Ballard P (2000) The dynamics of discrete mechanical systems with perfect unilateral constraints. Arch Ration Mech Anal 154:199–274
8. Ballard P (2001) Formulation and well-posedness of the dynamics of rigid-body systems with perfect unilateral constraints. Philos Trans R Soc Lond A 359:2327–2346
9. Cabannes H (1980) Mouvements périodiques d'une corde vibrante en présence d'un obstacle rectiligne. J de Mathématique et de Phys appliquées (ZAMP) 31:473–482
10. Cabannes H (1983) Mouvement d'une corde vibrante en présence d'un obstacle rectiligne fixe. Comptes rendus de l'Académie des sciences Série IIb Mécanique 296:1367–1371
11. Cabannes H (1984) Periodic motions of a string vibrating against a fixed point-mass obstacle: II. Math Methods Appl Sci 6:55–67
12. Cabannes H (1984) Cordes vibrantes avec obstacles. Acta Acust United Acust 55:14–20
13. Cabannes H (1997) Presentation of software for movies of vibrating strings with obstacles. Appl Math Lett 10:79–84
14. Cabannes H, Haraux A (1981) Mouvements presque-périodiques d'une corde vibrante en présence d'un obstacle fixe, rectiligne ou ponctuel. Int J Non-Linear Mech 16:449–458
15. di Bernardo M, Budd CJ, Champneys AR, Kowalczyk P (2008) Piecewise-smooth dynamical systems: theory and applications. Applied mathematical sciences. Springer, Berlin
16. Flach S, Gorbach A (2008) Discrete breathers: advances in theory and applications. Phys Rep 467:1–116
17. Gendelman OV (2013) Exact solutions for discrete breathers in a forced-damped chain. Phys Rev E 87:062911
18. Gendelman OV, Manevitch LI (2008) Discrete breathers in vibroimpact chains: analytic solutions. Phys Rev E 78:026609

19. Grinberg I, Gendelman OV (2016) Localization in finite vibroimpact chains: discrete breathers and multibreathers. Phys Rev E 94:032204
20. Haraux A, Cabannes H (1983) Almost periodic motion of a string vibrating against a straight fixed obstacle. Nonlinear Anal Theory Methods Appl 7:129–141
21. Hindmarsh AC (1983) ODEPACK, a systematized collection of ODE solvers. IMACS Trans Sci Comput 1:55–74
22. http://siconos.gforge.inria.fr/
23. Ibrahim RA (2009) In: Vibro-impact dynamics: modeling, mapping and applications. Lecture notes in applied and computational mechanics, vol 43, Springer, Berlin
24. Krupenin VL (2008) On the analysis of models of strongly nonlinear vibroimpact processes in equipped lattice systems. J Mach Manuf Reliab 37:552–557
25. Krupenin VL (2012) Simulation of vibroshock in two-dimensional systems with inherited properties. J Mach Manuf Reliab 41:361–368
26. Krupenin VL (2017) Vibration of string placed between extended and point limiters. J Mach Manuf Reliab 46:96–104
27. Le Thi H (2017) On some periodic solutions of discrete vibro-impact systems with a unilateral contact condition, PhD thesis, Université Côte d'Azur
28. Lebellego M (2011) Phénomènes ondulatoires dans un modèle discret de faille sismique, PhD thesis, Toulouse University
29. Legrand M, Junca S, Heng S (2017) Nonsmooth modal analysis of a N-degree-of-freedom system undergoing a purely elastic impact law. Commun Nonlinear Sci Numer Simul 45:190–219
30. MacKay RS, Aubry S (1994) Proof of existence of breathers for time-reversible or Hamiltonian networks of weakly coupled oscillators. Nonlinearity 7:1623–1643
31. Manevitch LI, Azeez MAF, Vakakis AF (1999) Exact solutions for a discrete systems undergoing free vibro-impact oscillations. In: Babitsky VI (ed) Dynamics of vibro-impact systems. Springer, Berlin pp 151–158
32. Nqi FZ, Schatzman M (2010) Computation of Lyapunov exponents for dynamical system with impact. Appl Math Sci 4:237–252
33. Perchikov N, Gendelman OV (2017) Dynamics of a discrete breather in a harmonically excited chain with vibro-impact on-site potential, Physica D 292–293: 8–28 (2015). Corrigendum: Physica D 340:26
34. Pilipchuk VN (2001) Impact modes in discrete vibrating systems with rigid barriers. Int J Non-Linear Mech 36:999–1012
35. Schatzman M (1978) A class of nonlinear differential equations of second order in time. Nonlinear Anal 2:355–373
36. Schatzman M (1980) A hyperbolic problem of second order with unilateral constraints: the vibrating string with a concave obstacle. J Math Anal Appl 73:138–191
37. Schatzman M, Bercovier M (1989) Numerical approximation of a wave equation with unilateral constraints. Math Comput 53:55–79
38. Sepulchre J-A, MacKay RS (1997) Localized oscillations in conservative or dissipative networks of weakly coupled autonomous oscillators. Nonlinearity 10:679–713
39. Shiroky IB, Gendelman OV (2016) Discrete breathers in an array of self-excited oscillators: exact solutions and stability. Chaos 26:103112
40. Thorin A, Delezoide P, Legrand M (2017) Nonsmooth modal analysis of piecewise-linear impact oscillators. SIAM J Appl Dyn Syst 16:1710–1747
41. Vakakis AF, Manevitch LI, Mikhlin YV, Pilipchuk VN, Zevin AA (1996). Normal modes and localization in nonlinear systems. Wiley series in nonlinear science
42. Vedenova EG, Manevich LI, Pilipchuk VN (1985) Normal oscillations of a string with concentrated masses on nonlinearly elastic supports. Prikl Mat Mekh 49:203–211
43. Whittlesey EF (1965) Analytic function in Banach spaces. Proc Am Math Soc 16:1077–1083

Mathematical Aspects of Vibro-Impact Problems

Laetitia Paoli

Abstract We consider in this chapter the dynamics of rigid multibody systems subjected to frictionless unilateral constraints. Starting from the mechanical description of the problem, we derive its formulation as a second-order Measure Differential Inclusion and we introduce the corresponding mathematical framework, namely functions of Bounded Variation and Stieltjes measures. Then, the main difficulties in the study of vibro-impact problems are described and an overview of the state of the art about existence results and relevant numerical methods (penalty approach, time-stepping schemes at the position or velocity level) is proposed. Throughout this chapter, the bouncing ball model problem is considered to highlight the key points of the mathematical analysis without too many technicalities.

1 Introduction

The dynamics of multibody systems subjected to perfect non-penetration conditions generates impact and vibrations, leading to unwanted noise and untimely wear of structures. Due to its wide range of applications – granular matter, robotics, aerospace, car engines – it is crucial to study these vibro-impact phenomena both from the qualitative and quantitative points of view.

At a first glance, the mathematical background in the framework of discrete mechanical systems seems quite elementary, since we may expect to deal with second-order Ordinary Differential Equations. Unfortunately, as soon as unilateral constraints are applied, impacts and velocity jumps may occur and the acceleration may contain Dirac masses. Hence, we have to consider Measure Differential Inclusions instead of Ordinary Differential Equations, and classical good properties, like

L. Paoli (✉)
University Lyon, UJM-Saint-Etienne, CNRS UMR 5208, Institut Camille Jordan, 23 rue Michelon, 42023 Saint-Etienne Cedex 2, France
e-mail: laetitia.paoli@univ-st-etienne.fr

R. I. Leine et al. (eds.), *Advanced Topics in Nonsmooth Dynamics*,
https://doi.org/10.1007/978-3-319-75972-2_4

uniqueness or continuity of data, are not always satisfied. As a consequence, reliable simulations of such systems cannot be performed easily and the choice of a numerical algorithm has to be substantiated by a convergence result.

So, the aim of this chapter is to highlight these difficulties and to describe the main tools that may be used to overcome them. In order to give the flavour of the proofs without too many technicalities, a model problem will be considered throughout the sections and some classical properties of functions of Bounded Variation and Stieltjes measures are recalled in Sect. 7, Appendix.

The chapter is organized as follows. In Sect. 2, starting from the mechanical description of the problem, its formulation as a second-order Measure Differential Inclusion is derived. Then, in Sect. 3, a list of bad mathematical properties is presented. Section 4 is devoted to existence results for smooth or convex constraints, based on a penalty approach. Finally, time-stepping schemes formulated at the position or velocity level are introduced in Sects. 5 and 6, leading to existence results also for non-smooth and/or non-convex constraints as well.

2 Description of the Problem and Mathematical Framework

We consider a discrete mechanical system with d degrees-of- freedom. We denote by $q \in \mathbf{R}^d$ its representative point in generalized coordinates. Starting from Lagrangian formalism, the dynamics is given by a second-order Ordinary Differential Equation (ODE)

$$\frac{d}{dt}\left(\frac{\partial \mathscr{L}}{\partial \dot{q}}\right) = \frac{\partial \mathscr{L}}{\partial q} + F(t, q, \dot{q}), \quad \mathscr{L} = \frac{1}{2}\langle \dot{q}, \mathsf{M}(q)\dot{q}\rangle - \mathscr{V}(q), \tag{1}$$

where $\langle \cdot, \cdot \rangle$ denotes the Euclidean inner product of $\mathbf{R}^d$, $\mathsf{M}(q)$ is the inertia operator of the system, $\mathscr{V}(q)$ is a smooth convex potential and $F(t, q, \dot{q})$ describes the forces acting on the system that do not derive from a potential, leading to some possible dissipation during the motion.

Let us assume now that the system is subjected to unilateral constraints characterized by geometrical inequalities

$$f_\alpha\big(q(t)\big) \geq 0, \quad \alpha \in \{1, \dots, \nu\}, \quad \nu \geq 1,$$

with smooth functions f_α such that ∇f_α does not vanish in a neighbourhood of $\{q \in \mathbf{R}^d;\ f_\alpha(q) = 0\}$. We define the set of admissible configurations as

$$K = \big\{q \in \mathbf{R}^d;\ f_\alpha(q) \geq 0 \ \ \forall \alpha \in \{1, \dots, \nu\}\big\}$$

and we assume that $\mathrm{Int}(K) \neq \emptyset$.

Whenever $q(t) \in \text{Int}(K)$, the motion is described by (1), but when $q(t) \in \partial K$, a reaction force due to the constraints is applied and we get

$$\frac{d}{dt}\left(\frac{\partial \mathscr{L}}{\partial \dot{q}}\right) = \frac{\partial \mathscr{L}}{\partial q} + F(t, q, \dot{q}) + R,$$

which can be rewritten as

$$\mathsf{M}(q)\ddot{q} = f(t, q, \dot{q}) + R, \quad \text{Supp}(R) \subset \big\{t;\ q(t) \in \partial K\big\}$$

with

$$f(t, q, \dot{q}) = -\mathscr{V}'(q) + F(t, q, \dot{q}) - \frac{1}{2}\big(\mathrm{d}\mathsf{M}(q) \cdot \dot{q}\big)\dot{q}.$$

Let us observe that, when $t \in (0, \tau)$ with $q(t) \in \partial K$, the velocity may be discontinuous. Indeed, let $\alpha \in \{1, \ldots, \nu\}$ such that $f_\alpha\big(q(t)\big) = 0$. Then,

$$\frac{f_\alpha\big(q(t \pm h)\big) - f_\alpha\big(q(t)\big)}{h} \geq 0, \quad h > 0,$$

and thus, if $\dot{q}(t \pm 0)$ is defined, we obtain

$$\big\langle \nabla f_\alpha\big(q(t)\big), \dot{q}(t + 0)\big\rangle \geq 0, \quad \big\langle \nabla f_\alpha\big(q(t)\big), \dot{q}(t - 0)\big\rangle \leq 0. \tag{2}$$

It follows that the appropriate mathematical framework for the generalized velocities is the space of *functions of Bounded Variation*. Thus, $\ddot{q}$ should be understood as the Stieltjes measure $d\dot{q}$ and the reaction force R is a measure with values in $\mathbf{R}^d$ (see Sect. 7, Appendix).

We assume, moreover, that the constraints are perfect, i.e.:

- contact is *without friction*

$$\forall v \in T_K(q) \cap \big(-T_K(q)\big) : \ \langle R, v\rangle = 0,$$

- there is *no adhesion*

$$\forall v \in T_K(q) : \ \langle R, v\rangle \geq 0,$$

where $T_K(q)$ is the set of kinematically admissible right velocities at q defined as

$$T_K(q) = \big\{v \in \mathbf{R}^d;\ \big\langle \nabla f_\alpha(q), v\big\rangle \geq 0 \ \ \forall \alpha \in J(q)\big\},$$

with $J(q) = \big\{\alpha \in \{1, \ldots, \nu\};\ f_\alpha(q) \leq 0\big\}$. We infer that $R \in -N_K(q)$ with

$$N_K(q) = \{w \in \mathbf{R}^d;\ \langle w, v\rangle \le 0\ \forall v \in T_K(q)\} \text{ if } q \in K, \quad N_K(q) = \emptyset \text{ otherwise}.$$

Let us recall Farkas-Minkowski's lemma.

Lemma 1 (Farkas-Minkowski's lemma [16]) *Let H be a Hilbert space endowed with the inner product $\langle \cdot, \cdot\rangle_H$, $r \in H$ and $a_\alpha \in H$, $\alpha \in J$, where J is a finite family of indexes. Then, the following inclusion holds:*

$$\{v \in H;\ \langle a_\alpha, v\rangle \ge 0\ \forall \alpha \in J\} \subset \{v \in H;\ \langle r, v\rangle \ge 0\}$$

if and only if there exist non-negative real numbers $(\lambda_\alpha)_{\alpha\in J}$ such that $r = \sum_{\alpha\in J} \lambda_\alpha a_\alpha$.

With $a_\alpha = \nabla f_\alpha(q)$ and $J = J(q)$, we obtain

$$R = \sum_{\alpha\in J(q)} \lambda_\alpha \nabla f_\alpha(q), \quad \lambda_\alpha \ge 0.$$

Hence, the motion of the system is described by a function $q : [0, \tau] \to \mathbf{R}^d$, with $\tau > 0$, such that

$$q(t) = q(0) + \int_0^t u(s)\, ds \in K \quad \forall t \in [0, \tau], \tag{3}$$

with $u \in BV([0, \tau]; \mathbf{R}^d)$ satisfying the following Measure Differential Inclusion (MDI),

$$\mathsf{M}(q)\mathrm{d}u - f(\cdot, q, u)\mathrm{d}t \in -N_K(q), \tag{4}$$

i.e., there exist non-negative real measures λ_α, $\alpha \in \{1, \dots, \nu\}$ such that

$$\mathsf{M}(q)\mathrm{d}u - f(\cdot, q, u)\mathrm{d}t = \sum_{\alpha=1}^{\nu} \lambda_\alpha \nabla f_\alpha(q), \tag{5}$$

with

$$\mathrm{Supp}(\lambda_\alpha) \subset \{t \in [0, \tau];\ f_\alpha(q(t)) = 0\}. \tag{6}$$

Let us observe that (3) implies that q is continuous, $\dot q(t \pm 0) = u(t \pm 0)$ for all $t \in (0, \tau)$ and $\mathrm{d}\dot q = \mathrm{d}u$ on $(0, \tau)$. Furthermore,

$$\dot q(t + 0) \in T_K(q(t)), \quad \dot q(t - 0) \in -T_K(q(t)) \quad \forall t \in (0, \tau).$$

It follows that, whenever $t \in (0, \tau)$ with $q(t) \in \partial K$ and $\dot q(t - 0) \notin T_K(q(t))$, the velocity is discontinuous at t. Furthermore, with (4),

$$\mathsf{M}\big(q(t)\big)\big(\dot{q}(t+0)-\dot{q}(t-0)\big)\in -N_K\big(q(t)\big), \tag{7}$$

and we should expect that the kinetic energy does not increase at t (mechanical consistency). But these properties do not define $\dot{q}(t+0)$ uniquely, and we need to introduce an *impact law*.

Let us consider first the simple case of a single active constraint at t, i.e., $J\big(q(t)\big) = \{\alpha\}$. Then, we may decompose the right and left velocities at t into normal and tangent parts as follows:

$$\dot{q}(t\pm 0)=\lambda^{\pm}\mathsf{M}^{-1}\big(q(t)\big)\nabla f_\alpha\big(q(t)\big)+\dot{q}_T(t\pm 0),$$

with $\lambda^{\pm}\in\mathbf{R}$ and $\big\langle\nabla f_\alpha\big(q(t)\big),\dot{q}_T(t\pm 0)\big\rangle=0$. Thus,

$$\lambda^{\pm}=\frac{\big\langle\nabla f_\alpha\big(q(t)\big),\dot{q}(t\pm 0)\big\rangle}{\big\langle\nabla f_\alpha\big(q(t)\big),\mathsf{M}^{-1}\big(q(t)\big)\nabla f_\alpha\big(q(t)\big)\big\rangle}\in\mathbf{R}^{\pm}.$$

Moreover, since $N_K\big(q(t)\big)=\mathbf{R}^{-}\nabla f_\alpha\big(q(t)\big)$, we infer from (7) that there exists $\mu\in\mathbf{R}$ such that

$$\mathsf{M}\big(q(t)\big)\big(\dot{q}_T(t+0)-\dot{q}_T(t-0)\big)=\mu\nabla f_\alpha\big(q(t)\big).$$

But

$$\big\langle\nabla f_\alpha\big(q(t)\big),\dot{q}_T(t+0)-\dot{q}_T(t-0)\big\rangle=0=\mu\underbrace{\big\langle\nabla f_\alpha\big(q(t)\big),\mathsf{M}^{-1}\big(q(t)\big)\nabla f_\alpha\big(q(t)\big)\big\rangle}_{>0}.$$

Thus, $\mu=0$ and $\dot{q}_T(t-0)=\dot{q}_T(t+0)$, which means that the tangent part is continuous.

We define the kinetic metric at $q(t)$ as

$$\langle v,w\rangle_{q(t)}=\big\langle v,\mathsf{M}\big(q(t)\big)w\big\rangle\quad\forall(v,w)\in\mathbf{R}^d\times\mathbf{R}^d$$

and the corresponding norm

$$\|v\|_{q(t)}=\big\langle v,\mathsf{M}\big(q(t)\big)v\big\rangle^{1/2}\quad\forall v\in\mathbf{R}^d.$$

The kinetic energy is given by

$$\begin{aligned}\mathscr{E}(t\pm 0)&=\frac{1}{2}\big\|\dot{q}(t)\big\|^2_{q(t)}=\frac{1}{2}\big\langle\dot{q}(t\pm 0),\mathsf{M}\big(q(t)\big)\dot{q}(t\pm 0)\big\rangle\\&=\frac{1}{2}(\lambda^{\pm})^2\big\langle\nabla f_\alpha\big(q(t)\big),\mathsf{M}^{-1}\big(q(t)\big)\nabla f_\alpha\big(q(t)\big)\big\rangle\\&\quad+\frac{1}{2}\big\langle\dot{q}_T(t\pm 0),\mathsf{M}\big(q(t)\big)\dot{q}_T(t\pm 0)\big\rangle.\end{aligned}$$

Hence, we have $\mathscr{E}(t+0) \le \mathscr{E}(t-0)$ (*mechanical consistency*) if and only if $|\lambda^+| \le |\lambda^-|$. Owing to the fact that $\lambda^+ \ge 0$ and $\lambda^- \le 0$, we obtain that there exists $e \in [0,1]$ such that $\lambda^+ = -e\lambda^-$ and

$$\dot{q}(t+0) = \dot{q}(t-0) - (1+e)\frac{\langle \nabla f_\alpha(q(t)), \dot{q}(t-0)\rangle}{\langle \nabla f_\alpha(q(t)), \mathsf{M}^{-1}(q(t))\nabla f_\alpha(q(t))\rangle}\mathsf{M}^{-1}(q(t))\nabla f_\alpha(q(t)),$$

i.e., we get a *family of impact laws* characterized by the choice of a parameter $e \in [0,1]$.

We observe that the projection of $\dot{q}(t-0)$ on the cone $\mathsf{M}^{-1}(q(t))N_K(q(t)) = \mathbf{R}^-\mathsf{M}^{-1}(q(t))\nabla f_\alpha(q(t))$ relatively to the kinetic metric at $q(t)$ is given by

$$\begin{aligned}&\mathrm{Proj}_{q(t)}\big(\mathsf{M}^{-1}(q(t))N_K(q(t)), \dot{q}(t-0)\big)\\ &= \frac{\langle \nabla f_\alpha(q(t)), \dot{q}(t-0)\rangle}{\langle \nabla f_\alpha(q(t)), \mathsf{M}^{-1}(q(t))\nabla f_\alpha(q(t))\rangle}\mathsf{M}^{-1}(q(t))\nabla f_\alpha(q(t)).\end{aligned}$$

Indeed, let

$$y = \frac{\langle \nabla f_\alpha(q(t)), \dot{q}(t-0)\rangle}{\langle \nabla f_\alpha(q(t)), \mathsf{M}^{-1}(q(t))\nabla f_\alpha(q(t))\rangle}\mathsf{M}^{-1}(q(t))\nabla f_\alpha(q(t)).$$

Since $\dot{q}(t-0) \in -T_K(q(t))$, we immediately have $y \in \mathbf{R}^-\mathsf{M}^{-1}(q(t))\nabla f_\alpha(q(t))$ and

$$\begin{aligned}&\langle \dot{q}(t-0) - y, v - y\rangle_{q(t)} = \langle \dot{q}_T(t-0), \underbrace{\mathsf{M}(q(t))(v-y)}_{\in \mathbf{R}\nabla f_\alpha(q(t))}\rangle \le 0\\ &\forall v \in \mathsf{M}^{-1}(q(t))N_K(q(t)).\end{aligned}$$

Then, we may apply

Lemma 2 (Lemma of the two cones [36]) *Let H be a real Hilbert space endowed with an inner product $\langle \cdot, \cdot\rangle_H$. Let P and Q be two mutually polar cones i.e.*

$$Q = \{x \in H;\ \langle x, y\rangle_H \le 0\ \forall y \in P\},\quad P = \{y \in H;\ \langle y, x\rangle_H \le 0\ \forall x \in Q\}.$$

Then, for all $(x,y,z) \in H^3$, the following properties are equivalent:
(i) $z = x + y,\ x \in P,\ y \in Q,\ \langle x, y\rangle_H = 0$,
(ii) $x = \mathrm{Proj}_H(P, z),\ y = \mathrm{Proj}_H(Q, z)$.

With $H = \mathbf{R}^d$ endowed with the kinetic metric at $q(t)$, $P = T_K(q(t))$ and $Q = \mathsf{M}^{-1}(q(t))N_K(q(t))$, we infer that

$$\dot{q}(t-0) = \underbrace{\mathrm{Proj}_{q(t)}\big(T_K(q(t)), \dot{q}(t-0)\big)}_{\dot{q}_T(t-0)} + \mathrm{Proj}_{q(t)}\big(\mathsf{M}^{-1}(q(t))N_K(q(t)), \dot{q}(t-0)\big)$$

and

$$\begin{aligned}\dot{q}(t+0) = \operatorname{Proj}_{q(t)}\big(T_K\big(q(t)\big), \dot{q}(t-0)\big) \\ -e\operatorname{Proj}_{q(t)}\big(\mathsf{M}^{-1}\big(q(t)\big)N_K\big(q(t)\big), \dot{q}(t-0)\big).\end{aligned} \tag{8}$$

(*Newton's law*).

This impact law may also be considered when several constraints are active at instant t (i.e., when $\operatorname{Card}\big(J\big(q(t)\big)\big) \geq 2$). In the general case, we can still check the mechanical consistency of this model of impact, since

$$\begin{aligned}\mathscr{E}(t+0) &= \frac{1}{2}\langle\dot{q}(t+0), \mathsf{M}\big(q(t)\big)\dot{q}(t+0)\rangle = \frac{1}{2}\left\|\dot{q}(t+0)\right\|^2_{q(t)} \\ &= \frac{1}{2}\left\|\operatorname{Proj}_{q(t)}\big(T_K\big(q(t)\big), \dot{q}(t-0)\big)\right\|^2_{q(t)} \\ &+\frac{1}{2}e^2\left\|\operatorname{Proj}_{q(t)}\big(\mathsf{M}^{-1}\big(q(t)\big)N_K\big(q(t)\big), \dot{q}(t-0)\big)\right\|^2_{q(t)} \leq \mathscr{E}(t-0),\end{aligned}$$

and equality holds if $e = 1$ (elastic shocks). Contrastingly, dissipation of energy at impacts is maximal when $e = 0$ (inelastic shocks), and then

$$\dot{q}(t+0) = \operatorname{Proj}_{q(t)}\big(T_K\big(q(t)\big), \dot{q}(t-0)\big) = \operatorname{Argmin}_{u\in T_K(q(t))}\|u - \dot{q}(t-0)\|_{q(t)}.$$

Let us emphasize that, whenever $\operatorname{Card}\big(J\big(q(t)\big)\big) = 1$, (8) is the unique impact law satisfying both the kinematic properties (2)–(7) and the mechanical consistency condition $\mathscr{E}(t+0) \leq \mathscr{E}(t-0)$. In this case, it simply means that the tangential part of the velocity is conserved while the normal part (relative to the kinetic metric) is reversed and multiplied by a restitution parameter $e \in [0, 1]$, i.e., it corresponds to a kind of optical reflexion rule at impacts.

Unfortunately, if $\operatorname{Card}\big(J\big(q(t)\big)\big) \geq 2$, (8) is no longer the unique kinematically and mechanically consistent model. As an example, let us consider the motion of a material point of mass $m = 1$ in the part of the plane $K = \mathbf{R}^- \times \mathbf{R}^+$. Without external forces, with an initial position $q_0 = (-1, 0)$ and an initial velocity $u_0 = (1, 0)$, the following two trajectories satisfy the MDI with conservation of energy:

$$\begin{aligned}q(t) &= (-1+t, 0) \quad \forall t \in [0, 1], \quad q(t) = (1-t, 0) \quad \forall t \geq 1, \\ \tilde{q}(t) &= (-1+t, 0) \quad \forall t \in [0, 1], \quad \tilde{q}(t) = (0, t-1) \quad \forall t \geq 1.\end{aligned}$$

Indeed, we may define $f_1(q) = -q_1$ and $f_2(q) = q_2$ for all $q = (q_1, q_2) \in \mathbf{R}^2$. Then, we get

$$q(t) = q(0) + \int_0^t u(s)\,ds \quad \left(\text{resp. } \tilde{q}(t) = q(0) + \int_0^t \tilde{u}(s)\,ds\right) \quad \forall t \geq 0,$$

with

$$u(s) = (1,0) \quad (\text{resp. } \tilde{u}(s) = (1,0)) \quad \text{if } s \in [0,1),$$
$$u(s) = (-1,0) \quad (\text{resp. } \tilde{u}(s) = (0,1)) \quad \text{if } s \geq 1.$$

It follows that $du = \big(u(1+0) - u(1-0)\big)\delta_{t=1}$ (respectively $d\tilde{u} = \big(\tilde{u}(1+0) - \tilde{u}(1-0)\big)\delta_{t=1}$), where $\delta_{t=1}$ is the unit Dirac mass measure at $t = 1$, and

$$J\big(q(t)\big) = \begin{cases} \{2\} & \text{if } t \in [0,1), \\ \{1,2\} & \text{if } t = 1, \\ \{2\} & \text{if } t > 1, \end{cases} \qquad \left(\text{respectively} \tilde{J}\big(\tilde{q}(t)\big) = \begin{cases} \{2\} & \text{if } t \in [0,1), \\ \{1,2\} & \text{if } t = 1, \\ \{1\} & \text{if } t > 1. \end{cases} \right)$$

Hence, the MDI is satisfied by both q and $\tilde{q}$. Moreover, for all $t \in (0,1) \cup (1,+\infty)$, we have $\dot{q}(t \pm 0) \in T_K\big(q(t)\big)$ (respectively $\dot{\tilde{q}}(t \pm 0) \in T_K\big(\tilde{q}(t)\big)$). Finally, for $t = 1$, we can check that $\dot{q}(t-0) = \dot{\tilde{q}}(t-0) = (1,0) \in N_K\big(q(t)\big)$, thus

$$\begin{aligned} &\mathrm{Proj}_{q(t)}\big(\mathsf{M}^{-1}(q(t))N_K\big(q(t)\big), \dot{q}(t-0)\big) = \dot{q}(t-0) \\ &= \mathrm{Proj}_{q(t)}\big(\mathsf{M}^{-1}(q(t))N_K\big(q(t)\big), \dot{\tilde{q}}(t-0)\big) = \dot{\tilde{q}}(t-0), \end{aligned}$$

and

$$\mathrm{Proj}_{q(t)}\big(T_K\big(q(t)\big), \dot{q}(t-0)\big) = 0_{\mathbf{R}^2} = \mathrm{Proj}_{q(t)}\big(T_K\big(q(t)\big), \dot{\tilde{q}}(t-0)\big).$$

So, only the first trajectory satisfies Newton's impact law (with $e = 1$).

In the rest of this chapter, we will focus on frictionless vibro-impact problems satisfying Newton's impact law. By gathering (3)–(5)–(6) and (8), we obtain the following mathematical formulation.

Problem (P) Let $q_0 \in K$, $u_0 \in T_K(q_0)$. Find $u : [0,\tau] \to \mathbf{R}^d$, $\tau > 0$, such that:

(P1) $u \in BV\big([0,\tau]; \mathbf{R}^d\big)$, $u(0+0) = u_0$,

(P2) $q(t) = q_0 + \int_0^t u(s)\, ds \in K$ for all $t \in [0,\tau]$,

(P3) the measure $\mathsf{M}(q)du - f(\cdot,q,u)dt$ takes its values in $-N_K(q)$, i.e., there exist non-negative real measures λ_α, $\alpha \in \{1,\ldots\nu\}$ such that

$$\mathsf{M}(q)du - f(\cdot,q,u)dt = \sum_{\alpha=1}^{\nu} \lambda_\alpha \nabla f_\alpha(q),$$

with

$$\mathrm{Supp}(\lambda_\alpha) \subset \big\{t \in [0,\tau];\ f_\alpha\big(q(t)\big) = 0\big\},$$

(P4) for all $t \in (0,\tau)$ Newton's impact law is satisfied, i.e.,

$$\dot{q}(t+0) = \mathrm{Proj}_{q(t)}\big(T_K\big(q(t)\big), \dot{q}(t-0)\big) - e\mathrm{Proj}_{q(t)}\big(M^{-1}\big(q(t)\big)N_K\big(q(t)\big), \dot{q}(t-0)\big).$$

Remark 1 It is also possible to consider *energy conservative solutions* (or *energy dissipative solutions*) by replacing property (P4) with the following *balance energy equation*:
(P'4) for almost every $t \in (0, \tau)$, we have

$$\frac{1}{2}\langle u(t), u(t)\rangle_{q(t)} = \frac{1}{2}\langle u_0, u_0\rangle_{q_0} + \int_0^t \langle f\big(s, q(s), u(s)\big), u(s)\rangle\, ds + \frac{1}{2}\int_0^t \langle u(s), \big(d\mathsf{M}\big(q(s)\big)u(s)\big)u(s)\rangle\, ds$$

(respectively the following *dissipativity property:*
(P"4) for all $t \in (0, \tau)$, we have

$$\left\|\dot q(t+0)\right\|_{q(t)} \leq \left\|\dot q(t-0)\right\|_{q(t)}$$

(see [3, 9, 11, 12, 43, 55, 60]). The former property implies that $\mathscr{E}(t+0) = \mathscr{E}(t-0)$ for all $t \in (0, \tau)$, which yields to Newton's impact law with $e = 1$ whenever $\mathrm{Card}\big(J\big(q(t)\big)\big) = 1$, while the latter property allows us to consider any mechanically consistent impact law.

3 Some Bad Mathematical Properties

From the mathematical point of view, natural questions arise: does problem (P) admit solutions? Do we have uniqueness? Are the solutions continuous with respect to data? How can we compute them exactly or approximately?

At a first glance, vibro-impact problems do not seem so difficult. Indeed, as long as the constraints are not saturated, problem (P) reduces to a second-order ODE and one may think that it is enough to solve this ODE, to detect when contact occurs by determining τ_c such that $q(\tau_c) \in \partial K$, and to use the impact law to define new initial data for the ODE at $\tau_c + 0$.

The main advantage of this strategy is its conceptual simplicity, but it relies on the assumption that the impacts are isolated, i.e., the time interval $[0, \tau]$ can be decomposed into a finite union of intervals $[\tau_i, \tau_{i+1}]$ such that $q(t) \in \mathrm{Int}(K)$ for all $t \in (\tau_i, \tau_{i+1})$. For this kind of motion, sometimes called *motions of finite sort*, any ODE solver combined with an impact detection procedure (leading to an event-driven algorithm) may be used to compute approximate solutions of problem (P).

Unfortunately, it has been known since the beginning of the 20th century that other kinds of solutions exists (see [17]), and in particular, impact instants may accumulate towards a finite limit impact instant τ_∞. As an example, let us consider a material point falling vertically on a horizontal plane (*bouncing ball example*). Then, $d = 1$, $K = \mathbf{R}^+$, $\mathsf{M}(q) \equiv 1$ and $f(t, q, v) = -g$ for all $(t, q, v) \in \mathbf{R}^3$. Assume that $e \in (0, 1)$. With $q_0 = 1$ and $u_0 = 0$, we obtain

$$q(t) = 1 - \frac{g}{2}t^2 \quad \text{if } t \in [0, \tau_1], \quad \tau_1 = \sqrt{\frac{2}{g}}.$$

Then, at $t = \tau_1$, the material point hits the obstacle, and we get

$$q(t) = v_1(t - \tau_1) - \frac{g}{2}(t - \tau_1)^2 \quad \text{if } t \in [\tau_1, \tau_2],$$

with $v_1 = e\sqrt{2g}$ and $\tau_2 = \tau_1 + \frac{2v_1}{g} = (1 + 2e)\sqrt{\frac{2}{g}}$. With an immediate induction, we obtain

$$q(t) = v_i(t - \tau_i) - \frac{g}{2}(t - \tau_i)^2, \quad \text{if } t \in [\tau_i, \tau_{i+1}],$$

with $v_i = e^i\sqrt{2g}$ and $\tau_{i+1} = \tau_i + \frac{2v_i}{g}$ for all $i \geq 1$. It follows that

$$\tau_{i+1} = \tau_i + 2e^i\sqrt{\frac{2}{g}} \longrightarrow_{i \to +\infty} \tau_\infty = \frac{1+e}{1-e}\sqrt{\frac{2}{g}}.$$

Thus, we get an infinite number of distinct impact instants within the bounded time-interval $[0, \tau_\infty)$, and afterwards, the material point remains at rest.

So, it appears that even-driven algorithms suffer a major drawback: they are well-suited only when we are able to justify that the solutions of the problem are motions of the finite sort.

The second major drawback is the possible *non-uniqueness* of solutions. Such an assertion may look strange, since the formulation of problem (P) derives from deterministic mechanical properties, but some counter-examples may be exhibited, even in very simple settings.

For instance, let us consider again the bouncing ball model problem, i.e., $d = 1$, $K = \mathbf{R}^+$ and $\mathsf{M}(q) \equiv 1$. Let us assume now that $e = 1$ and $q_0 = u_0 = 0$. Then, problem (P) reduces to

Problem (P) Find $u : [0, \tau] \to \mathbf{R}$, $\tau > 0$, such that:

(P1) $u \in BV\big([0, \tau]; \mathbf{R}\big)$, $u(0 + 0) = 0$,

(P2) $q(t) = \int_0^t u(s)\,ds \geq 0$ for all $t \in [0, \tau]$,

(P3) there exists a non-negative real measure λ such that

$$du - f(t, q, u)dt = \lambda, \quad \text{Supp}(\lambda) \subset \big\{t \in [0, \tau];\ q(t) = 0\big\},$$

(P4) for all $t \in (0, \tau)$ such that $q(t) = 0$, we have $\dot{q}(t + 0) = -\dot{q}(t - 0)$.

For any continuous non-positive function f of the time variable, the stationary motion $u \equiv 0$, $q \equiv 0$ is a solution to the problem. But we can also find a function

f of the time variable such that problem (P) admits another non-trivial solution with an infinite number of arches with grazing bounces. Such a counter-example to uniqueness was proposed first in [60], then in [4]. Let us describe it briefly.

Let f be the non-positive continuous function defined by $f(0) = 0$ and for all $n \geq 0$

$$f(t) = 0 \quad \text{if } t \in [\alpha_{n+1}, \alpha_{n+1} + \delta_n),$$
$$f(t) = -\frac{1}{2(n!)}\rho\left(\frac{t - \alpha_{n+1} - \delta_n}{\alpha_n - \alpha_{n+1} - \delta_n}\right) \quad \text{if } t \in [\alpha_{n+1} + \delta_n, \alpha_n),$$

with

$$\alpha_n = \sum_{i \geq n} \frac{(i+5)^2}{(i+1)(i+2)(i+3)(i+4)}, \qquad \delta_n = \frac{(n+5)}{(n+1)(n+2)(n+4)} \qquad \forall n \geq 0$$

and

$$\begin{cases} \rho(x) = 0 \quad \text{if} x = 0 \text{ or } x = 1, \\ \rho(x) = \dfrac{\exp\left(\frac{1}{x(x-1)}\right)}{\displaystyle\int_0^1 \exp\left(\frac{1}{t(t-1)}\right) dt} \quad \text{if} x \in (0, 1). \end{cases}$$

Next, we consider $\tau \in (\alpha_1, \alpha_0)$ and $u : [0, \tau] \to \mathbf{R}$ defined by $u(0) = 0$ and for all $n \geq 0$

$$u(t) = \frac{1}{(n+4)!} \text{ if } t \in [\alpha_{n+1}, \alpha_{n+1} + \delta_n) \cap [0, \tau],$$
$$u(t) = \frac{1}{(n+4)!} - \frac{1}{2(n!)} \int_{\alpha_{n+1}+\delta_n}^{t} \rho\left(\frac{s - \alpha_{n+1} - \delta_n}{\alpha_n - \alpha_{n+1} - \delta_n}\right) ds \text{ if } t \in [\alpha_{n+1} + \delta_n, \alpha_n) \cap [0, \tau].$$

It follows that u is a non-increasing function of class C^1 on each interval (α_{n+1}, α_n) with $\dot{u}(t) = f(t)$ for all $t \in (\alpha_{n+1}, \alpha_n)$ and

$$u(\alpha_{n+1} + 0) = u(\alpha_{n+1}) = \frac{1}{(n+4)!},$$
$$u(\alpha_n - 0) \quad = -\frac{1}{(n+3)!} = -u(\alpha_n + 0).$$

It follows that $\lim_{t \to 0^+} u(t) = 0$.

Moreover, u is a function of bounded variation on $[0, \tau]$. Indeed, let $S : 0 = t_0 < \cdots < t_p = \tau$ be a subdivision of $[0, \tau]$. Possibly adding new points, we may assume without loss of generality that $\{\alpha_n;\ n \geq 0\} \subset \{t_i;\ 0 \leq i \leq p\}$. Then, if, for some indexes $j_0, j_1 \in \{0, \ldots, p\}$ such that $j_0 < j_1 + 1$, we have $\{t_i;\ j_0 \leq i \leq j_1\} \subset [\alpha_{n+1}, \alpha_n)$ with $n \geq 1$, the monotonicity of u on $[\alpha_{n+1}, \alpha_n)$ implies that

$$\sum_{j=j_0}^{j_1-1} \left| u(t_{j+1}) - u(t_j) \right| = u(t_{j_0}) - u(t_{j_1}),$$

and an analogous property is valid if $\{t_i;\ j_0 \le i \le j_1\} \subset [\alpha_1, \tau)$. Hence,

$$\sum_{j=0}^{p} \left| u(t_{j+1} - u(t_j) \right| \\ \le \left| u(\alpha_1) - u(\tau) \right| + \sum_{n \ge 1} \Big(\left| u(\alpha_{n+1}) - u(\alpha_n - 0) \right| + \left| u(\alpha_n) - u(\alpha_n - 0) \right| \Big)$$

and

$$\mathrm{Var}\big(u, [0, \tau]\big) \le \left| u(\alpha_1) - u(\tau) \right| + \sum_{n \ge 1} \left(\frac{1}{(n+4)!} + \frac{3}{(n+3)!} \right) < +\infty.$$

Furthermore,

$$\mathrm{d}u - f\,\mathrm{d}t = \sum_{n \ge 1} \big(u(\alpha_n) - u(\alpha_n - 0) \big) \delta_{t=\alpha_n} = \sum_{n \ge 1} \frac{2}{(n+3)!} \delta_{t=\alpha_n} \in \mathscr{M}\big([0, \tau]; \mathbf{R}^+\big).$$

Finally, we let $q : [0, \tau] \to \mathbf{R}$ given by $q(t) = \displaystyle\int_0^t u(s)\,ds$ for all $t \in [0, \tau]$.

Since $u \in BV\big([0, \tau]; \mathbf{R}\big)$, u is bounded in $[0, \tau]$ and q is continuous on $[0, \tau]$ with $q(0) = 0$. Moreover, q is of class C^2, with a non-increasing derivative $\dot{q} = u$, on each subinterval (α_{n+1}, α_n) with $n \ge 1$ and on $(\alpha_1, \tau]$.

Let $n \ge 1$. By definition of $q(t)$, we have

$$q(\alpha_n) - q(\alpha_{n+1}) = \frac{\alpha_n - \alpha_{n+1}}{(n+4)!} - \frac{(\alpha_n - \alpha_{n+1} - \delta_n)^2}{2(n!)} \underbrace{\int_0^1 \left(\int_0^y \rho(\sigma)\,d\sigma \right) dy}_{= \int_0^1 (1-x)\rho(x)\,dx = \frac{1}{2}}$$

$$= \frac{\alpha_n - \alpha_{n+1}}{(n+4)!} - \frac{(\alpha_n - \alpha_{n+1} - \delta_n)^2}{4(n!)} = 0.$$

Since $\lim_{n \to +\infty} q(\alpha_n) = q(0) = 0$, we obtain $q(\alpha_n) = 0$ for all $n \ge 1$. Finally, since u decreases from $\dfrac{1}{(n+4)!}$ to $-\dfrac{1}{(n+3)!}$ on $(\alpha_{n+1} + \delta_n, \alpha_n)$, there exists $\xi_n \in (\alpha_{n+1} + \delta_n, \alpha_n)$ such that q is monotone increasing on (α_{n+1}, ξ_n), then monotone decreasing on (ξ_n, α_n). Hence, $q(t) > 0$ for all $t \in (\alpha_{n+1}, \alpha_n) \cap [0, \tau]$ for all $n \ge 0$ and q is *another solution* to problem (P).

The key point in this example is the fact that f possesses a flat point at $t = 0$. Indeed, $\rho \in C^\infty\big([0, 1]; \mathbf{R}\big)$, so we immediately have $f \in C^\infty\big((0, \alpha_0); \mathbf{R}\big)$. Moreover, for all $n \ge 0$ and for all $k \ge 1$, we have

$$f^{(k)}(t) = -\frac{1}{2(n!)}\frac{1}{(\alpha_n - \alpha_{n+1} - \delta_n)^k}\rho^{(k)}\left(\frac{t - \alpha_{n+1} - \delta_n}{\alpha_n - \alpha_{n+1} - \delta_n},\right)$$
$$\forall t \in [\alpha_{n+1} + \delta_n, \alpha_n] \cap [0, \tau]$$

thus

$$\begin{aligned}\left|f^{(k)}(t)\right| &\leq \frac{1}{2(n!)}\frac{1}{(\alpha_n - \alpha_{n+1} - \delta_n)^k}\max_{x\in[0,1]}\left|\rho^{(k)}(x)\right| \\ &\leq \frac{1}{2(n!)}\frac{1}{(\alpha_n - \alpha_{n+1} - \delta_n)^k}\frac{t}{\alpha_{n+1}}\max_{x\in[0,1]}\left|\rho^{(k)}(x)\right| \quad \forall t \in [\alpha_{n+1}, \alpha_n] \cap [0, \tau]\end{aligned}$$

and the same inequality also holds for $k = 0$. But

$$\alpha_n - \alpha_{n+1} - \delta_n \simeq_{+\infty} \frac{2}{n^3}, \quad \alpha_{n+1} \simeq_{+\infty} \frac{1}{n},$$

and thus

$$\lim_{t\to 0^+}\frac{1}{t}f^{(k)}(t) = 0 \quad \forall k \in \mathbf{N}.$$

By induction on k, we may conclude that f is infinitely differentiable at $t = 0$, with $f^{(k)}(0) = 0$ for all $k \in \mathbf{N}$.

Hence, we can only expect uniqueness results in the *analytical case* (see [54, 56] in the special case of energy conservative solutions to the bounce problem for a material point subjected to a single constraint, [61] for any value of the restitution coefficient and $d = 1$ and [4] for the general case) or as a generic property (see [13–15] for the one-dimensional elastic bounce problem and [9] for energy conservative solutions to the bounce problem for a material point subjected to a single constraint).

As a third mathematical drawback, *continuity of data* does not hold in general in the multi-constrained case, i.e., $\nu > 1$. Let us illustrate this with the example of a material point moving in the planar angular domain K given by

$$K = \left\{q = (q_1; q_2) \in \mathbf{R}^2;\ f_1(q) = q_2 \geq 0,\ f_2(q) = 2q_2 - q_1 \geq 0\right\}.$$

With $f \equiv 0$, $e = 1$, $q_0 = (\varepsilon, 1)$, $\varepsilon \in (-\infty, 2)$, $u_0 = (0, -1)$, we obtain

$$q_\varepsilon(t) = \begin{cases} (\varepsilon, 1 - t) & \text{if } t \in [0, 1], \\ (\varepsilon, t - 1) & \text{if } t \geq 1, \end{cases} \qquad \text{if } \varepsilon \leq 0,$$

and

$$q_\varepsilon(t) = \begin{cases} (\varepsilon, 1 - t) \quad \text{if } t \in [0, \tau_\varepsilon] \text{ with } \tau_\varepsilon = 1 - \dfrac{\varepsilon}{2}, \\ \left(\varepsilon - \dfrac{4}{5}(t - \tau_\varepsilon), \dfrac{\varepsilon}{2} + \dfrac{3}{5}(t - \tau_\varepsilon)\right) \quad \text{if } t \geq \tau_\varepsilon, \end{cases} \qquad \text{if } \varepsilon > 0.$$

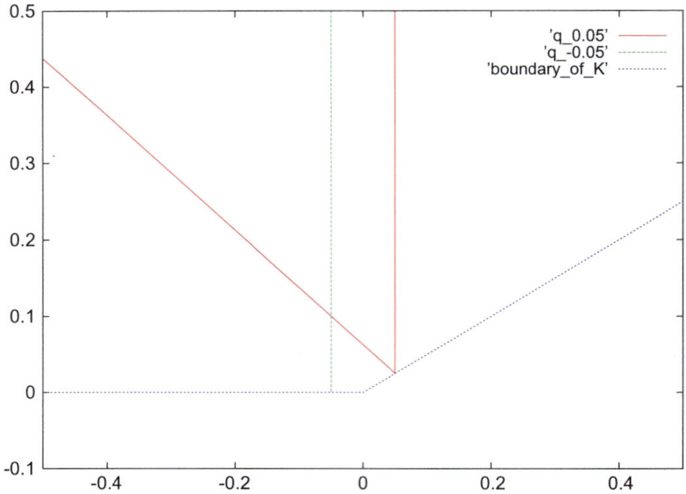

Fig. 1 Trajectories of a material point in an angular domain of $\mathbf{R}^2$, with $e = 1$

See, for instance, Fig. 1 with $\varepsilon = 0.05$. Thus, for any $t > 1$,

$$\lim_{\varepsilon \to 0^+} \frac{\left| q_\varepsilon(t) - q_{-\varepsilon}(t) \right|}{\left| q_\varepsilon(0) - q_{-\varepsilon}(0) \right|} = +\infty.$$

Such a bad property may lead to some numerical unpredictability due to round-up errors. Nevertheless, we can observe that, for this model problem, continuity of data holds if the vertex is right or acute whenever $e = 0$ or right whenever $e \in (0, 1]$. In a more general setting, the following proposition can be proved.

Proposition 1 (Continuity of data) *Continuity of data holds if the following geometrical "angle condition" on the active constraints is satisfied:*

$$\begin{aligned} \langle \nabla f_\alpha(q), \mathsf{M}(q)^{-1} \nabla f_\beta(q) \rangle &\le 0 \quad \text{if } e = 0 \\ \langle \nabla f_\alpha(q), \mathsf{M}(q)^{-1} \nabla f_\beta(q) \rangle &= 0 \quad \text{if } e \neq 0 \end{aligned}$$

for all $(\alpha, \beta) \in J(q)^2$ *such that* $\alpha \neq \beta$, *for all* $q \in \partial K$.

See [4] for a first proof giving a sufficient condition of continuity with respect to initial data and [44] for the general case.

Finally, as a last mathematical drawback, we may point out possible *finite time-explosions*. This bad property is not specific to vibro-impact problems and is somehow classical for ODE. For instance, let us consider once again the bouncing ball model problem, i.e., $d = 1$, $K = \mathbf{R}^+$, $\mathsf{M}(q) \equiv 1$, and let us assume that

$f(t, q, v) = v^2$ for all $(t, q, v) \in \mathbf{R}^3$. The motion without constraint is described by the ODE $\ddot{q} = \dot{q}^2$. With the initial condition $\dot{q}(0) = u_0 > 0$, we obtain

$$\dot{q}(t) = \frac{1}{\frac{1}{u_0} - t} > 0 \quad \forall t \in [0, \tau_1), \ \tau_1 = \frac{1}{u_0}.$$

Thus, for any $q_0 \geq 0$ and $u_0 > 1$, we get

$$q(t) = q_0 - \ln\left(\frac{1}{u_0} - t\right) > 0 \quad \forall t \in \left[0, \frac{1}{u_0}\right),$$

with

$$\lim_{t \to \frac{1}{u_0}^-} q(t) = +\infty = \lim_{t \to \frac{1}{u_0}^-} \dot{q}(t).$$

Hence, we have an explosion at $t = \dfrac{1}{u_0}$ without any impact during the time-interval $\left(0, \dfrac{1}{u_0}\right)$.

Nevertheless, owing to that the kinetic energy does not increase at impacts, we obtain that any solution to problem (P) satisfies the following *energy estimate*:

$$\begin{aligned}\mathscr{E}(t+0) \leq \mathscr{E}(0+0) &+ \int_0^t \langle f\big(s, q(s), \dot{q}(s)\big), \dot{q}(s)\rangle \, ds \\ &+ \frac{1}{2}\int_0^t \langle \dot{q}(s), \big(\mathrm{dM}\big(q(s)\big)\dot{q}(s)\big)\dot{q}(s)\rangle \, ds \quad \forall t \in [0, \tau)\end{aligned}$$

It follows that

Proposition 2 (Energy estimate [46]) *Let* $C > \|u_0\|_{q_0}$. *Then, there exists* $\tau(C) > 0$ *such that, for any solution* (q, u) *to problem (P) defined on* $[0, \tau]$, *we have*

$$\|q(t) - q_0\| \leq C \quad \text{and} \quad \|u(t)\|_{q(t)} \leq C \quad \text{for a.e. } t \in \big[0, \min\big(\tau(C), \tau\big)\big].$$

With such a long list of mathematical drawbacks, which kind of result can we expect? First of all, we may prove the existence of solutions. Moreover, we may propose some numerical methods, substantiated with appropriate convergence results, allowing us to compute approximate solutions. Of course, non-uniqueness will yield only to the convergence of subsequences of approximate solutions in general, but as soon as uniqueness holds (for instance, if all the data are analytical), we will recover the convergence of the whole sequence of approximate solutions.

4 Penalty Approach

A quite natural idea for solving constrained problems consists in relaxing the constraints in order to consider unconstrained problems $(P_n)_{n\geq 1}$ constructed in such a way that the constraints are better and better satisfied when n tends to $+\infty$.

Such a technique is especially suited to solving minimization problems. As an example, let us consider the problem

$$\text{Find } u \in U \quad \text{such that} \quad J(u) = \min_{v\in U} J(v)$$

where U is a closed, non-empty subset of $\mathbf{R}^d$ and $J : \mathbf{R}^d \to \mathbf{R}$ is a continuous mapping such that $\lim_{\|v\|\to+\infty} J(v) = +\infty$. Now, let $\psi : \mathbf{R}^d \to \mathbf{R}^+$ be a continuous mapping such that $\psi(v) = 0$ if and only if $v \in U$, and for all $n \in \mathbf{N}^*$, let us introduce the penalized problem

$$\text{Find } u_n \in \mathbf{R}^d \quad \text{such that} \quad J_n(u_n) = \min_{v\in\mathbf{R}^d} J_n(v),$$

with

$$J_n(v) = J(v) + n\psi(v) \quad \text{for all } v \in \mathbf{R}^d.$$

Both the constrained and the penalized problems admit a solution. Indeed, since $U \neq \emptyset$, there exists $v_0 \in U$, and since $\lim_{\|v\|\to+\infty} J(v) = +\infty$, there exists $r > 0$ such that $J_n(v) \geq J(v) > J(v_0) = J_n(v_0)$ for all $v \in \mathbf{R}^d$ such that $\|v\| > r$. It follows that the constrained (respectively penalized) problem is equivalent to

$$\text{Find } u \in U \cap \overline{B}(0_{\mathbf{R}^d}, r) \quad \text{such that} \quad J(u) = \min\nolimits_{v\in U\cap\overline{B}(0_{\mathbf{R}^d},r)} J(v)$$
$$(\text{resp. Find } u_n \in \mathbf{R}^d \cap \overline{B}(0_{\mathbf{R}^d}, r) \quad \text{such that} \quad J_n(u) = \min\nolimits_{v\in\mathbf{R}^d\cap\overline{B}(0_{\mathbf{R}^d},r)} J_n(v))$$

and the compactness of the closed ball $\overline{B}(0_{\mathbf{R}^d}, r)$ combined with the continuity of J (resp. J_n) allows us to conclude.

Moreover, for all $n \geq 1$, we have

$$J(u_n) \leq J_n(u_n) = J(u_n) + n\psi(u_n) \leq J_n(v) \quad \forall v \in \mathbf{R}^d.$$

It follows that $J(u_n) \leq J_n(u_n) \leq J_n(v_0) = J(v_0)$ for all $n \geq 1$ and $(u_n)_{n\geq 1}$ and $\big(n\psi(u_n)\big)_{n\geq 1}$ are bounded. Hence, possibly extracting a subsequence, still denoted $(u_n)_{n\geq 1}$, there exists $u \in \mathbf{R}^d$ such that

$$u_n \longrightarrow_{n\to+\infty} u.$$

Since ψ is continuous, $\lim_{n\to+\infty} \psi(u_n) = \psi(u) = 0$, which implies that $u \in U$, and by continuity of J, $J(u) = \lim_{n\to+\infty} J(u_n) \leq J(v)$ for all $v \in U$. Thus, the sequence $(u_n)_{n\geq 1}$ converges to a solution of the constrained problem.

This method is proposed for solving Lagrangian problems with bilateral constraints by adding a convex potential nW such that W is smooth and $W \equiv 0$ if and only if the constraints are satisfied (see, for instance, [2, 59]).

Let us now adapt this technique to our model problem of the bouncing ball, i.e., consider the vibro-impact problem (P) with $d = 1$, $K = \mathbf{R}^+$, $\mathsf{M}(q) \equiv 1$. In order to relax the constraint $q(t) \in \mathbf{R}^+$, we introduce the potential $W : \mathbf{R} \to \mathbf{R}^+$ given by $W(q) = \frac{1}{2}\big(\min(q, 0)\big)^2$ for all $q \in \mathbf{R}$. Clearly, W is Fréchet-differentiable at any $q \in \mathbf{R}$ and

$$W'(q) = \min(q, 0) \quad \forall q \in \mathbf{R}.$$

Then, we consider the unconstrained problems (P_n)

Find $q_n : [0, \tau] \to \mathbf{R}$, $\tau > 0$, such that $q_n(0) = q_0$, $\dot{q}_n(0) = u_0$ and

$$\ddot{q}_n(t) + nW'\big(q_n(t)\big) = f\big(t, q_n(t), \dot{q}_n(t)\big) \quad \text{in}(0, \tau), \tag{9}$$

with $n \in \mathbf{N}^*$. If f is continuous on $[0, \tau] \times \mathbf{R}^2$ and Lipschitz continuous with respect to its last two arguments, uniformly with respect to the first one, then the classical Cauchy-Lipschitz existence theorem for ODE implies that, for any $(q_0, u_0) \in \mathbf{R}^2$, problem (P_n) admits an unique solution $q_n \in C^2\big([0, \tau]; \mathbf{R}\big)$. Moreover, q_n satisfies the following energy estimate:

$$\begin{aligned}
&\frac{1}{2}\big(\dot{q}_n(t)\big)^2 + \frac{n}{2}W\big(q_n(t)\big) \\
&= \frac{1}{2}(u_0)^2 + \frac{n}{2}W(q_0) + \int_0^t f\big(s, q_n(s), \dot{q}_n(s)\big)\dot{q}_n(s)\, ds \quad \forall t \in [0, \tau].
\end{aligned}$$

Let us denote as L_f the Lipschitz constant of f. We obtain

$$\begin{aligned}
&\frac{1}{2}\big(\dot{q}_n(t)\big)^2 + \frac{n}{2}W\big(q_n(t)\big) \\
&\leq \frac{1}{2}(u_0)^2 + \frac{n}{2}W(q_0) + \int_0^t \big|f(s, q_0, 0)\big|\big|\dot{q}_n(s)\big|\, ds \\
&+L_f \int_0^t \big(\big|q_n(s) - q_0\big| + \big|\dot{q}_n(s)\big|\big)\big|\dot{q}_n(s)\big|\, ds \quad \forall t \in [0, \tau].
\end{aligned}$$

By using Cauchy-Schwarz's inequality and Grönwall's lemma, we infer that

$$
\begin{aligned}
&\left(\dot{q}_n(t)\right)^2 + nW\left(q_n(t)\right) \\
&\leq \left((u_0)^2 + nW(q_0) + \int_0^\tau \left| f(s, q_0, 0) \right|^2 ds \right) \exp\left(\left(2L_f(1+T)+1\right) t \right) \quad \forall t \in [0, \tau].
\end{aligned}
$$

As a consequence, if $q_0 \in \mathbf{R}^+$, we obtain that the sequences $(\dot{q}_n)_{n\geq 1}$ and $\left(nW(q_n)\right)_{n\geq 1}$ are bounded in $L^\infty(0, \tau; \mathbf{R})$. Next, we observe that

Lemma 3 *There exists a constant* C *(independent of* n*) such that* $\mathrm{Var}\left(\dot{q}_n, [0, \tau]\right) \leq C$.

Proof Let $n \geq 1$. Since $\dot{q}_n \in C^1\left([0, \tau]; \mathbf{R}\right)$ we have

$$
\mathrm{Var}\left(\dot{q}_n, [0, \tau]\right) = \int_0^\tau |\ddot{q}_n(s)|\, ds \leq \int_0^\tau \left| f\left(s, q_n(s), \dot{q}_n(s)\right) \right| ds + n \int_0^\tau \left| W'\left(q_n(s)\right) \right| ds.
$$

With the previous energy estimate we obtain

$$
\int_0^\tau \left| f\left(s, q_n(s), \dot{q}_n(s)\right) \right| ds \leq M\tau,
$$

where

$$
M = \sup\left\{ \left| f(s, q, v) \right|;\ (s, q, v) \in [0, \tau] \times [q_0 - R\tau, q_0 + R\tau] \times [-R, R] \right\}
$$

with

$$
R = \left((u_0)^2 + \int_0^\tau \left| f(s, q_0, 0) \right|^2 ds \right)^{1/2} \exp\left(\frac{2L_f(1+T)+1}{2} \tau \right).
$$

Now, let z be a continuous function from $[0, \tau]$ to $[-1, 1]$. We get

$$
\begin{aligned}
W'\left(q_n(s)\right)\left(1 \pm z(s) - q_n(s)\right) &= \left(1 \pm z(s)\right) \min\left(q_n(s), 0\right) - \left(\min\left(q_n(s), 0\right)\right)^2 \\
&\leq -\frac{1}{2}\left(\min\left(q_n(s), 0\right)\right)^2 \leq 0 \quad \forall s \in [0, \tau].
\end{aligned}
$$

Hence,

$$
\begin{aligned}
&\pm \int_0^\tau nW'\left(q_n(s)\right) z(s)\, ds \\
&\leq -\int_0^\tau nW'\left(q_n(s)\right)\left(1 - q_n(s)\right) ds \\
&= \int_0^\tau \left(\ddot{q}_n(s) - f\left(s, q_n(s), \dot{q}_n(s)\right)\right)\left(1 - q_n(s)\right) ds \\
&= \dot{q}_n(\tau) - u_0 - \dot{q}_n(\tau) q_n(\tau) + u_0 q_0 + \int_0^\tau \left(\left(\dot{q}_n(s)\right)^2 - f\left(s, q_n(s), \dot{q}_n(s)\right)\left(1 - q_n(s)\right) \right) ds \\
&\leq (R + |u_0|)(q_0 + 1) + 2R^2\tau + M(1 + q_0 + R\tau)\tau
\end{aligned}
$$

and we conclude that $\left(nW'(q_n)\right)_{n\geq 1}$ is bounded in $L^1(0, \tau; \mathbf{R})$ [19].

By applying Ascoli's theorem [64] and Helly's theorem ([27], see also Sect. 7, Appendix), we determine that there exists a subsequence, still denoted $(q_n)_{n\geq 1}$, such that

$$\begin{aligned}
&q_n(t) \longrightarrow_{n\to+\infty} q(t) \quad \text{uniformly in } [0,\tau],\\
&\dot{q}_n(t) \longrightarrow_{n\to+\infty} u(t) \quad \text{for all } t\in[0,\tau],\\
&\mathrm{d}\dot{q}_n - f(\cdot, q_n, \dot{q}_n)\mathrm{d}t \rightharpoonup_{n\to+\infty} \lambda = \mathrm{d}u - f(\cdot, q, u)\mathrm{d}t \quad \text{weakly} * \text{ in } \mathscr{M}\big([0,\tau];\mathbf{R}\big),
\end{aligned}$$

with $q \in C^0([0,\tau];\mathbf{R})$ and $u \in BV\big([0,\tau];\mathbf{R}\big)$. By using the continuity of W, we infer that

$$W\big(q_n(t)\big) \longrightarrow_{n\to+\infty} 0 = W\big(q(t)\big) \quad \text{uniformly in } [0,\tau],$$

i.e., $q(t) \geq 0$ for all $t \in [0,\tau]$. Moreover,

$$q_n(t) = q_0 + \int_0^t \dot{q}_n(s)\, ds \quad \forall t \in [0,\tau],$$

so at the limit, we have

$$q(t) = q_0 + \int_0^t u(s)\, ds \quad \forall t \in [0,\tau].$$

Furthermore, $\mathrm{d}\dot{q}_n - f(\cdot, q_n, \dot{q}_n)\mathrm{d}t = -nW'(q_n)\mathrm{d}t$ is a non-negative measure on $[0,\tau]$ for all $n \geq 1$ thus λ is also a non-negative measure. Let $v \in C^0\big([0,\tau];\mathbf{R}\big)$ such that $\mathrm{Supp}(v) \subset \big\{t \in [0,\tau];\ q(t) > 0\big\}$. By compacity of Supp($v$), we infer that, for all n big enough, we have $q_n(t) > 0$ for all $t \in \mathrm{Supp}(v)$. It follows that

$$\begin{aligned}
\langle \lambda, v\rangle_{\mathscr{M}([0,\tau];\mathbf{R}), C^0([0,\tau];\mathbf{R})} &= \lim_{n\to+\infty} \big\langle \mathrm{d}\dot{q}_n - f(\cdot, q_n, \dot{q}_n)\mathrm{d}t, v\big\rangle_{\mathscr{M}([0,\tau];\mathbf{R}), C^0([0,\tau];\mathbf{R})}\\
&= \lim_{n\to+\infty} -n \int_0^\tau \underbrace{W'\big(q_n(t)\big)v(t)}_{=0}\, dt = 0,
\end{aligned}$$

where $\langle\cdot,\cdot\rangle_{\mathscr{M}([0,\tau];\mathbf{R}), C^0([0,\tau];\mathbf{R})}$ denotes the duality product between the space of real (Borel) measures $\mathscr{M}([0,\tau];\mathbf{R})$ and the space of continuous functions $C^0([0,\tau];\mathbf{R})$. Thus, we may conclude that $\mathrm{Supp}(\lambda) \subset \big\{t \in [0,\tau];\ q(t) = 0\big\}$.

Finally, we obtain an energy estimate for the limit motion.

Proposition 3 *For almost every $t \in [0,\tau]$, we have*

$$\frac{1}{2}\big(u(t)\big)^2 = \frac{1}{2}(u_0)^2 + \int_0^t f\big(s, q(s), u(s)\big)u(s)\, ds.$$

Proof For all $n \geq 1$ and for all $t \in [0,\tau]$, we have

$$\frac{1}{2}\big(\dot{q}_n(t)\big)^2 + \frac{n}{2}W\big(q_n(t)\big) = \frac{1}{2}(u_0)^2 + \int_0^t f\big(s, q_n(s), \dot{q}_n(s)\big)\dot{q}_n(s)\,ds$$

and

$$\frac{1}{2}\big(\dot{q}_n(t)\big)^2 \longrightarrow_{n\to+\infty} \frac{1}{2}\big(u(t)\big)^2,$$
$$\int_0^t f\big(s, q_n(s), \dot{q}_n(s)\big)\dot{q}_n(s)\,ds \longrightarrow_{n\to+\infty} \int_0^t f\big(s, q(s), u(s)\big)u(s)\,ds.$$

It follows that $\left(\frac{n}{2}W\big(q_n(t)\big)\right)_{n\geq 1}$ admits a limit, denoted $g(t)$, for all $t \in [0, \tau]$.

Then, we observe that

$$\int_0^\tau \frac{n}{2}W\big(q_n(t)\big)\,dt \leq \frac{1}{\sqrt{2}}\max_{t\in[0,\tau]}\sqrt{W\big(q_n(t)\big)}\int_0^\tau n\big|W'\big(q_n(t)\big)\big|\,dt = \mathcal{O}\left(\frac{1}{\sqrt{n}}\right).$$

By using Fatou's lemma, we infer that $g \in L^1(0, \tau; \mathbf{R})$ and

$$0 = \liminf_{n\to+\infty}\int_0^\tau \frac{n}{2}W\big(q_n(t)\big)\,dt \geq \int_0^\tau \underbrace{g(t)}_{\geq 0}\,dt \geq 0.$$

Thus,

$$\frac{n}{2}W\big(q_n(t)\big) \longrightarrow_{n\to+\infty} g(t) = 0 \quad \text{a.e.} t \in [0, \tau].$$

With Proposition 3, we infer that

$$\big|\dot{q}(t+0)\big| = \big|u(t+0)\big| = \big|u(t-0)\big| = \big|\dot{q}(t-0)\big| \quad \forall t \in (0, \tau)$$

and we get an energy conservative solution to our vibro-impact problem. In this simple example, we have $\mathrm{Card}\big(J\big(q(t)\big)\big) = 1$ for all $t \in [0, \tau]$ such that $q(t) \in \partial K$, so we may conclude that q is a solution of problem (P) with a restitution coefficient $e = 1$.

Several existence results rely on a penalty approach, allowing us to consider either convex constraints, i.e., convex sets of admissible configurations or, more generally, dynamics driven by a convex (non-smooth) potential (see, for instance, [3, 9, 12, 18, 31, 43, 55, 60]). But, as in the model problem of the bouncing ball presented above, we always obtain energy conserving solutions, since the penalty term in the approximate problems derives from a potential and does not lead to dissipation of energy at the limit. So if we want to apply this kind of approach to study problem (P) with $e \in [0, 1)$, we have to introduce in the approximate problems some dissipation when the constraints are not satisfied. From the heuristic point of view, the simplest way to add dissipation during the motion consists in adding a viscous friction term. Indeed, if we consider the ODE

$$\ddot{y} + 2\varepsilon\dot{y} + y = 0,$$

with $\varepsilon \in (0, 1)$ and the initial conditions $y(0) = 0$, $\dot{y}(0) < 0$, we obtain

$$y(t) = \frac{\dot{y}(0)}{\sqrt{1-\varepsilon^2}} \sin\big(\sqrt{1-\varepsilon^2}t\big) \exp(-\varepsilon t) \quad \forall t \in \mathbf{R}^+.$$

It follows that $y(t) < 0$ on $(0, t_1)$, with $t_1 = \dfrac{\pi}{\sqrt{1-\varepsilon^2}}$ and $\dot{y}(t_1) = -\dot{y}(0)$ $\exp\left(-\dfrac{\pi\varepsilon}{\sqrt{1-\varepsilon^2}}\right)$. Observing that $\exp\left(-\dfrac{\pi\varepsilon}{\sqrt{1-\varepsilon^2}}\right) \in (0, 1)$, we choose ε such that $e = \exp\left(-\dfrac{\pi\varepsilon}{\sqrt{1-\varepsilon^2}}\right)$, which leads to

$$\varepsilon = -\frac{\ln(e)}{\sqrt{\pi^2 + \big(\ln(e)\big)^2}} \quad \text{if } e \in (0, 1).$$

Then, having in mind that we expect an immediate reflexion of the normal velocity at impact, we rescale the time-variable and we let $z_n(t) = \dfrac{1}{\sqrt{n}} y\big(\sqrt{n}(t - \tau_0)\big)$, which yields $\dot{z}_n\left(\tau_0 + \dfrac{\pi}{\sqrt{n(1-\varepsilon^2)}}\right) = -e\dot{z}_n(\tau_0)$.

Going back to the bouncing ball model problem, we now use these ideas to propose now the following sequence of approximate problems:

$$\ddot{q}_n(t) + 2\varepsilon\sqrt{n}G\big(q_n(t), \dot{q}_n(t)\big) + nW'\big(q_n(t)\big) = f\big(t, q_n(t), \dot{q}_n(t)\big) \quad \text{in}(0, \tau), \tag{10}$$

with the initial conditions $q_n(0) = q_0$, $\dot{q}_n(0) = u_0$ and

$$G(q, v) = \begin{cases} v & \text{if } q < 0, \\ 0 & \text{if } q \geq 0, \end{cases} \qquad \varepsilon = -\frac{\ln(e)}{\sqrt{\pi^2 + \big(\ln(e)\big)^2}}, \quad e \in (0, 1).$$

Let us observe that ε tends to zero when e tends to 1, and we recover (9). Otherwise, we have an additional penalty term which acts as a viscous friction force and is activated only when the constraint $q_n(t) \geq 0$ is not satisfied. Let us emphasize that (10) does not satisfy the usual assumptions, allowing us to apply classical existence results for ODE. Indeed, the mapping $(q, v) \mapsto G(q, v)$ is not continuous on $\{0\} \times \mathbf{R}^*$. Nevertheless, if f is Lipschitz continuous with respect to its last two arguments, uniformly with respect to the first one, we can establish that, for any $\tau > 0$, $q_0 \in \mathbf{R}^+$ and $u_0 \in \mathbf{R}$, (10) admits a solution $q_n \in C^1\big([0, \tau]; \mathbf{R}\big)$ such that $\dot{q}_n$ is absolutely continuous on $[0, \tau]$, $\ddot{q}_n \in L^\infty(0, \tau; \mathbf{R})$ and (10) holds for almost every $t \in (0, \tau)$ [49]. Moreover, we have

$$G(q,v)v \geq 0 \quad \forall (q,v) \in \mathbf{R}^2,$$

so with the same computations as earlier, we obtain the following energy inequality:

$$\frac{1}{2}\big(\dot{q}_n(t)\big)^2 + \frac{n}{2}W\big(q_n(t)\big) \leq \frac{1}{2}(u_0)^2 + \int_0^t f\big(s, q_n(s), \dot{q}_n(s)\big)\dot{q}_n(s)\,ds \quad \forall t \in [0,\tau]$$

and at the limit when n tends to $+\infty$, we may expect that $(q_n)_{n\geq 1}$ converges to an energy dissipative solution to the vibro-impact problem (see Remark 1). But we can also easily check that Newton's impact law will be satisfied at the limit. Indeed, let $\tau_{0n} \in (0,\tau)$ such that $q_n(\tau_{0n}) = 0$ and $\dot{q}_n(\tau_{0n}) < 0$. Then, q_n behaves like z_n, with $z_n(\tau_{0n}) = 0$ and $\dot{z}_n(\tau_{0n}) = \dot{q}_n(\tau_{0n})$ on a right neighbourhood of τ_{0n}. Indeed, q_n remains negative on some non-trivial interval $(\tau_{0n}, \widetilde{\tau}_{0n})$ and the function r_n given by

$$r_n(s) = \sqrt{n}(q_n - z_n)\left(\tau_{0n} + \frac{s}{\sqrt{n}}\right) \quad \forall s \in \big(0, \sqrt{n}(\widetilde{\tau}_{0n} - \tau_{0n})\big),$$

satisfies the ODE

$$\ddot{r}_n(s) + 2\varepsilon\dot{r}_n(s) + r_n(s) = \frac{1}{\sqrt{n}}\widetilde{f}(s) \quad \text{for a.e. } s \in \big(0, \sqrt{n}(\widetilde{\tau}_{0n} - \tau_{0n})\big)$$

with $r_n(0) = 0$, $\dot{r}_n(0) = 0$ and

$$\widetilde{f}(s) = f\left(\tau_{0n} + \frac{s}{\sqrt{n}}, q_n\left(\tau_{0n} + \frac{s}{\sqrt{n}}\right), \dot{q}_n\left(\tau_{0n} + \frac{s}{\sqrt{n}}\right)\right) \quad \forall s \in \big[0, \sqrt{n}(\widetilde{\tau}_{0n} - \tau_{0n})\big].$$

Then, with an energy inequality and Grönwall's lemma, we obtain

$$\big|\dot{r}_n(s)\big|^2 + \big|r_n(s)\big|^2 \leq \mathcal{O}\left(\frac{1}{n}\right)\exp(s) \quad \forall s \in \big[0, \sqrt{n}(\widetilde{\tau}_{0n} - \tau_{0n})\big],$$

i.e.,

$$\big|\dot{q}_n(t) - \dot{z}_n(t)\big|^2 + n\big|q_n(t) - z_n(t)\big|^2 \leq \mathcal{O}\left(\frac{1}{n}\right)\exp\big(\sqrt{n}(t - \tau_{0n})\big) \quad \forall t \in [\tau_{0n}, \widetilde{\tau}_{0n}].$$

Since $z_n(t) > 0$ for all $t \in \left(\tau_{0n} + \frac{\pi}{\sqrt{n(1-\varepsilon^2)}}, \tau_{0n} + \frac{2\pi}{\sqrt{n(1-\varepsilon^2)}}\right)$, we infer that

$$
\begin{aligned}
0 > q(t) &\geq z_n(t) - \mathcal{O}\left(\frac{1}{n}\right)\exp\left(\frac{\sqrt{n}}{2}(t-\tau_{0n})\right)\\
&\geq \frac{\dot q(\tau_{0n})}{\sqrt{n(1-\varepsilon^2)}}\sin\left(\sqrt{n(1-\varepsilon^2)}(t-\tau_{0n})\right)\exp\bigl(-\varepsilon\sqrt{n}(t-\tau_{0n})\bigr)\\
&\quad -\mathcal{O}\left(\frac{1}{n}\right)\exp\left(\frac{\sqrt{n}}{2}(t-\tau_{0n})\right)\\
&\quad \forall t\in(\tau_{0n},\widetilde{\tau}_{0n})\cap\left(\tau_{0n}+\frac{\pi}{\sqrt{n(1-\varepsilon^2)}},\tau_{0n}+\frac{2\pi}{\sqrt{n(1-\varepsilon^2)}}\right).
\end{aligned}
$$

Let us assume that there exists $\delta\in(0,\pi)$ and $n_*\geq 1$ such that

$$
\widetilde{\tau}_{0n} > \tau_{0n} + \frac{\pi+\delta}{\sqrt{n(1-\varepsilon^2)}} \quad \forall n\geq n_*.
$$

Then, with $t=\tau_{0n}+\dfrac{\pi+\delta}{\sqrt{n(1-\varepsilon^2)}}\in(\tau_{0n},\widetilde{\tau}_{0n})\cap\left(\tau_{0n}+\dfrac{\pi}{\sqrt{n(1-\varepsilon^2)}},\tau_{0n}+\dfrac{2\pi}{\sqrt{n(1-\varepsilon^2)}}\right)$, we obtain

$$
0\geq \underbrace{\frac{\dot q(\tau_{0n})}{\sqrt{1-\varepsilon^2}}\sin(\pi+\delta)\exp\left(-\varepsilon\frac{\pi+\delta}{\sqrt{1-\varepsilon^2}}\right)}_{>0} - \mathcal{O}\left(\frac{1}{\sqrt{n}}\right)\exp\left(\frac{\pi+\delta}{\sqrt{2(1-\varepsilon^2)}}\right)
$$

for all $n\geq n_*$, which is absurd. It follows that $\widetilde{\tau}_{0n}-\tau_{0n}\simeq_{n\to+\infty}\dfrac{\pi}{\sqrt{n(1-\varepsilon^2)}}$ and $\dot q_n(\widetilde{\tau}_{0n})\simeq_{n\to+\infty}\dot z_n\left(\tau_{0n}+\dfrac{\pi}{\sqrt{n(1-\varepsilon^2)}}\right)=-e\dot q_n(\tau_{0n})$. Of course, this does not allow us to conclude immediately that $(q_n)_{n\geq1}$ converges to a solution to problem (P). With the energy inequality, we already know that $(q_n)_{n\geq1}$ and $(\dot q_n)_{n\geq1}$ are uniformly bounded in $L^\infty(0,\tau;\mathbf{R})$, and we can check with the same kind of computations as in Lemma 3 that $\bigl(nW'(q_n)\bigr)_{n\geq1}$ is uniformly bounded in $L^1(0,\tau;\mathbf{R})$. Hence, we may apply Ascoli's theorem, but unfortunately, this is not enough to apply also Helly's compactness theorem as well and we have to prove the following Lemma.

Lemma 4 *The sequence* $\bigl(\sqrt{n}G(q_n,\dot q_n)\bigr)_{n\geq1}$ *is uniformly bounded in* $L^1(0,\tau;\mathbf{R})$.

Proof Let $n\geq 1$. We define $U_n^-=\{t\in[0,\tau];\ q_n(t)<0\}$ and we denote as I_{nj} the connex components of U_n^- i.e. $U_n^-=\bigcup_{j\in J_n}I_{nj}$ where J_n is at most countable and $I_{nj}\cap(0,\tau)=(\alpha_{nj},\beta_{nj})$ for all $j\in J_n$. For all $j\in J_n$ we have

$$
\ddot q_n(t)+2\varepsilon\sqrt{n}\dot q_n(t)=w_n(t)=f\bigl(t,q_n(t),\dot q_n(t)\bigr)-nq_n(t) \quad \text{for a.e.} t\in I_{nj},
$$

and thus

$$\dot{q}_n(t) = \dot{q}_n(t_{nj}) \exp\bigl(-2\varepsilon\sqrt{n}(t - t_{nj})\bigr) + \int_{t_{nj}}^{t} \exp\bigl(-2\varepsilon\sqrt{n}(t - s)\bigr) w_n(s)\, ds \quad \forall t \in I_{nj},$$

where $t_{nj} \in (\alpha_{nj}, \beta_{nj})$. It follows that $\dot{q}_n$ admits a finite (right) limit at α_{nj}. Let us assume now that $\alpha_{nj} \neq 0$. Then, $q_n(\alpha_{nj}) = 0$ and $\dot{q}_n(\alpha_{nj}) = \dot{q}_n(\alpha_{nj} + 0) \leq 0$, since $q_n(t) < 0$ on $(\alpha_{nj}, \beta_{nj})$. Let us assume, moreover, that $\dot{q}_n(\alpha_{nj} + 0) \neq 0$. Then, q_n remains positive on a left neighbourhood $(\gamma_{nj}, \alpha_{nj})$ of α_{nj} and

$$\ddot{q}_n(t) = f\bigl(t, q_n(t), \dot{q}_n(t)\bigr) \quad \text{for a.e. } t \in (\gamma_{nj}, \alpha_{nj}),$$

which yields

$$\begin{aligned}
&\dot{q}_n(\alpha_{nj})(\alpha_{nj} - t) + \bigl(q(t) - q(\alpha_{nj})\bigr) \\
&= \int_t^{\alpha_{nj}} \left(\int_s^{\alpha_{nj}} f\bigl(\sigma, q_n(\sigma), \dot{q}_n(\sigma)\, d\sigma \right) ds \quad \forall t \in (\gamma_{nj}, \alpha_{nj})
\end{aligned}$$

and thus $\dot{q}_n(\alpha_{nj}) = \dot{q}_n(\alpha_{nj} + 0) \leq \mathcal{O}(\alpha_{nj} - \gamma_{nj})$ if $\gamma_{nj} > 0$. Hence,

$$\begin{aligned}
\int_0^{\tau} \sqrt{n} \bigl| G\bigl(q_n(t), \dot{q}_n(t)\bigr) \bigr|\, dt &= \int_{U_n^-} \sqrt{n} |\dot{q}_n(t)|\, dt \\
&\leq \sum_{j \in J_n} |\dot{q}_n(\alpha_{nj} + 0)| \int_{\alpha_{nj}}^{\beta_{nj}} \sqrt{n} \exp\bigl(-2\varepsilon\sqrt{n}(t - \alpha_{nj})\bigr)\, dt \\
&+ \sqrt{n} \sum_{j \in J_n} \int_{\alpha_{nj}}^{\beta_{nj}} \left(\int_{\alpha_{nj}}^{t} \exp\bigl(-2\varepsilon\sqrt{n}(t - s)\bigr) |w_n(s)|\, ds \right) dt \\
&\leq \frac{1}{2\varepsilon} \left(\sum_{j \in J_n} |\dot{q}_n(\alpha_{nj} + 0)| + \int_{U_n^-} |w_n(t)|\, dt \right) \\
&\leq \frac{1}{2\varepsilon} \left(\sum_{j \in J_n} |\dot{q}_n(\alpha_{nj} + 0)| + \int_0^{\tau} |f\bigl(t, q_n(t), \dot{q}_n(t)\bigr)|\, dt \right. \\
&\qquad \left. + \int_0^{\tau} n \bigl| W'\bigl(q_n(t)\bigr) \bigr|\, dt \right),
\end{aligned}$$

which allows us to conclude.

So, by using Ascoli's and Helly's theorems, and possibly extracting a subsequence, still denoted $(q_n)_{n \geq 1}$, we infer that there exists $q \in C^0\bigl([0, \tau]; \mathbf{R}^d\bigr)$ and $u \in BV\bigl([0, \tau]; \mathbf{R}^d\bigr)$ such that

$$\begin{aligned}
&q_n(t) \longrightarrow_{n \to +\infty} q(t) \quad \text{uniformly in } [0, \tau], \\
&\dot{q}_n(t) \longrightarrow_{n \to +\infty} u(t) \quad \text{for all } t \in [0, \tau], \\
&\mathrm{d}\dot{q}_n - f(\cdot, q_n, \dot{q}_n)\mathrm{d}t \rightharpoonup_{n \to +\infty} \lambda = \mathrm{d}u - f(\cdot, q, u)\mathrm{d}t \quad \text{weakly} * \text{in } \mathcal{M}\bigl([0, \tau]; \mathbf{R}\bigr),
\end{aligned}$$

with

$$q(t) = q_0 + \int_0^t u(s)\,ds \geq 0 \quad \forall t \in [0,\tau]$$

and $\mathrm{Supp}(\lambda) \subset \{t \in [0,\tau];\ q(t) = 0\}$.

It remains to prove that λ is a non-negative measure, i.e.,

$$\langle \lambda, v\rangle_{\mathscr{M}([0,\tau];\mathbf{R}),C^0([0,\tau];\mathbf{R})} \geq 0 \quad \forall v \in C^0\big([0,\tau];\mathbf{R}^+\big).$$

Let us consider first $v \in C^1([0,\tau];\mathbf{R}^+)$. Then, for all $n \geq 1$, we have

$$\begin{aligned}
&\langle \mathrm{d}\dot{q}_n - f(\cdot, q_n, \dot{q}_n)\mathrm{d}t, v\rangle_{\mathscr{M}([0,\tau];\mathbf{R}),C^0([0,\tau];\mathbf{R})} \\
&= \int_{U_n^-} \underbrace{\big(-nq_n(t)\big)v(t)}_{\geq 0}\,dt - 2\varepsilon\sqrt{n}\int_{U_n^-} \dot{q}_n(t)v(t)\,dt \\
&\geq -2\varepsilon\sqrt{n}\sum_{j\in J_n}\big[q_n(t)v(t)\big]_{\alpha_{nj}}^{\beta_{nj}} + 2\varepsilon\sqrt{n}\sum_{j\in J_n}\int_{\alpha_{nj}}^{\beta_{nj}} q_n(t)\dot{v}(t)\,dt.
\end{aligned}$$

In the first sum, all the terms vanish, except perhaps the very last one, if $\beta_{nj} = \tau$, and then $q_n(\tau) \leq 0$. Thus,

$$\begin{aligned}
&\langle \mathrm{d}\dot{q}_n - f(\cdot, q_n, \dot{q}_n)\mathrm{d}t, v\rangle_{\mathscr{M}([0,\tau];\mathbf{R}),C^0([0,\tau];\mathbf{R})} \\
&\geq -\frac{2\varepsilon}{\sqrt{n}}\max_{t\in[0,\tau]}\big|\dot{v}(t)\big|\int_0^\tau n\big|W'\big(q_n(t)\big)\big|\,dt = -\mathscr{O}\left(\frac{1}{\sqrt{n}}\right).
\end{aligned}$$

It follows that

$$\langle \mathrm{d}\dot{q}_n - f(\cdot, q_n, \dot{q}_n)\mathrm{d}t, v\rangle_{\mathscr{M}([0,\tau];\mathbf{R}),C^0([0,\tau];\mathbf{R})} \longrightarrow_{n\to+\infty} \langle \lambda, v\rangle_{\mathscr{M}([0,\tau];\mathbf{R}),C^0([0,\tau];\mathbf{R})} \geq 0$$

for all $v \in C^1\big([0,\tau];\mathbf{R}^+\big)$ and by density we get finally $\langle \lambda, v\rangle_{\mathscr{M}([0,\tau];\mathbf{R}),C^0([0,\tau];\mathbf{R})} \geq 0$ for all $v \in C^0\big([0,\tau];\mathbf{R}^+\big)$.

By observing that $W'(q) = q - \mathrm{Proj}(K, q)$ and

$$G(q, v) = \big\langle v, \mathbf{n}\big(\mathrm{Proj}(K,q)\big)\big\rangle \mathbf{n}\big(\mathrm{Proj}(K,q)\big) \quad \text{if} q \notin K,$$

where $\mathbf{n}\big(\mathrm{Proj}(K,q)\big)$ is the outward unit normal to K at $\mathrm{Proj}(K,q)$, we may extend the previous penalty approach to any set K of admissible configurations such that K is convex and ∂K is smooth. Indeed, by applying some diffeomorphism, we may define local coordinates that transform the set K into a half-space, and the normal component of q in these new coordinates will satisfy an ODE of the same form as (10). Hence, the main ideas to prove the convergence of the approximate solutions remain the same as in the simple previous case of the bouncing ball example, but the technical aspects are more involved, since we have to deal, in general, with a curved boundary ∂K. For the complete proof when $\mathsf{M}(q) \equiv \mathrm{Id}_{\mathbf{R}^d}$ and ∂K of class C^2, the reader is referred to [49] for $e \in (0, 1]$.

Let us emphasize that the definition of ε as $\varepsilon = -\dfrac{\ln(e)}{\sqrt{\pi^2 + (\ln(e))^2}}$ does not allow us to consider the case $e = 0$. Nevertheless, we have $\lim_{e \to 0^+} \varepsilon = 1$, and we may wonder if the penalized problems

$$\begin{aligned}&\ddot{q}_n(t) + 2\sqrt{n}G(q_n(t), \dot{q}_n(t)) + n(q_n(t) - \text{Proj}(K, q_n(t)))\\&= f(t, q_n(t), \dot{q}_n(t)) \quad \text{in}(0, \tau),\end{aligned}$$

with $n \in \mathbf{N}^*$, provide approximate solutions of (P) when $e = 0$. For the bouncing ball model problem, we can compare q_n to the solution z_n to the following ODE:

$$\ddot{z}_n + 2\sqrt{n}\dot{z}_n + nz_n = 0.$$

More precisely, let us consider $\tau_{0n} \in (0, \tau)$ such that $q_n(\tau_{0n}) = 0$ and $\dot{q}_n(\tau_{0n}) < 0$. Let $(\tau_{0n}, \widetilde{\tau}_{0n})$ be a right neighbourhood of τ_{0n} such that $q_n(t) < 0$ for all $t \in (\tau_{0n}, \widetilde{\tau}_{0n})$. With $z_n(\tau_{0n}) = 0$ and $\dot{z}_n(\tau_{0n}) = \dot{q}_n(\tau_{0n})$, we get

$$0 \geq z_n(t) = \dot{q}_n(\tau_{0n})(t - \tau_{0n}) \exp(-\sqrt{n}(t - \tau_{0n})) \quad \forall t \in [\tau_{0n}, +\infty)$$

and

$$|\dot{q}_n(t) - \dot{z}_n(t)|^2 + n|q_n(t) - z_n(t)|^2 \leq \mathscr{O}\left(\frac{1}{n}\right) \exp(\sqrt{n}(t - \tau_{0n})) \quad \forall t \in (\tau_{0n}, \widetilde{\tau}_{0n})$$

and the mathematical analysis of the behaviour of the approximate trajectories can still be performed, leading to a convergence result. The general case with $\mathsf{M}(q) \not\equiv \text{Id}_{\mathbf{R}^d}$ and smooth enough ∂K (i.e., ∂K at least of class C^3) is considered in [62].

Of course, in the multi-constrained case (i.e., $\nu \geq 2$), the boundary of K is not smooth anymore, and when q is a "corner" of ∂K, i.e., $q \in \partial K$ with $\text{Card}(J(q)) \geq 2$, the normal cone to K at q is not reduced to a half-line and K can no longer be transformed into a half-space by using a local diffeomorphism. Thus, a natural idea consists in regularizing K in a neighbourhood of all such points of its boundary, for instance, by replacing K with $K^\varepsilon = K + \varepsilon \mathbf{B}_p$, where $\mathbf{B}_p$ is the closed unit ball of $\mathbf{R}^d$ for the usual p-norm, with $p \in (1, +\infty)$, and $\varepsilon > 0$. If K is convex, then K^ε is also convex, and the corresponding bounce problem admits energy conservative solutions that satisfy Newton's impact law with $e = 1$. When ε tends to zero, the reflexion of the velocity when an impact occurs in a corner of K depends on p and we recover Newton's impact law with $e = 1$ only for the choice $p = 2$ [10]. Moreover, if we solve the bounce problem with different kinds of penalty approaches, the limit impact law at corners also depends on the choice of penalty approach [3]!

Hence, the penalty approach provides a good theoretical (but heavy and technical) tool for proving existence results when ∂K is smooth or for convex sets K with energy conservative solutions, but it does not seem well-suited in the other cases.

Since this approach relies on the construction of a sequence of approximate trajectories that may be computed numerically (since q_n is the solution to an ODE), it also provides, as a by-product, a tool for approximately soving problem (P). Observing that $n\big(q_n - \mathrm{Proj}(K, q_n)\big)$ acts as an elastic drawback force and $2\varepsilon\sqrt{n}G(q_n, \dot{q}_n)$ as a viscous friction force applied to the system when the constraints are violated, problems (P_n) admit a nice mechanical interpretation: the boundary ∂K of the set of admissible configurations no longer behaves as a rigid obstacle but as a (visco)-elastic one, which seems more realistic from the physical point of view.

In the framework of deformable bodies, this approach, also called *normal compliance* approximation, has been extensively used to relax Signorini's complementary conditions (see, for instance, [25, 30] and references therein or [26] for some recent variants of this technique). Unfortunately, from the numerical point of view, it suffers from two major drawbacks.

First, as has been seen in the previous proof, the approximate trajectories are not feasible and the constraints are violated on time-intervals $I_{nj} = (\alpha_{nj}, \beta_{nj})$ with $\beta_{nj} - \alpha_{nj} \simeq_{n\to+\infty} \dfrac{\pi}{\sqrt{n(1-\varepsilon)^2}}$ when $e \in (0, 1]$. Hence, in order to compute q_n accurately enough to "catch" the reflexion of velocity at β_{nj}, we need to choose a time-step $\Delta t << \dfrac{\pi}{\sqrt{n(1-\varepsilon)^2}}$, and it is costly.

Moreover, we have to choose the value of the penalty parameter n, and then we solve numerically (thus only approximately) the penalized problem (P_n), which means that we have two kinds of approximation error: the approximation error due to the penalty approach, namely $\|q - q_n\|_{C^0([0,\tau];\mathbf{R}^d)}$, and the approximation error due to the numerical solver applied to (P_n). Of course, we can choose a very accurate ODE solver, and we may expect a bigger n the smaller $\|q - q_n\|_{C^0([0,\tau];\mathbf{R}^d)}$. But we also have to deal with the condition $\Delta t << \dfrac{\pi}{\sqrt{n(1-\varepsilon)^2}}$. Keeping in mind the physical interpretation of the term $n\big(q_n - \mathrm{Proj}(K, q_n)\big)$ as an elastic drawback force with stiffness n, we may hope to find some characteristic values of the stiffness associated to some material properties. Unfortunately, the range of these values given in the literature goes from 5.5 10^7 Nm for an impacting bar [63] to 10^{10} Nm for systems with joint clearance [57]. Furthermore, we can also observe a great sensitivity of the approximate solutions q_n to the penalty parameter [51].

So, we may conclude that the penalty approach does not provide any efficient procedure for simulating vibro-impact problems. As a consequence, we have to find other techniques for proving the existence of solutions and solving problem (P) numerically when K is not convex and/or ∂K is non-smooth.

An answer to the first question may be given by combining both existence results for ODE and for variational inequalities when the data are analytical (which means that no flat points like those in the example presented in Sect. 3 may appear along the trajectory). This idea has been developed by P.Ballard in [4], but it does not yield to the construction of approximate solutions, and thus does not give any numerical tool for solving problem (P).

Motivated now by both existence results and efficient numerical techniques that allow us to encompass the long list of drawbacks enumerated in Sect. 3, we will consider time-stepping approximations of problem (P) in the rest of this chapter.

5 Time-Discretization at the Position Level

Having in mind all the difficulties listed in the previous sections that we have to encompass in order to solve problem (P), it is clear that we have to propose an approach that

- avoids systematic impact detection,
- is able to deal with non-smooth sets of constraints and/or non-analytical data.

So, we may try to apply to the MDI (4) time-discretization techniques inspired by classical methods for ODE. In this framework, we may replace the acceleration term with finite difference approximations $\dfrac{u^n - u^{n-1}}{\Delta t}$, with $u^n = \dfrac{q^{n+1} - q^n}{\Delta t}$, where the q^j's are some approximation of q at the discrete instants $t_j = j\Delta t$, with $\Delta t > 0$. Then the simplest idea consists in considering an explicit time-discretization of the MDI given by

$$\mathsf{M}(q^n)\frac{u^n - u^{n-1}}{\Delta t} - f(t_n, q^n, u^{n-1}) \in -N_K(q^n) \quad \forall n \in \left\{1, \dots, \left\lfloor \frac{\tau}{\Delta t} \right\rfloor\right\}.$$

Unfortunately, if $q^n \notin K$, we have $N_K(q^n) = \emptyset$, and this inclusion does not admit any solution. The next simplest idea consists in considering the semi-implicit time-discretization of the MDI given by

$$\mathsf{M}(q^n)\frac{u^n - u^{n-1}}{\Delta t} - f(t_n, q^n, u^{n-1}) \in -N_K(q^{n+1}) \quad \forall n \in \left\{1, \dots, \left\lfloor \frac{\tau}{\Delta t} \right\rfloor\right\} \tag{11}$$

Of course, we need to check first that this inclusion always admits a solution.

Lemma 5 *Let us assume that $K \neq \emptyset$ and that $\left(\nabla f_\alpha(q)\right)_{\alpha \in J(q)}$ is linearly independent for all $q \in \partial K$. Then, for all $n \in \left\{1, \dots \left\lfloor \frac{\tau}{\Delta t} \right\rfloor\right\}$, $q^n \in \mathbf{R}^d$ and $q^{n-1} \in \mathbf{R}^d$, (11) admits at least one solution q^{n+1}.*

Proof Let $n \in \left\{1, \dots \left\lfloor \frac{\tau}{\Delta t} \right\rfloor\right\}$, $q^n \in \mathbf{R}^d$ and $q^{n-1} \in \mathbf{R}^d$. We define

$$W^n = 2q^n - q^{n-1} + \Delta t^2 \mathsf{M}^{-1}(q^n) f\left(t_n, q^n, \frac{q^n - q^{n-1}}{\Delta t}\right).$$

Then, (11) is equivalent to $\mathsf{M}(q^n)(W^n - q^{n+1}) \in N_K(q^{n+1})$.

If $W^n \in K$, then $q^{n+1} = W^n$ is the solution. Otherwise, since K is a closed non-empty subset of $\mathbf{R}^d$, we know that $\mathrm{Argmin}_{z \in K} \|W^n - z\|_{q^n}$ is not empty and we let

$q^{n+1} \in \mathrm{Argmin}_{z \in K} \|W^n - z\|_{q^n}$. Then, for all $z \in K$,

$$\|W^n - q^{n+1}\|_{q^n}^2$$
$$\leq \|W^n - z\|_{q^n}^2 = \|W^n - q^{n+1}\|_{q^n}^2 + 2\langle W^n - q^{n+1}, q^{n+1} - z\rangle_{q^n} + \|q^{n+1} - z\|_{q^n}^2,$$

i.e.,

$$\langle W^n - q^{n+1}, z - q^{n+1}\rangle_{q^n} \leq \frac{1}{2}\|z - q^{n+1}\|_{q^n}^2.$$

If $q^{n+1} \in \mathrm{Int}(K)$, then $z = q^{n+1} + r(W^n - q^{n+1})$ belongs to K for all positive small enough number r, which yields

$$\|W^n - q^{n+1}\|_{q^n}^2 \leq \frac{r}{2}\|W^n - q^{n+1}\|_{q^n}^2.$$

and at the limit as r tends to zero, we obtain $q^{n+1} = W^n \in \mathrm{Int}(K)$, which is absurd. So, $q^{n+1} \in \partial K$. Let $v \in T_K(q^{n+1}) = \left\{v \in \mathbf{R}^d;\ \langle \nabla f_\alpha(q^{n+1}), v\rangle \geq 0\ \forall \alpha \in J(q^{n+1})\right\}$. Since $\left(\nabla f_\alpha(q^{n+1})\right)_{\alpha \in J(q^{n+1})}$ is linearly independent, it may be completed as a basis of $\mathbf{R}^d$, and we denote as $(\varepsilon_j)_{1 \leq j \leq d}$ the dual basis. Then, for all $\delta > 0$, we define $v_\delta = v + \delta \sum_{\beta \in J(q^{n+1})} \varepsilon_\beta$. We get

$$\left\langle \nabla f_\alpha(q^{n+1}), v_\delta\right\rangle = \left\langle \nabla f_\alpha(q^{n+1}), v\right\rangle + \sum_{\beta \in J(q^{n+1})} \underbrace{\delta\left\langle \nabla f_\alpha(q^{n+1}), \varepsilon_\beta\right\rangle}_{=0 \text{ if } \beta \neq \alpha; = \delta \text{ if } \beta = \alpha} \geq \delta > 0$$

for all $\alpha \in J(q^{n+1})$. Then, $z(t) = q^{n+1} + v_\delta t \in K$ for all t in a right neighbourhood of 0, and we get

$$\langle W^n - q^{n+1}, v_\delta\rangle_{q^n} \leq \frac{t}{2}\|v_\delta\|_{q^n}^2,$$

which implies that $\langle W^n - q^{n+1}, v_\delta\rangle_{q^n} \leq 0$ for all $\delta > 0$. At the limit as δ tends to zero, we obtain $\langle W^n - q^{n+1}, v\rangle_{q^n} = \left\langle \mathsf{M}(q^n)(W^n - q^{n+1}), v\right\rangle \leq 0$ for all $v \in T_K(q^{n+1})$, which allows us to conclude.

This proof yields also a way to define q^0 and q^1 in order to initialize the algorithm: since (11) is satisfied by $q^{n+1} \in \mathrm{Argmin}_{z \in K} \|W^n - z\|_{q^n}$ for all $n \in \left\{1, \ldots, \left\lfloor \frac{\tau}{\Delta t} \right\rfloor\right\}$, we may choose

$$q^0 = q_0, \quad q^1 \in \mathrm{Argmin}_{z \in K} \|q_0 + \Delta t u_0 - z\|_{q_0}. \tag{12}$$

Clearly, (11) provides a time-discretization of the MDI (4) with feasible approximate positions q^n's, but we may wonder how the discrete velocities behave when the constraints are saturated. Let us check first what happens in the case of the bouncing ball model problem, i.e., $d = 1$, $K = \mathbf{R}^+$, $\mathsf{M}(q) \equiv 1$. Then, (11) reduces to

$$q^{n+1} = \max(W^n, 0), \; W^n = 2q^n - q^{n-1} + \Delta t^2 f\left(t_n, q^n, \frac{q^n - q^{n-1}}{\Delta t}\right) \tag{13}$$

for all $n \in \left\{1, \dots, \left\lfloor \frac{\tau}{\Delta t} \right\rfloor\right\}$. For simplicity, let us assume that $f \equiv 0$, $q_0 = 1$, $u_0 = -1$ and let us choose $\Delta t \in \left(0, \frac{1}{2}\right)$. With (12), we get

$$q^0 = 1, \quad q^1 = \max(1 - \Delta t, 0) = 1 - \Delta t > 0$$

and

$$q^{n+1} = 1 - (n+1)\Delta t \quad \text{for all } n \in \{0, \dots, n_*\} \text{with} n_* = \left\lfloor \frac{1}{\Delta t} \right\rfloor - 1.$$

At the next time-step, we get

$$W^{n_*+1} = 2q^{n_*+1} - q^{n_*} = 1 - (n_* + 2)\Delta t < 0$$

and $q^{n_*+2} = 0$. It follows that

$$W^{n_*+2} = 2q^{n_*+2} - q^{n_*+1} = -q^{n_*+1} \le 0,$$

thus $q^{n_*+3} = 0$, and by an immediate induction, $q^n = 0$ for all $n \ge n_* + 3$. The discrete velocities satisfy

$$u^n = -1 \quad \forall n \in \{1, \dots, n_*\}, \quad u^{n_*+1} = -\frac{q^{n_*+1}}{\Delta t} \in (-1, 0], \quad u^n = 0 \quad \forall n \ge n_* + 2$$

and we have an approximate solution to problem (P) with $e = 0$ (see Fig. 2).

Hence, we have to modify (11) in order to get approximate solutions of problem (P) with $e \in (0, 1]$ as well. We observe that (13) implies that $q^{n+1} = 0$ whenever $W^n \le 0$, and if it occurs for two successive time-steps, we get automatically $e = 0$. So, from the heuristic point of view, we may try to modify (13) as

$$q^{n+1} = -eq^{n-1} + (1 + e)\max(W_e^n, 0) \quad \forall n \ge 1, \tag{14}$$

with an appropriate choice of W_e^n, i.e., with W_e^n such that

$$q^{n+1} = -eq^{n-1} + (1 + e)W_e^n = 2q^n - q^{n-1} = 1 - (n+1)\Delta t,$$

as long as $(n + 1)\Delta t$ is not too close to $t_* = 1$. We obtain

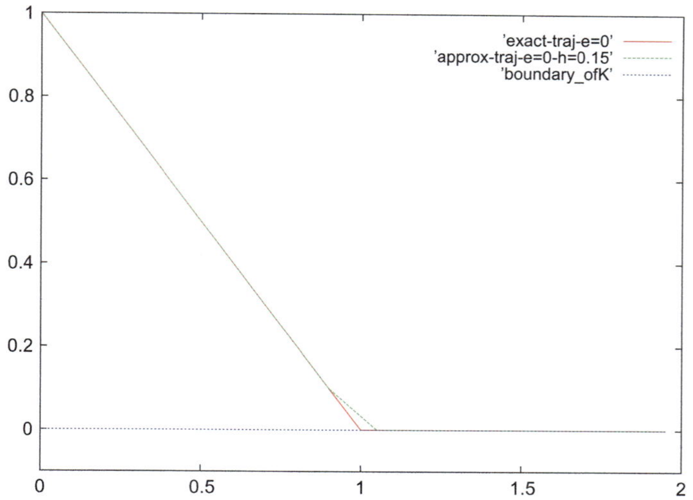

Fig. 2 Exact and approximate trajectories for $\Delta t = 0.15$ and $e = 0$

$$W_e^n = \frac{2q^n - (1-e)q^{n-1}}{1+e},$$

and (13) is replaced by

$$q^{n+1} = -eq^{n-1} + \max\big(2q^n - (1-e)q^{n-1}, 0\big) \quad \forall n \geq 1.$$

We obtain again that $q^{n+1} = 1 - (n+1)\Delta t$ for all $n \in \{1, \dots, n_{e*}\}$, with

$$n_{e*} = \max\left\{n \geq 1;\ 1 + \frac{2e\Delta t}{1+e} \geq (n+1)\Delta t\right\}.$$

Then,

$$W_e^{n_{e*}+1} < 0, \quad q^{n_{e*}+2} = -eq^{n_{e*}}$$

and

$$W_e^{n_{e*}+2} = \frac{2q^{n_{e*}+2} - (1-e)q^{n_{e*}+1}}{1+e} = \frac{-2eq^{n_{e*}} - (1-e)(2q^{n_{e*}} - q^{n_{e*}-1})}{1+e} = -W^{n_{e*}} \leq 0,$$

so $q^{n_{e*}+3} = -eq^{n_{e*}+1}$ and $u^{n_{e*}+2} = -eu^{n_{e*}} = e$. We infer that $W_e^{n_{e*}+3} = -eW_e^{n_{e*}+1} \geq 0$, thus $q^{n_{e*}+4} = 2q^{n_{e*}+3} - q^{n_{e*}+2}$ and $u^{n_{e*}+3} = u^{n_{e*}+2}$. The discrete velocities now satisfy

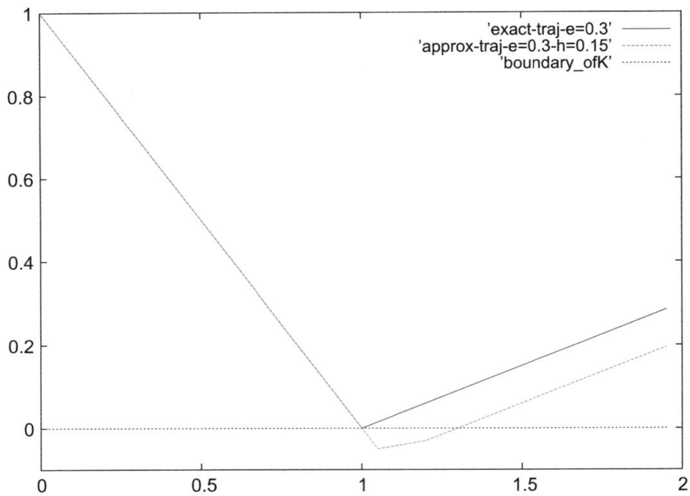

Fig. 3 Exact and approximate trajectories for $\Delta t = 0.15$ and $e = 0.3$

$$
\begin{aligned}
u^n &= -1 \quad \forall n \in \{1, \dots, n_{e*}\}, \\
u^{n_{e*}+1} &= \frac{-eq^{n_{e*}} - q^{n_{e*}+1}}{\Delta t} \in (-1, e], \\
u^n &= e \quad \forall n \geq n_{e*} + 2
\end{aligned}
$$

and we have obtained an approximate solution to problem (P) with $e \in (0, 1]$ (see Fig. 3).

Let us assume now that $f \not\equiv 0$. Then, we get

$$
q^{n+1} = -eq^{n-1} + \max\left(2q^n - (1-e)q^{n-1} + \Delta t^2 f\left(t_n, q^n, \frac{q^n - q^{n-1}}{\Delta t}\right), 0\right) \tag{15}
$$

and we let

$$
W_e^n = \frac{1}{1+e}\left(2q^n - (1-e)q^{n-1} + \Delta t^2 f\left(t_n, q^n, \frac{q^n - q^{n-1}}{\Delta t}\right)\right)
$$

for all $n \in \left\{1, \dots, \left\lfloor \frac{\tau}{\Delta t} \right\rfloor\right\}$. As long as $W_e^n \geq 0$ (15), reduces to

$$
q^{n+1} = 2q^n - q^{n-1} + \Delta t^2 f\left(t_n, q^n, \frac{q^n - q^{n-1}}{\Delta t}\right),
$$

i.e.,

$$\frac{q^{n+1} - 2q^n + q^{n-1}}{\Delta t^2} = f\left(t_n, q^n, \frac{q^n - q^{n-1}}{\Delta t}\right),$$

which is simply a centered time-discretization of the ODE $\ddot{q} = f(t, q, \dot{q})$ that describes the unconstrained dynamics. On the contrary, if there exists $n_{e*} \geq 1$ such that $W^{n_{e*}} \geq 0$ and $W^{n_{e*}+1} < 0$, we obtain

$$q^{n_{e*}+2} = -eq^{n_{e*}}, \quad q^{n_{e*}+1} = -eq^{n_{e*}-1} + (1+e)W_e^{n_{e*}} \geq -eq^{n_{e*}-1}.$$

Hence,

$$\begin{aligned}
W_e^{n_{e*}+2} &= \frac{2q^{n_{e*}+2} - (1-e)q^{n_{e*}+1} + \Delta t^2 f(t_{n_{e*}+2}, q^{n_{e*}+2}, u^{n_{e*}+1})}{1+e} \\
&= \frac{-e\big(2q^{n_{e*}} - (1-e)q^{n_{e*}-1}\big)}{1+e} - (1-e)W_e^{n_{e*}} + \frac{\Delta t^2}{1+e} f(t_{n_{e*}+2}, q^{n_{e*}+2}, u^{n_{e*}+1}) \\
&= -W_e^{n_{e*}} + \frac{e\Delta t^2}{1+e} f(t_{n_{e*}}, q^{n_{e*}}, u^{n_{e*}-1}) + \frac{\Delta t^2}{1+e} f(t_{n_{e*}+2}, q^{n_{e*}+2}, u^{n_{e*}+1}) \\
&\leq \frac{e\Delta t^2}{1+e} f(t_{n_{e*}}, q^{n_{e*}}, u^{n_{e*}-1}) + \frac{\Delta t^2}{1+e} f(t_{n_{e*}+2}, q^{n_{e*}+2}, u^{n_{e*}+1}).
\end{aligned}$$

If $W_e^{n_{e*}+2} \leq 0$, then $q^{n_{e*}+3} = -eq^{n_{e*}+1}$ and $u^{n_{e*}+2} = -eu^{n_{e*}}$. Otherwise,

$$0 < W_e^{n_{e*}+2} \leq \frac{\Delta t^2}{1+e} \big(ef(t_{n_{e*}}, q^{n_{e*}}, u^{n_{e*}-1}) + f(t_{n_{e*}+2}, q^{n_{e*}+2}, q^{n_{e*}+1})\big)$$

and

$$q^{n_{e*}+3} = -eq^{n_{e*}+1} + (1+e)W_e^{n_{e*}+2} = -eq^{n_{e*}+1} + \mathcal{O}(\Delta t^2),$$

which yields $u^{n_{e*}+2} = -eu^{n_{e*}} + \mathcal{O}(\Delta t)$. So, the discrete velocities are reversed and multiplied by the restitution coefficient e, up to some additional terms of order $\mathcal{O}(\Delta t)$.

Moreover, as long as $W_e^n \geq 0$, we have

$$u^n = u^{n-1} + \Delta t f\left(t_n, q^n, \frac{q^n - q^{n-1}}{\Delta t}\right),$$

so

$$|u^n| \leq |u^{n-1}| + \Delta t \big|f\left(t_n, q^n, u^{n-1}\right)\big|.$$

Whenever $n \geq 2$ and $W_e^n < 0$, we have $q^{n+1} + eq^{n-1} = 0$ and $q^n + eq^{n-2} \geq 0$, so

$$u^n = \frac{q^{n+1} - q^n}{\Delta t} = \frac{q^{n+1} + eq^{n-1} - (q^n + eq^{n-2}) - e(q^{n-1} - q^{n-2})}{\Delta t} \leq -eu^{n-2}$$

and

$$\frac{q^{n+1} + eq^{n-1} - (1+e)W_e^n}{\Delta t} = u^n - u^{n-1} - \Delta t f\left(t_n, q^n, \frac{q^n - q^{n-1}}{\Delta t}\right) \geq 0.$$

It follows that

$$|u^n| \leq \max\left(|u^{n-1}| + \Delta t \left|f\left(t_n, q^n, u^{n-1}\right)\right|, e|u^{n-2}|\right),$$

and by an immediate induction,

$$|u^n| \leq \max\left(|u^0|, |u^1|\right) + \sum_{k=2}^{n} \Delta t \left|f\left(t_k, q^k, u^{k-1}\right)\right| \quad \forall n \in \left\{2, \ldots, \left\lfloor \frac{\tau}{\Delta t} \right\rfloor\right\}.$$

Reminding ourselves that $q^0 = q_0 \in \mathbf{R}^+$, we obtain

$$|u^0| = \left|\frac{q^1 - q^0}{\Delta t}\right| = \frac{\left|\max(q_0 + \Delta t\, u_0, 0) - q_0)\right|}{\Delta t} \leq |u_0|$$

and

$$\begin{aligned} |u^1| &= \left|\frac{q^2 - q^1}{\Delta t}\right| = \left|\frac{(1+e)\max(W_e^1, 0) - (1+e)q^0 + q^0 - q^1}{\Delta t}\right| \\ &\leq \left|\frac{(1+e)W_e^1 - (1+e)q^0}{\Delta t}\right| + |u^0| \leq 3|u^0| + \Delta t \left|f\left(t_1, q^1, u^0\right)\right|. \end{aligned}$$

We infer that

$$\begin{aligned} |u^n| &\leq 3|u_0| + \sum_{k=1}^{n} \Delta t \left|f\left(t_k, q^k, u^{k-1}\right)\right| \\ &\leq 3|u_0| + \sum_{k=1}^{n} \Delta t \left|f\left(t_k, q_0, 0\right)\right| + \sum_{k=1}^{n} L_f \Delta t\left(|q^k - q^0| + |u^{k-1}|\right) \\ &\leq 3|u_0| + \tau \max_{t \in [0,\tau]} \left|f(t, q_0, 0)\right| + L_f(\tau + 1)\Delta t \sum_{k=0}^{n-1} |u^k| \quad \forall n \in \left\{1, \ldots, \left\lfloor \frac{\tau}{\Delta t} \right\rfloor\right\}, \end{aligned}$$

where we recall that L_f denotes the Lipschitz constant of f with respect to its last two arguments. By using the discrete Grönwall's lemma, we determine that there exists $h_* > 0$ and $C_* > 0$ such that, for all $\Delta t \in (0, h_*)$,

$$|u^n| \leq C_* \quad \forall n \in \left\{0, \ldots, \left\lfloor \frac{\tau}{\Delta t} \right\rfloor\right\}.$$

We define the approximate solutions q_h by linear interpolation of the q^n's, i.e.,

$$q_h(t) = q^n + (t - t_n)\frac{q^{n+1} - q^n}{h} \quad \forall t \in [t_n, t_{n+1}) \cap [0, \tau], \quad \forall n \in \left\{0, \dots, \left\lfloor \frac{\tau}{\Delta t} \right\rfloor\right\}$$

with $h = \Delta t$ and $t_n = nh$ for all $n \in \left\{0, \dots, \left\lfloor \frac{\tau}{\Delta t} \right\rfloor\right\}$. In order to prove that $(q_h)_{h>0}$ converges to a solution of the bouncing ball problem, we reproduce the same kind of mathematical analysis as in the previous section.

By construction, q_h is continuous and affine by parts on $[0, \tau]$ and

$$\dot{q}_h(t) = u^n \quad \forall t \in (t_n, t_{n+1}) \cap [0, \tau] \quad \forall n \in \left\{0, \dots, \left\lfloor \frac{\tau}{\Delta t} \right\rfloor\right\}.$$

So,

$$\begin{aligned}
TV\big(\dot{q}_h, [0, \tau]\big) &= \sum_{n=0}^{\lfloor \tau/h \rfloor - 1} |u^{n+1} - u^n| = |u^1 - u^0| + \sum_{n=1}^{\lfloor \tau/h \rfloor - 1} |u^{n+1} - u^n| \\
&\le |u^1 - u^0| + \sum_{n=1}^{\lfloor \tau/h \rfloor - 1} \underbrace{\left|u^{n+1} - u^n - hf(t_{n+1}, q^{n+1}, u^n)\right|}_{\ge 0} \\
&\quad + h \sum_{n=1}^{\lfloor \tau/h \rfloor - 1} \left|f(t_{n+1}, q^{n+1}, u^n)\right| \\
&\le |u^1 - u^0| + u^{\lfloor \tau/h \rfloor} - u^1 + 2h \sum_{n=1}^{\lfloor \tau/h \rfloor - 1} \left|f(t_{n+1}, q^{n+1}, u^n)\right| \\
&\le 4(C_* + \tau M_*),
\end{aligned}$$

with $M_* = \sup\big\{\left|f(t, q, v)\right|;\ (t, q, v) \in [0, \tau] \times [q_0 - C_*\tau, q_0 + C_*\tau] \times [-C_*, C_*]\big\}$.

Let us choose, from now on, $h = \Delta t = \dfrac{\tau}{N}$, with $N \in \mathbf{N}^*$. By using Ascoli's and Helly's theorem and possibly extracting a subsequence, still denoted $(q_h)_{h>0}$, we obtain

$$\begin{aligned}
&q_h(t) \longrightarrow_{h \to 0} q(t) \quad \text{uniformly in } [0, \tau], \\
&\dot{q}_h(t) \longrightarrow_{h \to 0} u(t) \quad \text{for all } t \in [0, \tau], \\
&\mathrm{d}\dot{q}_h - f(\cdot, q_h, \dot{q}_h)\mathrm{d}t \rightharpoonup_{h \to 0} \lambda = \mathrm{d}u - f(\cdot, q, u)\mathrm{d}t \quad \text{weakly} * \text{ in } \mathscr{M}\big([0, \tau]; \mathbf{R}\big),
\end{aligned}$$

with $q \in C^0\big([0, \tau]; \mathbf{R}\big)$ and $u \in BV\big([0, \tau]; \mathbf{R}\big)$ such that

$$q(t) = q_0 + \int_0^t u(s)\, ds \quad \forall t \in [0, \tau].$$

Moreover, for all $t \in (0, \tau)$, let $n = \left\lfloor \dfrac{t}{h} \right\rfloor$. For all h small enough, we have $n \in \left\{1, \dots, \left\lfloor \frac{\tau}{\Delta t} \right\rfloor - 1\right\}$ and

$$
\begin{aligned}
q(t) &= \frac{1}{1+e}\left(q(t_{n+1}) + eq(t_{n-1}) + e\int_{t_{n-1}}^{t} u(s)\,ds - \int_{t}^{t_{n+1}} u(s)\,ds\right)\\
&\geq \frac{1}{1+e}\underbrace{(q^{n+1}+eq^{n-1})}_{\geq 0} - \|q-q_h\|_{C([0,\tau];\mathbf{R})} - 2C_* h.
\end{aligned}
$$

Thus, $q(t) \geq 0$ for all $t \in [0,\tau]$.

Finally, we decompose the measure $\mathrm{d}\dot{q}_h - f(\cdot, q_h, \dot{q}_h)\mathrm{d}t$ as

$$
\mathrm{d}\dot{q}_h - f(\cdot, q_h, \dot{q}_h)\mathrm{d}t = \sum_{n=1}^{\lfloor \tau/h \rfloor - 1} \big(u^n - u^{n-1} - hf(t_n, q^n, u^{n-1})\big)\delta_{t=t_n} + \lambda_h,
$$

with

$$
\lambda_h = \sum_{n=1}^{\lfloor \tau/h \rfloor - 1} hf(t_n, q^n, u^{n-1})\delta_{t=t_n} - f(\cdot, q_h, \dot{q}_h)\mathrm{d}t,
$$

where $\delta_{t=t_n}$ is the Dirac measure of mass 1 at t_n. For all $v \in C^0\big([0,\tau];\mathbf{R}^+\big)$, we have

$$
\big\langle \mathrm{d}\dot{q}_h - f(\cdot, q_h, \dot{q}_h)\mathrm{d}t, v\big\rangle_{\mathscr{M}([0,\tau];\mathbf{R}), C^0([0,\tau];\mathbf{R})} \geq \langle \lambda_h, v\rangle_{\mathscr{M}([0,\tau];\mathbf{R}), C^0([0,\tau];\mathbf{R})}
$$

and

$$
\begin{aligned}
&\big|\langle \lambda_h, v\rangle_{\mathscr{M}([0,\tau];\mathbf{R}), C^0([0,\tau];\mathbf{R})}\big|\\
&= \Big|\sum_{n=1}^{\lfloor \tau/h \rfloor - 1} hf(t_n, q^n, u^{n-1})v(t_n) - \int_0^\tau f\big(t, q_h(t), \dot{q}_h(t)\big)v(t)\,dt\Big|\\
&\leq \sum_{n=1}^{\lfloor \tau/h \rfloor - 1} \int_{t_n}^{t_{n+1}} \big|f(t_n, q^n, u^{n-1}) - f\big(t, q_h(t), \dot{q}_h(t)\big)\big|\big|v(t)\big|\,dt\\
&\quad + \sum_{n=1}^{\lfloor \tau/h \rfloor - 1} \int_{t_n}^{t_{n+1}} \big|f(t_n, q^n, u^{n-1})\big|\big|v(t_n) - v(t)\big| + \int_0^h \big|f\big(t, q_h(t), \dot{q}_h(t)\big)v(t)\big|\,dt\\
&\leq \sum_{n=1}^{\lfloor \tau/h \rfloor - 1} L_f \int_{t_n}^{t_{n+1}} \big(\big|q^n - q_h(t)\big| + |u^n - u^{n-1}|\big)\|v\|_{C^0([0,\tau];\mathbf{R})}\,dt\\
&\quad + \tau M_* \omega_v(h) + hM_*\|v\|_{C^0([0,\tau];\mathbf{R})}\\
&\leq L_f h\big(C_*\tau + TV\big(\dot{q}_h, [0,\tau]\big)\big)\|v\|_{C^0([0,\tau];\mathbf{R})} + \tau M_*\omega_v(h) + hM_*\|v\|_{C^0([0,\tau];\mathbf{R})},
\end{aligned}
$$

where ω_v denotes the continuity modulus of v. At the limit as h tends to zero, we obtain

$$
\big\langle \mathrm{d}\dot{q}_h - f(\cdot, q_h, \dot{q}_h)\mathrm{d}t, v\big\rangle_{\mathscr{M}([0,\tau];\mathbf{R}), C^0([0,\tau];\mathbf{R})} \longrightarrow_{h\to 0} \langle \lambda, v\rangle_{\mathscr{M}([0,\tau];\mathbf{R}), C^0([0,\tau];\mathbf{R})} \geq 0
$$

for all $v \in C^0\big([0,\tau];\mathbf{R}^+\big)$. So, λ is a non-negative measure.

Moreover, let $v \in C^0\big([0,\tau];\mathbf{R}\big)$ such that $\mathrm{Supp}(v) \subset \big\{t \in [0,\tau];\ q(t) > 0\big\}$. By compacity of $\mathrm{Supp}(v)$, there exists $N_* \in \mathbf{N}^*$ such that, for all $h = \dfrac{\tau}{N}$ with $N \geq N_*$, we have $q_h(t) > 0$ for all $t \in \mathrm{Supp}(v)$ and $W_e^n > 0$ for all $nh \in \mathrm{Supp}(v)$. It follows that

$$\big\langle \mathrm{d}\dot{q}_h - f(\cdot, q_h, \dot{q}_h)\mathrm{d}t, v\big\rangle_{\mathscr{M}([0,\tau];\mathbf{R}),C^0([0,\tau];\mathbf{R})} = \langle \lambda_h, v\rangle_{\mathscr{M}([0,\tau];\mathbf{R}),C^0([0,\tau];\mathbf{R})} \longrightarrow_{h\to 0} \langle \lambda, v\rangle_{\mathscr{M}([0,\tau];\mathbf{R}),C^0([0,\tau];\mathbf{R})} = 0$$

which implies that $\mathrm{Supp}(\lambda) \subset \big\{t \in [0,\tau];\ q(t) = 0\big\}$, and we may conclude that q is a solution of the vibro-impact problem.

In the general case, with $d \geq 1$, $e \in [0,1]$ and a set of admissible configurations defined as

$$K = \big\{q \in \mathbf{R}^d;\ f_\alpha(q) \geq 0\ \forall \alpha \in \{1,\ldots,\nu\}\big\},$$

the time-stepping algorithm is given as a natural generalization of (14), i.e.,

- $q^0 = q_0,\ q^1 \in \mathrm{Argmin}_{z\in K}\left\|q_0 + \Delta t u_0 - z\right\|_{q^0}$,
- for all $n \in \left\{1,\ldots,\left\lfloor \dfrac{\tau}{\Delta t}\right\rfloor\right\}$

$$q^{n+1} \in -eq^{n-1} + (1+e)\mathrm{Argmin}_{z\in K}\|W_e^n - z\|_{q^n}, \tag{16}$$

with

$$W_e^n = \frac{1}{1+e}\left(2q^n - (1-e)q^{n-1} + \Delta t^2 \mathsf{M}^{-1}(q^n) f\left(t_n, q_n, \frac{q^n - q^{n-1}}{\Delta t}\right)\right). \tag{17}$$

By replacing q^{n+1} with $\dfrac{q^{n+1} + eq^{n-1}}{1+e}$ in the proof of Lemma 5, we determine that (16)–(17) imply that

$$\mathsf{M}(q^n)\frac{u^n - u^{n-1}}{\Delta t} - f\left(t_n, q^n, \frac{q^n - q^{n-1}}{\Delta t}\right) \in -N_K\left(\frac{q^{n+1} + eq^{n-1}}{1+e}\right)$$

for all $n \in \left\{1,\ldots,\left\lfloor \dfrac{\tau}{\Delta t}\right\rfloor\right\}$. Let us observe that whenever $W_e^n \in K$, we get

$$\mathsf{M}(q^n)\frac{q^{n+1} - 2q^n + q^{n-1}}{\Delta t^2} = f\left(t_n, q^n, \frac{q^n - q^{n-1}}{\Delta t}\right),$$

which is a centered time-discretization of the ODE describing the unconstrained dynamics. Moreover, the average approximate position $\dfrac{q^{n+1} + eq^{n-1}}{1+e}$ belongs to

K for all $n \in \left\{1, \ldots, \left\lfloor \frac{\tau}{\Delta t} \right\rfloor\right\}$, which implies that $\text{dist}(q^n, K) \le \mathcal{O}(\Delta t)$ for all $n \in \left\{0, \ldots, \left\lfloor \frac{\tau}{\Delta t} \right\rfloor\right\}$.

We define once again the approximate solution q_h as a linear interpolation of the q^n's. Let us assume:

(H1) for all $\alpha \in \{1, \ldots, \nu\}$, the function f_α belongs to $C^1(\mathbf{R}^d; \mathbf{R})$, ∇f_α is locally Lipschitz continuous, and does not vanish in a neighbourhood of $\{q \in \mathbf{R}^d : f_\alpha(q) = 0\}$,
(H2) the interior of K is not empty and the active constraints along ∂K are functionally independent, i.e., for all $q \in \partial K$, the vectors $\big(\nabla f_\alpha(q)\big)_{\alpha \in J(q)}$ are linearly independent,
(H3) M is a mapping of class C^1 from $\mathbf{R}^d$ to the set of symmetric positive definite $d \times d$ matrices,
(H4) f is a continuous function from $[0, \tau] \times \mathbf{R}^d \times \mathbf{R}^d$ $(\tau > 0)$ to $\mathbf{R}^d$,
(H5) for all compact subsets $\mathscr{B}$ of $\mathbf{R}^d$, there exist $C_{\mathscr{B}} > 0$ and $r_{\mathscr{B}} > 0$ such that, for all $(q_1, q_2) \in (K \cap \mathscr{B})^2$ such that $\|q_1 - q_2\| \le r_{\mathscr{B}}$, we have

$$
\begin{aligned}
&\langle e_\alpha(q_1), e_\beta(q_2)\rangle \le C_{\mathscr{B}} \|q_1 - q_2\| \quad \text{if } e = 0,\\
&\big|\langle e_\alpha(q_1), e_\beta(q_2)\rangle\big| \le C_{\mathscr{B}} \|q_1 - q_2\| \quad \text{if } e \in (0, 1],
\end{aligned}
$$

for all $(\alpha, \beta) \in J(q_1) \times J(q_2)$ such that $\alpha \ne \beta$, where $e_\alpha(q_i) = \dfrac{\mathsf{M}^{-1/2}(q_i)\nabla f_\alpha(q_i)}{\|\mathsf{M}^{-1/2}(q_i)\nabla f_\alpha(q_i)\|}$ for all $\alpha \in J(q_i)$, $i = 1, 2$.

Let us observe that, by choosing $q_1 = q_2 = q \in \partial K$, this last assumption reduces to the "angle condition" given in Proposition 1, i.e.,

$$
\begin{aligned}
&\langle \nabla f_\alpha(q), \mathsf{M}(q)^{-1} \nabla f_\beta(q)\rangle \le 0 \quad \text{if } e = 0\\
&\langle \nabla f_\alpha(q), \mathsf{M}(q)^{-1} \nabla f_\beta(q)\rangle = 0 \quad \text{if } e \ne 0
\end{aligned}
$$

for all $(\alpha, \beta) \in J(q)^2$ such that $\alpha \ne \beta$, for all $q \in \partial K$.
We obtain the following theorem.

Theorem 1 *Let $q_0 \in K$, $u_0 \in T_K(q_0)$ and $e \in [0, 1]$. For any $C > \|u_0\|_{q_0}$, let $\tau(C) > 0$ be defined by Proposition 2. Then, there exist $\tau_* \in \big[\min\big(\tau(C), \tau\big), \tau\big]$ and a subsequence of $(q_h)_{h>0}$, still denoted $(q_h)_{h>0}$, such that*

$$
\begin{aligned}
&\dot q_h(t) \longrightarrow_{h\to 0^+} u(t) \quad \text{for all } t \in [0, \tau_*],\\
&q_h(t) \longrightarrow_{h\to 0^+} q(t) = q_0 + \int_0^t u(s)\, ds \quad \text{uniformly in } [0, \tau_*],
\end{aligned}
$$

with $u \in BV\big([0, \tau_*]; \mathbf{R}^d\big)$, and u is a solution to problem (P).

The mathematical analysis follows the same steps as in the bouncing ball model problem: we establish first a uniform estimate of the approximate velocities $\dot q_h$ and of their total variation, then we pass to the limit as h tends to zero by applying

Ascoli's and Helly's theorems, and finally, we study the reflexion of the velocities when the constraints are saturated. Nevertheless, the details of the proofs are much more technical, even in the single constraint case ($\nu = 1$), and we refer the reader to [52] when $\nu = 1$ and [45, 46] when $\nu \geq 2$.

6 Time-Discretization at the Velocity Level

With the results of the previous section, it seems that there is nothing to add about the mathematical aspects of vibro-impact problems, since we have obtained a numerical method and existence results that allow us to consider rather weak regularity assumptions for the data and any value of d, ν and e.

Nevertheless, the time-stepping scheme proposed in the previous section requires to find, at each time-step, the proximal point of W_e^n in K relatively to the kinetic metric at q^n, which is not necessarily an easy task. Indeed, if $W_e^n \in K$, the solution to (16) is obvious and we simply have $q^{n+1} = -eq^{n-1} + (1+e)W_e^n$. Otherwise, if K is convex, (16) admits an unique solution that can be computed by standard procedures like gradient methods or relaxation (see [16], for instance). But when K is not convex, (16) is much more difficult to solve and may even admit several solutions.

So, the previous time-stepping scheme is a good theoretical tool for proving existence results in a more general framework than in [3, 4, 9, 12, 14, 43, 49, 55, 60, 62] but it is a useful numerical tool, mainly when K is convex, and in such a case, the efficiency of this algorithm has been clearly showned through several examples of implementation (see, for instance, [20, 21, 50, 53]).

Motivated by computational issues, it does, however, seem necessary to propose another numerical method allowing us to handle the non-convex case as well. In order to overcome the difficulty due to the lack of convexity of the set of admissible configurations, we observe that the set of admissible right velocities $T_K(q)$ is always convex. More precisely, we have the following properties: the constraint at the position level, i.e., the condition $q(t) \in K$ for all $t \in [0, \tau]$, yields $\dot q(t+0) \in T_K\big(q(t)\big)$ for all $t \in [0, \tau)$, which can be interpreted as a constraint at the velocity level. It follows that, for any function u of Bounded Variation such that $q(t) = q_0 + \int_0^t u(s)\,ds$ for all $t \in [0, \tau]$, we have $u(t) \in T_K\big(q(t)\big)$ for almost every $t \in (0, \tau)$. Indeed, we have $\dot q(t+0) = u(t+0) \in T_K\big(q(t)\big)$ for all $t \in [0, \tau)$. Since u is continuous except on a (at most) countable subset of $[0, \tau]$, the conclusion follows.

The converse property is also true and we have the Lemma.

Lemma 6 *Let $u \in BV\big([0, \tau]; \mathbf{R}^d\big)$ with $\tau > 0$, $q_0 \in K$ and q be defined by*

$$q(t) = q_0 + \int_0^t u(s)\,ds \quad \forall t \in [0, \tau].$$

Assume, moreover, that $u(t) \in T_K\big(q(t)\big)$ *for almost every* $t \in (0, \tau)$*. Then,* $q(t) \in K$ *for all* $t \in [0, \tau]$.

Proof Let us assume that there exists $\tau_* \in [0, \tau]$ such that $q(\tau_*) \notin K$. Then, $\tau_* > 0$ and there exists $\alpha \in \{1, \ldots, \nu\}$ such that $f_\alpha\big(q(\tau_*)\big) < 0$. We define

$$\tau_0 = \inf\big\{t \in [0, \tau];\ t \le \tau_*,\ f_\alpha\big(q(s)\big) < 0\ \forall s \in [t, \tau_*]\big\}.$$

By continuity of q, we get $\tau_0 \in [0, \tau_*)$ and $f_\alpha\big(q(\tau_0)\big) = 0$. Thus,

$$f_\alpha\big(q(t)\big) = \underbrace{f_\alpha\big(q(\tau_0)\big)}_{=0} + \int_{\tau_0}^{t} \underbrace{\big\langle \nabla f_\alpha\big(q(s)\big), u(s)\big\rangle}_{\ge 0}\, ds \quad \forall t \in (\tau_0, \tau_*],$$

which gives a contradiction.

So, we may wonder if it is possible to obtain another formulation of vibro-impact problems at the velocity level. Let us recall the impact law:

$$\dot q(t+0) = \mathrm{Proj}_{q(t)}\big(T_K\big(q(t)\big), \dot q(t-0)\big) - e\mathrm{Proj}_{q(t)}\big(\mathsf{M}^{-1}\big(q(t)\big)N_K\big(q(t)\big), \dot q(t-0)\big).$$

By definition of the projection on a convex set, it is equivalent to

$$\frac{\dot q(t+0) + e\dot q(t-0)}{1+e} \in T_K\big(q(t)\big)$$

and

$$\left\langle \mathsf{M}\big(q(t)\big)\underbrace{\left(\dot q(t-0) - \frac{\dot q(t+0) + e\dot q(t-0)}{1+e}\right)}_{=\frac{\dot q(t-0)-\dot q(t+0)}{1+e}}, v - \frac{\dot q(t+0) + e\dot q(t-0)}{1+e}\right\rangle \le 0$$

$\forall v \in T_K\big(q(t)\big)$,

i.e.,

$$\mathsf{M}\big(q(t)\big)\big(\dot q(t-0) - \dot q(t+0)\big) \in \partial\psi_{T_K(q(t))}\left(\frac{\dot q(t+0) + e\dot q(t-0)}{1+e}\right),$$

where $\psi_{T_K(q)}$ is the indicatrix function of $T_K(q)$ defined as

$$\psi_{T_K(q)}(v) = \begin{cases} 0 & \text{if } v \in T_K(q), \\ +\infty & \text{otherwise,} \end{cases}$$

and $\partial\psi_{T_K(q)}$ is its subdifferential [58] given by

$$\partial\psi_{T_K(q)}(v) = \begin{cases} \{w \in \mathbf{R}^d;\ 0 \geq \langle w, z - v\rangle\ \forall z \in T_K(q)\} & \text{if } v \in T_K(q), \\ \emptyset\ \ \text{otherwise.} \end{cases}$$

Next, we define the average velocity u_e by

$$u_e(t) = \frac{u(t+0) + eu(t-0)}{1+e} \quad \forall t \in (0,\tau), \quad u_e(0) = u(0), \quad u_e(\tau) = u(\tau).$$

Reminding ourselves that u is continuous except on a (at most) countable subset of $[0,\tau]$, we infer that $u_e(t \pm 0) = u(t \pm 0)$ for all $t \in (0,\tau)$. Thus, $u_e \in BV\big([0,\tau]; \mathbf{R}^d\big)$ with $\mathrm{d}u_e = \mathrm{d}u$. It follows that $\mathsf{M}(q)\mathrm{d}u - f(\cdot, q, u)\mathrm{d}t = \mathsf{M}(q)\mathrm{d}u_e - f(\cdot, q, u_e)\mathrm{d}t$ and

$$q(t) = q_0 + \int_0^t u(s)\,ds = q_0 + \int_0^t u_e(s)\,ds \quad \forall t \in [0,\tau].$$

Moreover, the impact law (P4) is equivalent to

$$u_e(t) = \frac{u(t+0) + eu(t-0)}{1+e} = \frac{\dot{q}(t+0) + e\dot{q}(t-0)}{1+e} \in T_K\big(q(t)\big)$$

and

$$\mathsf{M}\big(q(t)\big)\big(\dot{q}(t-0) - \dot{q}(t+0)\big) \in \partial\psi_{T_K(q(t))}\big(u_e(t)\big)$$

for all $t \in (0,\tau)$. Next, we observe that for any $q \in \partial K$ and $v \in T_K(q)$, we have $\partial\psi_{T_K(q)}(v) = N_{T_K(q)}(v)$, since $T_K(q)$ is convex [58] and $N_{T_K(q)}(v) \subset N_K(q)$ with equality if and only if $\big\langle \nabla f_\alpha(q), v\big\rangle = 0$ for all $\alpha \in J(q)$. Indeed,

$$T_K(q) = \Big\{v \in \mathbf{R}^d;\ \underbrace{\big\langle \nabla f_\alpha(q), v\big\rangle}_{=\varphi_\alpha(v)} \geq 0\ \forall \alpha \in J(q)\Big\}.$$

and, for all $v \in T_K(q)$, we have

$$\partial\psi_{T_K(q)}(v) = N_{T_K(q)}(v) = \left\{ w = -\sum_{\alpha \in J'(v)} \lambda_\alpha \underbrace{\nabla\varphi_\alpha(v)}_{=\nabla f_\alpha(q)},\ \lambda_\alpha \geq 0 \right\},$$

with

$$J'(v) = \big\{\alpha \in J(q);\ \varphi_\alpha(v) = \big\langle \nabla f_\alpha(q), v\big\rangle \leq 0\big\}.$$

Let $t \in (0,\tau)$. We can distinguish the following cases:

- if $q(t) \in \mathrm{Int}(K)$, then $T_K\big(q(t)\big) = \mathbf{R}^d$ and $N_K\big(q(t)\big) = \{0\} = \partial\psi_{T_K(q(t))}(v)$ for all $v \in T_K\big(q(t)\big)$;

• if $q(t) \in \partial K$ and $\dot q(t-0) = \dot q(t+0)$, then

$$\big\langle \nabla f_\alpha\big(q(t)\big), \dot q(t+0)\big\rangle = \big\langle \nabla f_\alpha\big(q(t)\big), \dot q(t-0)\big\rangle = 0 \quad \forall \alpha \in J\big(q(t)\big),$$

thus

$$\partial\psi_{T_K(q(t))}\left(\frac{\dot q(t+0) + e\dot q(t-0)}{1+e}\right) = \partial\psi_{T_K(q(t))}\Big(\underbrace{\frac{u(t+0)+eu(t-0)}{1+e}}_{=u_e(t)}\Big) = N_K\big(q(t)\big).$$

• if $q(t) \in \partial K$ and $\dot q(t-0) \neq \dot q(t+0)$, then

$$\begin{aligned}&\big(\mathsf{M}(q)\mathrm{d}u_e - f(\cdot, q, u_e)\mathrm{d}t\big)\big(\{t\}\big) = \mathsf{M}\big(q(t)\big)\big(u_e(t+0) - u_e(t-0)\big)\\ &= -\mathsf{M}\big(q(t)\big)\big(\dot q(t-0) - \dot q(t+0)\big).\end{aligned}$$

So, we can gather properties (P3) and (P4) of problem (P) into a single condition: the measure $f(\cdot, q, u_e)\mathrm{d}t - \mathsf{M}(q)\mathrm{d}u_e$ takes its values in $\partial\psi_{T_K(q)}(u_e)$. In order to give a precise mathematical meaning, we introduce the measure $\mu = |du_e| + dt$ where $|du_e|$ is defined by

$$|\mathrm{d}u_e|(A) = \sup \sum \big\|\mathrm{d}u_e(B_k)\big\|$$

for any Borel subset $A \subset [0, \tau]$, where the supremum is taken over all the finite families $(B_k)_k$ of disjoint Borel sets included in A (see [8] I-3.10, for instance).

Then, we can check that, for any Borel subset $A \subset [0, \tau]$, such that $\mu(A) = 0$ we have $|\mathrm{d}u_e|(A) = 0$ and $\mathrm{d}t(A) = 0$, which means that the measures $\mathrm{d}u_e$ and $\mathrm{d}t$ are *absolutely continuous* with respect to the measure μ. Using Radon-Nikodym's theorem, we infer that there exist $u'_\mu \in L^1\big([0, \tau]; \mathbf{R}^d, \mu\big)$ and $t'_\mu \in L^1\big([0, \tau]; \mathbf{R}, \mu\big)$ such that, for any Borel subset $A \subset [0, \tau]$,

$$\mathrm{d}u_e(A) = \int_A u'_\mu \mathrm{d}\mu, \quad \mathrm{d}t(A) = \int_A t'_\mu \mathrm{d}\mu.$$

([8] I-4.9)

So, we can now introduce another formulation of our vibro-impact problem.

Problem (P') Let $q_0 \in K$, $u_0 \in T_K(q_0)$. Find a function $u_e : [0, \tau] \to \mathbf{R}^d$, with $\tau > 0$, such that:

(P'1) $u_e \in BV\big([0, \tau]; \mathbf{R}^d\big)$, $u_e(0+0) = u_0$,

(P'2) $u_e(t) = \dfrac{u_e(t+0) + eu_e(t-0)}{1+e}$ for all $t \in (0, \tau)$,

(P'3) there exists a non-negative measure μ such that the Stieltjes measure $\mathrm{d}u_e$ and Lebesgue's measure $\mathrm{d}t$ admit densities relatively to μ, denoted, respectively, u'_μ and t'_μ, and

$$f\big(t, q(t), u_e(t)\big)t'_\mu(t) - \mathsf{M}\big(q(t)\big)u'_\mu(t) \in \partial\psi_{T_K(q(t))}\big(u_e(t)\big) \quad \mu \text{ a.e.on}(0, \tau),$$

with

$$q(t) = q_0 + \int_0^t u_e(s)\,ds \quad \forall t \in [0, \tau].$$

Let us emphasize that any μ-negligible subset of $[0, \tau]$ is also negligible with respect to Lebesgue's measure and (P'3) implies that $u_e(t) \in T_K\big(q(t)\big)$ for almost every $t \in (0, \tau)$. By using Lemma 6, we obtain $q(t) \in K$ for all $t \in [0, \tau]$.

Moreover, for any $t \in (0, \tau)$ such that $\dot{q}(t-0) \neq \dot{q}(t+0)$, we have

$$u_e(t-0) = \dot{q}(t-0) \neq \dot{q}(t+0) = u_e(t+0)$$

and thus $|\mathrm{d}u_e|\big(\{t\}\big) \neq 0$ and $\mu\big(\{t\}\big) > 0$. Recalling that $\partial\psi_{T_K(q(t))}\big(u_e(t)\big)$ is a cone, we infer from (P'3) that

$$\begin{aligned}&\big(f(\cdot, q, u_e)\mathrm{d}t - \mathsf{M}(q)\mathrm{d}u_e\big)\big(\{t\}\big) = \mathsf{M}\big(q(t)\big)\big(u_e(t-0) - u_e(t+0)\big)\\ &= \mathsf{M}\big(q(t)\big)\big(\dot{q}(t-0) - \dot{q}(t+0)\big) \in \partial\psi_{T_K(q(t))}\big(u_e(t)\big) = \partial\psi_{T_K(q(t))}\left(\frac{\dot{q}(t+0) + e\dot{q}(t-0)}{1+e}\right),\end{aligned}$$

which is equivalent to $\dfrac{\dot{q}(t+0) + e\dot{q}(t-0)}{1+e} = \mathrm{Proj}_{q(t)}\big(T_K\big(q(t)\big), \dot{q}(t-0)\big)$, i.e.,

$$\dot{q}(t+0) = \mathrm{Proj}_{q(t)}\big(T_K\big(q(t)\big), \dot{q}(t-0)\big) - e\mathrm{Proj}_{q(t)}\big(\mathsf{M}\big(q(t)\big)N_K\big(q(t)\big), \dot{q}(t-0)\big).$$

Remark 2 Since $\partial\psi_{T_K(q(t))}(v)$ is a cone for any $v \in \mathbf{R}^d$, we obtain that (P'3) is independent of the choice of any non-negative measure μ such that $\mathrm{d}u_e$ and $\mathrm{d}t$ are absolutely continuous with respect to μ [37, 38].

Starting from this new formulation, we propose another time-discretization of our problem. More precisely, for a given time-step $\Delta t > 0$, we define the approximate positions and velocities by

$$q^0 = q_0, \quad u^0 = u_0,$$

and for all $n \in \left\{0, \ldots, \left\lfloor \dfrac{\tau}{\Delta t} \right\rfloor - 1\right\}$

$$q^{n+1} = q^n + \Delta t\, u^n, \tag{18}$$

$$f\big(t_{n+1}, q^{n+1}, u^n\big) - \mathsf{M}(q^{n+1})\left(\frac{u^{n+1} - u^n}{\Delta t}\right) \in \partial\psi_{T_K(q^{n+1})}\left(\frac{u^{n+1} + eu^n}{1+e}\right). \tag{19}$$

By using the definition of $\partial \psi_{T_K(q)}(\cdot)$, we can deduce that this inclusion always admits an unique solution given by

$$u^{n+1} = -eu^n + (1+e)\mathrm{Proj}_{q^{n+1}}\left(T_K(q^{n+1}), u^n + \frac{\Delta t}{1+e}\mathsf{M}^{-1}(q^{n+1})f\big(t_{n+1}, q^{n+1}, u^n\big)\right).$$

Moreover, if $q^{n+1} \in \mathrm{Int}(K)$, we simply obtain

$$\frac{q^{n+2} - 2q^{n+1} + q^n}{\Delta t^2} = \mathsf{M}^{-1}(q^{n+1})f\big(t_{n+1}, q^{n+1}, u^n\big)$$

and we recover a centered time-discretization of the ODE describing the unconstrained dynamics. Otherwise, there exist non-negative real numbers $\big(\lambda_\alpha^{n+1}\big)_{\alpha \in J(q^{n+1})}$ such that

$$\mathsf{M}(q^{n+1})(u^{n+1} - u^n) - \Delta t f\big(t_{n+1}, q^{n+1}, u^n\big) = \sum_{\alpha \in J(q^{n+1})} \lambda_\alpha^{n+1} \nabla f_\alpha(q^{n+1}),$$

and we get a quite natural time-discretization of the MDI (5).

In order to have an idea of how this scheme behaves, let us consider once again the bouncing ball model problem, i.e., $d = 1$, $K = \mathbf{R}^+$ and $\mathsf{M}(q) \equiv 1$. Then, (19) reduces to

$$u^{n+1} = \begin{cases} u^n + \Delta t f\big(t_{n+1}, q^{n+1}, u^n\big) \ \text{ if } q^{n+1} > 0, \\ -eu^n + (1+e)\max\left(u^n + \dfrac{\Delta t}{1+e} f\big(t_{n+1}, q^{n+1}, u^n\big), 0\right) \ \text{ if } q^{n+1} \le 0, \end{cases}$$

or equivalently,

$$u^{n+1} = \begin{cases} u^n + \Delta t f\big(t_{n+1}, q^{n+1}, u^n\big) \ \text{ if } q^{n+1} > 0 \\ \qquad \text{or if } q^{n+1} \le 0 \text{ and } u^n + \dfrac{\Delta t}{1+e} f\big(t_{n+1}, q^{n+1}, u^n\big) \ge 0, \\ -eu^n \ \text{ otherwise.} \end{cases}$$

It follows that

$$|u^{n+1}| \le |u^n| + \Delta t\big|f\big(t_{n+1}, q^{n+1}, u^n\big)\big| \le |u^0| + \sum_{k=1}^{n+1} \Delta t\big|f\big(t_k, q^k, u^{k-1}\big)\big|$$

for all $n \in \left\{0, \ldots, \left\lfloor \frac{\tau}{\Delta t} \right\rfloor - 1\right\}$. Then, with the same kind of computations as in the previous section, we deduce that there exists $h_* > 0$ and $C_* > 0$ such that, for all $\Delta t \in (0, h_*)$,

$$|u^n| \le C_* \quad \forall n \in \left\{0, \ldots, \left\lfloor \frac{\tau}{\Delta t} \right\rfloor\right\}.$$

Next, we observe that

$$u^{n+1} - u^n - \Delta t f\left(t_{n+1}, q^{n+1}, u^n\right) = \begin{cases} 0 \ \text{ if } q^{n+1} > 0 \text{ or if} \\ q^{n+1} \leq 0 \text{ and } u^n + \dfrac{\Delta t}{1+e} f\left(t_{n+1}, q^{n+1}, u^n\right) \geq 0, \\ -(1+e)\left(u^n + \dfrac{\Delta t}{1+e} f\left(t_{n+1}, q^{n+1}, u^n\right)\right) \\ \text{otherwise,} \end{cases}$$

and thus

$$u^{n+1} - u^n - \Delta t f\left(t_{n+1}, q^{n+1}, u^n\right) \geq 0 \quad \forall n \in \left\{0, \ldots, \left\lfloor \frac{\tau}{\Delta t} \right\rfloor - 1\right\}.$$

We infer that

$$\sum_{n=0}^{\lfloor \tau/\Delta t \rfloor - 1} |u^{n+1} - u^n| \leq u^{\lfloor \tau/\Delta t \rfloor} - u^0 + 2\Delta t \sum_{n=1}^{\lfloor \tau/\Delta t \rfloor} \left| f(t_n, q^n, u^{n-1}) \right| \leq 2(C_* + \tau M_*)$$

with $M_* = \max\left\{\left|f(t, q, v)\right|; \quad (t, q, v) \in [0, \tau] \times [q_0 - C_*\tau, q_0 + C_*\tau] \times [-C_*, C_*]\right\}$.

Let us define the approximate solutions q_h by linear interpolation of the q^n's, i.e.,

$$q_h(t) = q^n + (t - t_n)\frac{q^{n+1} - q^n}{h} \quad \forall t \in [t_n, t_{n+1}], \quad \forall n \in \left\{0, \ldots, \left\lfloor \frac{\tau}{h} \right\rfloor - 1\right\},$$

where $h = \Delta t = \dfrac{\tau}{N}$ with $N \in \mathbf{N}^*$ and $t_n = nh$ for all $n \in \{0, \ldots, N\}$.

We pass to the limit as Δt tends to zero, with Ascoli's and Helly's theorems: there exists a subsequence, still denoted $(q_h)_{h>0}$, such that

$$\begin{aligned} q_h(t) &\longrightarrow_{h\to 0} q(t) \quad \text{uniformly in } [0, \tau], \\ \dot{q}_h(t) &\longrightarrow_{h\to 0} u(t) \quad \text{for all } t \in [0, \tau], \end{aligned}$$

with $q \in C^0\big([0, \tau]; \mathbf{R}\big)$ and $u \in BV\big([0, \tau]; \mathbf{R}\big)$ such that

$$q(t) = q_0 + \int_0^t u(s)\, ds \quad \forall t \in [0, \tau].$$

Next, we prove the following proposition.

Proposition 4 *For all $t \in [0, \tau]$ we have $q(t) \geq 0$.*

Proof We use a contradiction argument. Let us assume that there exists $\tau_0 \in (0, \tau)$ such that $q(\tau_0) < 0$ and let

$$\tau_1 = \inf\left\{s \in [0, \tau_0); \ q(t) \leq \frac{1}{2} q(\tau_0) \ \forall t \in [s, \tau_0]\right\}.$$

Since $q(0) = q_0 \geq 0$, we obtain $\tau_1 > 0$ and $q(\tau_1) = \frac{1}{2}q(\tau_0)$. With the uniform convergence of $(q_h)_{h>0}$ to q on $[0, \tau]$, we obtain that

$$q_h(t) \leq \frac{1}{4}q(\tau_0) < 0 \quad \forall t \in [\tau_1, \tau_0]$$

for all h small enough. Let $n_1 = \left\lfloor \frac{\tau_1}{h} \right\rfloor$ and $n_0 = \left\lfloor \frac{\tau_0}{h} \right\rfloor$. By definition of the scheme,

$$\begin{aligned} q_h(t_{n_0+1}) + eq_h(t_{n_0}) &= q_h(t_{n_1+1}) + eq_h(t_{n_1}) + h \sum_{n=n_1+1}^{n_0} \underbrace{(u^n + eu^{n-1})}_{\geq 0 \text{ since } q^n \leq 0} \\ &\geq q_h(t_{n_1+1}) + eq_h(t_{n_1}) \end{aligned}$$

for all h small enough. By passing to the limit as h tends to zero, we obtain

$$0 > q(\tau_0) \geq q(\tau_1) = \frac{1}{2}q(\tau_0),$$

which is absurd.

Let us observe that

$$u(t+0) = \dot{q}(t+0) \in T_K\big(q(t)\big), \quad u(t-0) = \dot{q}(t-0) \in -T_K\big(q(t)\big) \quad \forall t \in (0, \tau),$$

and we cannot infer that $u(t) = \dfrac{u(t+0) + eu(t-0)}{1+e} \in T_K\big(q(t)\big)$ for all $t \in (0, \tau)$. Hence, we modify u on a (at most) countable subset of $[0, \tau]$ and we define $u_e \in BV\big([0, \tau]; \mathbf{R}^d\big)$ by

$$u_e(t) = \frac{u(t+0) + eu(t-0)}{1+e} \quad \forall t \in [0, \tau],$$

with the convention that $u(0-0) = u(0)$ and $u(\tau+0) = u(\tau)$. It follows that $u_e \in BV\big([0, \tau]; \mathbf{R}^d\big)$ and $u_e(t \pm 0) = u(t \pm 0)$ for all $t \in (0, \tau)$. Thus,

$$u_e(t) = \frac{u_e(t+0) + eu_e(t-0)}{1+e} \quad \forall t \in (0, \tau)$$

and

$$q(t) = q_0 + \int_0^t u_e(s)\, ds \quad \forall t \in [0, \tau].$$

In order to check that u_e is a solution to (P'), it remains to prove (P'3). This is achieved in two steps.

Let $\mu = |\mathrm{d}u_e| + \mathrm{d}t$ and $t \in (0, \tau)$. First, we observe that, for all $n \in \left\{0, \ldots, \left\lfloor \frac{\tau}{\Delta t} \right\rfloor - 1\right\}$,

$$\left(u^n + \frac{h}{1+e} f(t_{n+1}, q^{n+1}, u^n) - \frac{u^{n+1} + eu^n}{1+e}\right)\left(x - \frac{u^{n+1} + eu^n}{1+e}\right) \le 0 \quad \forall x \in \mathbf{R}^+,$$

i.e.,

$$\begin{aligned} hf(t_{n+1}, q^{n+1}, u^n)(x - u^n) &\le (u^{n+1} - u^n)x - \frac{1}{2}\left(|u^{n+1}|^2 - |u^n|^2\right) \\ &+ \left(\frac{1}{2} - \frac{1}{1+e}\right)|u^{n+1} - u^n|^2 + hf(t_{n+1}, q^{n+1}, u^n)\left(\frac{u^{n+1} - u^n}{1+e}\right). \end{aligned}$$

Hence, for any subinterval $[t, s] \subset [0, \tau]$, we have

$$\begin{aligned} \sum_{n=j-1}^{k-1} hf(t_{n+1}, q^{n+1}, u^n)(x - u^n) &\le \big(u_h(t_k) - u_h(t_{j-1})\big)x - \frac{1}{2}\left(\left|u_h(t_k)\right|^2 - \left|u_h(t_{j-1})\right|^2\right) \\ &\quad + \frac{hM_*}{1+e} \sum_{n=j}^{k-1} |u^{n+1} - u^n|, \end{aligned}$$

with $t_{j-1} \le t < t_j < \cdots < t_k \le s < t_{k+1}$. If $q(t) > 0$, then, for any small enough Δt and $s - t$, we have $q^{n+1} = q_h(t_{n+1}) > 0$, which implies, $\frac{u^{n+1} + eu^n}{1+e} = u^n + \frac{h}{1+e} f(t_{n+1}, q^{n+1}, u^n)$ for all $n \in \{j-1, \ldots, k-1\}$. Thus, the same inequality is valid for any $x \in \mathbf{R} = T_K\big(q(t)\big)$. So, whatever the value of $q(t)$, when Δt tends to zero, we get

$$\begin{aligned} &\int_t^s f\big(\sigma, q(\sigma), u(\sigma)\big)\big(x - u(\sigma)\big)\, d\sigma \\ &\le \big(u(s) - u(t)\big)x - \frac{1}{2}\left(\left|u(s)\right|^2 - \left|u(t)\right|^2\right) \quad \forall x \in T_K\big(q(t)\big). \end{aligned}$$

By using Jeffery's theorem [29] and differentiation rules for functions of Bounded Variation, we infer that there exists a μ-negligible subset A of $[0, \tau]$ such that, for all $t \in (0, \tau) \setminus A$ such that u_e is continuous at t, we have

$$f\big(t, q(t), u_e(t)\big)t'_\mu(t)\big(x - u_e(t)\big) \le u'_\mu(t)\big(x - u_e(t)\big) \quad \forall x \in T_K\big(q(t)\big),$$

which yields

$$f\big(t, q(t), u_e(t)\big)t'_\mu(t) - u'_\mu(t) \in \partial\psi_{T_K(q(t))}\big(u_e(t)\big).$$

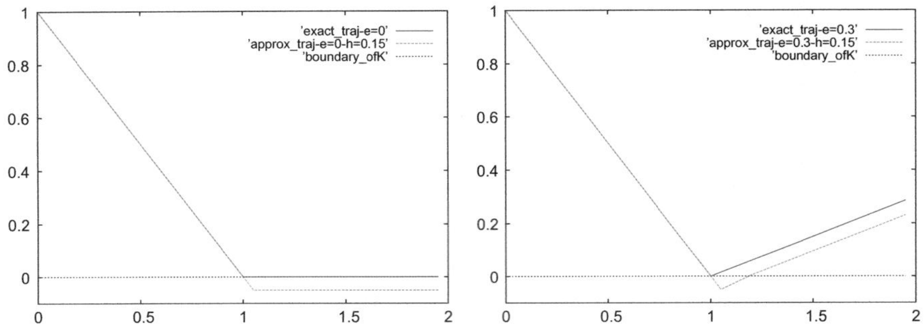

Fig. 4 Exact and approximate trajectories for $\Delta t = 0.15$ and $e = 0$ (left) or $e = 0.3$ (right)

Finally, there remains to prove that (P'3) holds at discontinuity points of u_e. But in such a case, the measures du_e and μ have a Dirac mass and (P'3) reduces to the impact law (P4).

Let us assume for simplicity that $f \equiv 0, q_0 = 1$ and $u_0 = -1$. Then, if $\Delta t \in (0, 1)$, we have $q^1 = 1 - \Delta t > 0$, so $u^1 = u^0 = -1$, and we get

$$q^{n+1} = 1 - (n+1)\Delta t, \quad u^n = -1 \quad \forall n \in \{0, \ldots, n_*\},$$

with

$$n_* = \max\{n \geq 1;\ 1 - n\Delta t > 0\} = \begin{cases} \left\lfloor \dfrac{1}{\Delta t} \right\rfloor & \text{if} \dfrac{1}{\Delta t} \notin \mathbf{N}^*, \\ \left\lfloor \dfrac{1}{\Delta t} \right\rfloor - 1 & \text{if} \dfrac{1}{\Delta t} \in \mathbf{N}^*. \end{cases}$$

Then, $q^{n_*+1} \leq 0$, $u^{n_*+1} = -eu^{n_*} + (1+e)\max(u^{n_*}, 0) = e$, and by induction,

$$q^n = 1 - (n_* + 1)h + e(n - n_* - 1)\Delta t, \quad u^n = e \quad \forall n \geq n_* + 1.$$

The convergence of this velocity-based time-stepping scheme has been proved first in the single constraint case (i.e., $\nu = 1$, which implies that $T_K(q)$ is given as $\mathbf{R}^d$ if $q \in \text{Int}(K)$ or as an half-line otherwise) for a trivial inertia operator and inelastic shocks ([35], see also [32]). This proof has been extended to partially or totally elastic shocks but still with a trivial kinetic metric [33], and then to a non-trivial inertia operator [23, 24, 34]. Finally the general case, with $\nu \geq 1$, $e \in [0, 1]$ and $\mathsf{M}(q) \not\equiv \text{Id}_{\mathbf{R}^d}$, is considered in [47] (see also [48]) and we have the following theorem.

Theorem 2 *Assume that (H1)–(H3) holds and*
(H'4) the function $f : [0, \tau] \times \mathbf{R}^d \times \mathbf{R}^d \to \mathbf{R}^d$ $(\tau > 0)$ *is continuous and locally Lipschitz continuous with respect to its second and third arguments,*
(H'5) for all $q \in \partial K$, *we have*

$$\langle \nabla f_\alpha(q), \mathsf{M}(q)^{-1} \nabla f_\beta(q) \rangle \le 0 \quad \text{if } e = 0$$
$$\langle \nabla f_\alpha(q), \mathsf{M}(q)^{-1} \nabla f_\beta(q) \rangle = 0 \quad \text{if } e \neq 0$$

for all $(\alpha, \beta) \in J(q)^2$ *such that* $\alpha \neq \beta$.
Let $q_0 \in K$, $u_0 \in T_K(q_0)$ *and* $e \in [0, 1]$. *For any* $C > \|u_0\|_{q_0}$, *let* $\tau(C) > 0$ *be defined by Proposition 2. Then, there exist* $\tau_* \in \big[\min\big(\tau(C), \tau\big), \tau\big]$, $u_e \in BV\big([0, \tau_*]; \mathbf{R}^d\big)$ *and a subsequence of* $(q_h, u_h)_{h>0}$, *still denoted* $(q_h, u_h)_{h>0}$, *such that*

$$u_h(t) \longrightarrow_{h\to 0^+} u_e(t) \quad \text{except on a (at most) countable subset of } [0, \tau_*],$$
$$q_h(t) \longrightarrow_{h\to 0^+} q(t) = q_0 + \int_0^t u_e(s)\, ds \quad \text{uniformly in } [0, \tau_*],$$

and u_e *is a solution to problem (P').*

Of course, one may wonder if this numerical method is capable of the complex dynamical behaviour of vibro-impact problems efficiently. Indeed, with the bouncing ball model problem, it seems that this velocity-based time-stepping scheme gives a better approximation of the impact law (P4) than the position-based algorithm described in Sect. 5, since reflexion of the velocity occurs immediately when the discrete position saturates the constraints (we obtain $u^{n_*+1} = -eu^{n_*} = e$ when $q^{n_*+1} \le 0$), while two time-steps are needed to obtain the reflexion of the discrete velocities with (14) (we got $u^{n_{e*}+1} = \dfrac{-eq^{n_{e*}} - q^{n_{e*}+1}}{\Delta t} \in (-1, e]$ and $u^{n_{e*}+2} = -eu^{n_{e*}} = e$ when $W^{n_{e*}+1} = \dfrac{2q^{n_{e*}+1} - (1-e)q^{n_{e*}}}{1+e} < 0$). But, in the general case, the approximate average positions $\dfrac{q^{n+1} + eq^{n-1}}{1+e}$ always satisfy the constraints, since $\dfrac{q^{n+1} + eq^{n-1}}{1+e} \in K$ by definition with the position-based algorithm (16)–(17), which implies that $\operatorname{dist}(q^n, K) \le \mathcal{O}(\Delta t)$, and it can be checked that the reflexion of the velocities occurs in (at most) $\nu + 1$ time-steps (in the general case). On the contrary with (18)–(19), only the constraints at the velocity level are satisfied at each time-step, since we have $\dfrac{u^{n+1} + eu^n}{1+e} \in T_K(q^{n+1})$, but it may lead to some discrepancy at the position level in the case of grazing impacts or when $e = 0$. Indeed, let us go back to the bouncing ball example. If $q^n > 0$ and $q^{n+1} \le 0$, then

$$u^n = \frac{q^{n+1} - q^n}{\Delta t} < 0$$

and

$$u^{n+1} = -eu^n + (1+e) \max\left(u^n + \frac{\Delta t}{1+e} f\big(t_{n+1}, q^{n+1}, u^n\big), 0\right).$$

If $u^n + \dfrac{\Delta t}{1+e} f\left(t_{n+1}, q^{n+1}, u^n\right) \geq 0$, then

$$u^{n+1} = u^n + \Delta t f\left(t_{n+1}, q^{n+1}, u^n\right) \leq \mathcal{O}(\Delta t)$$

and

$$q^{n+2} = q^{n+1} + \Delta t\, u^{n+1} \leq q^{n+1} + \mathcal{O}(\Delta t^2).$$

If $u^{n+1} + \dfrac{\Delta t}{1+e} f\left(t_{n+2}, q^{n+2}, u^{n+1}\right)$ is still non-negative, we get

$$u^{n+2} = u^{n+1} + \Delta t f\left(t_{n+2}, q^{n+2}, u^{n+1}\right) \leq \mathcal{O}(\Delta t),$$

and so on.

Contrastingly, if $u^n + \dfrac{\Delta t}{1+e} f\left(t_{n+1}, q^{n+1}, u^n\right) < 0$, we get $u^{n+1} = -eu^n$ and $q^{n+2} = q^{n+1} - e\Delta t\, u^n = q^n + (1-e)\Delta t\, u^n$, which still may be non-positive and which is certainly non-positive if $e = 0$ (see Fig. 4). Hence, it may be useful in numerical simulations to modify (18)–(19) by adding a post-stabilization procedure [39]. See also the chapter by Brüls et al. in this book [7] related to numerical time-integration schemes in which stabilization of the constraints at position and acceleration levels is implemented. Nevertheless, the mathematical proof of convergence is still open.

For examples of implementation, the reader is referred to [22, 28, 39, 41, 42].

7 Conclusion

Despite their importance in many industrial applications and some pioneering works [17], vibro-impact problems, i.e., dynamics of rigid multibody systems with perfect unilateral constraints, were not really investigated before the end of the '50s [5] and their formulation in the appropriate mathematical framework of functions of Bounded Variation was introduced for the first time in [60] at the end of the '70s. So, this part of non-smooth dynamics is a rather recent and very active research field.

In this chapter, an overview of the state-of-the-art about existence, (non-) uniqueness and continuity of data has been proposed, as well as the main difficulties to be overcome in numerical simulations. Different kinds of approximation – penalty approach, time-stepping scheme at the position or velocity level – have also been described and their behaviour has been illustrated with the bouncing ball model problem.

This chapter focuses on mathematical aspects, so it was not within its scope to present examples of implementations; for numerical aspects, the reader is referred [6] or [1] and the references therein.

Appendix: Functions of Bounded Variation

Definition 1 Let $I = [a, b]$ be a real interval and $u : I \to \mathbf{R}^d$, $d \geq 1$, be a function. The total variation of u on I is given by

$$\mathrm{Var}(u, I) = \sup \sum_{i=1}^{n} \left\| u(t_i) - u(t_{i-1}) \right\| \in \mathbf{R}^+ \cup \{+\infty\},$$

where the supremum is taken over all increasing finite sequences $S : t_0 < t_1 < \cdots < t_n$ of points of I. If $\mathrm{Var}(u, I) < +\infty$, then we say that u is a *function of Bounded Variation* on I and we denote $u \in BV(I; \mathbf{R}^d)$.

Examples: step functions, functions of class C^1, monotone functions when $d = 1$.

Remark 3 If $u \in BV(I; \mathbf{R}^d)$, then u is bounded on I. Indeed, let $t \in (a, b)$ and $S : a = t_0 < t < t_2 = b$. We have

$$\begin{aligned} \left\| u(t) \right\| &\leq \|u(a)\| + \|u(t) - u(a)\| \\ &\leq \|u(a)\| + \|u(t) - u(a)\| + \|u(b) - u(t)\| \\ &\leq \|u(a)\| + \mathrm{Var}(u, I). \end{aligned}$$

Proposition 5 ([27] II-7 or [40] Proposition 4.2 and Corollary 4.4)

If $u \in BV(I; \mathbf{R}^d)$, then u possesses a left limit (respectively right limit), denoted $u(t - 0)$ (respectively $u(t + 0)$) for all $t \in (a, b]$ (respectively $t \in [a, b)$). Furthermore, the set of discontinuity points of u in I, is at most, countable.

By convention, we will denote $u(a - 0) = u(a)$, $u(b + 0) = u(b)$.

Let $u \in BV(I; \mathbf{R})$ and $S : a = t_0 < t_1 < \cdots < t_n = b$ a finite sequence of points of I. For all $i \in \{1, \ldots, n\}$, we consider $\theta_S^i \in [t_{i-1}, t_i]$. Then, for any function $\varphi \in C^0(I; \mathbf{R})$, we define

$$\Sigma(S, \theta, \varphi) = \sum_{i=1}^{n} \varphi(\theta_S^i)\big(u(t_i) - u(t_{i-1})\big).$$

Obviously, we have

$$\left| \Sigma(S, \theta, \varphi) \right| \leq \max_{t \in I} \left| \varphi(t) \right| \mathrm{Var}(u, I).$$

Remark 4 If $u(t) = t$ for all $t \in [a, b]$, then $\Sigma(S, \theta, \varphi)$ is the Riemann sum of φ associated with the pointed subdivision (S, θ).

As in Riemann's theory of integration, it can be proved that $\Sigma(S, \theta, \varphi)$ converges to a limit when $\sigma(S) = \max_{1 \leq i \leq n}(t_i - t_{i-1})$ tends to zero, uniformly with respect to θ ([27] II.9). With the previous inequality, we obtain

$$\lim_{\sigma(S)\to 0}\left|\Sigma(S,\theta,\varphi)\right| \le \max_{t\in I}\left|\varphi(t)\right|\mathrm{Var}(u,I).$$

Thus, the mapping $\varphi \mapsto \lim_{\sigma(S)\to 0}\Sigma(S,\theta,\varphi)$ is linear and continuous from the space $C^0(I;\mathbf{R})$, endowed with the uniform norm on I, to $\mathbf{R}$. It allows us to define a measure, denoted $\mathrm{d}u \in \mathscr{M}^1(I;\mathbf{R})$ such that, for all $\varphi \in C(I;\mathbf{R})$,

$$\lim_{\sigma(S)\to 0}\Sigma(S,\theta,\varphi) = \int_I \varphi\, du \quad \text{(Riemann − Stieltjes integral)}.$$

Definition 2 The measure $\mathrm{d}u$ is called the *Stieltjes measure* associated with u.

Examples:
If $u \in C^1(I;\mathbf{R})$, we obtain $\mathrm{d}u = u'\mathrm{d}t$ where $\mathrm{d}t$ denotes Lebesgue's measure on I. So, $\mathrm{d}u$ generalizes the notion of a derivative for functions of Bounded Variation.
If u is a step function that is discontinuous at t_j, $j = 1,\dots,p$, we obtain $\mathrm{d}u = \sum_{j=1}^{p}\big(u(t_j+0) - u(t_j-0)\big)\delta_{t=t_j}$, where $\delta_{t=t_j}$ denotes the Dirac measure at t_j.

If $u \in BV(I;\mathbf{R}^d)$ with $d > 1$, we define

$$\mathrm{d}u = \sum_{i=1}^{d}\mathrm{d}u_i\, e_i, \qquad \int_I \varphi\, du = \sum_{i=1}^{d}\int_I \varphi_i\, du_i \quad \forall\varphi \in C^0(I;\mathbf{R}^d),$$

where $(e_i)_{1\le i\le d}$ is the canonical basis of $\mathbf{R}^d$ and $(u_i)_{1\le i\le d}$ (respectively $(\varphi_i)_{1\le i\le d}$) are the coordinates of u (respectively φ) in the basis $(e_i)_{1\le i\le d}$.

Proposition 6 ([40] Proposition 8.1 and Corollary 8.2) *Let $u \in BV(I;\mathbf{R}^d)$. Then, for any $(a',b') \in I\times I$ such that $a' < b'$, we have*

$$\begin{aligned}
&\mathrm{d}u\big(\{a'\}\big) = u(a'+0) - u(a'-0), \quad \mathrm{d}u\big([a',b']\big) = u(b'+0) - u(a'-0),\\
&\mathrm{d}u\big([a',b'[\big) = u(b'-0) - u(a'-0), \quad \mathrm{d}u\big(]a',b']\big) = u(b'+0) - u(a'+0),\\
&\mathrm{d}u\big(]a',b'[\big) = u(b'-0) - u(a'+0).
\end{aligned}$$

Remark 5 The previous relations uniquely characterize the measure $\mathrm{d}u$, since the Borel tribute on I is generated by compact subintervals of I. As a consequence, if u and v are two functions of Bounded Variation on I such that $u(t\pm 0) = v(t\pm 0)$ for all $t \in I$, we have $\mathrm{d}u = \mathrm{d}v$.

Finally, let us recall a compactness result for functions of Bounded Variation.

Theorem 3 (Helly's theorem [27] II-8.9 and II-15.3) *Let $(u_n)_{n\ge 1}$ be a sequence of functions of $BV\big([a,b];\mathbf{R}^d\big)$ such that $(u_n)_{n\ge 1}$ is uniformly of bounded variation and $\big(u_n(a)\big)_{n\ge 1}$ is bounded. Thus, there exists a subsequence, still denoted $(u_n)_{n\ge 1}$, and $u \in BV\big([a,b];\mathbf{R}^d\big)$ such that*

$$u_n(x) \longrightarrow_{n\to+\infty} u(x) \quad \forall x \in [a,b],$$

and

$$\lim_{n\to+\infty}\int_a^b \varphi\, du_n = \int_a^b \varphi\, du \quad \forall\varphi \in C\big([a,b];\mathbf{R}^d\big),$$

i.e.,

$$\mathrm{d}u_n \rightharpoonup_{n\to+\infty} \mathrm{d}u \quad \textit{weakly} * \textit{ in } \mathcal{M}\big([a,b];\mathbf{R}^d\big).$$

References

1. Acary V, Brogliato B (2008) Numerical methods for nonsmooth dynamical systems. Applications in mechanics and electronics. Springer, Berlin
2. Arnold VI (1978) Mathematical methods of classical mechanics. Springer, Berlin
3. Attouch H, Cabot A, Redont P (2002) The dynamics of elastic shocks via epigraphical regularization of a differential inclusion. Barrier and penalty approximations. Adv Math Sci Appl 12(1):273–306
4. Ballard P (2000) The dynamics of discrete mechanical systems with perfect unilateral constraints. Arch Ration Mech Anal 154(3):199–274
5. Bressan A (1959) Questioni de regolarita e di unicita del moto in presenza di vincoli olonomi unilaterali. Rend Sem Mat Univ Padova 29:271–315
6. Brogliato B, ten Dam AA, Paoli L, Genot F, Abadie M (2002) Numerical simulation of finite dimensional multibody nonsmooth mechanical systems. ASME Appl Mech Rev 55(2):107–149
7. Brüls O, Acary V, Cardona A (2018) On the constraints formulation in the nonsmooth generalized-α method. In: Leine R, Acary V, Brüls O (eds) Advanced topics in nonsmooth dynamics, transactions of the European network for nonsmooth dynamics. Springer, Berlin (2018)
8. Buchwalter H (1992) Variations sur l'analyse. Ellipses, Paris
9. Buttazzo G, Percivale D (1983) On the approximation of the elastic bounce problem on Riemannian manifolds. J Differ Equ 47(2):227–245
10. Cabot A (2004) Bounce law at the corners of convex billiards. Nonlinear Anal TMA 57(4):597–614
11. Cabot A, Paoli L (2007) Asymptotics for some vibro-impact problems with a linear dissipation term. J Math Pures Appl 87(3):291–323
12. Carriero M, Pascali E (1981) The one-dimensional rebound problem and its approximations with nonconvex penalties. Rend Mat 13(4):541–553
13. Carriero M, Pascali E (1982) Uniqueness of the one-dimensional bounce problem as a generic property in $L^1([0,T];I\!R)$. Boll Un Mat Ital A 6(1):87–91
14. Carriero M, Leaci A, Pascali E (1983) Convergence for the first-integral equation associated with the one-dimensional elastic bounce problem. Ann Mat Pura Appl 133:227–256
15. Carriero M, Leaci A, Pascali E (1993) Convergence for the first integral equation associated with the bounce problem. Atti Accad Naz Lincei Rend Cl Sci Fis Mat Nat 72(4):209–216
16. Ciarlet PG (1990) Introduction à l'analyse numérique matricielle et à l'optimisation. Masson, Paris
17. Delassus E (1917) Mémoire sur la théorie des liaisons finies unilatérales. Ann Sci Ecole Norm Sup 34:95
18. Deryabin MV (1995) On the realization of unilateral constraints. J Appl Math Mech 58(6):1079–1083
19. Dieudonné J (1969) Eléments d'analyse. Gauthier-Villars, Paris

20. Dumont Y, Paoli L (2006) Vibrations of a beam between obstacles, convergence of a fully discretized approximation. M2AN Math Model Numer Anal 40(4):705–734
21. Dumont Y, Paoli L (2008) Numerical simulation of a model of vibrations with joint clearance. Int J Comput Appl Technol 33(1):41–53
22. Dumont Y, Paoli L (2015) Dynamic contact of a beam against rigid obstacles: convergence of a velocity-based approximation and numerical results. Nonlinear Anal RWA 22:520–536
23. Dzonou R, Monteiro Marques M (2007) A sweeping process approach to inelastic contact problems with general inertia operators. Eur J Mech A/Solids 26(3):474–490
24. Dzonou R, Monteiro Marques M, Paoli L (2009) A convergence result for a vibro-impact problem with a general inertia operator. Nonlinear Dyn 58(1–2):361–384
25. Eck C, Jarusek J, Krbec M (2005) Unilateral contact problems in mechanics. Variational methods and existence theorems. Monographs and textbooks in pure and applied mathematics, vol 270. Chapman & Hall/CRC, Boca Raton (2005)
26. Eck C, Jaruek J, Star J (2013) Normal compliance contact models with finite interpenetration. Arch Ration Mech Anal 208(1):25–57
27. Hildebrandt TH (1963) Introduction to the theory of integration. Academic Press, New-York
28. Jean M (1999) The nonsmooth contact dynamics method. Comput Methods Appl Mech Eng 177:235–257
29. Jeffery RL (1932) Non-absolutely convergent integrals with respect to functions of bounded variations. Trans AMS 34: 645–675 (1932)
30. Kikuchi N, Oden JT (1988) Contact problems in elasticity: a study of variational inequalities and finite element methods. SIAM, Philadelphia
31. Kozlov VV (1990) A constructive method for justifying the theory of systems with nonretaining constraints. J Appl Math Mech 52(6):691–699
32. Kunze M, Monteiro Marques M (2000) An introduction to Moreau's sweeping process. In: Brogliato B (ed) Impacts in mechanical systems. Lecture notes in physics, vol 551. Springer, Berlin, pp 1–60
33. Mabrouk M (1998) A unified variational model for the dynamics of perfect unilateral constraints. Eur J Mech A/Solids 17:819–842
34. Maury B (2006) A time-stepping scheme for inelastic collisions. Numerical handling of the nonoverlapping constraint. Numer Math 102(4):649–679
35. Monteiro Marques M (1993) Differential inclusions in nonsmooth mechanical problems. Birkhauser, Basel
36. Moreau JJ (1962) Decomposition orthogonale d'un espace hilbertien selon deux cônes mutuellement polaires. Comptes Rendus Acad Sci Paris 255:238–240
37. Moreau JJ (1983) Liaisons unilatérales sans frottement et chocs inélastiques. Comptes Rendus Acad Sci Paris, Série II 296:1473–1476
38. Moreau JJ (1985) Standard inelastic shocks and the dynamics of unilateral constraints. In: Del Piero G, Maceri F (eds) Unilateral problems in structural analysis. CISM courses and lectures vol 288. Springer, New-York, pp 173–221
39. Moreau JJ (1986) Dynamique des systèmes à liaisons unilatérales avec frottement sec eventuel. Technical note no. 1-85, LMGC, Université des Sciences et Techniques du Languedoc
40. Moreau JJ (1988) Bounded variation in time. In: Topics in nonsmooth mechanics. Birkhauser, Basel
41. Moreau JJ (1994) Some numerical methods in multibody dynamics: application to granular ma- terials. Eur J Mech A/Solids 13(4):93–114
42. Moreau JJ (1999) Numerical aspects of sweeping process. Comput Methods Appl Mech Eng 177:329–349
43. Paoli L (2000) An existence result for vibrations with unilateral constraints: case of a nonsmooth set of constraints. Math Models Methods Appl Sci 10(6):815–831
44. Paoli L (2005) Continuous dependence on data for vibro-impact problems. Math Models Methods Appl Sci 15(1):53–93
45. Paoli L (2005) An existence result for non-smooth vibro-impact problems. J Differ Equ 211(2):247–281

46. Paoli L (2010) Time-stepping approximation of rigid-body dynamics with perfect unilateral constraints. I and II. Arch Ration Mech Anal 198(2):457–503 and 505–568 (2010)
47. Paoli L (2011) A proximal-like method for a class of second order measure-differential inclusions describing vibro-impact problems. J Differ Equ 250(1):476–514
48. Paoli L (2011) A proximal-like algorithm for vibro-impact problems with a non-smooth set of constraints. AIMS Proc Disc Control Dyn Syst II:1186–1195
49. Paoli L, Schatzman M (1993) Mouvement à un nombre fini de degrés de liberté avec contraintes unilatérales: cas avec perte d'énergie. RAIRO Modél Math Anal Numér (M2AN) 27(6):673–717
50. Paoli L, Schatzman M (1998) Resonance in impact problems. Math Comput Model 28(4–8):385–406
51. Paoli L, Schatzman M (2000) Ill-posedness in vibro-impact and its numerical consequences. In: Proceedings of European congress on computational methods in applied sciences and engineering (ECCOMAS), 11–14 September 2000, CDRom
52. Paoli L, Schatzman M (2002) A numerical scheme for impact problems, I and II. SIAM J Numer Anal 40(2):702–733 and 734–768 (2002)
53. Paoli L, Schatzman M (2007) Numerical simulation of the dynamics of an impacting bar. Comput Methods Appl Mech Eng 196(29–30):2839–2851
54. Percivale D (1985) Uniqueness in the elastic bounce problem. J Differ Equ 56(2):206–215
55. Percivale D (1986) Bounce problem with weak hypotheses of regularity. Ann Mat Pura Appl 143:259–274
56. Percivale D (1991) Uniqueness in the elastic bounce problem. II. J Differ Equ 90(2):304–315
57. Ravn P (1998) A continuous analysis method for planar multibody systems with joint clearance. Multibody Syst Dyn 2(1):1–24
58. Rockafellar RT (1970) Convex analysis. Princeton University Press, Princeton
59. Rudin H, Ungar P (1957) Motion under a strong constraining force. Commun Pure Appl Math 10:431–447
60. Schatzman M (1978) A class of nonlinear differential equations of second order in time. Nonlinear Anal TMA 2(3):355–373
61. Schatzman M (1998) Uniqueness and continuous dependence on data for one-dimensional impact problems. Math Comput Model 28(4–8):1–18
62. Schatzman M (2001) Penalty method for impact in generalized coordinates. Philos Trans R Soc Lond A 359(1789):2429–2446
63. Stoianovici D, Hurmuzlu Y (1996) A critical study of the applicability of rigid body collision theory. ASME J Appl Mech 63:307–316
64. Tisseron C (1996) Notions de topologie, introduction aux espaces fonctionnels. Hermann, Paris

Nonsmooth Modal Analysis: From the Discrete to the Continuous Settings

Anders Thorin and Mathias Legrand

Abstract This chapter addresses the prediction of vibratory resonances in nonsmooth structural systems via *Nonsmooth Modal Analysis*. Nonsmoothness in the trajectories is induced by unilateral contact conditions in the governing (in)equations. Semi-analytical and numerical state-of-the-art solution methods are detailed. The significance of nonsmooth modal analysis is illustrated in simplified one-dimensional space semi-discrete and continuous frameworks whose theoretical and numerical discrepancies are explained. This contribution establishes clear evidence of correlation between periodically forced and autonomous unilaterally constrained oscillators. It is also shown that strategies using semi-discretization in space are not suitable for nonsmooth modal analysis. The spectrum of vibration exhibits an intricate network of backbone curves with no parallel in nonlinear smooth systems.

The purpose of this chapter is to provide a general picture of the state-of-the-art vibratory analysis of nonsmooth systems. This topic lies at the interface between *modal analysis* of smooth nonlinear systems and *nonsmooth contact dynamics* dedicated to the time-evolution of nonsmooth systems, undergoing impact or dry friction, for instance. Some elementary concepts are succinctly recalled for the purpose of completeness.

Terminology

Unless otherwise stated, the epithet *discrete* (as in "discrete systems" or "discrete setting") designates *semi-discretization in space*, while *continuous* refers to everything else.

The original version of this chapter was revised: Incorrect author name has been changed. The erratum to this chapter is available at https://doi.org/10.1007/978-3-319-75972-2_11

A. Thorin (✉) · M. Legrand
Structural Dynamics and Vibration Laboratory, McGill University, Montreal, Canada
e-mail: anders.thorin@mcgill.ca

M. Legrand
e-mail: mathias.legrand@mcgill.ca

R. I. Leine et al. (eds.), *Advanced Topics in Nonsmooth Dynamics*,
https://doi.org/10.1007/978-3-319-75972-2_5

1 Introduction to Nonlinear Modal Analysis

Mechanical systems, from those of large scale (buildings) to those of small scale (MEMS switches), commonly undergo forced vibrations. The efficient and accurate characterization of the response of such systems to an external periodic loading is essential to ensure safe designs. It also has various other applications, such as retrofit, damage detection or model reduction, to name a few. In this context, frequency-response curves play a key role for the dynamicist: they indicate the energy level of a periodic solution produced by an external periodic forcing of (angular) frequency ω, as a function of ω. For nonlinear systems, computing these frequency-response curves is not a straightforward task. Actually, they are known to depend, in a possibly intricate manner, on the forcing amplitude, the forcing frequency, and the forcing shape [48]. A brute-force time-domain approach consisting in solving the governing equations for various external forces and initial conditions is, in practice, not conceivable for large-scale systems. Instead, modal analysis provides a means of computing, for a much more reasonable cost, the so-called *backbone curves* that shape the forced response curves. Such backbone curves correspond to the underlying autonomous (i.e., unforced) and conservative (i.e., undamped) periodic solutions of the governing differential equation.[1] Autonomous periodic solutions of conservative systems may seem "unrealistic" in the sense that no undamped systems are observable in the physical world. Their investigation can yet provide germane information on periodically-forced and slightly damped systems. Essentially, they extend the concept of spectrum, defined for linear systems, to the nonlinear framework. In particular, they show the energy-dependence of vibration frequencies. The above statements are illustrated by considering a finite-dimensional system relevant to structural dynamics and governed by a linear Ordinary Differential Equation (ODE) of the form

$$\mathbf{M}\ddot{\mathbf{u}}(t) + \mathbf{C}\dot{\mathbf{u}}(t) + \mathbf{K}\mathbf{u}(t) = \mathbf{f}_{\text{ext}}(t), \tag{1}$$

where $\mathbf{u}$ is the vector of generalized displacements, $\mathbf{M}$ is its positive-definite mass matrix, $\mathbf{C}$ its damping matrix, $\mathbf{K}$ its positive-definite stiffness matrix and $\mathbf{f}_{\text{ext}}$ its vector of external loadings. The backbone curves are trivially obtained by considering the autonomous conservative counterpart $\mathbf{M}\ddot{\mathbf{u}} + \mathbf{K}\mathbf{u} = \mathbf{0}$ and its periodic solutions, yielding vertical lines in the energy–frequency diagram at the eigenfrequencies of the system defined as the square roots of the eigenvalues of $\mathbf{M}^{-1}\mathbf{K}$: in linear dynamics, the frequency of vibration is independent of the magnitude of vibration. This can be seen in Fig. 1 (top), which illustrates a typical Frequency-Energy Plot (FEP) for a two-degrees-of-freedom (dof) linear oscillator. The forced frequency-response curves are clearly aligned on the two backbone curves, which completely characterize the spectrum of vibration. Let us now consider a smooth nonlinear system of the form

$$\mathbf{M}\ddot{\mathbf{u}}(t) + \mathbf{f}(\dot{\mathbf{u}}(t), \mathbf{u}(t)) = \mathbf{f}_{\text{ext}}(t), \tag{2}$$

[1] Modal analysis can also be defined in the autonomous damped case, which is more complicated and not further discussed here.

where *smooth* refers to the smoothness of $\mathbf{f}$ with respect to $\mathbf{u}$ and $\dot{\mathbf{u}}$. Its dynamics is unsurprisingly more subtle than the previous linear case, and systematic solution methods for characterizing the vibrations globally are not available [40, 90, Sect. 1.3]. However, it is known that fixed points, periodic and quasi-periodic limit cycles may exist, in the vicinity of which it may be possible to approximate the nonlinear dynamical response. In particular, the centre manifold theorem [40, 47], together with Lyapunov's centre theorem [9, p. 5], show that under sufficient regularity[2] and non-internal resonance conditions, two-dimensional invariant manifolds exist locally in the phase space and are tangent to the linear modes of the system linearized at the fixed points. Such two-dimensional invariant manifolds were later defined as *nonlinear normal modes* of vibration [93, 111] in the vibration community. They can be understood as curved extensions of linear modes that correspond to flat two-dimensional invariant manifolds defined by one-parameter continuous families of elliptic trajectories. However, the nonlinear framework encompasses many phenomena that are not observed in linear systems, such as internal resonance, frequency–energy dependence, emergence of subharmonics or chaos, and existence of isolated loops in the FEP [49]. Again, the relevance of nonlinear modal analysis is illustrated in Fig. 1 (bottom), depicting the forced response of a two-dof Duffing oscillator. The response curves warp around the backbone curves. As opposed to the linear spectrum, the nonlinear backbone curves are frequency-dependent and stiffening is exhibited here. The kink of the forced response in the neighborhood of $\omega_2/3$ corresponds to a subharmonic resonance. The loop near ω_1 corresponds to an internal resonance, where the first nonlinear mode and the third subharmonic of the second nonlinear mode interact.

Among all nonlinearities found in mechanics, unilateral and frictional contact nonlinearities form a specific class in which nonsmoothness arises in the dynamics [94]. Typically, the impact between two bodies induces velocity discontinuities and acceleration impulses [2]. The present chapter focuses on the frictionless framework. The governing equation can no longer be written in the form (2), where $\mathbf{f}$ is a smooth function of $\mathbf{u}$ and $\dot{\mathbf{u}}$. However, classical analytical techniques available for computing nonlinear modes [49] require smoothness of the governing equation. Indeed, the invariant manifold approach is based on the Taylor series of the solution written as a function of a pair of master coordinates [93]; the method of multiple scales [73], as a subclass of perturbation methods, requires asymptotic expansions; normal forms rely on the nonlinearity being an analytic function [45]. When it comes to nonsmoothness, such strategies no longer apply.

Nonsmooth modal analysis is the extension of nonlinear modal analysis to nonsmooth systems. This is accomplished by computing *nonsmooth modes*, that is, families of nonsmooth periodic solutions of the autonomous and conservative dynamics. Even simplistic nonsmooth oscillators exhibit intricate responses [20, 103, 106]. The regularizing approach, consisting in replacing nonsmoothness with smooth strong nonlinearities [7, 15, 26, 44, 58, 92, 101, 114], has the adverse effect of introduc-

[2]Notably, the linearized flow should be invertible. For example, the equation $\ddot{u} + u^3 = 0$ has nontrivial periodic solutions, but its linearized ODE $\ddot{u} = 0$ does not.

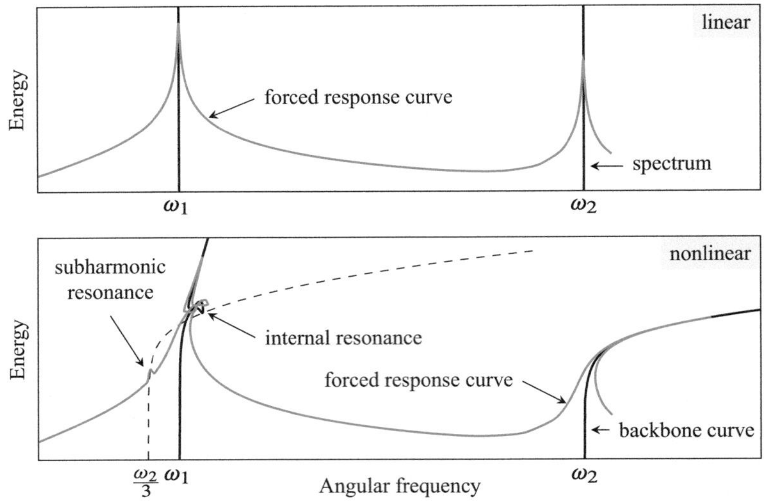

Fig. 1 Frequency–energy plot of a two-dof Duffing oscillator, linearized (top) and nonlinear (bottom). [—] Backbone curves. [—] Forced-response. [- -] Subharmonic of the second mode

ing issues such as numerical stiffness [19, 77, 78] and is not further discussed in this work. Another approach is to include nonsmoothness as such. Many investigations on the dynamics of forced vibro-impact oscillators [35, 83, 85, 117, 120] and grazing bifurcations [20, 20, 30, 75, 81] or stability issues [60] are available. The specific target of families of periodic solutions of a conservative nonsmooth problem has emerged recently for space-discretized systems [59, 104, 107] or continuous ones [41, 125].

Multiple applications which could benefit from nonsmooth modal analysis can be listed: rotor-stator contact interactions in rotating machinery involving unilateral contact occurrences between blades and casings [116], boiler tube dynamics with a loose support [26, 76], grid-to-rod fretting [42], percussive drilling systems [79, 80], cutting tools [121] or, on a smaller scale, capsule systems (capsubots) [63, 64], and electrostatically-driven and piezoelectric actuators [31, 69].The Sensitivity of an atomic force microscope, in tapping mode, can be improved through understanding of the response of impact oscillators [113]. Additional examples include impact dampers implemented to reduce vibrations [62, 91] or fret-string contact interactions within musical instruments [12, 16, 43]. More applications can be found in [7]. All applications have in common the need to properly characterize nonsmooth vibratory resonances.

The purpose of this chapter is to give a picture of the state-of-the-art nonsmooth modal analysis. While the standard procedure in mechanical engineering is to approximate continuous systems by n-dof systems, complications arise when contact is involved. Nonsmooth modes of a continuous system have intricate relationships with that of their semi-discretized counterparts, which raises open-ended questions. The available analytical and numerical methods for nonsmooth modal analysis are first presented for finite-dimensional systems (Sect. 2) and continuous systems (Sect. 3). The relationships between forced response and Nonsmooth Modes (NSMs) are then illustrated in Sect. 4. The comparison between modal analysis of continuous systems and semi-discretized counterparts is addressed in Sect. 5, which concludes the chapter.

2 Nonsmooth Modal Analysis of Discrete Oscillators

Consider the dynamics governed by a differential equation of the form (2), where $\mathbf{f}_{\text{ext}} = \mathbf{0}$. A contact condition, which prevents penetration between two colliding bodies, is commonly expressed as a unilateral constraint $g(\mathbf{u}, t) \geq 0$, where g stands for *gap*, that is, the distance between the bodies. This constraint is incorporated into the dynamics via a Lagrange multiplier λ corresponding to the reaction force in the outward normal direction of the contact surface. The non-sticking condition implies that $\lambda \geq 0$ and λ can be non-zero only if the gap is closed: $g(\mathbf{u}, t)\lambda(t) = 0$. These three conditions, known as the Signorini conditions [2], are commonly written in the synthetic form $0 \leq \lambda \perp g(\mathbf{u}, t) \geq 0$. In the case of multiple unilateral constraints, each gap function and its corresponding Lagrange multiplier can be stacked in vectors $\mathbf{g}$ and $\boldsymbol{\lambda}$, respectively; the inequalities and the orthogonality operator $\perp$ are then defined component-wise. Altogether, the autonomous dynamics now writes

$$
\begin{cases}
\mathbf{M}\ddot{\mathbf{u}} + \mathbf{f}(\mathbf{u}, \dot{\mathbf{u}}) - \nabla_{\mathbf{u}}\, \mathbf{g}(\mathbf{u}, t)\boldsymbol{\lambda} = \mathbf{0} & \text{(3a)} \\
\mathbf{0} \leq \boldsymbol{\lambda} \perp \mathbf{g}(\mathbf{u}, t) \geq \mathbf{0} & \text{(3b)}
\end{cases}
$$

and nonsmooth modal analysis consists in finding continuous families of periodic solutions to this problem. Equation (3a) should be read in a weak sense, since $\mathbf{u}$ is only of regularity $\mathcal{C}^0$ because of the complementarity condition (3b). Various other formalisms are available to describe the dynamics [2].

2.1 *Necessity of an Impact Law*

An aspect that does not always seem to be understood is that Problem (3), together with some initial conditions $\mathbf{u}(0)$ and $\dot{\mathbf{u}}(0)$, does not *uniquely* determine a solution. For instance, consider a punctual ball of mass m located above a rigid ground and subjected to gravity. When dropped from a given height, the ball first undergoes a free flight uniquely determined by its initial position and initial velocity, together with an ODE of the form $m\ddot{u} + mg = 0$ (Cauchy problem). It then reaches the ground: from there, infinitely many solutions are possible, all satisfying Eq. (3) adapted to the problem at hand. The ball could remain on the ground: $\dot{u} = 0$ and $\lambda = mg$. It could also bounce with the same kinetic energy: $\dot{u}^+ = -\dot{u}^-$ and $\lambda = -2m\dot{u}^-$ at the impact

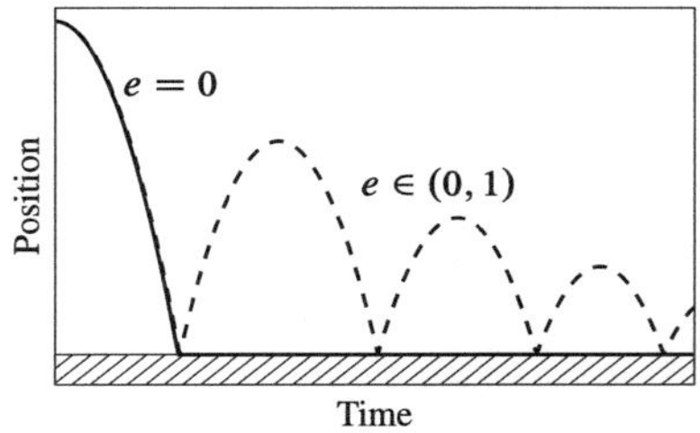

Fig. 2 Two distinct solutions to the problem of a bouncing ball of the form (3): uniqueness is not guaranteed when an impact law is not specified

time, where $\dot{u}^-$ (respectively $\dot{u}^+$) denotes the normal pre-impact (respectively post-impact) contact velocity. These two acceptable solutions are depicted in Fig. 2. This non-uniqueness indicates that information is missing.[3] To ensure well-posedness, Eq. (3) is complemented with a constitutive *impact law*. If the latter does not lead to an increase of kinetic energy, uniqueness is guaranteed as soon as the unilateral constraints, the smooth nonlinear terms and the smooth external forces are analytic functions [8, 87]. Nevertheless, even with impact laws, the continuity of the solutions with respect to the initial conditions is not guaranteed in the case of multiple unilateral constraints [8].

The necessity of an impact law holds for any unilateral constraint arising in systems semi-discretized in space, unless special treatment is enforced [51]. Numerical strategies which do not explicitly include an impact law, such as [13, 122], produce only one among infinitely many possible solutions.

Among possible impact laws, only conservative ones should be considered in the framework of nonsmooth modal analysis, since autonomous *periodic* solutions are sought. The most common choice[4] is Newton's purely elastic impact law, $\dot{u}^+ = -\dot{u}^-$ at impact times. This choice, dictated by the periodicity condition, is incompatible with lasting contact phases observed during collisions in the continuous framework. This can be illustrated by considering the position of the contacting end of a one-dimensional bar colliding with a rigid obstacle, as depicted in Fig. 3. In the continuous framework, contact phases last a finite amount of time, while the energy is preserved (left plot). When the bar is discretized, the conservative impact law implies instantaneous bounces (right plot). For n-dof systems, lasting contact phases necessitate a purely inelastic impact law of the form $\dot{u}^+ = 0$, leading to a loss of kinetic energy incompatible with the conservative framework of modal analysis. Also, it is worth mentioning that when subjected to an external load, a unilaterally-constrained system can exhibit lasting contact phases after a countable infinity of impacts occurring in finite time, for non-purely elastic impact laws [17, 66]. This phenomenon is called *chattering* and is illustrated in Sect. 5.

For very specific initial conditions, systems governed by (3), together with a purely elastic impact law, may have solutions with lasting contact phases, also called *sticking*

[3]As explained in Sect. 3, this results from the fact that shock waves, emanating from the contact interface where bodies collide, are not properly described in the semi-discrete setting.

[4]Other strategies, consisting in redistributing energy or mass, have also been explored, (see Sect. 5.2).

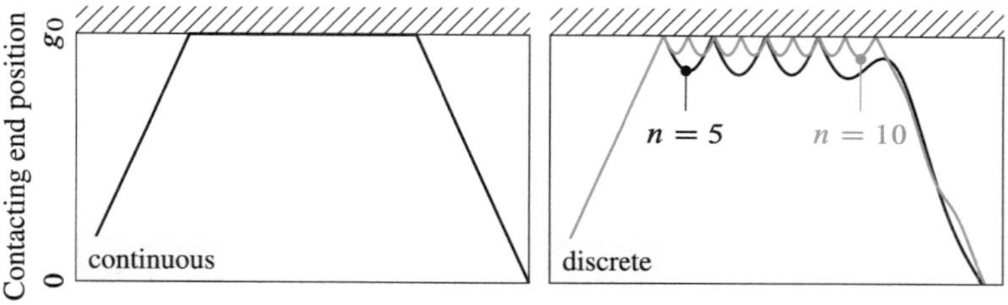

Fig. 3 Displacement of the contacting with end of a bar colliding a rigid obstacle with no external force. In the continuous framework, no impact law is needed for well-posedness and the contact is lasting, even for energy-preserving motions. The discretized bar with a conservative impact law exhibits chattering instead

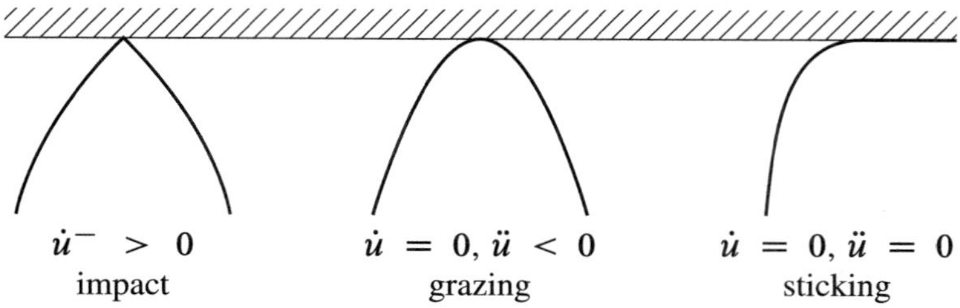

Fig. 4 Possible gap-closing trajectories for conservative autonomous systems, in terms of the normal velocity $\dot{u}$

phases, despite the non-sticking condition on the contact force. Such trajectories can be seen as one specific type of contact, as impact or grazing see Fig. 4. They were investigated in [56] for a linear two-dof spring-mass system. An extension to n degrees of freedom, general mass matrices and a single unilateral constraint is proposed in [105]. In both cases, T-periodic trajectories with one lasting contact phase were shown to exist only for isolated values of T. While they may seem of purely theoretical interest, it was recently demonstrated that such trajectories play an important role in the response spectrum of piecewise-linear impact oscillators [106, Fig. 4]. No systematic results are presently available in the literature on periodic motions with lasting contact phases of systems with additional smooth nonlinearities or multiple unilateral constraints.

2.2 Quasi-analytical Techniques in Simple Cases

The systematic analytic derivation of NSM for n-dof systems has recently been provided for a piecewise-linear spring–mass system with one Impact Per Period (IPP) [59], as well as for any piecewise-linear system with a single linear unilateral constraint and an arbitrary number of IPPs [104]. Preliminary investigations show that there exist strong relationships between the forced response of piecewise-linear impact oscillators and backbone curves obtained using NSM, as detailed in Sect. 4. Additional weak smooth nonlinearities do not seem to change the overall picture, as succinctly discussed in Sect. 4.1. Extension to multiple unilateral constraints quickly becomes tedious, because of the combinatorial nature of the sequence of unilateral constraint activations.

We now derive the main ideas on how to carry out nonsmooth modal analysis on n-dof piecewise-linear impact oscillators [104]. The generalized displacements and velocities are denoted by $\mathbf{u}$ and $\dot{\mathbf{u}}$; the state $\mathbf{x}$ is such that $\mathbf{x}^\top = [\mathbf{u},\ \dot{\mathbf{u}}]^\top \in \mathbb{R}^{2n}$. The unilateral constraint is assumed to be a linear function of the $\mathbf{u}$. As a consequence, there exists a vector $\mathbf{w} \in \mathbb{R}^n$ and a constant $g_0 \in \mathbb{R}$ such that $g(\mathbf{u}) = \mathbf{w}^\top \mathbf{u} + g_0$. The elastic impact law can be written as [104, Sect. 4.2]

$$g(\mathbf{u}) = 0 \quad \Longrightarrow \quad \mathbf{x}^+ = \mathbf{N}\mathbf{x}^-, \tag{4}$$

where $\mathbf{N}$ is similar to a reflection matrix with respect to a hyperplane of $\mathbb{R}^{2n}$ (also known as a Householder matrix), which depends only on $\mathbf{w}$ and the mass matrix $\mathbf{M}$. This describes the impact as a simple relationship in terms of the system state $\mathbf{x}$. In the same spirit, let $\mathbf{S}(\sigma)\mathbf{x}$ denote the state after a free flight of duration σ from a state $\mathbf{x}$. A k IPP motion ($k \in \mathbb{N}^*$) is the succession of one free flight of duration $\sigma_1 > 0$, one impact, one free flight of duration $\sigma_2 > 0$, one impact, and so on, k times. Such a motion is depicted in Fig. 5. Starting from a post-impact state, the periodicity condition reads as

$$\mathbf{x}(0) = \mathbf{x}(T) = \mathbf{N}\mathbf{S}(\sigma_k)\mathbf{N}\mathbf{S}(\sigma_{k-1})\mathbf{N}\ldots\mathbf{N}\mathbf{S}(\sigma_1)\mathbf{x}(0), \tag{5}$$

where $T = \sigma_1 + \cdots + \sigma_k$. This condition comes with the k gap closure conditions at impact times, that is,

$$g(\mathbf{x}(0)) = 0,\ g(\mathbf{x}(\sigma_1)) = 0,\ g(\mathbf{x}(\sigma_1 + \sigma_2)) = 0,\ \ldots,\ g(\mathbf{x}(\sigma_1 + \cdots + \sigma_{k-1})) = 0. \tag{6}$$

The initial conditions $\mathbf{x}_0$, determining a motion $\mathbf{x}$ that satisfies conditions (5) and (6) for some $\mathbf{s} = (\sigma_1, \ldots, \sigma_k)$, define an autonomous periodic motion, provided the gap remains non-negative, in line with (3b). Finding such $\mathbf{x}_0$ reduces to determining a vector $\boldsymbol{\lambda} \in \mathbb{R}^k$ that satisfies [104]

$$\boldsymbol{\Pi}(\mathbf{s})\boldsymbol{\lambda} = \mathbf{0} \quad \text{and} \quad \boldsymbol{\Sigma}(\mathbf{s})\boldsymbol{\lambda} = g_0\mathbf{j}, \tag{7}$$

with $\mathbf{j} = [1,\ \ldots,\ 1]^\top \in \mathbb{R}^k$ and where $\boldsymbol{\Pi}$ and $\boldsymbol{\Sigma}$ are two $k \times k$ matrices, whose expressions are known explicitly [104, Sect. 3.1] and depend on the parameters $\mathbf{M}$, $\mathbf{K}$ and $\mathbf{w}$. The physical interpretation of vector $\boldsymbol{\lambda}$ is that it is proportional to the pre-impact contact velocities. Several major consequences follow from (7):

- it suffices to find the k components of $\boldsymbol{\lambda}$ instead of the $2n$ unknown components of $\mathbf{x}_0$;

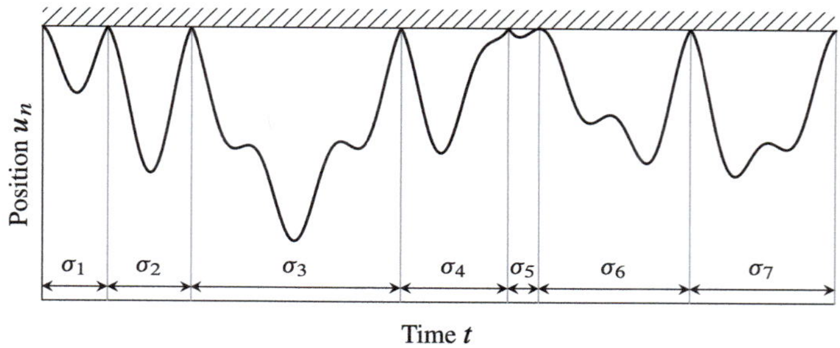

Fig. 5 Example of motion with 7 IPPs for a 5-dof system

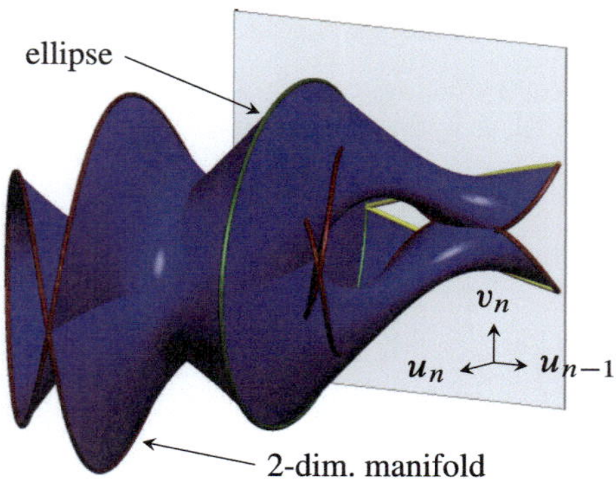

Fig. 6 Projection of a 1 IPP NSM in the (u_{n-1}, u_n, v_n) space for $n = 5$ (from [104]). This NSM is a continuum of periodic nonsmooth trajectories with 1 IPP continuously connected to a linear grazing mode (green ellipse). This two-dimensional manifold is invariant: if a motion starts on it, it will remain on it as time unfolds. In particular, this manifold cannot be intersected by other trajectories in the phase space

- $\mathbf{\Sigma}$ is invertible almost everywhere in $\mathbb{R}^k$, so $\boldsymbol{\lambda}$ can be eliminated by combining (5) and (6). As a result, all the periodic solutions are governed by the equation $\mathbf{\Pi}(\mathbf{s})\mathbf{\Sigma}(\mathbf{s})^{-1}\mathbf{j} = \mathbf{0}$. The first step is to solve for $\mathbf{s}$. Then, the corresponding initial state is recovered via $\mathbf{x}_0(\mathbf{s}) = \boldsymbol{\varphi}(g_0\mathbf{\Sigma}(\mathbf{s})^{-1}\mathbf{j})$, where $\boldsymbol{\varphi}$ is a known function (see [104], not recalled here for conciseness);
- the skew-symmetry of $\mathbf{\Pi}$ is such that $\mathbf{\Pi}(\mathbf{s})\mathbf{\Sigma}(\mathbf{s})^{-1}\mathbf{j} = \mathbf{0}$ generically leads to $k-1$ independent equations. As a result, the set of solutions is a curve in $\mathbb{R}^k$ and periodic orbits with k IPPs belong to a one-parameter continuous family, corresponding to a two-dimensional manifold in the phase space (see Fig. 6). This feature is shared by smooth nonlinear systems away from internal resonances.

The above methodology is summarized in Fig. 7. Each NSM corresponds to a backbone curve in terms of FEP. An example of such FEP is provided in Fig. 8 for a two-dof spring-mass system (see Fig. 14), with up to seven IPPs. For one to three IPPs, the spectrum was computed with the quasi-analytical method described above.

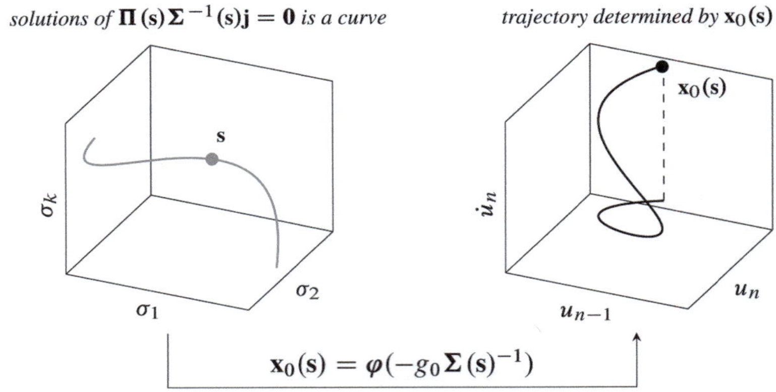

Fig. 7 Summary of analytical nonsmooth modal analysis in the generic case for $g_0 \neq 0$. The dependency of $\mathbf{x}_0$ to $\mathbf{s}$ is highlighted by the notation $\mathbf{x}_0(\mathbf{s})$

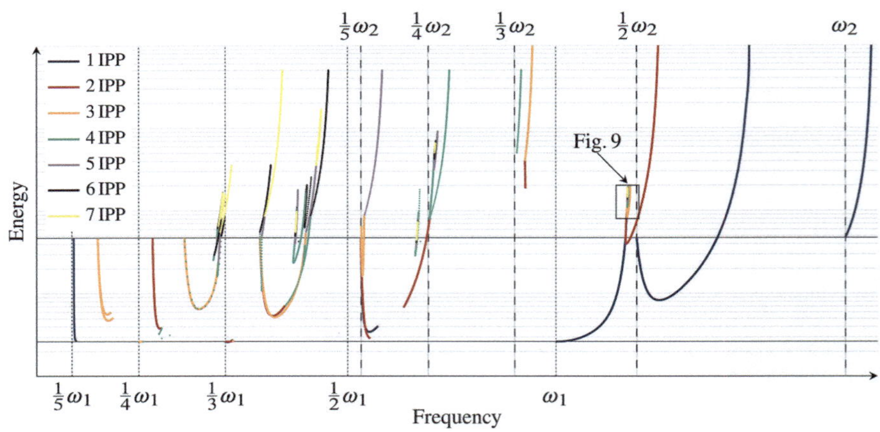

Fig. 8 Backbone curves of a two-dof impact oscillator with up to 7 IPPs. The two horizontal lines correspond to the two linear grazing modes. Axes in log scale

A multiple shooting method was used for four to seven IPPs (see Sect. 2.3.2). Figure 8 displays no isolated branches. Indeed, all backbone curves can either [106]:

- diverge to unbounded energy, which corresponds to a singularity of $\boldsymbol{\Sigma}(\mathbf{s})$;
- be connected to a linear grazing mode (this is true in the case for 1 IPP);
- be connected to another backbone curve, with the junction then corresponding, to a nonsmooth trajectory with impacts and grazing;
- converge to a motion with one Sticking Per Period (SPP).

In the neighborhood of a 1 SPP, backbone curves seem to converge to the SPP as the number of IPPs increases. This phenomenon is illustrated in Fig. 9. Convergence to trajectories combining 1 IPP and 1 SPP have also been observed. While very likely to be true, there is no formal proof of such convergence.

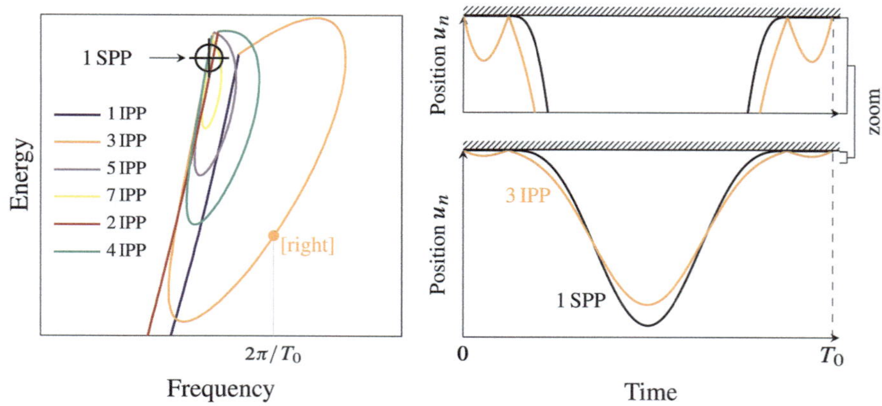

Fig. 9 Parallel sequences of NSMs with increasing IPP ($1 \to 3 \to 5 \to 7$ and $2 \to 4$) converging to a 1 SPP motion. Convergence is shown via backbone curves (left) and in time domain (right) (from [106])

As already reported [104, 107], seemingly independent backbone curves might be connected through a vertical backbone curve: this additional non-generic feature was referred to as a *bridge*. This occurs for isolated $\mathbf{s}$, making $\boldsymbol{\Sigma}(\mathbf{s})$ singular. However, such $\mathbf{s}$ and those leading to unbounded energy are distinct.

Stability analysis of k IPP motions is carried out in a straightforward fashion by linearizing the kth return map on the hyperplane $g(\mathbf{u}) = 0$. A perturbation of an initial condition $\mathbf{x}_0$ propagates through the mapping

$$\mathbf{x}_0 + \delta\mathbf{x}_0 \mapsto \mathbf{NS}(\sigma_k + \delta\sigma_k)\mathbf{N} \dots \mathbf{NS}(\sigma_1 + \delta\sigma_1)(\mathbf{x}_0 + \delta\mathbf{x}_0), \tag{8}$$

where $\delta\sigma_i$ is an unknown yet small change of duration of the ith free flight. The first-order Taylor expansion of this assumed smooth mapping yields an equation of the form

$$\delta\mathbf{x} = \mathbf{NS}(\sigma_k)\mathbf{N} \dots \mathbf{NS}(\sigma_1)\delta\mathbf{x}_0 + \Big(\sum_{\ell=1}^{k} \mathbf{NS}(\sigma_k) \dots \mathbf{NS}'(\sigma_\ell)\mathbf{N} \dots \mathbf{NS}(\sigma_1)\delta\sigma_\ell\Big)\mathbf{x}_0. \tag{9}$$

The unknowns $\delta\sigma_1, \dots, \delta\sigma_k$ are found by solving the linearized system $g(\mathbf{u}((\sigma_1 + \delta\sigma_1) + \dots + (\sigma_\ell + \delta\sigma_\ell))) = 0$ for $\ell \in [\![1, k]\!]$. Ultimately, there exists a linear mapping between $\delta\mathbf{x}_0$ and $\delta\mathbf{x}$ through a matrix $\mathbf{A}(\mathbf{x}_0)$ such that

$$\delta\mathbf{x} = \mathbf{A}(\mathbf{x}_0)\delta\mathbf{x}_0. \tag{10}$$

The eigenvalues of $\mathbf{A}(\mathbf{x}_0)$ determine the spectral stability of the periodic solutions emanating from $\mathbf{x}_0$ [90, Summary 7.5].

2.3 Numerical Techniques

The above (semi-)analytical developments provide essential insight in understanding nonsmooth modes. They are inevitable for proving mathematical results, but are limited to piecewise-linear systems. Numerical techniques take over for more challenging vibro-impact systems, for instance, with multiple unilateral constraints or polynomial nonlinearities.

In the following, we restrict ourselves to two well-known procedures devoted to periodic solutions: Harmonic Balance Method (HBM) and Shooting Method (SM). HBM enforces periodicity exactly by construction, while contact conditions are only approximated. In contrast, SM handles contact conditions accurately, to the detriment of periodicity. Other methods, such as multiple scales, invariant manifold approach and alike, are not considered, as they essentially apply to smooth nonlinearities.

2.3.1 Harmonic Balance Method and Its Variants

For n-dof systems, setting the unilateral constraints apart, smooth dynamics is described by ODEs in the form

$$\mathbf{f}(\mathbf{u}, \dot{\mathbf{u}}, \ddot{\mathbf{u}}, t) = \mathbf{0}, \tag{11}$$

where $\mathbf{f}$ is a nonlinear function of the displacements $\mathbf{u}$ and velocities $\dot{\mathbf{u}}$. The unknown displacement $\mathbf{u}$ is approximated by $\mathbf{u}_h$, which is defined as a linear combination of N chosen shape functions stacked in a vector $\boldsymbol{\varphi}$ so that

$$\mathbf{u}(t) \approx \mathbf{u}_h(t) = \mathbf{A}\boldsymbol{\varphi}(t), \tag{12}$$

where $\mathbf{A}$ is a $n \times N$ matrix of unknown coefficients. Equation (11) is approximately solved by making the residual $\mathbf{f}(\mathbf{A}\boldsymbol{\varphi}, \mathbf{A}\dot{\boldsymbol{\varphi}}, \mathbf{A}\ddot{\boldsymbol{\varphi}}, t)$ orthogonal to a well-chosen set of M test functions $\boldsymbol{\phi}$ for the usual inner product

$$\forall k \in [\![1, M]\!], \quad \int_0^T \phi_k(t)\mathbf{f}(\mathbf{A}\boldsymbol{\varphi}, \mathbf{A}\dot{\boldsymbol{\varphi}}, \mathbf{A}\ddot{\boldsymbol{\varphi}}, t)\,\mathrm{d}t = \mathbf{0}. \tag{13}$$

Such integrals collectively form a system of nonlinear equations and can be evaluated numerically if the integrand does not easily simplify. Choosing $M = N$, the nN coefficients in $\mathbf{A}$ are then found using a root-finding algorithm such as Newton-Raphson to solve the nN Eq. (13).

This method can be used to compute periodic responses to either forced or autonomous ODEs. Equation (12) shows that the periodicity condition is transferred to a condition on $\boldsymbol{\varphi}$, which must therefore be periodic. In the case of a periodic external force of angular frequency Ω, periodic solutions are expected to have a frequency multiple of Ω, so $T = 2\pi/\Omega$ can be chosen, or $T = 2p\pi/\Omega$, $p \in \mathbb{N}^*$ to accommodate possible subharmonics [38, 53, 123]. In the autonomous case, T is unknown and continuation procedures must be used [3].

The HBM is a well-established technique [54, 118] for finding approximations of periodic solutions to (13). It is obtained by specifying

$$\boldsymbol{\varphi} = \boldsymbol{\phi} = \left[1 \ \exp \mathrm{i}\omega t \ \ldots \ \exp \mathrm{i}N\omega t\right]^{\top}, \tag{14}$$

with $\omega = 2\pi/T$. While commonly producing accurate results for weak nonlinearities, HBM is mostly used heuristically, and there is no proof that a truncated series is a valid approximation of the exact solution [32]. Other shape and test functions $\boldsymbol{\varphi}$ and $\boldsymbol{\phi}$ shall be adopted. Another well-known method is the *collocation* method, which corresponds to a low-order piecewise periodized polynomial for $\boldsymbol{\varphi}$ and

$$\forall t \in [0, T], \quad \boldsymbol{\phi}(t) = \left[\delta(t - t_1) \ \ldots \ \delta(t - t_N)\right]^{\top}, \tag{15}$$

where $t_1, \ldots, t_N$ are the collocations points. The Dirac deltas have the property to transform the computation of the inner product (13) into the simple evaluation of $\mathbf{u}_h$ at the collocation points. The derivatives of $\mathbf{u}_h$ are computed from the shape functions if they are differentiable, through a finite difference scheme, for instance, or via a conservative Simo scheme [5, 98]. When orthogonal polynomials are chosen as shape functions and the collocation points are the roots of one of the orthogonal polynomials, the method is called *orthogonal collocation* or *pseudospectral* [10, 33] and is reported to be efficient for dealing with sharp fronts [22].

For unilateral contact problems, HBM has mostly been implemented in conjunction with regularizing techniques [26, 38, 53, 116, 123] and the contact forces are directly included in the governing ODE (11). A variant of HBM in which the truncated Fourier series is replaced by wavelets has been proposed to compute periodic solutions of a turbine blade with regularized contact conditions [99]. HBM with regularized friction has been investigated in [46].

The unilateral contact conditions can also be treated without regularization, and the problem reads as

$$\begin{cases} \mathbf{f}(\mathbf{u}, \dot{\mathbf{u}}, \ddot{\mathbf{u}}, t, \boldsymbol{\lambda}) = \mathbf{0} & \text{(16a)} \\ \mathbf{0} \leq \mathbf{g}(\mathbf{u}) \perp \boldsymbol{\lambda} \geq \mathbf{0} & \text{(16b)} \\ \mathbf{u}(0) = \mathbf{u}(T), \ \dot{\mathbf{u}}(0) = \dot{\mathbf{u}}(T). & \text{(16c)} \end{cases}$$

Above, no impact law is specified. It is instead replaced by the periodicity conditions. We are not aware of any formal proof of this supposed equivalence. However, within HBM, the impact law with $e = 1$ is implied by the conservation of the total energy in an autonomous problem with no simultaneous impacts, but it is unclear which solutions are picked by the numerical procedure in other cases, such as in the presence of external forces.

The Signorini conditions are transformed by means of a max operator, observing that for any $\boldsymbol{\alpha} > \mathbf{0}$ [88], in the component-wise sense,

$$\mathbf{0} \leq \boldsymbol{\lambda} \perp \mathbf{g}(\mathbf{u}) \geq \mathbf{0} \iff \boldsymbol{\lambda} - \mathbf{max}(\boldsymbol{\lambda} - \alpha\mathbf{g}(\mathbf{u}), \mathbf{0}) = \mathbf{0}, \tag{17}$$

and can be readily included in (11) at the cost of reducing the regularity of $\mathbf{f}$ [96]. The inner product (13) is computed numerically and the solution is found through a semi-smooth Newton solver. An alternative is to implement HBM together with an augmented Lagrangian in a case of unilateral conditions only [55] or a variation of the augmented Lagrangian in the case friction [71]. Another possibility is to approximate $\mathbf{u}$ and $\boldsymbol{\lambda}$ with adapted periodic shape functions and satisfy the Signorini conditions at discrete times (collocation points), leading to a Linear or Nonlinear Complementary Problem [68].

Another possible strategy could consist in adding a chosen nonsmooth function with the same regularity as the expected solution ($\mathcal{C}^0$ in the case of impacts) as a shape function; a faster convergence would then be expected, as in the dry frictional case [52]. Irrespective of the chosen discretization, contact-induced nonlinearities require a large number of harmonics (see Fig. 20).

2.3.2 Shooting Method

The Shooting Method is a well-known procedure capable of tracking periodic solutions of ODEs in the form (11) [6, 70]. It consists in finding initial conditions $(\mathbf{u}_0, \dot{\mathbf{u}}_0)$ such that they are recovered after a time integration over some interval $[0, T]$ for some $T > 0$. The analytical method presented in Sect. 2.2 can be understood for one IPP as a SM in which exact integration is performed through matrix exponentials. In more general cases, time integration can be carried out either by event-driven schemes or time-stepping methods [2]. Enforcing periodicity conditions reduces to finding the roots of a vector function $\mathbf{z}(\mathbf{u}_0, \dot{\mathbf{u}}_0, T)$, bearing in mind that $\mathbf{z}$ might be nonsmooth. In modal analysis, T is unknown and there are a priori $2n + 1$ unknowns for $2n$ independent equations: the solution space is a curve, which can be found, as in the HBM, via numerical continuation. The multiple shooting method enlarges the domain of attraction of the root-finding algorithm by splitting the integration domain, increasing the robustness of the numerical procedure [90, 102].

This approach was applied to contact problems with regularized nonsmoothness [84, 112]. It was also used to locate grazing [110]. The merits of SM for nonsmooth modal analysis rely on the fact that efficient numerical schemes dedicated to nonsmoothness, such as the Moreau–Jean scheme [2, 89], can be employed. Convergence proofs exist for a few schemes [25].

For solutions with multiple impacts per period, period T can be replaced by the succession of unknown free flight durations $\sigma_1, \ldots, \sigma_k$ (see Sect. 2.2). Complementing the set of equations with $k - 1$ additional conditions of gap closure ($g(\sigma_1) = 0, \ldots, g(\sigma_1 + \cdots + \sigma_k) = 0$) imposes prescribed times of impact, which has the advantage of eliminating the nonsmoothness without regularizing the contact conditions. Again, continuation can be used to recover a backbone curve with a given number of IPPs. The robust features of Manlab could be explored in this context [11, 109].

One drawback of the shooting method is that it hardly captures unstable parts of backbone curves, because of the time integration [22].

HBM and SM have been combined in the context of forced nonsmooth dynamics in a hybrid method [89]. The linear part of the dynamics is captured by HBM, while SM deals with the nonlinearities.

2.3.3 Gauss' Principle

Another possibility would be to consider Gauss' principle for translating the problem of finding a periodic solution into an optimization problem [108]. This principle is known to be equivalent to d'Alembert's or Jourdain's in the nonsmooth dynamics framework [36]. The acceleration field $\ddot{\mathbf{u}}$ solution to an ODE of the form (3) obeys Gauss' principle with the unilateral constraints

$$\min G(\ddot{\mathbf{u}}) \quad \text{subject to} \quad g(\mathbf{u}) \geq 0, \tag{18}$$

with $G(\ddot{\mathbf{u}}) = (\ddot{\mathbf{u}} - \mathbf{a})^\top \mathbf{M}(\ddot{\mathbf{u}} - \mathbf{a})$ and $\mathbf{a} = -\mathbf{M}^{-1}\mathbf{f}(\mathbf{u}, \dot{\mathbf{u}})$. The idea is to seek periodic solutions by replacing $\mathbf{u}$ with a truncated Fourier series $\mathbf{u}_h$, as in Eq. (12), and to express Gauss' principle in a weak sense in which the cost function is $G(\ddot{\mathbf{u}}_h) \approx G_h(\mathbf{A}, t)$. This yields a problem of the form: find $\mathbf{A}$ solution to

$$\min_{\mathbf{A}} \left(\int_0^T G_h(\mathbf{A}, t)\, \mathrm{d}t \right) \quad \text{subject to} \quad \forall t_i \in S,\ g(\mathbf{A}\boldsymbol{\varphi}(t_i)) \geq 0, \tag{19}$$

where S is a chosen set of discrete times in the interval $[0, T]$. This approach has been adopted for a one-dof system in [74].

3 Nonsmooth Modal Analysis of Continuous Systems

Contact between two linear elastic media generates shock waves featuring discontinuous stress and velocity fronts. For example, when a bar hits the rigid ground, a shock wave emanates at the contact interface, propagates to the free surface of the bar and reflects. The bar departs from the ground when the reflection of the shock wave comes back to the contact interface. Mathematically, the dynamics is described by a Partial Differential Equation (PDE), a solution of which is completely determined by the initial displacement and velocity fields, even in the presence of unilateral constraints [57]: in contrast to the semi-discretized framework (see Sect. 2.1), no impact law is needed for well-posedness. The situation is already quite sophisticated for three-dimensional isotropic homogeneous linear elastic materials, where uncoupled longitudinal and transverse waves propagate at distinct velocities. When a nonlinear constitutive law is considered instead, the governing equations are still hyperbolic,

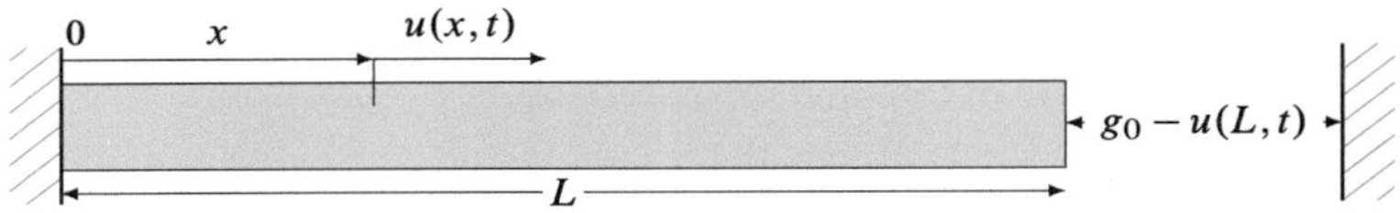

Fig. 10 Fixed–free bar subjected to a unilateral constraint

but the longitudinal and transverse waves are coupled [27, Chap. 4]. Here, we focus on one-dimensional homogeneous linear elastodynamics and explore solution methods that do not rely on space semi-discretization techniques exposed in Sect. 2.

3.1 One-Dimensional Problem of Interest

The system of interest is a fixed–free bar, whose free end is subjected to a unilateral constraint, as illustrated in Fig. 10. The displacement u is assumed to be small compared to the length L of the bar. The dynamics is governed by

$$\begin{cases} \forall x \in (0, L),\ t \in \mathbb{R},\ \partial^2_{tt} u(x,t) = c^2 \partial^2_{xx} u(x,t) & \text{wave equation} \quad (20a) \\ \forall t \geq 0,\ u(0,t) = 0 & \text{Dirichlet condition} \quad (20b) \\ \forall t \geq 0,\ 0 \leq -\partial_x u(L,t) \perp g_0 - u(L,t) \geq 0 & \text{Signorini condition} \quad (20c) \\ \forall x \in (0, L),\ u(x,0) = u_0(x),\ v(x,0) = v_0(x) & \text{initial conditions,} \quad (20d) \end{cases}$$

where g_0 denotes the gap at rest and $c = \sqrt{E/\rho}$ is the wave propagation speed, defined from the Young modulus E and the density ρ of the material. It is worth mentioning that the eigenfrequencies of the linear fixed–free bar are all multiples of the first one, that is, $\omega_k = k\omega_1$, $k \in \mathbb{N}^*$: any initial condition generates a periodic motion and all linear frequencies satisfy an internal resonance condition.

3.2 Analytical Solution

A few analytical solutions of (20) are available for colliding bars [37] or vibrating strings with an obstacle [12, 41] which share similar governing equations. New ingredients are introduced below.

The general solution to (20a) is of the form $u(x,t) = f(ct + x) + h(ct - x)$, for $x \in [0, L]$ and $t \in \mathbb{R}$. In the weak sense, it suffices to require continuity and piecewise $\mathcal{C}^1$-regularity for f and h. Condition (20b) implies that $f = -h$. Let φ denote the derivative of f. It follows that

$$\forall x \in [0, L],\ \forall t \in \mathbb{R}, \quad u(x,t) = f(ct + x) - f(ct - x) = \int_{ct-x}^{ct+x} \varphi(s)\,\mathrm{d}s. \quad (21)$$

Condition (20c) implies that $\partial_x u(L,t) = 0$ when the gap is open, in other words, $\varphi(x+L) = -\varphi(x-L)$, which means that φ is a $2L$-antiperiodic function on $\mathbb{R}$. When the gap closes, it remains closed as long as $\partial_x u \leq 0$. In particular, $\partial_t u(L,t) = 0$, which is equivalent to $\varphi(ct+L) - \varphi(ct-L) = 0$, or φ is $2L$-periodic. Consider a free phase over $[0, t_1]$. On this interval, the displacement field is associated with a $2L$-antiperiodic function φ. Assume the gap is then closed over $[t_1, t_1+t_2]$. The displacement field is then associated with a $2L$-periodic function φ_1. Introducing the function ϵ, defined over $\mathbb{R}$ by $2L$-antiperiodicity and the value 1 over $[-L, L)$, it can be shown that the periodicity condition reduces to the following condition on φ[5]:

$$\forall x \in \mathbb{R}, \quad \varphi(x) = \epsilon(x)\epsilon(x+ct_2)\varphi(x+c(t_1+t_2)). \tag{22}$$

The problem of finding (potential) periodic solutions with one contact phase per period for the unilaterally constrained bar hence reduces to finding φ solutions of (22). The period is given by $T := t_1 + t_2$. Three additional conditions apply, which can be understood as *admissibility conditions* [104, 105]:

- the contacting end of the bar must not penetrate the obstacle during the free flight:

$$\forall t \in [0, t_1], \quad g_0 - \int_{ct-L}^{ct+L} \varphi(s)\,\mathrm{d}s \geq 0; \tag{23a}$$

- at $x = L$, the bar must remain in compression during the contact phase:

$$\forall t \in [t_1, t_1+t_2], \quad \varphi(ct+L) + \varphi(ct-L) \leq 0; \tag{23b}$$

- the gap must be closed at t_1:

$$g_0 - \int_{ct_1-L}^{ct_1+L} \varphi(s)\,\mathrm{d}s = 0. \tag{23c}$$

Equations (22) and (23) can either be solved collectively to find periodic solutions or be used to check the correctness of a candidate periodic solution identified from numerical methods.

An interesting direct consequence follows from the absolute value of (22): $\forall x \in \mathbb{R}, |\varphi(x)| = |\varphi(x+cT)|$. Recall that φ is $2L$-antiperiodic, so $|\varphi|$ is $2L$-periodic, and also cT-periodic. This is possible only if cT/L is a rational, or if $|\varphi|$ is constant. Continua parametrized by T are hence possible only if $|\varphi|$ is constant, meaning that all backbone curves, which are not vertical lines, correspond to piecewise-linear displacement fields, that is, piecewise-constant velocity fields.

Not only does it provide a sound mathematical basis, this approach was proven successful for rediscovering the nonsmooth modes previously conjectured [125] (see Fig. 11). The main backbone curves emanate from the linear eigenfrequencies of the

[5]This formula was established by Pierre Delezoide.

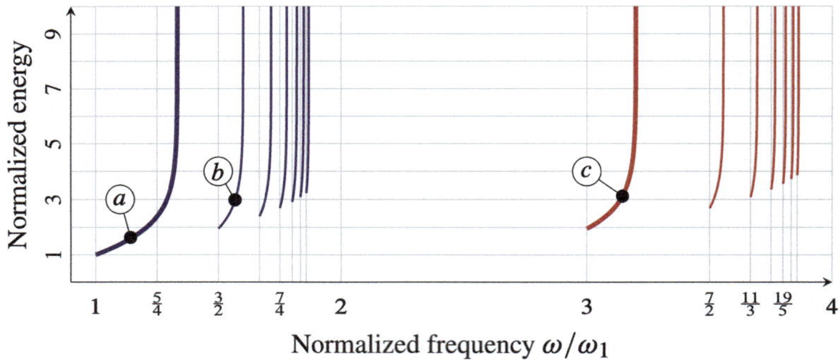

Fig. 11 Backbone curves in the vicinity of the first two linear modes of the bar. ω_1 is the first linear mode of the fixed–free bar. Labels ⓐ, ⓑ and ⓒ correspond to the first NSM, a subharmonic backbone curve and the second NSM, respectively

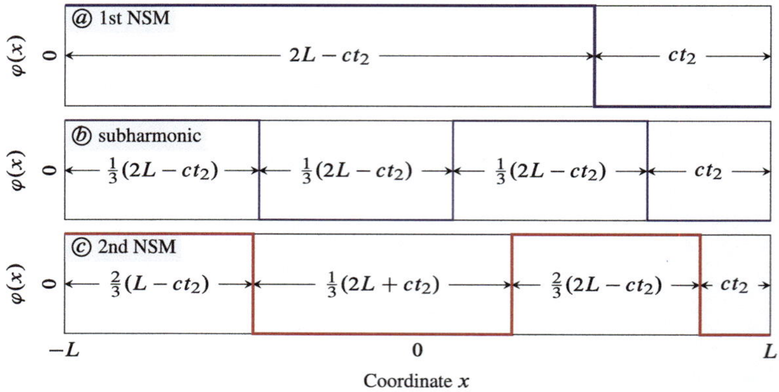

Fig. 12 Functions φ corresponding to the first NSM, the first subharmonic of the first NSM and the second NSM. The corresponding energies and frequencies are marked by the labels ⓐ, ⓑ and ⓒ in Fig. 11

fixed–free bar. The additional curves correspond to subharmonics of higher frequency modes. The functions φ labeled ⓐ, ⓑ and ⓒ in Fig. 11 are plotted in Fig. 12.

Among the solutions to (22) are the two main NSMs determined by φ_1 and φ_2. Each of these functions is defined by its value over $[-L, L]$ and its $2L$-antiperiodicity over $\mathbb{R}$. For the first one,

$$\varphi_1(x) = \alpha \begin{cases} +1 & x \in [-L, L - t_2) \\ -1 & x \in [2L - t_2, L], \end{cases} \tag{24}$$

where the duration of the contact phase t_2 relates to T through $T = 4L/c - t_2$. For the second mode,

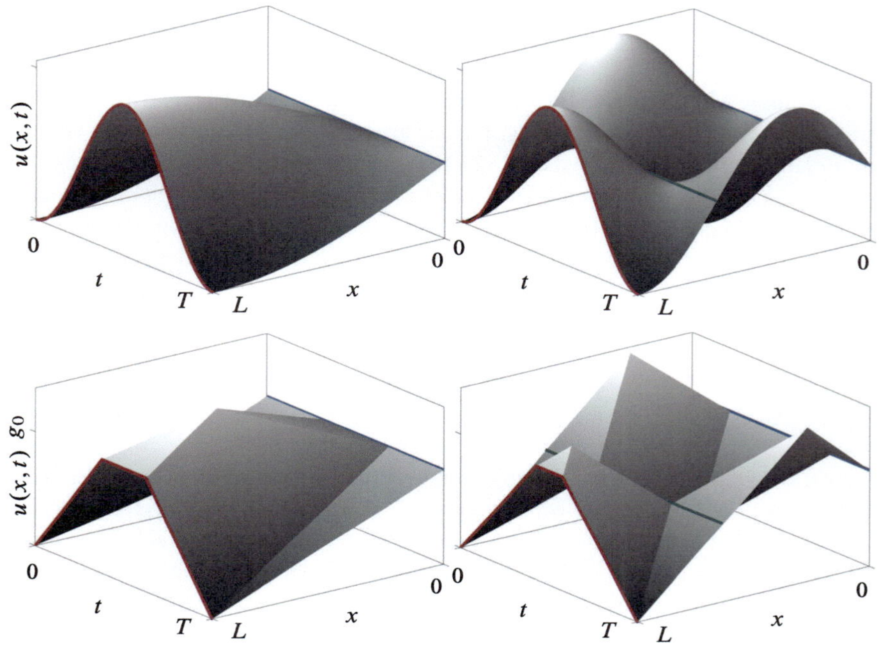

Fig. 13 Displacement field on the first and second linear modes (top), and first and second nonsmooth modes (bottom)

$$\varphi_2(x) = \alpha \begin{cases} +1 & x \in [-L, -\frac{1}{3}(L + 2ct_2)) \cup [\frac{1}{3}(L - ct_2), L - ct_2) \\ -1 & x \in [-\frac{1}{3}(L + 2ct_2), \frac{1}{3}(L - ct_2)) \cup [L - ct_2, L] \end{cases}, \tag{25}$$

where t_2 satisfies $T = (4L/c - t_2)/3$. In both cases, the mode is parametrized by t_2, or equivalently, T or ω. The coefficient α is such that $u(L, 0) = g_0$ and is not explained for the sake of conciseness. The displacement field u, calculated from φ using Eq. (21), is depicted for the first two linear and nonsmooth modes in Fig. 13 with appropriate labels. The first nonsmooth mode and its linear counterpart show similar features: this also holds for the second mode, where both exhibit nodes of vibration. However, standing waves in the linear setting become travelling waves in the unilateral setting, where the characteristic lines are clearly identified.

The analytical approach developed above is limited to simple systems such as the one considered. Numerical techniques capable of handling more general systems are now exposed.

3.3 Finite Volumes and the Wave Finite Element Method

Finite Volume Methods (FVMs) form a family of numerical methods widely used in fluid mechanics [61] to solve PDEs. By construction, they are designed to enforce conservation laws. They consist in discretizing the space domain into cells. As opposed to other well-known numerical techniques such as the Finite Element Method (FEM), the strong form of the PDE is considered and the unknown field is averaged in every cell through volume integrals. Time evolutions are calculated via fluxes on the cell boundaries. In the one-dimensional case, the wave equation (20a) is recast into a system of two hyperbolic conservation laws:

$$\begin{cases} \partial_t \sigma - E \partial_x v = 0 & (26) \\ \rho \partial_t v - \partial_x \sigma = 0, & (27) \end{cases}$$

where v and σ are the velocity and stress fields, respectively. The Wave Finite Element Method (WFEM) is a shock-capturing FVM, in which the time-discretization is coupled to space in such a way that waves propagate along the characteristics lines of the hyperbolic PDE [97]. The Dirichlet-type fixed boundary at $x = 0$ can be dealt with straightforwardly using ghost cells [61]. The treatment of the unilateral contact condition is more challenging: one possibility is to use the floating boundary condition technique [97], which can be understood as a conditional switch between free and fixed boundary conditions.

Finding periodic solutions of the colliding bar reduces to finding the initial stress and velocity fields, in the form of constant averaged values in every cell, which propagate along the characteristic lines, satisfy the clamped boundary condition at $x = 0$ and the switches between fixed and free boundaries at $x = L$ such that the initial state is recovered at time T after a prescribed number k of contact phases per period. The analytical backbone curves in Fig. 11 are retrieved with this approach [125]. A more complicated configuration in which, at $x = 0$, the Dirichlet condition is replaced with a Robin condition of the form $\partial_x u(0, t) = \alpha u(0, t)$ is also of interest, since the internal resonance condition previously mentioned no longer holds. No analytical results could be derived, but nonsmooth modes can be numerically computed.

Also, WFEM implies a projection step when penetration is predicted. This should not be confused with an impact law, since the exact solution of a bouncing bar [23] and the exact solutions in Fig. 13 are retrieved.[6] The main drawbacks of semi-discretization in space are not observed: in particular, there is no chattering, the velocity of the contacting end undergoes a jump at gap openings, and the energy is accurately preserved. Forced responses can be computed as well [125]. However, extension to higher dimensions looks challenging. Indeed, the description of how a discontinuity (between two finite volumes) propagates, the so-called Riemann problem, can no longer be solved exactly. Moreover, conservation laws in the multi-dimensional framework raise a number of issues that are not well-understood [65].

[6] Presumably, in agreement with the continuous framework, no impact law is needed, because information propagates accurately along the characteristic lines.

3.4 Boundary Element Method

Problem (20) can be solved using a variant of the Boundary Element Method (BEM) called the Time-Domain Boundary Element Method (TD-BEM) [100, 115]. BEM is a weighted residual method, with a different weighting function chosen as the fundamental solution u^* of the PDE of interest. For the wave equation in one dimension, u^* is defined as the displacement field in response to an impulse at an arbitrary position $\xi \in [0, L]$ and time $\tau \in \mathbb{R}$:

$$\begin{aligned}&\forall x \in [0, L],\ \forall t \in \mathbb{R},\\ &\qquad \partial^2_{xx} u^*(x, t, \xi, \tau) - \frac{1}{c^2}\partial^2_{tt} u^*(x, t, \xi, \tau) = \delta(x - \xi)\delta(t - \tau). \qquad (28)\end{aligned}$$

A fundamental solution to this PDE reads as [37, Sect. 1.1.8]:

$$u^*(x, t, \xi, \tau) = -\frac{c}{2} H\big(c(t - \tau) - |x - \xi|\big), \qquad (29)$$

where H is the Heaviside function. Using u^* as the weighting function in the space-time integral form of Eq. (20a) yields

$$c^2 \int_0^\tau \int_0^L \partial^2_{xx} u(x, t) u^*(x, t, \xi, \tau)\, \mathrm{d}x\mathrm{d}t - \int_0^\tau \int_0^L \partial^2_{tt} u(x, t) u^*(x, t, \xi, \tau)\, \mathrm{d}x\mathrm{d}t = 0 \qquad (30)$$

which, after integration by parts and a few manipulations [115], leads to

$$\begin{aligned} u(\xi, \tau) = {} & \frac{1}{2} u(L, \tau - (L - \xi)/c) + \frac{1}{2} u(0, \tau - \xi/c) && (31a)\\ & - \int_0^\tau \partial_x u(L, t) u^*(L, t, \xi, \tau)\, \mathrm{d}t - \int_0^\tau \partial_x u(0, t) u^*(0, t, \xi, \tau)\, \mathrm{d}t && (31b)\\ & + \frac{1}{c^2} \int_0^L v_0(x) u^*(x, t, \xi, 0)\, \mathrm{d}x - \frac{1}{c^2} \int_0^L u_0(x) \partial_t u^*(x, t, \xi, 0)\, \mathrm{d}x, && (31c)\end{aligned}$$

where the last integral stands in the distributional sense. This is the principle of the TD-BEM in one dimension. The general solution is a linear combination of u^* defined in (29), which is a progressive wave. The Convolution Quadrature-BEM (CQ-BEM) [1, 86] computes the integrals in (31) via the Convolution Quadrature Method. They can also be computed with piecewise-constant or piecewise-linear polynomials [14]. After discretizing space and time integrals, the sought solution u becomes a linear combination of the boundary conditions $u(0, \cdot)$, $\partial_x u(0, \cdot)$, $u(L, \cdot)$, $\partial_x u(L, \cdot)$ and the initial conditions u_0 and v_0. Due to clamping at $x = 0$, $u(0, \cdot)$ is known and $\partial_x u(0, \cdot)$ is unknown. The contact condition at $x = L$ corresponds to either a free or a fixed boundary condition, and the switch is triggered by monitoring the gap and the normal contact force. In either case, exactly half of the boundary conditions are known and half are unknown. The latter can be deduced from the evaluation of (31) at $\xi = 0$ and $\xi = L$, providing two equations in two unknowns at each prescribed time step. The displacement of internal prescribed nodes can then be recovered through (31).

When targeting periodic solutions, the shooting method (see Sect. 2.3.2) can be used, together with the discretized governing equations obtained from (31), providing $2n$ equations where n is the number of discretized space nodes, for $2n + 1$ unknowns (the initial conditions at the n nodes plus the period T). Again, numerical continuation techniques involving a semi-smooth Newton solver are employed to find nonsmooth modes of vibration.

This approach was successful in computing the first two main backbone curves, some subharmonics and internal resonance backbone curves, (see Fig. 11). The main challenge for the extension to the multi-dimensional framework is that the fundamental solutions are only known exactly for simple geometries. In such cases, Green's functions (which are fundamental solutions with specified boundary conditions) can, however, be approximated numerically [24, Chap. 7].

3.5 Space-Time Finite Differences

Many other discretization schemes relying on finite differences are available for hyperbolic PDEs [67]. We focus on numerical methods that simultaneously discretize space and time. When discontinuous solutions are expected, common methods include Lax-Friederich, Lax-Wendroff, MacCormack Upwind, Forward-Time-Centered-Space (FTCS), and Leapfrog. These schemes all stem from truncated Taylor series, and differ mostly according to their order in space and in time. For example, the FTCS method is second-order in space and first-order in time. Another method can be derived as follows. Writing the second-order Taylor series of u in time

$$u(x_j, t^{n+1}) = u(x_j, t^n) + \Delta t \partial_t u(x_j, t^n) + \frac{1}{2}\Delta t^2 \partial^2_{tt} u(x_j, t^n) + \mathcal{O}(\Delta t^3), \tag{32}$$

then replacing $\partial_t u$ with $-c\partial_x u$ (and thus $\partial^2_{tt} u$ with $c^2 \partial^2_{xx} u$) [82], and applying a first-order central difference for $\partial_x u$ and second-order central difference for $\partial^2_{xx} u$ yields

$$\begin{aligned} u(x_j, t^{n+1}) = u(x_j, t^n) &- \frac{c\Delta t}{2\Delta x}(u(x_{j+1}, t^n) - u(x_{j-1}, t^n) \\ &+ \frac{c^2 \Delta t^2}{2\Delta x^2}(u(x_{j+1}, t^n) - 2u(x_j, t^n) + u(x_{j-1}, t^n)), \end{aligned} \tag{33}$$

which is the well-known Lax-Wendroff scheme.

Stability is governed by the Courant–Friederich–Lewy (CFL) condition, which provides a necessary condition (sometimes sufficient) on the time step Δt, given the wave celerity c_i in direction i and the space discretization step Δx_i, taking the following form in N dimensions:

$$\Delta t \sum_{i=1}^{N} \frac{c_i}{\Delta x_i} \leq C_{\mathrm{CFL}}, \tag{34}$$

where C_{CFL} depends on the finite difference scheme. The interpretation of this condition is that numerical "information" should not propagate slower than physical "information". It is not a sufficient condition for stability.

The main issue with these finite difference schemes for propagating discontinuous fields is that they are either first-order accurate, thus numerical viscosity[7] smoothens the solution, or second-order accurate, in which case they are dispersive,[8] leading to numerical oscillations known as Gibbs phenomenon. Apart from Glimm's method, which suffers from inaccuracy during smooth phases [18], this is clearly illustrated by several examples in [34].

A common strategy for reduce spurious oscillations is to add numerical diffusion tuned to the Gibbs phenomenon. This approach is problem-dependent, and may therefore be tedious to accomplish, and is hardly compatible with periodic solutions. Possibly more promising are *limiters* [21, 28]. Signorini conditions and impact laws have yet to be incorporated into this formalism. As of now, it is unclear whether these approaches would be suitable for nonsmooth modal analysis. Other potentially relevant methods are listed in [28].

Mixed space-time HBM [119] or a time-space FEM with a discretization along the characteristics[9] for one-dimensional systems might also be useful for nonsmooth modal analysis, and are hence worthy of further investigation.

4 Relationships Between Forced Response and Nonsmooth Modes

Various analytical and numerical methods capable of performing nonsmooth modal analysis have been reviewed in the preceding sections, both in the discrete and continuous frameworks. Some are sufficiently mature for nonsmooth modal analysis, while others have yet to be thoroughly explored, as their usability has not been comprehensively assessed. In the following, the nonsmooth modal analysis of a FEM-like

[7]Numerical viscosity, or *diffusion*, arises when the numerical scheme introduces a velocity term with a positive prefactor.

[8]Numerical dispersion occurs when the numerically approximated propagation celerity of a wave depends on its frequency. Note that dispersion and numerical dispersion are two distinct concepts.

[9]The unknown displacement field would be expanded as

$$u(x,t) = \sum a_i \phi_i(x+ct) + b_i \phi_i(x-ct), \tag{35}$$

where the ϕ_i could be the usual hat functions, for instance.

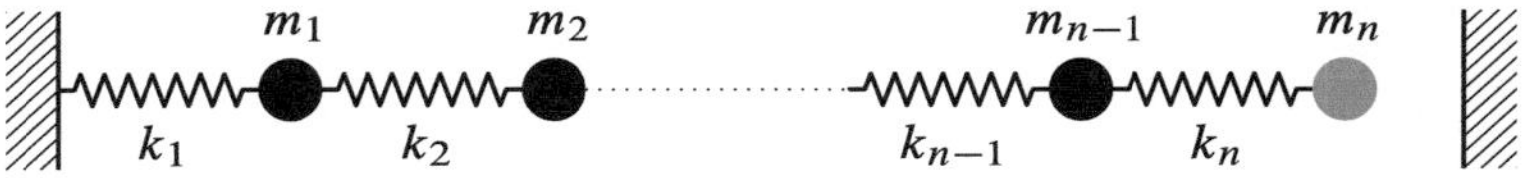

Fig. 14 Spring-mass system subjected to a unilateral constraint

semi-discretization of the colliding bar is explored by means of the analytical and the multiple shooting methods. In the continuous framework, nonsmooth modal analysis of the same system is carried out with WFEM and analytical techniques. The fact that peaks of resonance in the forced response emerge along the backbone curves in the FEP demonstrates the main purpose of nonsmooth modal analysis.

4.1 Discrete Oscillators

Recall that all numerical methods detailed in Sect. 2.3 are capable of computing periodic solutions of a forced system. The brute-force approach is another possibility, which does not work in the autonomous case. It consists in time-integrating Eq. (2), where $\mathbf{f}_{\text{ext}}$ is periodic in time, until a periodic response is obtained or a stopping criterion is reached [72]. This simple technique is CPU-intensive when damping is light and the detection of the steady-state may be delicate. Nevertheless, it was implemented to compute the forced response of the two-dof spring-mass system in Fig. 14 with $n = 2$.

Results are presented as a function of the forcing period in Fig. 15 (top), where colors indicate the number of impacts per period. For clarity, only the five lowest IPPs are shown, even though solutions with as high as 24 IPPs were found, an example of which is depicted in Fig. 16. The period of the response can differ from the period of the forcing. For instance, the period of the 24 IPP-response in Fig. 16 is $8T_0$, where T_0 is the forcing period. Accordingly, the results in Fig. 15 (top) can also be plotted as a function of the *response* period, (see Fig. 15 (bottom)). This results in a correlation between the number of IPP and the response period: IPP curves are clustered. It also shows that identical nonsmooth resonances can be obtained for distinct forcing periods: the two 1 IPP resonance peaks in the top plot seem to correspond to the same resonance in the bottom plot.

The purpose of modal analysis is to predict vibratory resonances. Using the analytical method described in Sect. 2.2, it appears that resonance peaks in the forced response mostly emanate in the vicinity of NSM backbone curves. This is illustrated for the main peaks in Figs. 15 (bottom) and 17, for 1, 3 and 5 IPPs. Note that several response curves are depicted, because the horizontal axis corresponds to the *response* period. A vast majority of the branches in Fig. 15 look like they were connected to NSMs. The way in which nonsmooth modes relate to forced responses is not restricted to peaks in the FEP, but rather extends to shapes. This is illustrated in Fig. 18 for a two-dof system in which the forced response trajectories of the masses are compared

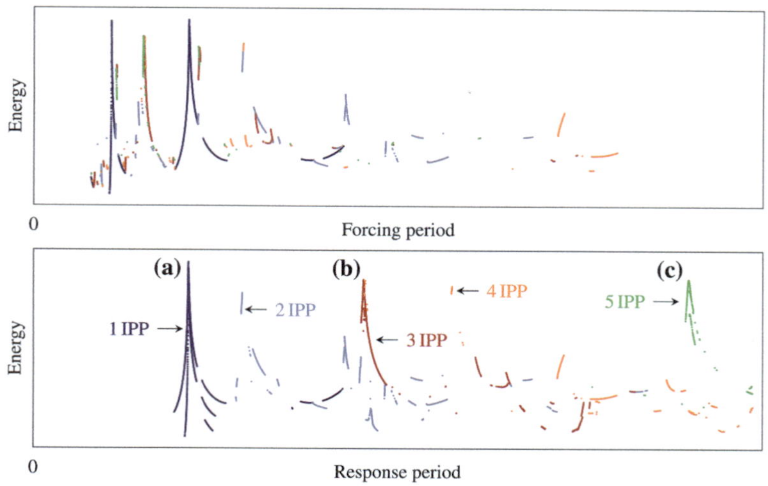

Fig. 15 FEP of a two-dof impact oscillator. Responses with 6 IPPs or more are excluded for clarity. Colors correspond to IPPs with labels used in Fig. 17

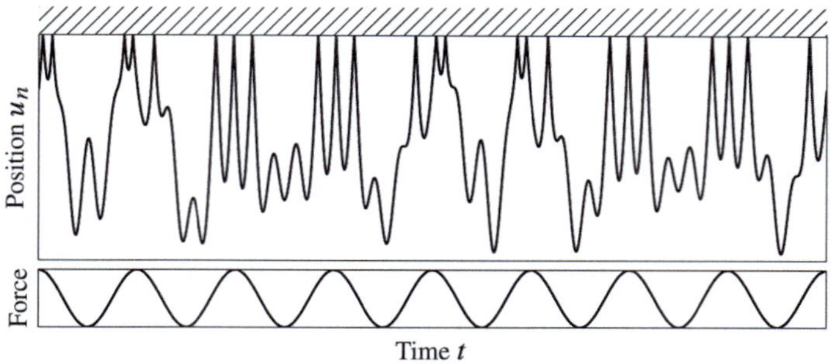

Fig. 16 24-IPP Periodic forced response of period 8 times the forcing period

to the nonsmooth modal shape of the same period. Though no longer symmetric, the forced response is strikingly similar to the periodic solution of the autonomous problem. The above observations extend, in part, to Duffing impact oscillators, as depicted in Fig. 14 for $n = 2$. The corresponding autonomous dynamics between impacts is governed by

$$\begin{bmatrix} m_1 & 0 \\ 0 & m_2 \end{bmatrix} \begin{bmatrix} \ddot{u}_1 \\ \ddot{u}_2 \end{bmatrix} + \begin{bmatrix} 2k & -k \\ -k & k \end{bmatrix} \begin{bmatrix} u_1 \\ u_2 \end{bmatrix} + \epsilon \begin{bmatrix} (u_2 - u_1)^3 \\ (u_1 - u_2)^3 \end{bmatrix} = \begin{bmatrix} 0 \\ 0 \end{bmatrix}. \qquad (36)$$

where ϵ is user-defined. The previous piecewise-linear system corresponds to $\epsilon = 0$, but no analytical techniques exist to compute the periodic solutions when $\epsilon \neq 0$, except for $n = 1$. Using the shooting method between time instants 0^+ and T^-, as described in Sect. 2.3.2, both the backbone curves and the forced response curves can

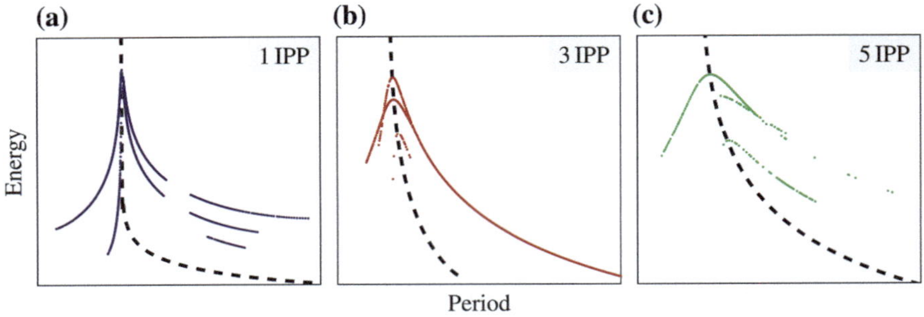

Fig. 17 Forced response resonances as a function of the response period. They perfectly match the backbone curves [- -]. Labels refer to Fig. 15

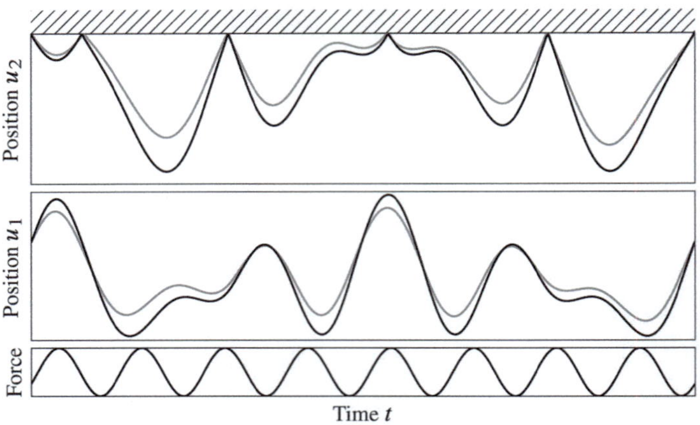

Fig. 18 Comparison between a forced response and the corresponding autonomous periodic solution on the NSM for 5 IPPs. [—] NSM and [—] forced response

be computed for several values of ϵ. They are exposed in Fig. 19, in the neighborhood of a backbone curve with 1 IPP. In this figure, the thick backbone curve is the one in Fig. 17 (left), plotted in terms of frequency. It continuously deforms as the cubic nonlinearity increases. The forced response changes accordingly, so that even in the piecewise-nonlinear case, nonsmooth modal analysis seems to provide backbone curves that perfectly support the forced response curves.

We now proceed with the illustration of HBM, as described in Sect. 2.3.1 for a one-dof impact oscillator [95]. Figure 20 shows the approximated backbone curves for an increasing number of harmonics: the backbone curve converges to the exact one. Also, the approximated forced response is seen to be perfectly organized around the backbone curves. The time evolution of position (right plot) shows that the residual penetration gets smaller as N increases. This very simple example establishes numerical evidence that when periodicity is enforced, constitutive impact laws are unnecessary.

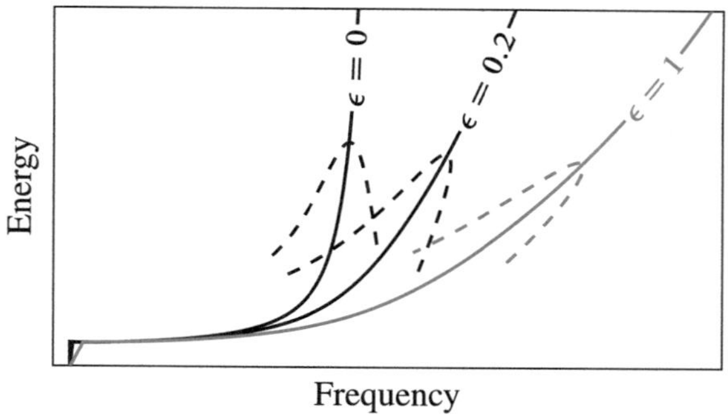

Fig. 19 Sensitivity to the cubic nonlinearity in a Duffing impact oscillator with ϵ defined in Eq. (36). [- -] Forced response and [—] backbone curves

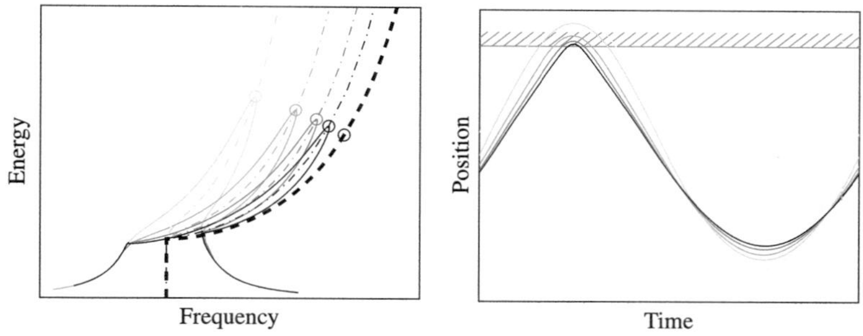

Fig. 20 Convergence of HBM for a one-dof oscillator and no impact law (from [95]). Forced responses are computed from Eq. (17). [- -] Exact backbone curve. [-··-] Backbone curves calculated with HBM. With $N \in \{1, 2, 5, 10, 20\}$ (from light gray to black)

4.2 *Continuous Oscillators*

This subsection succinctly extends the previous results to the continuous framework by exploring the autonomous and forced dynamics of a one-dimensional bar colliding with a rigid wall (see Fig. 10). As explained previously, backbone curves can be obtained analytically (Sect. 3.2), via WFEM (Sect. 3.3) or using TD-BEM (Sect. 3.4). The first four main backbone curves are depicted in Fig. 21 together with the periodically forced response at various energy levels. The top plot corresponds to an excitation induced by a harmonically moving obstacle, while the bottom plot considers an external periodic and distributed force along the bar. As in the discrete case, the main peaks of the forced response align, in both cases, with the main backbone curves. The additional minor peaks on the top plot might correspond to internal resonances. However, this point requires further work.

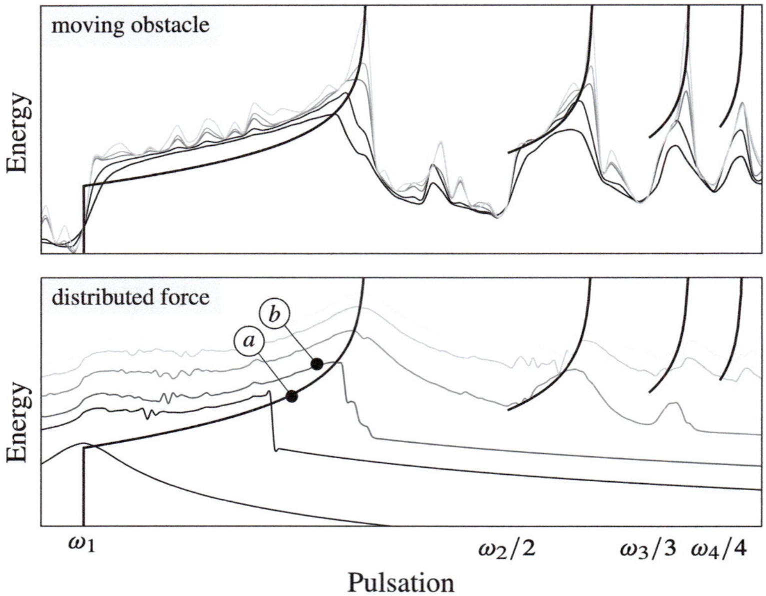

Fig. 21 Main backbone curves of the colliding bar [thick] and forced response curves [thin] (from [125]). External loading is either a harmonically moving obstacle (top) or a harmonic distributed force (bottom). Labels ⓐ and ⓑ refer to Fig. 22

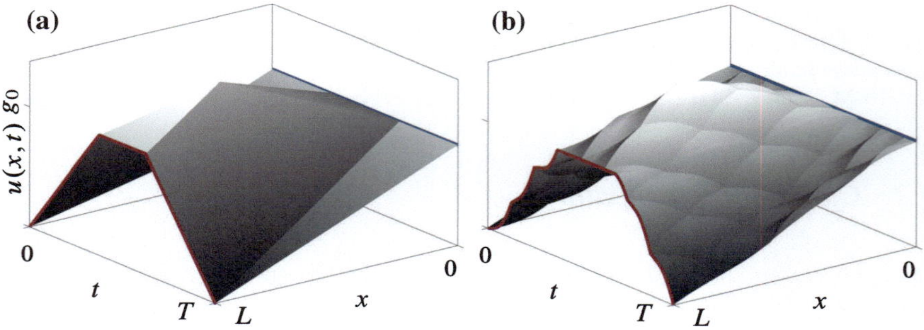

Fig. 22 Space-time forced response and comparison with the nonsmooth mode of the same frequency. Labels are reported in Fig. 21. Left plot is the one in Fig. 13 bottom left

Similarities between autonomous and forced responses also emerge in terms of frequency and modal shapes. For instance, Fig. 22 compares one periodic solution belonging to the first nonsmooth mode to a forced response arising in its vicinity. It is remarkable that the forced response is dominated by the resonant response, that is, the first mode shape (see Fig. 22, left), which is only slightly altered by the type of external forcing.

5 From Discrete to Continuous NSM: Similarities and Differences

We have seen that space-continuous and space-discrete models fall under two different paradigms. In the first category, contact is simply a constraint from which emanate shock waves propagating in the continuous solid. The second category introduces a number of pitfalls and difficulties. An impact law is required, propagation of a wave is difficult to approximate accurately, and lasting contact phases are hardly compatible with the conservation of energy required by the periodicity condition. Additionally, the regularity of the generalized positions is higher than in the continuous case, characterized by discontinuous velocity waves and not just the degree-of-freedom involved in the unilateral constraint.

This last section attempts to highlight the similarities and differences between the two "worlds" within the unidimensional framework presented in Sect. 3.1.

5.1 *Without Unilateral Contact Constraints*

Unilateral contact conditions are temporarily set aside. In structural dynamics, the Finite Element Method is widely used to discretize PDE (20a). Loosely speaking, the weak form of (20a) consists in finding u such that for all v in an appropriate space

$$\int_0^L v\, \partial_{tt}^2 u \,\mathrm{d}x + c^2 \int_0^L \partial_x v\, \partial_x u \,\mathrm{d}x - c^2 \big[v\, \partial_x u\big]_{x=0}^{x=L} = 0. \tag{37}$$

Posing $u_h(x,t) = \sum_{i=1}^n \phi_i(x) u_i(t)$, $v_h(x) = \sum_{i=1}^n \phi_i(x) v_i$ for some chosen shape functions $\phi_1, \dots, \phi_n$, approximating u and v by u_h and v_h in (37), respectively, leads to a system of ODEs standard in structural dynamics:

$$\forall t \in \mathbb{R}^+, \quad \mathbf{M}\ddot{\mathbf{u}}(t) + \mathbf{K}\mathbf{u}(t) = \mathbf{0}, \tag{38}$$

where $\mathbf{M}$ and $\mathbf{K}$ are calculated from (37). In the sequel, we consider, for simplicity, the space semi-discretization of the clamped–free bar with punctual masses (see Fig. 14). Accordingly, $\mathbf{M} = m\mathbf{I}_n$ and

$$\mathbf{K} = k \begin{bmatrix} 2 & -1 & \dots & \dots & 0 \\ -1 & 2 & -1 & \dots & 0 \\ \vdots & \ddots & \ddots & \ddots & \vdots \\ 0 & \dots & -1 & 2 & -1 \\ 0 & \dots & \dots & -1 & 1 \end{bmatrix}. \tag{39}$$

The Young modulus E, the length L and the cross-sectional area S of the bar are related to the stiffnesses and the masses through

$$k = \frac{nS}{L}E \quad \text{and} \quad m = \frac{\rho S}{n}L. \tag{40}$$

Illustrations are given for the following arbitrary values: $E = 1\text{Pa}$, $S = 1\text{m}^2$, $L = 1\text{m}$, $\rho = 1\text{kg m}^{-3}$ and the corresponding k and m given by (40).

Space-discretization formulations are not able to capture the progressive nature of shock waves properly and may lead to non-causal spurious oscillations in space [39]. In order to explain this, let us compare the modal properties of the continuous bar with those of the spring-mass system. The eigenfrequencies and corresponding mode-shapes of the continuous bar are given by [37]

$$\forall p \in \mathbb{N}^*, \quad \omega_p = \frac{(2p-1)\pi c}{2L} \quad \text{and} \quad \phi_p(x) = \sin\left(\frac{(2p-1)\pi x}{2L}\right). \tag{41}$$

In contrast, the eigenfrequencies of the discrete system are

$$\forall p \in [\![1, n]\!], \quad \tilde{\omega}_p = \sqrt{\frac{2k}{m}}\sqrt{1 - \cos\left(\frac{(2p-1)\pi}{2n+1}\right)}, \tag{42}$$

with the corresponding eigenvectors

$$\tilde{\boldsymbol{\phi}}_p = \left[\sin\left(j\frac{(2p-1)\pi}{2n+1}\right)\right]_{j=1,\dots,n}. \tag{43}$$

When $n \gg p$, using (40), the result is that the eigenfrequencies of the discrete and the continous bar are equivalent:

$$\tilde{\omega}_p \sim n\sqrt{\frac{2ES}{\rho SL^2}}\sqrt{\frac{1}{2}\left(\frac{(2p-1)\pi}{2n+1}\right)^2} \sim \sqrt{\frac{E}{\rho}}\frac{\pi(2p-1)}{2L} = \omega_p. \tag{44}$$

Relating the node j of the discrete system to the position x in the bar via $x = L(j-1)/(n-1)$, an analogous consequence holds for the mode shapes $\tilde{\boldsymbol{\phi}}_p$ and $\phi_p(x)$. This is shown in Fig. 23 where both the eigenfrequencies and the eigenmodes are in good agreement in the low-frequency range. However, when the index p is no longer negligible compared to n, the approximation becomes inaccurate. By injecting a progressive monochromatic wave of the form $u(x, t) = e^{\mathrm{i}(\omega t - \kappa x)}$ into the wave equation (here, i stands for the imaginary unit), it results that $\kappa = c/\omega$, which constitutes a linear dispersion relation: the phase and group velocities coincide and there is no dispersion. Now, let Δx denote the space discretization step such that $\Delta x = L/n$. A progressive monochromatic wave of the form $u_p(t) = e^{\mathrm{i}(\omega t - p\tilde{\kappa}\Delta x)}$ in (38) propagates by satisfying

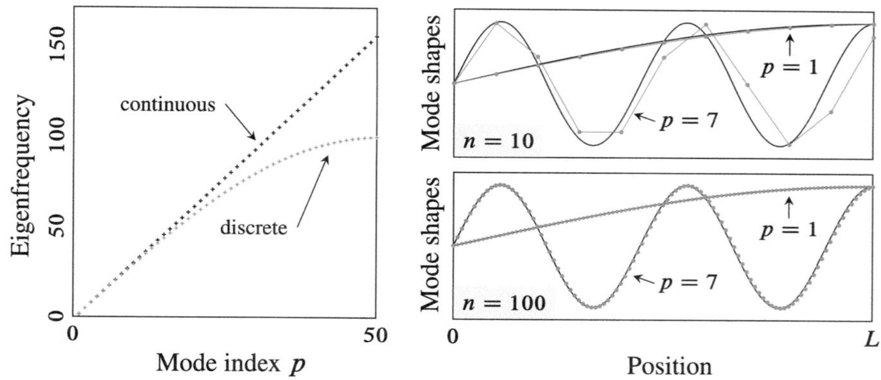

Fig. 23 Comparison between the linear modes of a continuous clamped bar and the linear modes of its discretized counterpart. [—] Mode shape of the continuous bar. [—] Mode shape of the discretized system

$$
\begin{aligned}
0 &= -\omega^2 e^{\mathrm{i}(\omega t - p\Delta\xi)} - \frac{k}{m} e^{\mathrm{i}(\omega t - p\tilde{\kappa}\Delta \mathbf{x})}\left(e^{-\mathrm{i}\tilde{\kappa}\Delta x} - 2 + e^{\mathrm{i}\tilde{\kappa}\Delta x}\right) \\
&= -\omega^2 u_p(t) - 2\frac{k}{m} u_p(t)(\cos(\tilde{\kappa}\Delta x) - 1),
\end{aligned} \tag{45}
$$

so that

$$
\tilde{\kappa}\Delta x = \arccos\left(1 - \frac{\omega^2 m}{2k}\right) = \arccos\left(1 - \frac{\omega^2 \Delta x^2}{2c^2}\right). \tag{46}
$$

When $\Delta x \ll \kappa = c/\omega$, then $\tilde{\kappa} \sim \kappa$, translating the fact that low-frequency waves propagate at the same velocity as in the continuous bar. Nevertheless, dispersion appears for higher frequencies, as illustrated in Fig. 24. This figure shows the time histories for zero initial displacements and velocities except a unit initial velocity on the free node n. Even with a relatively high number of degrees of freedom ($n = 100$), the solution displays spurious oscillations due to the dispersion of high frequency waves. This questions the relevance of the space semi-discretization formalism when shock waves are sought. A comparison between numerous different schemes is proposed in [23]. Even the most accurate of them yields significant discrepancies with the exact solution, even after only one (pseudo-)period of motion [4, 29].

5.2 With Unilateral Contact Constraints

The relationships between nonsmooth modes and forced response curves have been presented in Sect. 4 for discrete and continuous systems, separately. The relationships between discrete NSMs and continuous NSMs is now examined in an exploratory and qualitative manner.

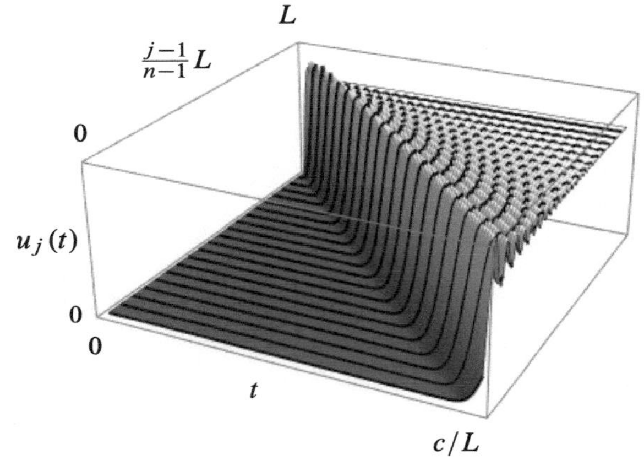

Fig. 24 Time evolution of a spring-mass system with $n = 100$. All initial displacements and velocities are zero, except a unit initial velocity for the free node n. The main wave propagates at the velocity c, but spurious oscillations become visible due to dispersion. Index j goes from 1 to n. The black curves correspond to the trajectory of every fifth degree of freedom. Trajectories are merged in a surface to facilitate visualization

As discussed in Sect. 2.1, the space semi-discretization of a PDE brings in the necessity of an impact law: modal analysis requires $e = 1$ for energy conservation, while $e = 0$ is needed if sticking phases are of interest. Sticking phases are meaningful, as they emerge naturally in the continuous framework (see Fig. 3). Some authors have proposed the *mass redistribution method*. It removes the mass of the contacting node and redistributes it to other nodes [29, 50], so that kinetic energy is not affected. However, it is not clear how it differs from a penalization approach. In the same vein, a recent exploratory work that incorporates an elastic law $e = 0$ proposed to redistributing the kinetic energy of the non-massless contacting node to the neighboring mass [124]. Let us now analyze the sensitivity of the responses to e with n fixed, and to n with e fixed, respectively.

It is observed that the sensitivity of the solution to e reduces when n increases. Figure 25 displays the periodic forced responses for various e and n, obtained using a Moreau–Jean scheme together with a θ-method ($\theta = 1/2$) [2].

For n as small as 20, displacements of the masses are not much affected by e, meaning the forced response curves computed for various e are very similar. Chattering obtained for $e > 0$ seems to have a negligible effect on the overall dynamics [78].

Interestingly, when scaled with respect to the length L, the local behavior of the contact node for large values of n is indistinguishable from that of the continuous bar. This is illustrated with $e = 1$ in Fig. 26 where the periodic solution with $n \in \{5, 20, 1000\}$ is compared with the continuous periodic solutions produced by WFEM (see Sect. 3.3). In particular, no elastic bounces are visible and the contact behaves very much like the lasting contact experienced in the continuous framework. Indeed, the solutions seem to converge with n, irrespective of e, to the solution of the continuous bar. Overall, this paradigmatic difference between the continuous and the discrete systems *with forcing and damping* vanishes as n becomes large. The chattering phenomenon appears to be the pivot between the discrete and the continuous

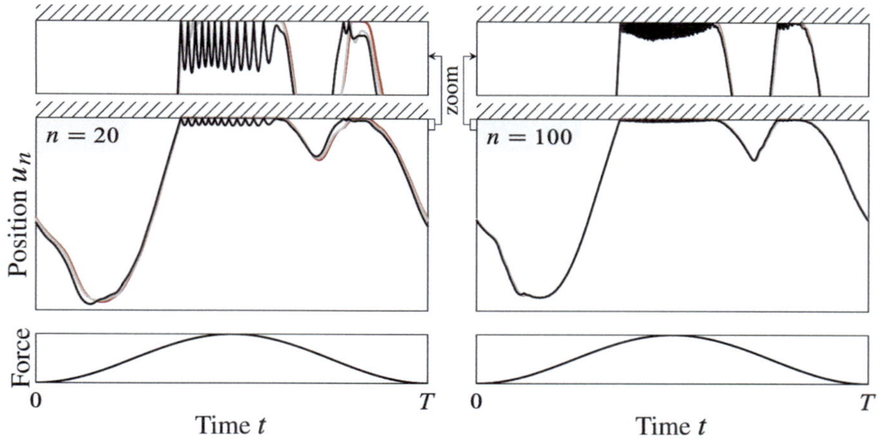

Fig. 25 Sensitivity of a forced periodic solution to the coefficient of restitution e with respect to n for $T = 5.9$ and $g_0 = 1$. [—] $e = 1$. [—] $e = 0.7$. [—] $e = 0$. When n is sufficiently large, the influence of e becomes negligible

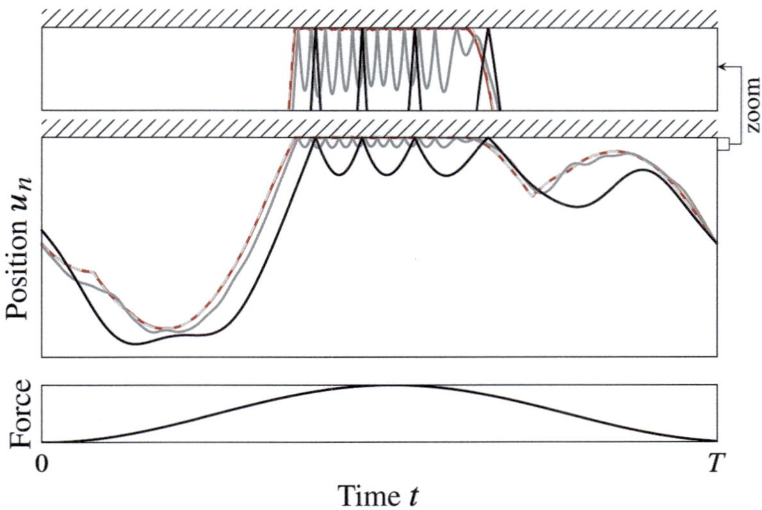

Fig. 26 Convergence to the continuous periodic solutions as n increases for $e = 1$ and $T = 5.7$. Time-integration with $n = 1000$ is indistinguishable from the WFEM solution [- -]. [—] $n = 5$. [—] $n = 20$. [—] $n = 1000$

frameworks. Damping is likely to play an important role as well, since it acts like a low-pass filter, and thus reduces the discrepancies between continuous and discrete models mentioned in Sect. 5.1.

Naturally, one may wonder, in the autonomous and conservative framework, how the backbone curves of the continuous bar compare to the ones of the semi-discretized bar. More explicitly, we would like to approach the backbone curves in Fig. 11 according to the ones exhibited in its n-dof counterpart, as in Fig. 8 for a sufficiently large

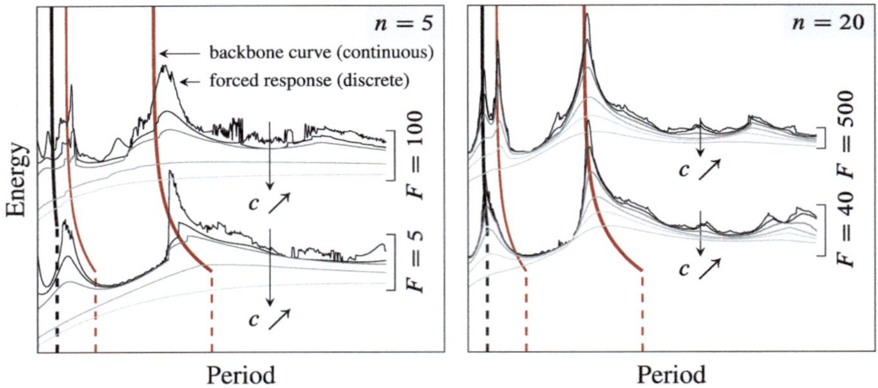

Fig. 27 Comparison between the forced response curve of the discrete system with $e = 1$ and the backbone curve of the continuous bar. [—] First continuous NSM. [—] Second continuous NSM. [—] Harmonic of the first NSM. Dashed parts correspond to the linear part. The damping is denoted by c

n. The challenge comes from the fact that when n becomes large, the spectrum is extremely dense and numerically demanding and currently not accessible. Nonetheless, we provide a few clarifications. In Fig. 27, the energy averaged over six forcing periods for $n = 5$ and $n = 20$ is plotted, for two levels of forcing and several levels of damping. For $n = 5$, the resonance peaks roughly correspond to the main backbone curves of the continuous bar. For $n = 20$, the agreement is clear, and thus, irrespective of the level of damping, for the first two modes as well.

Figure 27 also shows that the forced response curve is very jagged for low levels of damping (dark curves) and becomes a smooth function of the forcing period as damping increases. This can be understood by plotting the position as a function of time for distinct damping levels, as illustrated in Fig. 28. The forcing magnitude is tuned to approximately maintain the magnitude for the position u_n, to compensate for increasing damping levels. The three following types of forced response curves can be distinguished:

- For low damping, the forced response curve is governed by k IPP nonsmooth modes, as stated in Sect. 4.1. The backbone curves feature a number of small branches[10] (see Fig. 15 for $n = 2$). It follows that a forced response is very sensitive to the forcing frequency, as witnessed by the numerous irregularities in the forced response curves in Fig. 27. This situation corresponds to the top left plot in Fig. 28 where a $6T$-periodic response with 3 IPPs is observed.

[10]The number of linear modes increases with n, together with possible internal resonances. This gives the intuition behind the density of backbone curves, which quickly escalates with n. In the piecewise-linear framework, this can be understood in light of the matrices $\mathbf{\Pi}$ and $\mathbf{\Sigma}$ in Eq. (7), whose domains of definition always become more intricate.

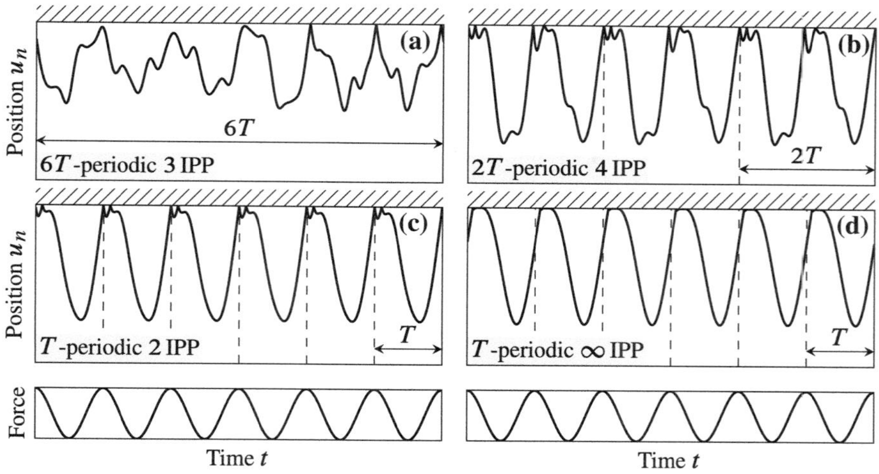

Fig. 28 As the damping (and external force) increases, the motion becomes more and more organized; eventually, chattering appears. ⓐ 3 IPP of period $6T$ $(F = 1, \mathbf{C} = 0.001\mathbf{K})$. ⓑ 4 IPP of period $2T$ $(F = 4, \mathbf{C} = 0.02\mathbf{K})$. ⓒ 2 IPP of period T $(F = 5, \mathbf{C} = 0.1\mathbf{K})$. ⓓ ∞ IPP (chattering) with period T $(F = 14, \mathbf{C} = \mathbf{K})$. With $n = 5$ and $T = 4.4$. Identical vertical scale for the position

- For moderate damping, the forced response curve is smoother and the trajectory is simpler. This corresponds to the $2T$-periodic response with 4 IPPs (top right) and T-periodic response with 2 IPPs (bottom left) in Fig. 28.
- For large damping, the response curves are very smooth, as shown in Fig. 27. Contact settles through chattering mechanisms, and the macroscopic coefficient of restitution, i.e., seen from the scale of the whole system, is $e = 0$, even though the computations were performed with $e = 1$.

Given that the motion of the discrete system converges to that of the continuous bar for sufficient damping, it is not surprising that, for medium or high levels of damping, the resonance peaks are close to those of the bar, for $n = 20$. More surprising is the fact that nonsmooth resonances for low damping also match the continuous backbone curves for large n. In other words, for low damping, as shown in Sect. 4, the forced response of a n-dof system appears to be driven by its (discrete) NSMs, at least for small n. Figure 27 shows that, for large n, this forced response resonates along the backbone curves corresponding to the NSMs of the continuous system. Accordingly, there must be a relationship between the backbone curves of the discrete system and those of the continuous one. This is presently to be clarified, even in the one-dimensional framework, because computing the FEP for the autonomous case with large n is challenging.

We close this chapter with two observations, which tend to confirm some degree of correlation between backbone curves in the continuous and discrete settings.

The first one is concerned with the non-existence of nonsmooth modes for the continuous bar within certain frequency ranges. The continuous bar does not seem to

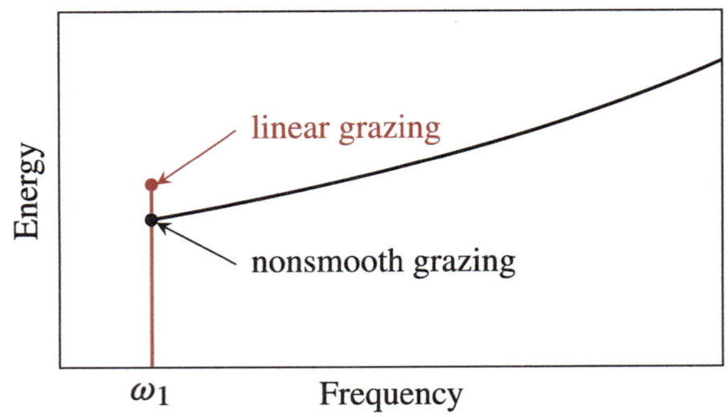

Fig. 29 Close-up view of the first backbone curve of the continuous bar. Two different grazing trajectories coexist at ω_1. [—] Backbone curve of the first linear mode. [—] Backbone curve of the first nonsmooth mode

possess any backbone curves within the range $\omega/\omega_1 \in [2, 3]$ in Fig. 11 and the same applies to the discretized bar for large n: nearly no periodic solutions are detected in this range, at least for 1 IPP, which is encouraging.

The second point relates to similarities in grazing motions:

- In the vicinity of ω_1, the continuous bar features two grazing modes, as illustrated in Fig. 29: the linear grazing mode of the clamped–free bar, which is a sinusoidal function in time, and the nonsmooth grazing mode, which is a triangular function in time (see Fig. 30 (left)). This triangular shape corresponds to the limit case when the mode shape shown in Fig. 13 (bottom left) has a contact duration approaching 0 and can be found exactly from φ given in (24), that is, $\varphi(x) = 1$ for $x \in [-L, L]$ and $2L$-antiperiodic, by evaluating integral (21).
- For the discrete bar, the corresponding linear grazing mode is a sine of frequency $\tilde{\omega}_1$ as well. There is a priori no equivalent for the triangular function found for the continuous bar, since the modal manifold is known to be continuous for any fixed n [59].

However, the triangular shape can be recovered in the discrete setting for large n, as a 1 IPP trajectory. Let us focus on the contacting end of the first nonsmooth mode, for a grazing amplitude. From [104, Eq. (93a)], the position of the nth mass with 1 IPP is

$$u_n(t) = -\frac{\lambda}{\sqrt{m}} \sum_{j=1}^{n} \frac{\cos(\tilde{\omega}_j(t - T/2))}{\tilde{\omega}_j \sin(\tilde{\omega}_j T/2)} \tilde{\phi}_{j,n}^2, \tag{47}$$

where $\tilde{\omega}_j$ and $\tilde{\phi}_{j,n}$ are given by (42) and (43). The value of λ is such that $u_n(0) = g_0$ (closed gap); to simplify, only the time-domain shape of $u_n(t)$ is studied, its magnitude being dropped. When T approaches $2\pi/\tilde{\omega}_1$, the first term of the sum dominates and the shape converges to $\cos((t - T/2)\pi/2)$. This situation corresponds to the first linear grazing mode. For the triangular shape, it should be observed that the sum is dominated by the first terms such that for some $n' \ll n$ and $j \leq n'$, $\tilde{\omega}_j \sim \pi(2j-1)/2$ and $\tilde{\phi}_{j,n} \sim \sin((2j-1)\pi/2) = (-1)^{j+1}$, $u_n(t)$ can be approximated by

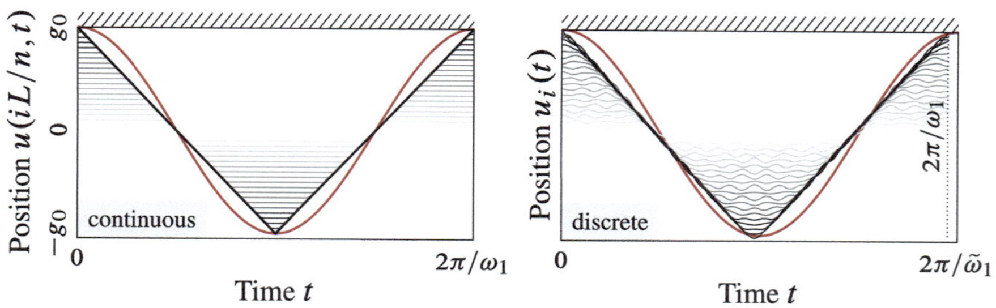

Fig. 30 In the continuous framework, the first linear grazing (a sine function in time) and first nonsmooth grazing (a triangular function in time) trajectories share the same frequency ω_1. For the discrete system, the linear grazing mode is also a sine. When n is sufficiently large, the triangular shape of the continuous nonsmooth grazing mode is retrieved. [—] Displacement of the contacting end $u(L,t)$ and $u_n(t)$, respectively, along the first linear grazing mode. Other positions within the bar/discrete oscillator are also indicated from $x \approx 0$ to $x \approx L$ (white to black)

$$\sum_{j=1}^{n'} \frac{\cos\big(\pi(2j-1)(t-T/2)/2\big)}{(2j-1)\sin(\tilde{\omega}_j T/2)}. \tag{48}$$

Now, concerning the period of the first grazing nonsmooth mode $T = 2\pi/\omega_1 = 4$, and since $\sin(2\tilde{\omega}_j) \sim (2j-1)\pi/(2n)$ when $j \ll n$, the shape is also similar to

$$\sum_{j=1}^{n'} \frac{\cos(\pi(2j-1)t/2)}{(2j-1)^2} = \sum_{j=1}^{n'} (-1)^j \frac{\sin((2j-1)\pi/2(t-1)}{(2j-1)^2}, \tag{49}$$

which is the truncated Fourier series of a triangular wave. As $n \to \infty$, $\tilde{\omega}_1 \to \omega_1$, which is in the neighborhood of ω_1 (and its multiples), both the exact grazing sine and the approximated grazing triangular function are found (see Fig. 30 (right)). The n-dof system thus mimics the continuous bar's nonsmooth grazing behaviour. The corresponding energies, for the discrete and continuous system, are also found to be comparable.

The (nonsmooth grazing) triangular displacement reported above emerges because it can be expressed as a combination of the linear modes of the clamped–free bar, whose time-domain participations in the nonsmooth periodic solution follow a Fourier sequence of fundamental frequency ω_1: this unique attribute stems from the full internal resonance condition enjoyed by the continuous time considered and no longer holds when this condition is not satisfied.

6 Conclusion

In the literature, nonlinear modal analysis is recognized as a matured tool for smooth nonlinear vibratory systems of small to moderate size. However, new methods of analysis are needed when vibro-impact dynamics and unilateral contact conditions are involved. Nonsmooth modal analysis is one such tool. It consists in finding continuous families of periodic solutions of unforced nonsmooth systems, as specified by the definition of modal analysis. Existing solution methods serving that purpose, including very recent developments, were presented in this chapter for simplified systems in the form of a one-dimensional continuous bar and a corresponding n-dof discrete spring–mass oscillator. Conceptual dissimilarities between these two frameworks are summarized as follows:

- For modal analysis purposes, the discrete setting necessitates an energy-preserving impact law with restitution coefficient e, while the continuous setting does not.
- The discrete setting with an energy-preserving impact law generates *chattering*, which manifests itself as k-IPP trajectories that are challenging to capture numerically when k and n grow. Chattering was found to be the pivot between the discrete and continuous worlds.
- It is not clear whether the backbone curves (which define the nonlinear spectrum of vibration) of the discrete oscillator converge towards the backbone curves of the continuous system as n increases.
- The sensitivity to the restitution coefficient e of the periodically forced displacement of the discrete oscillator with low damping decreases with n.
- The backbone curves calculated in the continuous setting accurately predict the vibratory resonances of the discrete oscillator for a sufficiently large n, irrespective of e.
- By virtue of the above comment, vibratory resonances of the continuous bar and discrete oscillator are in good agreement as soon as n is sufficiently large. The peaks of resonance are not much affected by the type of forcing (distributed, concentrated at the contacting end, or far from the contact zone).

In the long run, the aim is to settle Nonsmooth Modal Analysis as an attractive and standard engineering tool aiding in the the efficient prediction and comprehension of nonsmooth vibratory signatures, in replacement of tedious time-domain computations. Among the various possible avenues to be explored in the future, the following are pressing issues:

- In the finite element framework, removing the problematic chattering could be overcome by taking advantage of the vanishing influence of the impact laws for large n and choosing a purely inelastic impact law, that is, $e = 0$. The very small loss of energy should be compensated for in some way.
- Nonsmooth Modal Analysis of multi-dimensional systems should be tentatively performed employing Finite Volumes and the Time-domain Boundary Element Method.

- In the context of continuum mechanics with the assumptions of large displacements and strains, smooth nonlinearities emerge. The resulting dynamics involving unilateral contact constraints should be addressed.

Acknowledgements The authors would like to thank David Urman for providing the data for Figs. 1, 8 and 19, Charlélie Bertrand for Fig. 11, Carlos Yoong for Figs. 13 and 22, as well as Pierre Delezoide for his assistance in Sect. 3.2. The authors would also like to acknowledge the NSERC Discovery Grant program (421542-2012) and Fonds de recherche du Québec Nature et technologies Établissement de nouveaux chercheurs universitaires (2014-NC-173113) for their financial support.

References

1. Abreu AI, Carrer JAM, Mansur WJ (2003) Scalar wave propagation in 2D: a BEM formulation based on the operational quadrature method. Engineering analysis with boundary elements 27(2):101–105
2. Acary V, Brogliato B (2008) Numerical methods for nonsmooth dynamical systems: applications in mechanics and electronics, vol 35. Springer, Berlin
3. Allgower E, Georg K (2012) Numerical continuation methods: an introduction, vol 13. Springer Science and Business Media
4. Armero F, Pet?ocz E (1998) Formulation and analysis of conserving algorithms for frictionless dynamic contact/impact problems. Comput Methods Appl Mech Eng 158(3–4):269–300
5. Arquier R, Bellizzi S, Bouc R, Cochelin B (2006) Two methods for the computation of nonlinear modes of vibrating systems at large amplitudes. Comput Struct 84(24):1565–1576
6. Ascher U, Mattheij R, Russell R (1995) Numerical solution of boundary value problems for ordinary differential equations. SIAM, Philadelphia, USA
7. Attar M, Karrech A, Regenauer-Lieb K (2017) Non-linear modal analysis of structural components subjected to unilateral constraints. J Sound Vib 389:380–410
8. Ballard P (2000) The dynamics of discrete mechanical systems with perfect unilateral constraints. Arch Ration Mech Anal 154(3):199–274
9. Berti M (2007) Nonlinear oscillations of Hamiltonian PDEs. Progress in nonlinear differential equations and their applications. Birkhäuser, Boston
10. Boyd J (2001) Chebyshev and fourier spectral methods. Dover Publications Inc, Mineola
11. Bruno C, Christophe V (2009) An interactive path following software, Manlab
12. Cabannes H (1984) Cordes vibrantes avec obstacles [in French]. Acta Acust United Acust 55(1):14–20
13. Carpenter N, Taylor R, Katona M (1991) Lagrange constraints for transient finite element surface contact. Int J Numer Methods Eng 32(1):103–128
14. Carrer JAM, Costa VL (2015) Boundary element method formulations for the solution of the scalar wave equation in one-dimensional problems. J Braz Soc Mech Sci Eng 37(3):959–971
15. Chati M, Rand R, Mukerhjee S (1997) Modal analysis of a cracked beam. J Sound Vib 207(2):249–270
16. Chatziioannou V, van Walstijn M (2015) Energy conserving schemes for the simulation of musical instrument contact dynamics. J Sound Vib 339:262–279
17. Chris B, Felix D (1994) Chattering and related behaviour in impact oscillators. Philos Trans R Soc Lond A: Math Phys Eng Sci 347(1683):365–389
18. Colella P (1982) Glimm's method for gas dynamics. SIAM J Sci Stat Comput 3(1):76–110
19. David S (2000) Rigid-body dynamics with friction and impact. SIAM Rev 42(1):3–39
20. di Bernardo M, Budd C, Champneys A, Kowalczyk P (2008) Piecewise-smooth dynamical systems: theory and applications. Applied mathematical sciences. Springer Science and Business Media, London

21. Dinshaw B, Chi-Wang S (2000) Monotonicity preserving weighted essentially non-oscillatory schemes with increasingly high order of accuracy. J Comput Phys 160(2):405–452
22. Doedel E (1997) Nonlinear numerics. Journal of the Franklin Institute 334(5-6):1049–1073
23. Doyen D, Ern A, Piperno S (2011) Time-integration schemes for the finite element dynamic Signorini problem. SIAM J Sci Comput 33(1):223–249
24. Duffy D (2015) Green's functions with applications. CRC Press, Boca Raton
25. Dzonou R, Marques MM, Paoli L (2009) A convergence result for a vibro-impact problem with a general inertia operator. Nonlinear Dyn 58(1–2):361
26. El Hadi M, Bellizzi S, Cochelin B, Nistor I (2015) Nonlinear normal modes of a two degrees-of-freedom piecewise linear system. Mech Syst Signal Process 64:266–281
27. Engelbrecht J (1997) Nonlinear wave dynamics: complexity and simplicity. Kluwer, Dordrecht
28. Ewing R, Wang H (2001) A summary of numerical methods for time-dependent advection-dominated partial differential equations. J Comput Appl Math 128(1):423–445
29. Farshid D, Adrien P, Jérôme P, Yves R (2013) Numerical approximations of a one dimensional elastodynamic contact problem based on mass redistribution method. HAL preprint 00917450
30. Fredriksson M, Nordmark A (2000) On normal form calculations in impact oscillators. Proc R Soc Lond A: Math Phys Eng Sci 456(1994):315–329 (The Royal Society)
31. Fung R-F, Han C-F, Ha J-L (2008) Dynamic responses of the impact drive mechanism modeled by the distributed parameter system. Appl Math Model 32(9):1734–1743
32. García-Saldaña J, Gasull A (2013) A theoretical basis for the harmonic balance method. J Differ Equ 254(1):67–80
33. Garg D, Patterson M, Hager W, Rao A, Benson D, Huntington G (2009) An overview of three pseudospectral methods for the numerical solution of optimal control problems. Adv Astronaut Sci 135(1):475–487
34. Gary S (1978) A survey of several finite difference methods for systems of nonlinear hyperbolic conservation laws. J Comput Phys 27(1):1–31
35. Gendelman O (2013) Exact solutions for discrete breathers in a forced-damped chain. Phys Rev E 87(6):062911
36. Glocker C (1998) The principles of d'Alembert, Jourdain, and Gauss in nonsmooth dynamics part I: scleronomic multibody systems. ZAMM 78(1):21–37
37. Graff K (1991) Wave motion in elastic solids. Dover publications, New York
38. Griffin JH (1989) An alternating frequency/time domain method for calculating the steady-state response of nonlinear dynamic systems. J Appl Mech 56:149
39. Grosu E, Harari I (2007) Stability of semidiscrete formulations for elastodynamics at small time steps. Finite Elem Anal Des 43(6):533–542
40. Guckenheimer J, Holmes P (1983) Nonlinear oscillations, dynamical systems, and bifurcations of vector fields, vol 42. Springer Science and Business Media, New York
41. Haraux A, Cabannes H (1983) Almost periodic motion of a string vibrating against a straight fixed obstacle. Nonlinear Anal Theory Methods Appl 7(2):129–141
42. Hu Z, Thouless MD, Lu W (2016) Effects of gap size and excitation frequency on the vibrational behavior and wear rate of fuel rods. Nucl Eng Des 308:261–268
43. Issanchou C, Bilbao S, Le Carrou J-L, Touzé C, Doaré O (2017) A modal-based approach to the nonlinear vibration of strings against a unilateral obstacle: simulations and experiments in the pointwise case. J Sound Vib 393:229–251
44. Jayaprakash KR, Starosvetsky Y, Vakakis A, Peeters M, Kerschen G (2011) Nonlinear normal modes and band zones in granular chains with no pre-compression. Nonlinear Dyn 63(3):359–385
45. Jézéquel L, Lamarque C-H (1991) Analysis of non-linear dynamical systems by the normal form theory. J Sound Vib 149(3):429–459
46. Joannin C, Chouvion B, Thouverez F, Mbaye M, Ousty J-P (2016) Nonlinear modal analysis of mistuned periodic structures subjected to dry friction. J Eng Gas Turbines Power 138(7):072504

47. Kelley Al (1967) The stable, center-stable, center, center-unstable, unstable manifolds. J Differ Equ 3(4):546–570
48. Kerschen G (ed) (2014) Modal analysis of nonlinear mechanical systems, vol 7. Springer, CISM International Centre for Mechanical Sciences
49. Kerschen G, Peeters M, Golinval J-C, Vakakis A (2009) Nonlinear normal modes, part I: A useful framework for the structural dynamicist. Mech Syst Signal Process 23(1):170–194
50. Khenous HB, Laborde P, Renard Y (2006) On the discretization of contact problems in elastodynamics. Lecture notes in applied and computational mechanics. Springer, p 31–38
51. Khenous HB, Laborde P, Renard Y (2008) Mass redistribution method for finite element contact problems in elastodynamics. Eur J Mech A Solid 27(5):918-932
52. Kim W-J, Perkins NC (2003) Harmonic balance/Galerkin method for nonsmooth dynamic systems. J Sound Vib 261(2):213–224
53. Kim YB, Noah ST, Choi YS (1991) Periodic response of multi-disk rotors with bearing clearances. J Sound Vib 144(3):381–395
54. Laslett J (1959) Concerning the v/N → 1=3 resonance, IV: a trial function for the limiting amplitude solution of $d^2u/d\phi^2 + (a + b\cos 2\phi)u + B_1/2(\sin 2\phi)u^2 = 0$. Technical report
55. Laxalde D, Legrand M (2011) Nonlinear modal analysis of mechanical systems with frictionless contact interfaces. Comput Mech 47(4):469–478
56. Le Thi H, Junca S, Legrand M (2017) Periodic solutions of a two-degree-of-freedom autonomous vibro-impact oscillator with sticking phases. Nonlinear Anal Hybrid Syst [in press]
57. Lebeau G, Schatzman M (1984) A wave problem in a half-space with a unilateral constraint at the boundary. J Differ Equ 53(3):309–361
58. Lee Y, Nucera F, Vakakis A, McFarland D, Bergman L (2009) Periodic orbits, damped transitions and targeted energy transfers in oscillators with vibro-impact attachments. Phys D Nonlinear Phenom 238(18):1868–1896
59. Legrand M, Junca S, Heng S (2017) Nonsmooth modal analysis of a N-degree-of-freedom system undergoing a purely elastic impact law. Commun Nonlinear Sci Numer Simul 45:190–219
60. Leine RI, van de Wouw N (2007) Stability and convergence of mechanical systems with unilateral constraints. Lecture notes in applied and computational mechanics, vol 36. Springer Science and Business Media, Berlin
61. LeVeque R (2002) Finite volume methods for hyperbolic problems, vol 31. Cambridge University Press, Cambridge
62. Li K, Darby A (2009) Modelling a buffered impact damper system using a spring-damper model of impact. Struct Control Health Monit 16(3):287–302
63. Liu Y, Pavlovskaia E, Wiercigroch M (2016) Experimental verification of the vibro-impact capsule model. Nonlinear Dyn 83(1-2):1029–1041
64. Liu Y, Wiercigroch M, Pavlovskaia E, Hongnian Y (2013) Modelling of a vibro-impact capsule system. Int J Mech Sci 66:2–11
65. Loubere R, Dumbser M, Diot S (2014) A new family of high order unstructured MOOD and ADER finite volume schemes for multidimensional systems of hyperbolic conservation laws. Commun Comput Phys 16(3):718–763
66. Luo G, Ma L, Lv X (2009) Dynamic analysis and suppressing chaotic impacts of a two-degree-of-freedom oscillator with a clearance. Nonlinear Anal Real World Appl 10(2):756–778
67. Mazzia A (2010) Numerical methods for the solution of hyperbolic conservation laws. Science applicate, via Belzoni, Italy
68. Meingast M, Legrand M, Pierre C (2014) A linear complementarity problem formulation for periodic solutions to unilateral contact problems. Int J Non-Linear Mech 66:18–27
69. Mita M, Ataka M, Fujita H, Toshiyoshi H (2014) An inertia driven micro-actuator for space applications. Electron Commun Jpn 97(3):60–67
70. Morrison D, Riley J, Zancanaro J (1962) Multiple shooting method for two-point boundary value problems. Commun ACM 5(12):613–614

71. Nacivet S, Pierre C, Thouverez F, Jezequel L (2003) A dynamic Lagrangian frequency-time method for the vibration of dryfriction- damped systems. J Sound Vib 265(1):201–219
72. Nayfeh A, Balachandran B (2008) Applied nonlinear dynamics: analytical, computational and experimental methods. Wiley, New York
73. Nayfeh A, Mook D (2008) Nonlinear oscillations. Wiley
74. Nodoushan AA (2015) On the use of Gauss' principle in vibration analysis. Master Thesis, McGill University
75. Nordmark A (2001) Existence of periodic orbits in grazing bifurcations of impacting mechanical oscillators. Nonlinearity 14(6):1517
76. Paidoussis M, Li GX (1992) Cross-flow-induced chaotic vibrations of heat-exchanger tubes impacting on loose supports. J Sound Vib 152(2):305–326
77. Paoli L, Schatzman M (2000) Ill-posedness in vibro-impact and its numerical consequences. In: Proceedings of the European congress on computational methods in applied sciences and engineering (ECCOMAS), Barcelona, Spain
78. Paoli L, Schatzman M (2007) Numerical simulation of the dynamics of an impacting bar. Comput Methods Appl Mech Eng 196(29):2839–2851
79. Pavlovskaia E, Hendry D, Wiercigroch M (2015) Modelling of high frequency vibro-impact drilling. Int J Mech Sci 91:110–119
80. Pavlovskaia E, Wiercigroch M (2003) Periodic solution finder for an impact oscillator with a drift. J Sound Vib 267(4):893–911
81. Pilipchuk V (2001) Impact modes in discrete vibrating systems with rigid barriers. Int J Non-Linear Mech 36(6):999–1012
82. Pletcher R, Tannehill J, Anderson D (2012) Computational fluid mechanics and heat transfer. CRC Press, Boca Raton
83. Pun D, Lau SL, Law SS, Cao DQ (1998) Forced vibration analysis of a multidegree impact vibrator. J Sound Vib 213(3):447–466
84. Renson L, Noël J-P, Kerschen G (2015) Complex dynamics of a nonlinear aerospace structure: numerical continuation and normal modes. Nonlinear Dyn 79(2):1293–1309
85. Sami M (1973) Steady-state response of a multidegree systemwith an impact damper. J Appl Mech 40(1):127–132
86. Schanz M (2012) Wave propagation in viscoelastic and poroelastic continua: a boundary element approach, vol 2. Springer Science and Business Media, Berlin
87. Schatzman M (1998) Uniqueness and continuous dependence on data for one-dimensional impact problems. Math Comput Model 28(4-8):1–18
88. Schindler T, Nguyen B, Trinkle J (2011) Understanding the difference between prox and complementarity formulations for simulation of systems with contact. 2011 IEEE/RSJ International conference on intelligent robots and systems (IROS). IEEE, pp 1433–1438
89. Schreyer F, Leine RI (2016) A mixed shooting-harmonic balance method for unilaterally constrained mechanical systems. Arch Mech Eng 63(2):297–314
90. Seydel R (2009) Practical bifurcation and stability analysis, vol 5. Springer Science and Business Media, New York
91. Sharif-Bakhtiar M, Shaw S (1988) The dynamic response of a centrifugal pendulum vibration absorber with motion-limiting stops. J Sound Vib 126(2):221–235
92. Shaw S, Holmes P (1983) A periodically forced piecewise linear oscillator. J Sound Vib 90(1):129–155
93. Shaw S, Pierre C (1991) Non-linear normal modes and invariant manifolds. J Sound Vib 150(1):170–173
94. Shevitz D, Paden B (1994) Lyapunov stability theory of nonsmooth systems. IEEE Trans Autom Control 39(9):1910–1914
95. Shi Y (2016) Computation of nonlinear modes of vibration of systems undergoing unilateral contact through the semi-smooth Newton approach. Master Thesis, McGill University
96. Shi Y, Legrand M (2016) Semismooth Newton solver for periodicallyforced solutions to a unilateral contact formulation. In: 24th international congress of theoretical and applied mechanics

97. Shorr B (2004) The wave finite element method. Springer Science and Business Media, Berlin
98. Simo JC, Tarnow N (1992) The discrete energy-momentum method: conserving algorithms for nonlinear elastodynamics. Zeitschrift für Angewandte Mathematik und Physik 43(5):757–792
99. Simon J, Mathias L (2015) Forced vibrations of a turbine blade undergoing regularized unilateral contact conditions through the wavelet balance method. Int J Numer Methods Eng 101(5):351–374
100. Soares D, Carrer JAM, Mansur WJ (2005) Non-linear elastodynamic analysis by the BEM: an approach based on the iterative coupling of the D-BEM and TD-BEM formulations. Eng Anal Bound Elem 29(8):761–774
101. Sotirios N (1993) Dynamics of multiple-degree-of-freedom oscillators with colliding components. J Sound Vib 165(3):439–453
102. Stoer J, Bulirsch R (2013) Introduction to numerical analysis, vol 12. Springer Science and Business Media, New York
103. Thompson JMT, Ghaffari R (1983) Chaotic dynamics of an impact oscillator. Phys Rev A 27(3):1741
104. Thorin A, Delezoide P, Legrand M (2017) Nonsmooth modal analysis of piecewise-linear impact oscillators. SIAM J Appl Dyn Syst 16(3):1710–1747
105. Thorin A, Delezoide P, Legrand M (2017) Periodic solutions of n-dofs autonomous vibroimpact oscillators with one lasting contact phase. Nonlinear Dyn 90(3):1771–1783
106. Thorin A, Legrand M (2017) Spectrum of an impact oscillator via nonsmooth modal analysis. In: 9th European nonlinear dynamics conference, Budapest, Hungary
107. Thorin A, Legrand M, Junca S (2015) Nonsmooth modal analysis: investigation of a two-dof spring-mass system subject to an elastic impact law. In: Proceedings of the ASME international design engineering technical conferences and computers and information in engineering conference, Boston
108. Udwadia F, Kalaba R (2007) Analytical dynamics: a new approach. Cambridge University Press, Cambridge
109. Urman D, Legrand M (2016) Nonlinear modes of vibration of vibroimpact Duffing oscillators. In: International congress of theoretical and applied mechanics, Montreal, Canada
110. Vaibhav D, Ian H (2006) Shooting methods for locating grazing phenomena in hybrid systems. Int J Bifurc Chaos 16(03):671–692
111. Vakakis A, Manevitch L, Mikhlin Y, Pilipchuk V, Zevin A (1996) Normal modes and localization in nonlinear systems. Wiley, New York
112. Van de Vorst ELB, Van Campen DH, De Kraker A, Fey RHB (1996) Periodic solutions of a multi-DOF beam system with impact. J Sound Vib 192(5):913–925
113. van de Water W, Molenaar J (2000) Dynamics of vibrating atomic force microscopy. Nanotechnology 11(3):192
114. Vedenova E, Manevich L, Pilipchuk V (1985) Normal oscillations of a string with concentrated masses on non-linearly elastic supports. J Appl Math Mech 49(2):153–159
115. Venkatesh J, Thorin A, Legrand M (2017) Nonlinear modal analysis of a one-dimensional bar undergoing unilateral contact via the time-domain boundary element method. In: Proceedings of the ASME international design engineering technical conferences and computers and information in engineering conference, Cleveland
116. Von Groll G, Ewins D (2001) The harmonic balance method with arclength continuation in rotor/stator contact problems. J Sound Vib 241(2):223–233
117. Wagg D (2006) Multiple non-smooth events in multi-degree-of-freedom vibroimpact systems. Nonlinear Dyn 43(1–2):137–148
118. Wanda S-S (1978) The generalized harmonic balance method for determining the combination resonance in the parametric dynamic systems. J Sound Vib 58(3):347–361
119. Wang X, Zhu W (2017) The spatial and temporal harmonic balance method for obtaining periodic responses of a nonlinear partial differential equation with a linear complex boundary condition. In: Proceedings of the ASME international design engineering technical conferences and computers and information in engineering conference. Cleveland

120. Whiston GS (1987) Global dynamics of a vibro-impacting linear oscillator. J Sound Vib 118(3):395–424
121. Wiercigroch M, Budak E (2001) Sources of nonlinearities, chatter generation and suppression in metal cutting. Philos Trans R Soc Lond A Math Phys Eng Sci 359(1781):663–693
122. Wriggers P (2006) Computational contact mechanics. Springer Science and Business Media, Berlin
123. Yeon-Sun C, Sherif N (1988) Forced periodic vibration of unsymmetric piecewise-linear systems. J Sound Vib 121(1):117–126
124. Yoong C, Acary V, Legrand M (2017) Modification of Moreau-Jean's scheme for energy conservation in inelastic impact dynamics. In: Proceedings of the 9th European nonlinear dynamics conference. Budapest
125. Yoong C, Thorin A, Legrand M (2018) Nonsmooth modal analysis of an elastic bar subject to a unilateral contact constraint. Nonlinear Dyn 91(4):2453–2476

Variational and Numerical Methods Based on the Bipotential and Application to the Frictional Contact

Géry de Saxcé

Abstract First, we define the bipotential and discuss some fundamental aspects concerning the existence and construction of a bipotential generating a given constitutive law. After a quick review of applications to solid mechanics, we highlight, in particular, problems of unilateral contact with isotropic and anisotropic Coulomb dry friction. The second part is devoted to variational methods and numerical algorithms inspired by the bipotential, illustrated, in particular, to multi-body systems. Extended limit analysis techniques are used to determine the collapse load of structures with plasticity and friction contact.

1 Basic Tools

X and Y are topological, locally convex, real vector spaces of dual variables $\mathbf{x} \in X$ and $\mathbf{y} \in Y$, with the duality product $\langle \bullet, \bullet \rangle : X \times Y \to \mathbb{R}$. We shall suppose that X, Y have topologies compatible with the duality product. We use the notation: $\bar{\mathbb{R}} = \mathbb{R} \cup \{+\infty\}$. For any convex and closed set $A \subset X$, its indicator function, χ_A, is defined by

$$\chi_A(\mathbf{x}) = \begin{cases} 0 & \text{if } \mathbf{x} \in A \\ +\infty & \text{otherwise} . \end{cases}$$

The subgradient of a function $\phi : X \to \bar{\mathbb{R}}$ at a point $\mathbf{x} \in X$ is the (possibly empty) set

$$\partial\phi(\mathbf{x}) = \left\{ \mathbf{y} \in Y \mid \forall \mathbf{x}' \in X \ \langle \mathbf{x}' - \mathbf{x}, \mathbf{y} \rangle \le \phi(\mathbf{x}') - \phi(\mathbf{x}) \right\} .$$

Its Fenchel conjugate $\phi^* : Y \to \bar{\mathbb{R}}$ is defined by

G. de Saxcé (✉)
Laboratoire de Mécanique, Multiphysique, Multiéchelle (FRE CNRS 2016), Av. Paul Langevin, Cité scientifique, 59655 Villeneuve d'Ascq, France
e-mail: gery.desaxce@univ-lille1.fr

R. I. Leine et al. (eds.), *Advanced Topics in Nonsmooth Dynamics*,
https://doi.org/10.1007/978-3-319-75972-2_6

$$\phi^*(\boldsymbol{y}) = \sup\{\langle \boldsymbol{y}, \boldsymbol{x}\rangle - \phi(\boldsymbol{y}) \mid \boldsymbol{y} \in X\} \ .$$

The conjugate is always convex and lower semi-continuous (lsc).

We denote by $\Gamma(X)$ the class of convex and lower semi-continuous functions $\phi : X \rightarrow \bar{\mathbb{R}}$. The class of convex and lower semi-continuous functions $\phi : X \rightarrow \mathbb{R}$ different from the constant $+\infty$ is denoted by $\Gamma_0(X)$.

A graph M is **cyclically monotone** if, for all integer $m > 0$ and any finite family of couples $(\boldsymbol{x}_j, \boldsymbol{y}_j) \in M$, $j = 0, 1, \ldots, m$,

$$\langle \boldsymbol{x}_0 - \boldsymbol{x}_m, \boldsymbol{y}_m\rangle + \sum_{k=1}^{m} \langle \boldsymbol{x}_k - \boldsymbol{x}_{k-1}, \boldsymbol{y}_{k-1}\rangle \leq 0. \tag{1}$$

A cyclically monotone graph M is **maximal** if it does not admit a strict prolongation that is cyclically monotone. By reindexing the couples, we easily recast the previous inequality as

$$\langle \boldsymbol{x}_m, \boldsymbol{y}_0 - \boldsymbol{y}_m\rangle + \sum_{k=1}^{m} \langle \boldsymbol{x}_{k-1}, \boldsymbol{y}_k - \boldsymbol{y}_{k-1}\rangle \leq 0, \tag{2}$$

which shows that the graphs of a law and its dual law are simultaneously cyclically monotone. Rockafellar [26], Theorem 24.8 (see also Moreau [23], Proposition 12.2) proved a theorem that can be stated as follows.

Theorem 1.1 *Given a graph M, there exist potentials $\phi \in \Gamma_0(X)$ such that $M \subset Graph(\partial\phi)$ if, and only if, M is cyclically monotone. They are defined by*

$$\phi(\boldsymbol{x}) = \ \sup\left\{ \langle \boldsymbol{x} - \boldsymbol{x}_m, \boldsymbol{y}_m\rangle + \sum_{k=1}^{m} \langle \boldsymbol{x}_k - \boldsymbol{x}_{k-1}, \boldsymbol{y}_{k-1}\rangle \right\} + \phi(\boldsymbol{x}_0), \tag{3}$$

where $\boldsymbol{x}_0$ and $\phi(\boldsymbol{x}_0)$ are arbitrarily fixed and the 'sup' is extended to any $m > 0$ and to any couples $(\boldsymbol{x}_k, \boldsymbol{y}_k) \in M$, $k = 1, 2, \ldots, m$.

Because the dual law is also cyclically monotone, we can once again apply the construction of the previous Theorem, giving the function

$$\psi(\boldsymbol{y}) = \ \sup\left\{ \langle \boldsymbol{x}_m, \boldsymbol{y} - \boldsymbol{y}_m\rangle + \sum_{k=1}^{m} \langle \boldsymbol{x}_{k-1}, \boldsymbol{y}_k - \boldsymbol{y}_{k-1}\rangle \right\} + \psi(\boldsymbol{y}_0) \tag{4}$$

such that $M \subset M(\partial\psi^*)$. With the exception of when M is maximal, ϕ and ψ^* are, in general, a distinct function, as will be seen further in the application.

2 What Is a Bipotential?

The constitutive laws of the materials can be represented, as in Elasticity, by a univalued mapping $T : X \to Y$ or, as in Plasticity, can be generalized in the form of a multivalued mapping $T : X \to 2^Y$, but this representation is not necessarily convenient. Its graph $M = Graph(T)$ is a non-empty part of $X \times Y$. When the graph is maximal cyclically monotone, we can modelize it thanks to a convex and lower semi-continuous (l.s.c.) function $\phi : X \to \bar{\mathbb{R}}$, called a **superpotential** (or pseudo-potential), such that the graph M is the one of its subdifferential $Graph(\partial\phi)$. The dissipative materials admitting a superpotential of dissipation are often qualified as standard [17], and the law is said to be a normality law, a subnormality law or an associated law. However, many experimental laws proposed over these last decades, particularly in Plasticity, are non-associated. For such laws, we proposed in [12] a suitable modelization thanks to a function called a bipotential.

Definition 1 A **bipotential** is a function $b : X \times Y \to \bar{\mathbb{R}}$, with the properties:

(a) b is convex and lower semicontinuous in each argument;
(b) for any $\mathbf{x} \in X, \mathbf{y} \in Y$, we have $b(\mathbf{x}, \mathbf{y}) \geq \langle \mathbf{x}, \mathbf{y} \rangle$;
(c) for any $(\mathbf{x}, \mathbf{y}) \in X \times Y$, we have the equivalences:

$$\mathbf{y} \in \partial b(\bullet, \mathbf{y})(\mathbf{x}) \iff \mathbf{x} \in \partial b(\mathbf{x}, \bullet)(\mathbf{y}) \iff b(\mathbf{x}, \mathbf{y}) = \langle \mathbf{x}, \mathbf{y} \rangle \, . \tag{5}$$

The **graph** of b is

$$M(b) \;=\; \{(\mathbf{x}, \mathbf{y}) \in X \times Y \;\mid\; b(\mathbf{x}, \mathbf{y}) = \langle \mathbf{x}, \mathbf{y} \rangle\} \, . \tag{6}$$

If the graph M of a law is the graph of a bipotential b, we say that the law (the graph) admits a bipotential. In particular, for each superpotential ϕ, we can associate the **separable bipotential**

$$b(\mathbf{x}, \mathbf{y}) = \phi(\mathbf{x}) + \phi^*(\mathbf{y}) \tag{7}$$

where ϕ^* is the Fenchel conjugate of ϕ. Hence, the cornerstone inequality of the bipotential (Definition 1 (b)) is reduced to well-known Fenchel's inequality [13].

We also introduce, in [5, 6], the notion of a strong bipotential. Conditions (B1S) and (B2S) appear as relations (51), (52) in Laborde and Renard [21].

Definition 2 A function $b : X \times Y \to \bar{\mathbb{R}}$ is a **strong bipotential** if it satisfies the conditions:

(a) b is convex and lower semicontinuous in each argument;
(B1S) for any $\mathbf{y} \in Y$, $\inf \left\{ b(\mathbf{x}', \mathbf{y}) - \langle \mathbf{x}', \mathbf{y} \rangle : \mathbf{x}' \in X \right\} \in \{0, +\infty\}$;
(B2S) for any $\mathbf{x} \in X$, $\inf \left\{ b(\mathbf{x}, \mathbf{y}') - \langle \mathbf{x}, \mathbf{y}' \rangle : \mathbf{y}' \in Y \right\} \in \{0, +\infty\}$.

Proposition 2.1 *Any strong bipotential is a bipotential.*

The notion of a strong bipotential (introduced in relations (51), (52) [21]) is also motivated by the fact that all bipotentials considered in applications to mechanics are, in fact, strong bipotentials.

The introduction of non-separable bipotentials allows modelizing, in a more general way, the non-associated constitutive laws. The laws admitting a bipotential are called laws of implicit standard materials, because the relation $\mathbf{y} \in \partial b(\bullet, \mathbf{y})(\mathbf{x})$ is a subnormality law but the relation between $\mathbf{x}$ and $\mathbf{y}$ is implicit. Linked to the structural mechanics and, in particular, with the Calculus of Variation, the bipotential theory offers an elegant framework for modelizing a broad spectrum of non-associated laws. Examples of such non-associated constitutive laws are: non-associated Drücker-Prager [9] and Cam-Clay models [8] in soil mechanics, cyclic Plasticity ([2, 7]) and Viscoplasticity [18] of metals with non-linear kinematical hardening rule, Lemaitre's damage law [1], the coaxial laws ([10, 29]), and the Coulomb's friction law [3, 7, 9, 12, 14, 16, 20, 21]. A complete survey can be found in [10]. In the previous works, robust numerical algorithms were proposed for solving structural mechanics problems.

3 Existence and Non-uniqueness of the Bipotential

For all these particular constitutive laws, the bipotentials were heuristically constructed, without knowing beforehand the conditions under which the law admits a bipotential, nor a systematic algorithm for constructing this bipotential. This question was answered later. In [4], we solved two key problems: (a) when the graph of a given multivalued operator can be expressed as the graph (6) of a bipotential, and (b) a method of construction of a bipotential associated (in the sense of point (a)) with a multivalued, typically non-monotone, operator. The main tool was the notion of **convex Lagrangian cover** of the graph of the multivalued operator, and a related notion of implicit convexity of this cover. The results of [4] apply only to bi-convex, bi-closed graphs (for short BB-graphs) admitting at least one convex Lagrangian cover by **maximal cyclically monotone graphs**. This is a rather large class of graph of multivalued operators, but important applications to mechanics, such as the bipotential associated to contact with friction [12], are not in this class.

In more recent papers [5, 6], we proposed an extension of the method presented in [4] to a more general class of BB-graphs. This is done in two steps. In the first step, we proved that the intersection of two maximal cyclically monotone graphs is the critical set of a bipotential if, and only if, a condition formulated in terms of the inf convolution of a family of convex lsc functions is true [5]. In the second step, we extended the main result of [4] by replacing the notion of convex Lagrangian cover with the one of **bipotential convex cover** (Definition 5). In this way, we were able to apply our results to the bipotential for the Coulomb's friction law.

Definition 3 For any graph $M \subset X \times Y$, we can introduce the **sections**

$$M(\mathbf{x}) = \{\mathbf{y} \in Y \mid (\mathbf{x}, \mathbf{y}) \in M\}, \qquad M^*(\mathbf{y}) = \{\mathbf{x} \in X \mid (\mathbf{x}, \mathbf{y}) \in M\},$$

The **domain** of M is, by definition,

$$dom(M) = \{\mathbf{x} \in X \mid M(\mathbf{x}) \neq \emptyset\},$$

Hence, the law T assigns at each $\mathbf{x} \in X$ the section $M(\mathbf{x})$ and the inverse law assigns to each $\mathbf{y} \in Y$ the section $M^*(\mathbf{y})$. Let a constitutive law be given by a graph M. Does it admit a bipotential? The existence problem is easily settled by the following result.

Theorem 3.1 *Given a non-empty set $M \subset X \times Y$, there is a bipotential b such that $M = M(b)$ if, and only if, for any $\mathbf{x} \in X$ and $\mathbf{y} \in Y$, the sections $M(\mathbf{x})$ and $M^*(\mathbf{y})$ are convex and closed.*

The proof can be found in [4]. Then, we say that M is bi-convex and bi-closed, or in short, that M is a **BB-graph**. This criterion, simple to verify, allows for straightaway moving laws without bipotential aside. If the law is represented by a BB-graph, a closely related topic is to know whether the bipotential is unique. The answer is no. The proof of the previous result is based on the introduction of the bipotential

$$b_\infty(\mathbf{x}, \mathbf{y}) = \langle \mathbf{x}, \mathbf{y} \rangle + \chi_M(\mathbf{x}, \mathbf{y}).$$

Therefore, here is a counterexample. If M is cyclically monotone maximal, it admits at least two distinct bipotentials, the separable bipotential defined by (7) and b_∞. Therefore, the graph of the law alone is not sufficient to uniquely define the bipotential.

4 Bipotential Convex Cover

For a given multivalued constitutive law, Theorem 3.1 does not give a satisfactory bipotential, because the bipotential b_∞ is somehow degenerate. We would like to be able to find a bipotential b that is not everywhere infinite outside the graph M. We saw that the graph alone is not sufficient to construct interesting bipotentials. We need more information to start from. This is provided by the notion of bipotential convex cover.

Let $Bp(X, Y)$ be the set of all bipotentials $b : X \times Y \to \bar{\mathbb{R}}$. We shall need the following definitions.

Definition 4 Let Λ be an arbitrary non-empty set and V a real vector space. The function $f : \Lambda \times V \to \bar{\mathbb{R}}$ is **implicitly convex** if, for any two elements $(\lambda_1, \mathbf{z}_1), (\lambda_2, \mathbf{z}_2) \in \Lambda \times V$ and for any two numbers $\alpha, \beta \in [0, 1]$ with $\alpha + \beta = 1$, there exists $\lambda \in \Lambda$ such that

$$f(\lambda, \alpha \mathbf{z}_1 + \beta \mathbf{z}_2) \leq \alpha f(\lambda_1, \mathbf{z}_1) + \beta f(\lambda_2, \mathbf{z}_2) \quad . \tag{8}$$

Definition 5 A **bipotential convex cover** of the non-empty set M is a function $\lambda \in \Lambda \mapsto b_\lambda$ from Λ with values in the set $Bp(X, Y)$, with the properties:

(a) The set Λ is a non-empty compact topological space,
(b) Let $f : \Lambda \times X \times Y \to \mathbb{R} \cup \{+\infty\}$ be the function defined by

$$f(\lambda, \mathbf{x}, \mathbf{y}) = b_\lambda(\mathbf{x}, \mathbf{y}).$$

Then, for any $\mathbf{x} \in X$ and for any $\mathbf{y} \in Y$, the functions $f(\bullet, \mathbf{x}, \bullet) : \Lambda \times Y \to \bar{\mathbb{R}}$ and $f(\bullet, \bullet, \mathbf{y}) : \Lambda \times X \to \bar{\mathbb{R}}$ are lower semi-continuous on the product spaces $\Lambda \times Y$ and, respectively, $\Lambda \times X$ endowed with the standard topology,

(c) We have $M = \bigcup_{\lambda \in \Lambda} M(b_\lambda)$,
(d) With the notations from point (b), the functions $f(\bullet, \mathbf{x}, \bullet)$ and $f(\bullet, \bullet, \mathbf{y})$ are implicitly convex in the sense of Definition 4.

A bipotential convex cover is in some sense described by the collection $\{b_\lambda : \lambda \in \Lambda\}$. This is the point of view that we will adopt in the sequel. The next result defines the conditions under which the notion of bipotential convex cover is independent of the choice of the parametrization [6].

Proposition 4.1 *Let $\lambda \in \Lambda \mapsto b_\lambda \in Bp(X, Y)$ be a bipotential convex cover and $g : \Lambda \to \Lambda$ be a continuous, invertible, with continuous inverse, function. Then, $\lambda \in \Lambda \mapsto b_{g(\lambda)} \in Bp(X, Y)$ is a bipotential convex cover.*

The next theorem, proved in [6], is the key result needed further.

Theorem 4.2 *Let $\lambda \mapsto b_\lambda$ be a bipotential convex cover of the graph M and $b : X \times Y \to R$ defined by*

$$b(\mathbf{x}, \mathbf{y}) = \inf \{b_\lambda(\mathbf{x}, \mathbf{y}) \mid \lambda \in \Lambda\} \quad . \tag{9}$$

Then, b is a bipotential and $M = M(b)$.

The result is rather surprising, because an inferior envelop of functions, even convex, is not generally a convex function. The property (d) of the Definition 5 is essential for ensuring the convexity properties of b.

5 Bipotential for Cyclically Monotone Graphs

Maximal cyclically monotone graphs are critical sets of separable bipotentials. For a non-maximal cyclically monotone graph M, Rockafellar's theorem [26] claims only that there exists a superpotential ϕ such that $M \subset Graph(\partial\phi)$. Hence, ϕ is not sufficient to define M unambiguously. In this section, we show that M can be characterized unequivocally by a bipotential $b = max(b_1, b_2)$, where b_1 and b_2 are separable bipotentials.

Theorem 5.1 *Let b_1 and b_2 be separable bipotentials associated, respectively, with the convex and lsc functions $\phi_1, \phi_2 : X \to \bar{\mathbb{R}}$, that is,*

$$b_i(\mathbf{x}, \mathbf{y}) \;=\; \phi_i(\mathbf{x}) + \phi_i^*(\mathbf{y})$$

for any $i = 1, 2$ and $(\mathbf{x}, \mathbf{y}) \in X \times Y$. Consider the following assertions:

(i) $b = max(b_1, b_2)$ *is a strong bipotential.*
(ii') *For any $\mathbf{y} \in dom\, \phi_1^* \cap dom\, \phi_2^*$ and for any $\lambda \in [0, 1]$, we have*

$$(\lambda\, \phi_1 \,+\, (1-\lambda)\, \phi_2)^*\, (\mathbf{y}) \;=\; \lambda\, \phi_1^*(\mathbf{y}) \,+\, (1-\lambda)\, \phi_2^*(\mathbf{y}). \tag{10}$$

(ii") *For any $\mathbf{x} \in dom\, \phi_1 \cap dom\, \phi_2$ and for any $\lambda \in [0, 1]$, we have*

$$\left(\lambda\, \phi_1^* \,+\, (1-\lambda)\, \phi_2^*\right)^*\, (\mathbf{x}) \;=\; \lambda\, \phi_1(\mathbf{x}) \,+\, (1-\lambda)\, \phi_2(\mathbf{x}). \tag{11}$$

Then, the point (i) is equivalent with the conjunction of (ii'), (ii"), (for short: (i) $\Longleftrightarrow$ ((ii') AND (ii"))).

Remark 1 If b_1, b_2 are separable bipotentials and $b = max(b_1, b_2)$ is a bipotential, then $M(b) = M(b_1) \cap M(b_2)$, therefore $M(b)$ is the intersection of two maximal cyclically monotone graphs.

The demonstration of the Theorem is given in [6].

6 Example

For any BB-graph M, let us show how to construct b_∞. For any $\boldsymbol{u} \in dom(M)$, the graph $M_{\boldsymbol{u}} = \{\boldsymbol{u}\} \times M(\boldsymbol{u})$ is cyclically monotone. Indeed, for any finite family of couples $(\boldsymbol{x}_j, \boldsymbol{y}_j) \in M$, $j = 0, 1, \ldots, m$, we have $\boldsymbol{x}_j = \boldsymbol{u}$ for all j and

$$\langle \boldsymbol{x}_0 - \boldsymbol{x}_m, \boldsymbol{y}_m \rangle + \sum_{k=1}^{m} \langle \boldsymbol{x}_k - \boldsymbol{x}_{k-1}, \boldsymbol{y}_{k-1} \rangle = \langle \boldsymbol{u} - \boldsymbol{u}, \boldsymbol{y}_m \rangle + \sum_{k=1}^{m} \langle \boldsymbol{u} - \boldsymbol{u}, \boldsymbol{y}_{k-1} \rangle = 0.$$

The sets $M_{\boldsymbol{u}}$ obviously cover the graph M when $\boldsymbol{u}$ runs in $dom(M)$. As $\boldsymbol{x}_j = \boldsymbol{u}$ in (3), and putting $\mathbf{v} = \boldsymbol{y}_m$, the function defined by Theorem 1.1 is reduced to

$$\phi_{\boldsymbol{u}}(\boldsymbol{x}) = \;\sup\{\langle \boldsymbol{x} - \boldsymbol{u}, \mathbf{v} \rangle \mid \mathbf{v} \in M(\boldsymbol{u})\}$$

and its Fenchel conjugate is

$$\phi_u^*(y) = \langle u, y\rangle + \chi_{M(u)}(y).$$

Besides, choosing $(x_0, y_0) = (u, 0)$, and taking into account $x_j = u$ for all j, the function defined by (4) is

$$\psi_u(y) = \sup\left\{\langle u, y - y_m\rangle + \langle u, \sum_{k=1}^{m}(y_k - y_{k-1})\rangle\right\} = \langle u, y\rangle,$$

which is, in general, different from ϕ_u^*. Its Fenchel conjugate is

$$\psi_u^*(x) = \sup\{\langle x - u, y\rangle \mid y \in Y\} = \chi_{\{u\}}(x).$$

For $u \in dom(M)$, let us now calculate the bipotential of M_u by Theorem 5.1:

$$b_u(x, y) = \sup(\phi_u(x) + \phi_u^*(y), \psi_u(y) + \psi_u^*(x)) =$$

$$\sup\left(\sup\{\langle x - u, \mathrm{v}\rangle \mid \mathrm{v} \in M(u)\} + \langle u, y\rangle + \chi_{M(u)}(y), \langle u, y\rangle + \chi_{\{u\}}(x)\right).$$

This function is equal to $+\infty$ if $x \neq u$. Otherwise, it equal to

$$\langle u, y\rangle + \chi_{M(u)}(y).$$

Hence, for any $(x, y) \in X \times Y$, we have

$$b_u(x, y) = \langle u, y\rangle + \chi_{M(u)}(y) + \chi_{\{u\}}(x).$$

Finally, we construct the bipotential of M by Theorem 4.2:

$$b(x, y) = \inf\{b_u(x, y) \mid u \in dom(M)\}$$

$$b(x, y) = \inf\left\{\langle u, y\rangle + \chi_{M(u)}(y) + \chi_{\{u\}}(x) \mid u \in dom(M)\right\}.$$

If x does not belong to $dom(M)$, the function is equal to $+\infty$. Otherwise, we choose $u = x$ to minimise, which gives us

$$b(x, y) = \langle x, y\rangle + \chi_{M(x)}(y) = \langle x, y\rangle + \chi_M(x, y) = b_\infty(x, y)$$

Of course, as discussed earlier, this bipotential is rather trivial, but the method allows us to obtain more interesting ones by choosing suitable bipotential convex covers, as in the next section.

7 Application to Unilateral Contact with Coulomb's Dry Friction

To be brief, the space $X = \mathbb{R}^3$ is the one of relative velocities between points of two bodies, and the space Y, also identified to $\mathbb{R}^3$, is the one of the contact reaction stresses. The duality product is the usual scalar product. We put

$$-\dot{\mathbf{u}} = -(\dot{u}_n, \dot{\mathbf{u}}_t) \in X = \mathbb{R} \times \mathbb{R}^2, \quad (r_n, \mathbf{r}_t) \in Y = \mathbb{R} \times \mathbb{R}^2 ,$$

where $\dot{u}_n$ is the gap velocity, $\dot{\mathbf{u}}_t$ is the sliding velocity, r_n is the contact pressure and $\mathbf{r}_t$ is the friction stress. The friction coefficient is $\mu > 0$. The graph of the law of unilateral contact with Coulomb's dry friction is defined as the union of three sets, respectively corresponding to the 'body separation', the 'sticking' and the 'sliding'.

$$\begin{aligned} M = \{(-\dot{\mathbf{u}}, \mathbf{0}) \in X \times Y \ \mid \ -\dot{u}_n < 0\} \cup \{(\mathbf{0}, \mathbf{r}) \in X \times Y \ \mid \| \mathbf{r}_t \| \leq \mu r_n\} \cup \\ \cup \left\{(-\dot{\mathbf{u}}, \mathbf{r}) \in X \times Y \ \mid \ \dot{u}_n = 0, \ \dot{\mathbf{u}}_t \neq 0, \ \mathbf{r}_t = \mu r_n \frac{-\dot{\mathbf{u}}_t}{\| -\dot{\mathbf{u}}_t \|}\right\}. \end{aligned} \tag{12}$$

It is well known that this graph is not monotone, and thus not cyclically monotone. As usual, we introduce Coulomb's cone

$$K_\mu = \{(r_n, \mathbf{r}_t) \in Y \ \mid \| \mathbf{r}_t \| \leq \mu r_n\} \tag{13}$$

and its conjugate cone

$$K_\mu^* = \{(-\dot{u}_n, -\dot{\mathbf{u}}_t) \in X \ \mid \ \mu \| -\dot{\mathbf{u}}_t \| - \dot{u}_n \leq 0\} .$$

In particular, we have

$$K_0^* = \{-\dot{\mathbf{u}} \in X \ \mid \ -\dot{u}_n \leq 0\} .$$

Now, we define some sets useful in the sequel. Let us consider $p > 0$ and the closed convex disc obtained by cutting Coulomb's cone at the level $r_n = p$

$$D(p) = \left\{\mathbf{r}_t \in \mathbb{R}^2 \ \mid \| \mathbf{r}_t \| \leq \mu p\right\} .$$

Therefore, for each value of $p > 0$, we define a set of 'sticking couples'

$$M_p^{(a)} = \{(\mathbf{0}, (p, \mathbf{r}_t)) \in X \times Y \ \mid \ \mathbf{r}_t \in D(p)\}$$

and a set of 'sliding couples'

$$M_p^{(s)} = \{((0, -\dot{\mathbf{u}}_t), (p, \mathbf{r}_t)) \in X \times Y \ \mid \ \| \mathbf{r}_t \| = \mu p, \ \exists \lambda > 0, \ -\dot{\mathbf{u}}_t = \lambda \mathbf{r}_t\} .$$

So, we can cover the graph M with the set of subsequent subgraphs parameterized by $p \in [0, +\infty]$:

(a) $M_p = M_p^{(a)} \cup M_p^{(s)}, \quad p \in (0, +\infty)$,
(b) $M_0 = \{(-\dot{\mathbf{u}}, \mathbf{0}) \in X \times Y \mid -\dot{u}_n \leq 0\}$,
(c) $M_{+\infty} = \emptyset$, by convention.

All these subgraphs are cyclically monotone, but none of them is maximal. By applying Rockafellar's Theorem [26] to M_p twice, let us construct the corresponding superpotentials $\phi_p : X \to \bar{\mathbb{R}}$ such that $M_p \subset Graph(\partial\phi_p)$ and $\psi_p : Y \to \bar{\mathbb{R}}$ such that $M_p \subset Graph(\partial\psi_p)$. It is worthwhile observing that ψ_p is not the Fenchel conjugate of ϕ_p, because M_p is not maximal. To fix the arbitrary constant in the definition of the superpotentials, we suppose that $\phi_p(\mathbf{0}) = \psi_p(\mathbf{0}) = 0$. For $p \in (0, +\infty)$, the computations give us

$$\phi_p(-\dot{\mathbf{u}}) = -p\dot{u}_n + \mu p \parallel -\dot{\mathbf{u}}_t \parallel \qquad \psi_p(\mathbf{r}) = \chi_{D(p)}(\mathbf{r}_t).$$

Their Fenchel conjugates are

$$\phi_p^*(\mathbf{r}) = \chi_{\{p\}}(r_n) + \chi_{D(p)}(\mathbf{r}_t) \quad \psi_p^*(-\dot{\mathbf{u}}) = \mu p \parallel -\dot{\mathbf{u}}_t \parallel + \chi_{\{0\}}(\dot{u}_n).$$

For $p = 0$, we obtain

$$\phi_0(-\dot{\mathbf{u}}) = 0 \,, \quad \psi_0(\mathbf{r}) = \chi_{K_0}(\mathbf{r}).$$

Their Fenchel conjugates are

$$\phi_0^*(\mathbf{r}) = \chi_{\{0\}}(\mathbf{r}) \,, \quad \psi_0^*(-\dot{\mathbf{u}}) = \chi_{K_0^*}(-\dot{\mathbf{u}}).$$

For fixed p, define the bipotentials $b_{i,p}$, $i = 1, 2$, by

$$b_{1,p}(-\dot{\mathbf{u}}, \mathbf{r}) = \phi_p(-\dot{\mathbf{u}}) + \phi_p^*(\mathbf{r}) \quad ,$$

$$b_{2,p}(-\dot{\mathbf{u}}, \mathbf{r}) = \psi_p^*(-\dot{\mathbf{u}}) + \psi_p(\mathbf{r}) \quad .$$

As an application of Theorem 5.1, we obtain that $b_p = \max\left\{b_{1,p}, b_{2,p}\right\}$ is a bipotential. Indeed, we shall check only the point (ii') from Theorem (5.1) (the point (ii")) is true by a similar computation). For $\lambda \in [0, 1)$ and $p \neq 0$, we have

$$\lambda\phi_p(-\dot{\mathbf{u}}) + (1 - \lambda)\psi_p^*(-\dot{\mathbf{u}}) = \chi_{\{0\}}(\dot{u}_n) + \mu p \parallel -\dot{\mathbf{u}}_t \parallel,$$

therefore we get

$$\left(\lambda\phi_p + (1 - \lambda)\psi_p^*\right)^* (\mathbf{r}) = \chi_{D(p)}(\mathbf{r}_t).$$

Also, by computation, we obtain

$$\lambda\phi_p^*(\mathbf{r}) + (1-\lambda)\psi_p(\mathbf{r}) = \chi_{\{p\}}(r_n) + \chi_{D(p)}(\mathbf{r}_t).$$

If $\phi_p^*(\mathbf{r}) < +\infty$, $\psi_p(\mathbf{r}) < +\infty$, then, in particular, $r_n = p$, and we obtain (10) as an equality $0 = 0$. All other cases involving $\lambda = 1$ or $p = 0$ are solved in the same way.

The bipotential b_p has the expression

$$b_p(-\dot{\mathbf{u}}, \mathbf{r}) = \mu p \parallel -\dot{\mathbf{u}}_t \parallel + \chi_{D(p)}(\mathbf{r}_t) + \chi_{\{p\}}(r_n) + \chi_{\{0\}}(\dot{u}_n), \quad p \in (0, +\infty)$$

$$b_0(-\dot{\mathbf{u}}, \mathbf{r}) = \chi_{\{0\}}(\mathbf{r}) + \chi_{(-\infty,0]}(\dot{u}_n).$$

It is easy to check that the function $p \in [0, +\infty] \mapsto b_p$ is a bipotential convex cover, therefore, by Theorem 4.2, we obtain a bipotential for the set M. By direct computation, this bipotential, defined as

$$b(-\dot{\mathbf{u}}, \mathbf{r}) = \inf \left\{ b_p(-\dot{\mathbf{u}}, \mathbf{r}) : p \in [0, +\infty] \right\} \quad ,$$

has the following expression:

$$b(-\dot{\mathbf{u}}, \mathbf{r}) = \mu r_n \parallel -\dot{\mathbf{u}}_t \parallel + \chi_{K_\mu}(\mathbf{r}) + \chi_{K_0^*}(-\dot{\mathbf{u}}) \ . \tag{14}$$

Therefore, we recover the bipotential heuristically obtained in [12].

8 Unilateral Contact with Orthotropic Friction

For many industrial applications, the assumption of isotropic friction is unrealistic because of directional surface machining and finishing operations. For orthotropic frictional contact, Michałowski and Mróz have pointed out the non-associated nature of the sliding rule [22] (see also Mróz and Stupkiewicz [24]). In this model, the convex friction cone is defined by

$$K_\mu = \left\{ \mathbf{r} \in \mathbb{R}^3 \mid \|\mathbf{r}_t\|_\mu - r_n \leq 0 \right\}, \tag{15}$$

where $\| \bullet \|_\mu$ denotes the elliptic norm

$$\|\mathbf{r}_t\|_\mu = \sqrt{\left(\frac{r_{t_x}}{\mu_x}\right)^2 + \left(\frac{r_{t_y}}{\mu_y}\right)^2} = \|\mathbb{F}^{-1}\mathbf{r}_t\| \,, \tag{16}$$

with the Euclidean norm $\| \bullet \|$ and

$$\mathbb{F} = \begin{bmatrix} \mu_x & 0 \\ 0 & \mu_y \end{bmatrix}. \tag{17}$$

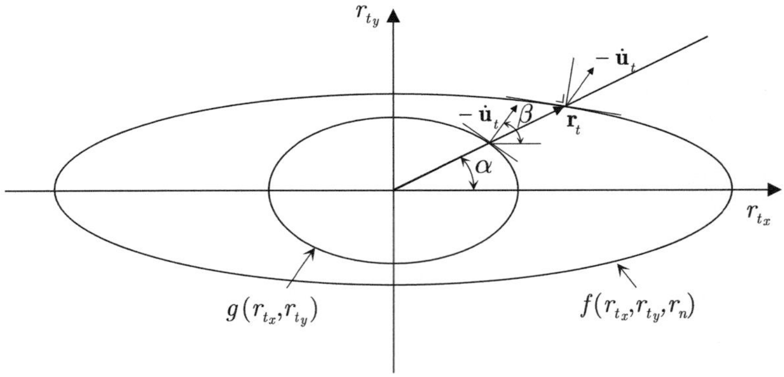

Fig. 1 Friction condition and sliding rule

The classical isotropic Coulomb's friction condition is recovered by setting $\mu_x = \mu_y = \mu$. The dual norm $\| \bullet \|_\mu^*$ associated with (16) is given by

$$\| - \dot{\mathbf{u}}_t \|_\mu^* = \sqrt{\mu_x^2 \, (-\dot{u}_{t_x})^2 + \mu_y^2 \, (-\dot{u}_{t_y})^2} = \| \mathbb{F} \, (-\dot{\mathbf{u}}_t) \| \, . \tag{18}$$

The Michałowski-Mróz model is twice non-associated because, as in isotropic friction, the normal velocity is absent when sliding occurs, but also because the tangential velocity is not normal to the elliptical level curves $\|\mathbf{r}_t\|_\mu = \text{constant}$ (Fig. 1). This additional lack of associativity can be described introducing a slip potential defined by

$$g(r_{t_x}, r_{t_y}) = \|\mathbf{r}_t\|_p, \tag{19}$$

in which $\|\mathbf{r}_t\|_p$ is given by

$$\|\mathbf{r}_t\|_p = \sqrt{\left(\frac{r_{t_x}}{p_x}\right)^2 + \left(\frac{r_{t_y}}{p_y}\right)^2} = \left\| \mathbb{P}^{-1} \mathbf{r}_t \right\| , \tag{20}$$

with

$$\mathbb{P} = \begin{bmatrix} p_x & 0 \\ 0 & p_y \end{bmatrix} . \tag{21}$$

It is convenient to introduce the *sliding non-associativity* matrix

$$\mathbb{Q} = \mathbb{P}\mathbb{F}^{-1} = \mathbb{F}^{-1}\mathbb{P} = \begin{bmatrix} \dfrac{p_x}{\mu_x} & 0 \\ 0 & \dfrac{p_y}{\mu_y} \end{bmatrix} . \tag{22}$$

The complete form of the frictional contact law involves three possible states, which are separating, contact with sticking, and contact with sliding. It can be described using two overlapped "*if...then...else*" statements:

$$
\boxed{
\begin{array}{ll}
\textit{if}\, r_n = 0 \,\textit{then} & \\
\quad \dot{u}_n \geq 0 & \text{!separating} \\
\textit{elseif}\, \mathbf{r} \in \operatorname{int} K_\mu \,\textit{then} & \\
\quad \dot{u}_n = 0 \,\textit{and}\, \dot{\mathbf{u}}_t = \mathbf{0} & \text{!sticking} \\
\textit{else}(\, \mathbf{r} \in \operatorname{bd} K_\mu \text{ and } r_n > 0) & \\
\quad \left\{ \dot{u}_n = 0 \;,\; \dot{\lambda} \geq 0 \text{ and } -\dot{\mathbf{u}}_t = \dot{\lambda} \dfrac{\partial g}{\partial \mathbf{r}_t} = \dot{\lambda} \dfrac{\mathbb{P}^{-2}\mathbf{r}_t}{\|\mathbf{r}_t\|_p} \right\} & \text{!sliding} \\
\textit{endif} &
\end{array}
}
\tag{23}
$$

where "int K_μ" and "bd K_μ" denote the interior and the boundary of K_μ, respectively.

It was proved in [19] that the graph of the previous law is the critical set of the following bipotential:

$$b(-\dot{\mathbf{u}}, \mathbf{r}) = \chi_{K_\mu}(\mathbf{r}) + \chi_{\mathbb{R}_-}(-\dot{u}_n) + \left(\mathbb{I} - \mathbb{Q}^2\right) \bullet \mathbf{r}_t + r_n \, \|\mathbb{Q}^2(-\dot{\mathbf{u}}_t)\|_\mu^* \tag{24}$$

For the isotropic case, we recover the bipotential (14) as a particular case.

9 Implicit Predictor-Corrector Scheme

The time interval $[0, T]$, within which the loading history is defined, is partitioned into N sub-intervals of size Δt, not necessarily equal. For the sake of clarity, we focus our attention on the first time increment $[t_0, t_1]$. The value of any quantity a at the beginning (respectively at the end) of the step is denoted by a_0 (respectively a_1). The corresponding increment is Δa. In order to ensure convergence and stability requirements, the implicit scheme is considered. Taking into account the fact that b is positively homogeneous of order one with respect to the first argument, the complete contact law is satisfied at the end of each time step:

$$-\Delta\mathbf{u} \in \partial b_c(-\Delta\mathbf{u}, \bullet)\,(\mathbf{r}_1), \tag{25}$$

where b_c is given by

$$
\begin{aligned}
b_c(-\Delta\mathbf{u}, \mathbf{r}_1) = {} & \chi_{K_\mu}(\mathbf{r}_1) + \chi_{\mathbb{R}_-}(-h_0 - \Delta u_n) + \left(\mathbb{I} - \mathbb{Q}^2\right)(-\Delta\mathbf{u}_t) \bullet \mathbf{r}_{t_1} \\
& + r_{n_1} \, \|\mathbb{Q}^2(-\Delta\mathbf{u}_t)\|_\mu^*,
\end{aligned}
\tag{26}
$$

which takes into account the gap $-h_0$ at the beginning of the time-step.

Unfortunately, unlike elastoplastic problems, the incremental bipotential is not differentiable. For this reason, the incremental bipotential is not directly used and a regularization technique is applied in order to replace the unpleasant inequation with an equivalent equation, simpler to solve. Thus, the implicit scheme (25) leads to

$$\mathbf{r}_1 \in \mathbb{R}^3 : \quad b_c(-\Delta\mathbf{u}, \mathbf{r}') - b_c(-\Delta\mathbf{u}, \mathbf{r}_1) \geq -\Delta\mathbf{u} \bullet (\mathbf{r}' - \mathbf{r}_1), \quad \forall \mathbf{r}' \in \mathbb{R}^3. \quad (27)$$

The variational inequality (27) is rewritten in an obvious manner as

$$\mathbf{r}_1 \in \mathbb{R}^3 : \ \rho b_c(-\Delta\mathbf{u}, \mathbf{r}') - \rho b_c(-\Delta\mathbf{u}, \mathbf{r}_1) + [\mathbf{r}_1 - (\mathbf{r}_1 - \rho\Delta\mathbf{u})] \bullet (\mathbf{r}' - \mathbf{r}_1) \geq 0, \quad \forall \mathbf{r}' \in \mathbb{R}^3 \quad (28)$$

where ρ is a positive number that needs to be chosen within a suitable range to ensure convergence. The variational inequality (28) means that $\mathbf{r}_1$ is the proximal point of the augmented force $\hat{\mathbf{r}} = \mathbf{r}_1 - \rho\,\Delta\mathbf{u}$, with respect to the function $\mathbf{r}_1 \mapsto \rho b_c(-\Delta\mathbf{u}, \mathbf{r}_1)$ (see [23])

$$\mathbf{r}_1 = prox\,(\mathbf{r}_1 - \rho\Delta\mathbf{u}, \rho b_c(-\Delta\mathbf{u}, \bullet))\,. \quad (29)$$

The solution of (29) can be obtained using Uzawa's algorithm, which involves two steps: prediction : $\hat{\mathbf{r}}^{i+1} = \mathbf{r}_1^i - \rho\,\Delta\mathbf{u}$. correction: $\mathbf{r}_1^{i+1} = prox\left(\hat{\mathbf{r}}^{i+1}, \rho b_c(-\Delta\mathbf{u}, \bullet)\right)$ By substituting $b_c(-\Delta\mathbf{u}, \mathbf{r}_1)$ with its expression (26) in (28), the inequality can be rearranged to obtain

$$\mathbf{r}_1 \in K_\mu : \quad \left(\mathbf{r}_{t_1} - \boldsymbol{\tau}_t\right) \bullet \left(\mathbf{r}'_t - \mathbf{r}_{t_1}\right) + \left(r_{n_1} - \tau_n\right)\left(r'_n - r_{n_1}\right) \geq 0 \quad \forall \mathbf{r}' \in K_\mu, \quad (30)$$

where

$$\boldsymbol{\tau}_t = \mathbf{r}_{t_1} - \rho\,\mathbb{Q}^2\Delta\mathbf{u}_t \quad \text{and} \quad \tau_n = r_{n_1} - \rho\left(\Delta u_n + \|\mathbb{Q}^2(-\Delta\mathbf{u}_t)\|_\mu^*\right) \quad (31)$$

are the components of the modified augmented surface traction. The last inequality means that the reaction $\mathbf{r}$ at the end of the time step is the projection of the augmented surface traction onto the convex friction cone K_μ. With respect to algorithms with separated treatment of the friction and the unilateral contact, the main advantage of the method is that only one predictor-corrector step is required for the discrete frictional contact problem. At each step, the iterative algorithm is the following:

(1) A value $\Delta\mathbf{u}$ of the displacement increment being known, the new value of the contact stress $\mathbf{r}_1$ at the end of the time step is obtained by computing the augmented stress $\boldsymbol{\tau}^{i+1}$ and the proximal point $\mathbf{r}_1^{i+1}$:

Predictor :

$$\boldsymbol{\tau}^{i+1} = \mathbf{r}_1^i - \rho\left[\mathbb{Q}^2\Delta\mathbf{u}_t + \left(\Delta u_n + \|\mathbb{Q}^2(-\Delta\mathbf{u}_t)\|_\mu^*\right)\mathbf{n}\right]$$

Corrector :

$$\mathbf{r}_1^{i+1} = \quad proj\,(\boldsymbol{\tau}^{i+1}, K_\mu)$$

(2) Next, the displacement increment is updated and the procedure is iteratively repeated.

In the solution of the projection problem, three different situations must be considered according to the position of the prediction $\boldsymbol{\tau}$ in $\mathbb{R}^3$. The first case corresponds to a prediction $\boldsymbol{\tau}$ located in the cone K_μ. Its projection is the prediction itself, i.e., $\mathbf{r}_1 = \boldsymbol{\tau}$. The second one relates to a prediction situated in the cone K_μ^*, where its projection turns out to be the origin ($\mathbf{r}_1 = \mathbf{0}$). In the last case, the prediction is neither in K_μ nor in K_μ^* and the corrector step requires computing the projection of the prediction. The projection of a point onto a convex set is equivalent to the minimization of the distance between this point and the convex set. Next, the problem is reformulated as an unconstrained minimization problem by means of the Lagrange multipliers technique and leads to finding the intersection of a quartic and a straight line. Details are given in [19].

10 Numerical Applications

In previous papers [11, 12], several applications involving frictional contact problems with isotropic friction condition and associated sliding rule have been carried out using an algorithm based on the bipotential approach. The examples treated have shown that the algorithm is very competitive, as the augmentation phase involves only one prediction-correction step.

However, although many works have been devoted to the isotropic friction, this hypothesis is not often realistic. In fact, most frictional contacts are anisotropic. The source of the roughness anisotropy is technological; the industrial process used to fabricate the bodies can create striations along preferential directions. In fact, most machining, finishing and superfinishing operations are directional, and machined surfaces have particular striation patterns unique to the type of machining. Also, specific techniques of manufacture produce a surface with anisotropic frictional properties. For a large number of machining processes, the striation directions are mutually orthogonal. For such surfaces, an orthotropic friction condition will provide a better description of the frictional behavior.

In [20], we proposed a benchmark test for validating the algorithm for a class of non-associated anisotropic friction laws. The test of such frictional contact laws requires a 3D finite element model. The problem under consideration is a deformable elastic cylinder in contact with a rigid surface (Fig. 2). The radius and the height of the cylinder are both equal to 10 mm. The Young modulus E of the cylinder is taken as equal to 210000 MPa and the Poisson ratio is 0.3. On the surface contact, the

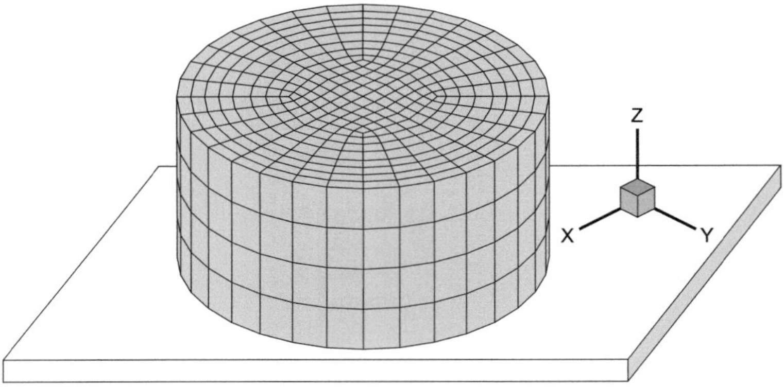

Fig. 2 Compression of a cylinder in contact with a rigid plate

Table 1 Frictional properties

Case	μ_x	μ_y	p_x	p_y
1	0.20	0.20	0.20	0.20
2	0.30	0.25	0.30	0.25
3	0.30	0.15	0.30	0.15
4	0.30	0.15	0.20	0.20
5	0.30	0.15	0.05	0.20

friction condition is assumed to be anisotropic. A vertical rigid motion is imposed on the upper surface of the cylinder by an amount of 0.1 mm. The displacement is applied in one step. The base of the cylinder is in contact with the rigid plate whose normal vector is (0, 0, 1). The cylinder is subdivided into 1280 eight-node brick-like elements, as shown in Fig. 2. Each element has 27 integration points.

Five different sets of frictional parameters, shown in Table 1, are considered.

The first case corresponds to a classical isotropic friction condition, considered here for comparison with anisotropic cases. Case 2 and case 3 represent an anisotropic frictional model with an associated sliding rule. The anisotropy is mild in case 2. The last two cases consider a non-associated sliding rule.

Figure 3 shows the contour plots of the slip and the relative displacements between the lower surface of the cylinder and the rigid plate in the x-direction. The slip corresponds to the Euclidean norm of the tangential displacement. The frictional model being isotropic, the iso-values of the slips are circular. As expected, the stick zone is located around the base center and sliding increases monotonically as we get closer the periphery. The magnitude of the tangential displacement in the x-direction is largest on the cylinder edge.

Figure 4 shows, for case 1, the iso-values of the normal component of the contact reaction and the tangential component of the contact reaction in the x-direction. In all figures, the x-axis is horizontal and the y-axis is vertical. Since the model is isotropic, a simple rotation about the z-axis of 90° gives the tangential component of the contact

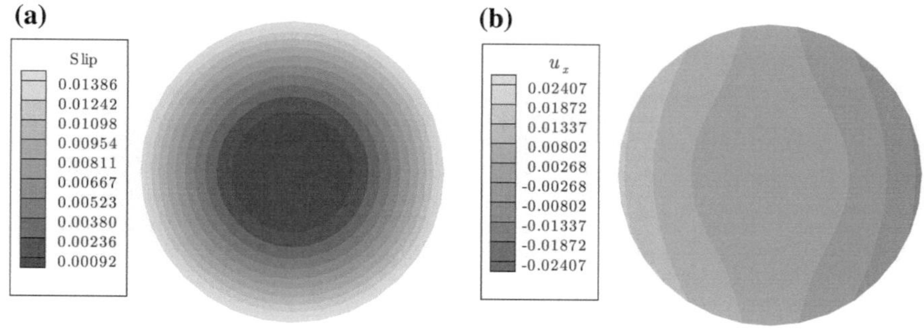

Fig. 3 Case 1: **a** Contour plot of slip - **b** Contour plot of u_{t_x}

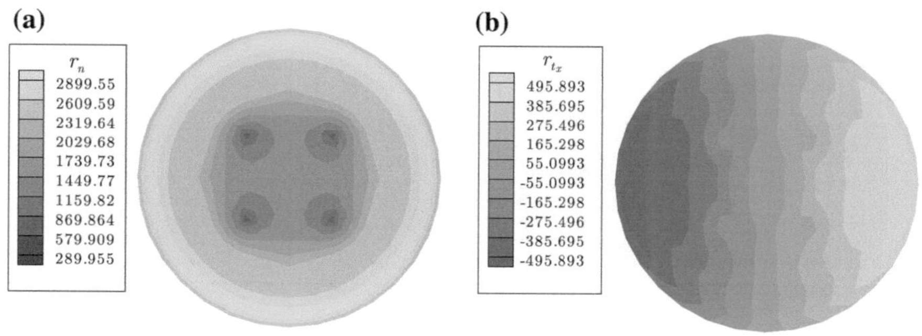

Fig. 4 Case 1: **a** Contour plot of r_n - **b** Contour plot of r_{t_x}

reaction in the y-direction. The normal reaction is higher in the periphery of the cylinder and decreases as we approach the center of the cylinder basis. However, the decrease is not monotonic along every radius. Indeed, the normal reaction attains its minimum in four tiny areas located at approximately the third of the radius from the base center. For all cases considered, contour plots of the normal component of the contact reaction have a similar pattern. The tangential reaction is higher in the periphery of the cylinder, as sliding is more important there. Now, the effect of anisotropy is considered. Slip contour plots for cases 2 and 3 are shown in Fig. 5. As can be seen, the anisotropy of the friction condition significantly influences the slip distribution pattern. The stick area is now an ellipse with a semi-axes ratio equal to the semi-axes ratio of the friction criterion. The slips increase gradually from the stick area and are maximum on the periphery in the y-direction, since, for both cases, μ_x is greater than μ_y. If the disparity between the friction coefficients μ_x and μ_y is significant (as it is for case 3), the distribution of r_{t_y} changes notably, but the iso-values pattern of r_{t_x} remains similar to the isotropic case 1 (see Fig. 6). The algorithm is still convergent if the sliding rule is non-associated, but requires a few more iterations. The number of iterations depends strongly on the degree of non-associativity. A ratio $\dfrac{\mu_x}{\mu_y}$

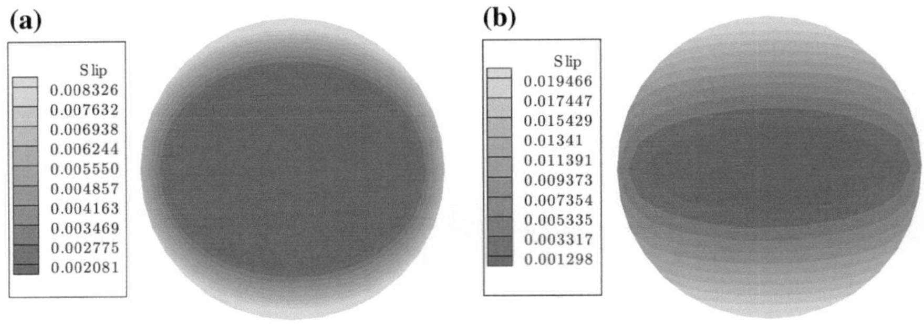

Fig. 5 Contour plots of slip: **a** Case 2 - **b** Case 3

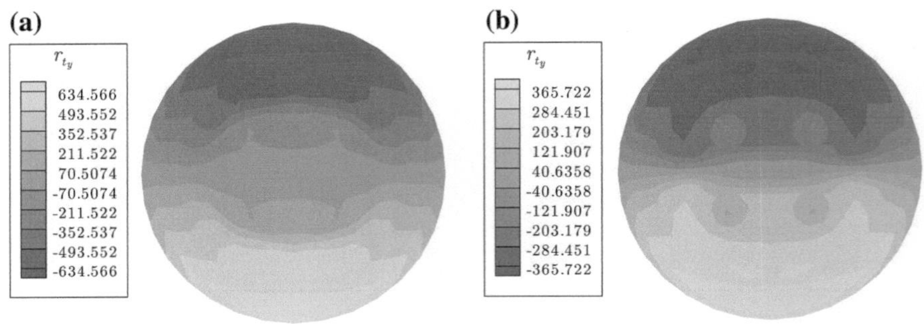

Fig. 6 Contour plots of r_{t_x}: **a** Case 2 - **b** Case 3

much larger or much smaller than the ratio $\frac{p_x}{p_y}$ gives a strong non-associated sliding rule. In Case 4, an isotropic sliding potential is considered that corresponds here to a mild non-associativity. In practice, high non-associativity (case 5) is often observed ([22, 24]). Once a non-associative sliding rule is considered, the slip distribution can change drastically according to the degree of non-associativity. Indeed, if the slip rule is strongly non-associated, the iso-values of the slip become non-convex, as shown in Fig. 7. With an isotropic sliding potential, the stick zone is elliptical and the maximum displacement is lower than the one obtained with an associated sliding rule. Although the principal friction coefficients are the same for both cases 4 and 5, the maximum slip is larger for case 5. The distribution pattern of r_{t_y} (Fig. 8) is similar to cases 2 and 3 (see Fig. 5a), but the iso-values distribution of r_{t_x} is totally different if the non-associativity is strong.

Moreover, it is worth observing that, unlike the isotropic case, a strong hysteretic behavior occurs when friction is anisotropic [15]. This effect is particularly important, because the wear rate is strongly coupled to the friction dissipation. Figure 9 shows the relationship between the equivalent applied force (resulting from imposed displacement) and the slip of point P at the intersection of the boundary of the contact zone and the bisectrix of the quadrant Oxy. It can be seen that the relationship is linear during loading. The unloading stage starts with a sharp drop in the applied

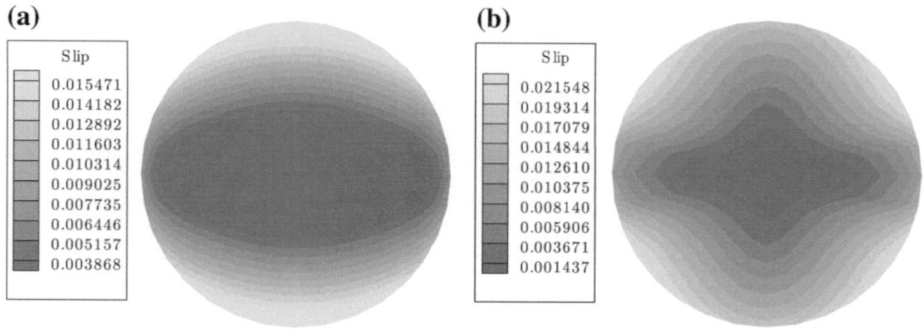

Fig. 7 Contour plots of slip: **a** Case 4 - **b** Case 5

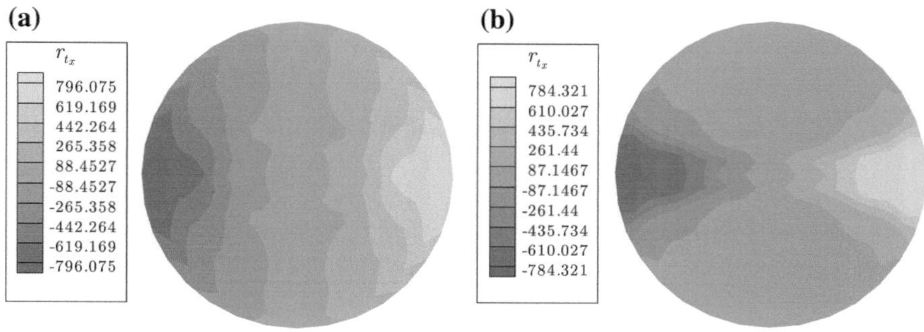

Fig. 8 Contour plots of r_{t_x}: **a** Case 4 - **b** Case 5

force, followed by a non-linear decrease. The difference between the loading path and the unloading path appears to be evident. Furthermore, there is a slight difference between the respective trajectories during the first and second cycles.

11 Limit Load of Plastic Frames with Frictional Contact Supports

Limit analysis is a method of direct calculation of the collapse mechanism of elasto-plastic structures under proportional loading and statical conditions, bypassing the complete study of the loading history [27]. It is based on the essential hypothesis of associated plasticity (with the normality law). The standard approach of limit analysis does not allow to take into account frictional contact at support, because of the non-associativity of Coulomb's dry friction law.

On the basis of the bipotential theory, we establish an extended limit analysis theory for frames in the presence of the unilateral contact with Coulomb's dry friction at supports. The kinematic and static approaches are formulated through the calculation of the total dissipation power of the frame. As it will be shown, on account of the

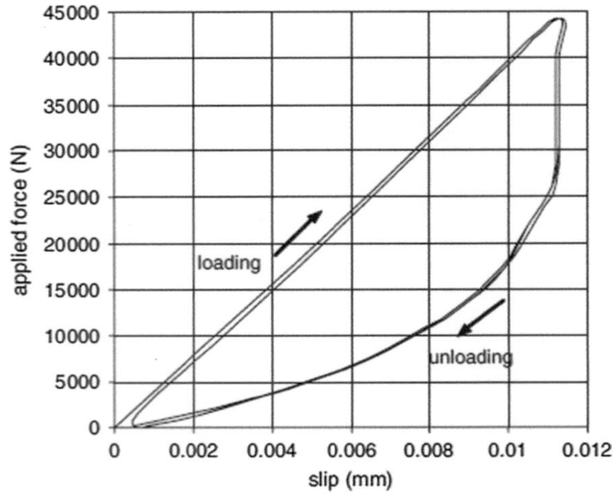

Fig. 9 Total force versus slip at point P

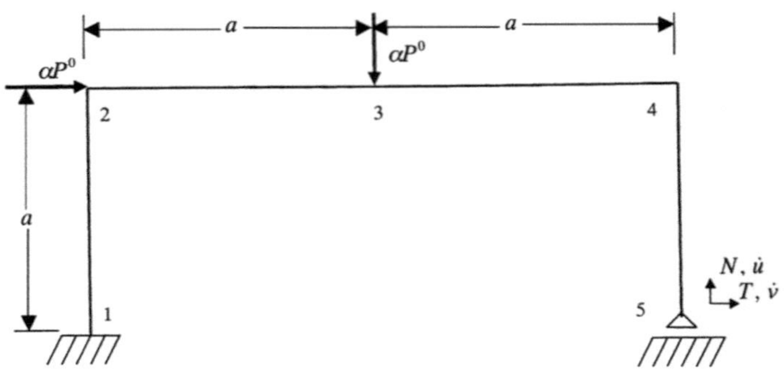

Fig. 10 PlaneFrame with frictional support

presence of contact with friction, the two approaches are coupled in the sense that the kinematic approach of limit analysis contains static variables and, conversely, the statical approach contains kinematic variables. To deal with this, an iterative algorithm, based on the successive approximations method, will be described here. The method is applied to the study of a simple example, consisting of a rectangular frame (Fig. 10), which demonstrates how the value of the friction coefficient affects the plastic limit state, and therefore the limit load and the corresponding collapse mechanism.

The frame structure is supposed to have n plastic hinges or critical sections. The ith hinge is characterised by bending moment M_i gathered in a vector m, angular velocity $\dot{\theta}_i$ gathered in a vector $\dot{q}_p$ and its plastic capacity M_{pi}. The total plastically dissipated power is given by

$$D(\dot{q}_p) = \sum_{i=1}^{n} M_{pi} \mid \dot{\theta}_i \mid ,$$

Its Fenchel conjugate is

$$D^*(m) = \chi_K(m),$$

where $K = \{m \;\; \text{s.t.} \mid M_i \mid \leq M_{pi} \text{ for } 1 \leq i \leq n\}$. Hence, the plastic yielding rule reads as

$$\dot{q}_p \in \partial D^*(m)$$

and the converse relation is

$$m \in \partial D(\dot{q}_p).$$

In detail, the plastic yielding rule is

$$\text{if} \;\; \mid M_i \mid < M_{pi}, \quad \text{then} \;\; \dot{\theta}_i = 0$$

$$\text{else, if} \;\; M_i = M_{pi}, \quad \text{then} \;\; \dot{\theta}_i \geq 0,$$

$$\text{else, if} \;\; M_i = -M_{pi}, \quad \text{then} \;\; \dot{\theta}_i \leq 0.$$

We suppose the frame contains some columns that can be pinned and unilaterally supported with possible frictional contact (for instance, the right-hand column in Fig. 10). Let N (respectively T) be the normal (respectively tangential) component of the reaction R at support. Coulomb's cone is defined by (13):

$$K_\mu = \{R = (T, N) \mid \mid T \mid \leq N\}.$$

Let V be the relative velocity at the unilateral support, $\dot{u}$ and $\dot{v}$ being, respectively, their horizontal and vertical components. Coulomb's dry friction law is derived from the contact bipotential (14):

$$b(-V, R) = \begin{cases} \mu N \mid \dot{u} \mid & \text{if} \quad \dot{v} \geq 0 \quad \text{and} \quad R \in K_\mu \\ +\infty & \text{otherwise}. \end{cases}$$

The total dissipative power due to plastic hinges and sliding friction, expressed in terms of the generalized velocities $\dot{q} = (\dot{q}_p, V)$ and the generalized stresses $Q = (m, R)$, is given by the bipotential

$$\beta(\dot{q}, Q) = D(\dot{q}_p) + D^*(m) + b_c(-V, R). \tag{32}$$

Let h be the redundancy degree of the plane structure with n plastic hinges and l unilateral contacts with friction. The number of independent mechanisms, denoted e, is given by

$$e = 2l + n - h. \tag{33}$$

The equilibrium equations corresponding to the independent mechanisms can be written in the following matricial form:

$$C\,Q = f = \alpha\, f^0, \tag{34}$$

where $f, \alpha > 0$, f^0 and C are, respectively, the actual load, the load factor, the reference load and the so-called rotation matrix. f^0 are arbitrarily fixed for convenience (for instance, with unitary values), and α, controlling the intensity of the overall loading, may be given or unknown. The corresponding compatibility relations relating the generalised strain velocities $\dot{q}$ to the vector $\dot{w}$ of generalised velocities of the independent mechanisms can be obtained through the usual duality:

$$\dot{q} = C^T\,\dot{w}. \tag{35}$$

A vector Q^s is said to be admissible if it is statically admissible (i.e., it satisfies the equilibrium equations (34)) and plastically admissible (i.e., it satisfies the plasticity condition and friction condition). A collapse mechanism $\dot{w}^k$ is said to be admissible if it is compatible with the compatibility conditions (35) and provides a positive power:

$$\dot{w}^{kT}\, f^0 > 0.$$

For given loads, an exact solution $(\dot{w}, Q)$ of the **structural problem** is such that:

- $\dot{w}$ is kinematically admissible (ka),
- Q is statically admissible (sa),
- $(\dot{q}, Q)$, where $\dot{q} = C^T\,\dot{w}$ satisfies the constitutive law:

$$Q \in \partial\beta(\bullet, Q)(\dot{q}) \quad \Leftrightarrow \dot{q} \in \partial\beta(\dot{q}, \bullet)(Q) \quad \Leftrightarrow \beta(\dot{q}, Q) = \dot{q}^T\,Q. \tag{36}$$

Let us introduce the following function, called **bifunctional**, representing the difference between the internal power and the external one:

$$B(\dot{w}, Q) = \beta(C^T\,\dot{w}, Q) - \dot{w}^T\, f.$$

The key result is given below.

Proposition 11.1 *A solution to the $(\dot{q}, Q)$ of the structural problem is simultaneously a solution to the variational principles:*

$$\inf\left\{B(\dot{w}^k, Q) : \dot{w}^k\ ca\right\}, \qquad \inf\left\{B(\dot{w}, Q^s) : Q^s\ sa\right\}. \tag{37}$$

Proof. The former differential inclusion (36) reads as

$$\beta(\dot{q}^k, Q) - \beta(\dot{q}, Q) \geq (\dot{q}^k - \dot{q})^T\,Q$$

which leads to, with $\dot{q}^k = C^T\,\dot{w}^k$,

$$B(\dot{w}^k, Q) - B(\dot{w}, Q) \geq (\dot{q}^k - \dot{q})^T\,Q - (\dot{w}^k - \dot{w})^T\, f.$$

Owing to (35) and (34), one obtains the virtual power principle

$$(\dot{q}^k - \dot{q})^T Q = (\dot{w}^k - \dot{w})^T f,$$

hence

$$B(\dot{w}^k, Q) \geq B(\dot{w}, Q),$$

which achieves the proof. ■

We would like now to determine the value α^l of the load factor for which a mechanism $\dot{w} \neq 0$ with non-negative dissipation

$$\beta(\dot{q}, Q) > 0 \tag{38}$$

is developed with unlimited displacements and strains under constant load. This is called the **limit factor** (or **limit load**). In this new problem, α^l is an unknown variable, in addition to the **collapse mechanism** $\dot{w}$ and the corresponding generalized stresses Q. Owing to (34), (35) and the latter relation (36), one has

$$\beta(\dot{q}, Q) = \dot{w}^T C\, Q = \alpha^l \dot{w}^T f^0.$$

Then, the minimum in the variational principles (37) is

$$B(\dot{w}, Q) = 0. \tag{39}$$

Also, taking into account (38) and $\alpha^l > 0$, one has $\dot{w}^T f^0 > 0$, and thus the collapse mechanism $\dot{w}$ is admissible. Hence, the limit factor is given by the ratio of the internal dissipation and the external one:

$$\alpha^l = \frac{\beta(\dot{q}, Q)}{\dot{w}^T f^0}.$$

By analogy, for any admissible mechanism $\dot{w}^k$, the corresponding **kinematical factor** is defined by

$$\alpha^k = \frac{\beta(\dot{q}^k, Q)}{\dot{w}^{kT} f^0}. \tag{40}$$

As the bipotential (32) is positively homogeneous of order one,

$$\forall \lambda > 0, \qquad \beta(\lambda\, \dot{q}, Q) = \lambda\, \beta(\dot{q}, Q),$$

the kinematical factor does not depend on the intensity of the mechanism. For convenience, it can be arbitrarily fixed by imposing the **normalization condition**

$$\dot{w}^{kT} f^0 = 1, \tag{41}$$

hence the kinematical factor becomes equal to the internal dissipation:

$$\alpha_k = \beta(\dot{q}^k, Q).$$

We are now able to state the **kinematical theorem**.

Proposition 11.2 *For any admissible mechanism $\dot{q}^k$, the corresponding kinematical factor majorizes the limit factor:*

$$\alpha^k \geq \alpha^l.$$

Proof. According to the former variational principle (37) and to (39), one has

$$B(\dot{w}^k, Q) = \beta(\dot{q}^k, Q) - \alpha^l \dot{w}^{kT} f^0 \geq B(\dot{w}, Q) = 0.$$

Owing to the definition (40) of the kinematical factor, one has

$$(\alpha^k - \alpha^l)\dot{w}^T f^0 \geq 0,$$

which achieves the proof, because $\dot{w}^k$ is admissible, hence the second factor is positive. ■

The dual proposition is the **statical theorem**.

Proposition 11.3 *For any admissible generalized stress Q^s in equilibrium with reference forces multiplied by α^s, one has*

$$\alpha^l - \mu N \mid \dot{u} \mid \geq \alpha^s - \mu N^s \mid \dot{u} \mid .$$

The proof can be found in [3].

The interest of Proposition 11.2 lies in the fact that it transforms the kinematic approach into an optimization problem:

$$\inf \left\{ \beta(C^T \dot{w}^k, Q) : \dot{w}^k \text{ admissible and } \dot{w}^{kT} f^0 = 1 \right\}. \tag{42}$$

In solving this problem, the major difficulty faced is that the functional to minimise contains the generalised stress vector Q at the limit state, which is not yet known. A general algorithm for solving this kind of problem, having a coupling term between dual variables, is based on the successive approximations method, taking into account the constitutive relations. Let $(\dot{w}_i, Q_i)$ be the approximate solution assumed to be

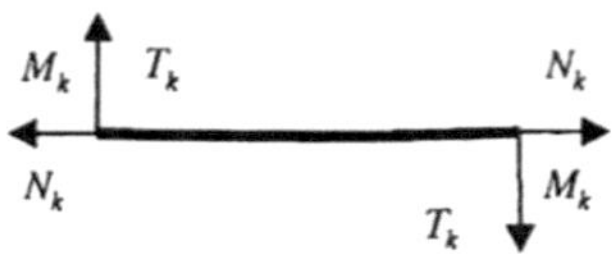

Fig. 11 Beam element: M_k: bending moment; N_k: normal force; and T_k: shear force

known at the ith iteration. The novel mechanism $\dot{w}_{i+1}$ is obtained by solving the optimization problem:

$$\inf \left\{\beta(C^T \dot{w}^k, Q_i) : \dot{w}^k \text{ admissible and } \dot{w}^{kT} f^0 = 1\right\}, \tag{43}$$

where the exact solution Q is replaced by the approximation Q_i in (42).

The generalised strain vector $\dot{w}_{i+1}$ being known, the new approximation of generalised stress Q_{i+1} is obtained by seeking a sub-gradient of the bifunctional:

$$Q_{i+1} \in \partial\beta(\bullet, Q_i)(C^T \dot{w}_{i+1}).$$

Finally, we can make a converging minimising sequence that, at the limit, provides the limit state $(\dot{w}, Q)$.

12 Application to a Plane Frame

In order to illustrate the effect of the presence of the unilateral contact with frictional support on the limit load and associated mechanism of the plane frames, a steel rectangular plane frame, built-in at the left-hand column, pinned and unilaterally supported at the right-hand column with possible Coulomb dry friction, is considered. The loading and dimensions are indicated in Fig. 10. The plastic capacity is constant on the structure and is equal to M_p. The joints are assumed infinitely rigid. Hence, the plastic hinge model can be used. Concerning the bending moment, shear and normal forces, the usual convention of beam theory is considered with respect to a given reference fibre, represented in Fig. 11.

Because the redundancy of the structure is $h = 2$, having one contact, the number of independent mechanisms is given by (33) and equal to 4. They are illustrated in Fig. 12. The vector of corresponding velocities is

$$\dot{w}^T = (\dot{u}_3, \dot{v}_3, \dot{u}_5, \dot{v}_5).$$

For convenience, the following dimensionless quantities are used:

$$\alpha = \frac{P\,a}{M_p}, \qquad m_i = \frac{M_i}{M_p}, \qquad t = \frac{T\,a}{M_p}, \qquad n = \frac{N\,a}{M_p}.$$

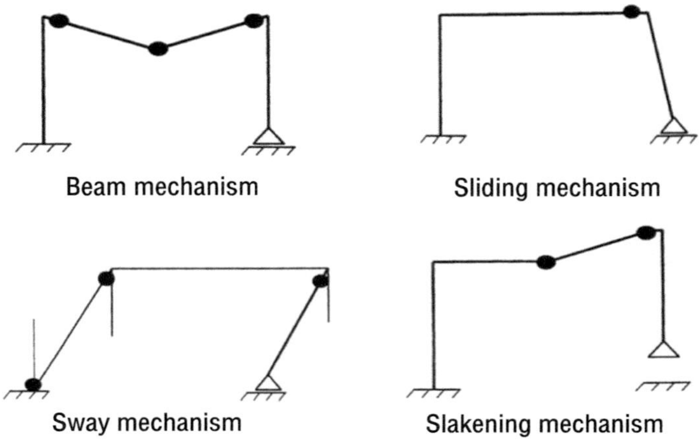

Fig. 12 Mechanisms

The equilibrium equation system (34) reads as

$$-m_2 + 2m_3 - m_4 = \alpha \tag{44}$$

$$-m_1 + m_2 - m_4 = \alpha \tag{45}$$

$$m_4 + t = 0 \tag{46}$$

$$m_3 - m_4 - n = 0. \tag{47}$$

The generalized variables are given in the following order:

$$Q^T = (t, n, m_1, m_2, m_3, m_4), \qquad \dot{q}^T = (\dot{u}, \dot{v}, \dot{\theta}_1, \dot{\theta}_2, \dot{\theta}_3, \dot{\theta}_4)$$

The compatibility condition system (35) reads as

$$\dot{u} = \dot{u}_5, \qquad \dot{v} = -\dot{v}_5$$

$$\dot{\theta}_1 = \frac{1}{a}(\dot{v}_5 - \dot{v}_3), \qquad \dot{\theta}_2 = \frac{1}{a}(\dot{v}_3 - \dot{u}_3)$$

$$\dot{\theta}_3 = \frac{1}{a}(2\dot{u}_3 + \dot{v}_5), \qquad \dot{\theta}_4 = \frac{1}{a}(-\dot{u}_3 - \dot{v}_3 + \dot{u}_5 - \dot{v}_5).$$

The normalization condition (41) is

$$\dot{u}_3 + \dot{v}_3 = 1.$$

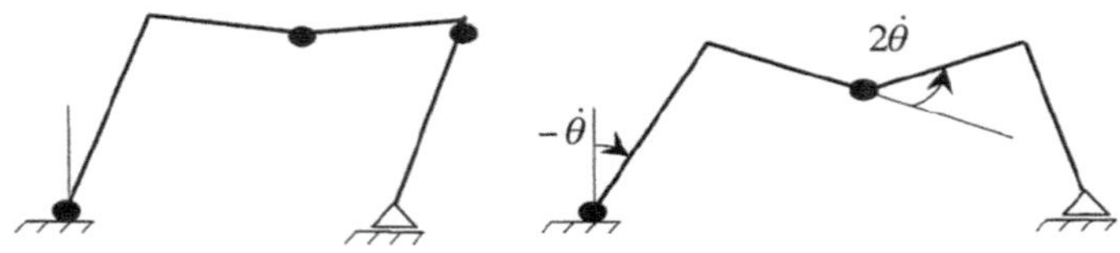

Fig. 13 Exact collapse mechanism. Case 1: for $\mu \geq 0.5$ (at the left). Case 2: for $\mu < 0.5$ (at the right)

A simple way to solve the kinematical problem (42) is to use Neals-Symonds method of mechanism combination ([25, 28]). The principle of the method consists in making a collapse mechanism by combination of some simple independent mechanisms. The kinematical theorem of limit analysis shows that the real mechanism is characterised by the smallest dissipation. As a general rule, the combination of two mechanisms does not give us an improved value of dissipation unless these mechanisms have in common opposed and equal rotations that cancel each other out. It can be remarked that because of the loading direction, the contact obviously occurs at the support that allows us to eliminate mechanism (47) of the combination.

12.1 *Iteration* 1

As an initial approximation, the problem with usual simple bilateral supports is considered. It is equivalent to assuming that the sticking contact occurs. It is easy to find the exact mechanism through a combination of the beam and sway mechanisms such that the plastic rotation of the second hinge is eliminated to minimise the total dissipation. The mechanism exactly involves $h + 1 = 3$ active variables $\dot{\theta}_1, \dot{\theta}_3$ and $\dot{\theta}_4$. The other ones are equal to zero. The equilibrium equation corresponding to the combined mechanism (44) + (45) is obtained by sum member to member:

$$-m_1 + 2m_3 - 2m_4 = 2\alpha. \tag{48}$$

The associated general velocity vector respecting the normalization condition is $\dot{w}^T = (0.5, 0.5, 0, 0)$. The corresponding mechanism is represented in Fig. 13 (Case 1).

Moreover, we have 6 statical unknowns that must be calculated. According to (48) and the plastic yielding rule, we choose $m_1 = -1$, $m_3 = 1$, and $m_4 = -1$. Using equilibrium equations (44)–(47), it is easy to compute the other variables m_2, n and t. Finally, we have to check whether the generalised stress vector $Q^T = (1, 2, 1, 0 : 5, 1, 1)$ is admissible. Although all the plasticity conditions are satisfied, Coulomb's friction criterion $|\ t\ | \leq \mu\, n$ is not verified if $\mu < 0.5$. Otherwise, if $\mu \geq 0.5$, Coulomb's sliding condition is satisfied and the generalized stresses are admissible, thus the exact collapse mechanism is obtained at the first iteration. Later on, the non-trivial case $\mu < 0.5$ is only considered (Fig. 13 (Case 1)). The sliding condition is violated and the vector Q is not admissible. For clarity, the first iteration gives us

$$\dot{w}_1^T = (0.5, 0.5, 0, 0), \qquad Q_1^T = (1, 2, -1, 0.5, 1, -1),$$

with a limit factor $\alpha = 2.5$. In summary, this mechanism is not valid for $\mu < 0.5$. So, other mechanisms must be looked for.

12.2 Iteration 2

Taking into account the existence of frictional contact, it is necessary to include the sliding mechanism (47) in the combination of mechanisms. Keeping the generalised stress vector constant and equal to the Q_1 found in the last iteration, let us calculate a new one. A new mechanism may be taken as the combination of (48) and (46) multiplied by λ:

$$-m_1 + 2\,m_3 + (\lambda - 2)\,m_4 + \lambda\,t = 2\,\alpha.$$

Assuming the sliding occurs, we put $m_1 = -1$, $m_3 = 1$ and $|\,t\,| = \mu\,n$, which leads to the following expression of the kinematical factor:

$$\alpha = 1.5 + \left|\,1 - \frac{\lambda}{2}\,\right| + \frac{\lambda\,\mu}{2}\,n.$$

The value of λ is obtained by minimizing the function $\alpha(\lambda)$, which gives us

$$\lambda = 2, \qquad \alpha = 1.5 + \mu\,n,$$

which corresponds to the elimination of the last hinge ($\lambda = 2$). So, it is clear that the kinematic load factor contains the static variables, which are not yet known. Based on the principle of the successive approximation algorithm, the value of normal reaction n is the one obtained in the previous iteration, i.e., $n = 2$. The active generalised velocities are, in this case, $\dot{\theta}_1$, $\dot{\theta}_3$ and $\dot{u}$. The other velocities are null: $\dot{\theta}_2 = \dot{\theta}_4 = \dot{u} = 0$. By using the compatibility equations and the normalization condition, one has

$$\dot{w}_2^T = (0.5, 0.5, 1, 0), \qquad \dot{q}_2^T = (2, 0, -1, 0, 2, 0).$$

The new approximation Q_2, which must satisfy the constitutive relations, is obtained by means of equilibrium equations:

$$Q_2^T = (2\,\mu, 1 + 2\,\mu, -1, 0.5, 1, -2\,\mu), \qquad \alpha_3 = 1.5 + 2\,\mu.$$

Once again, the sliding condition is violated for $\mu < 0.5$. So, we should try a new approximation.

Table 2 Values of α, n, t at iteration 3 for different values of μ

μ	0	0.1	0.2	0.3	0.4	0.5
α	1.5	1.512	1.556	1.644	1.788	2.50
n	1	1.012	1.056	1.144	1.288	2.00
t	0	0.012	0.056	0.144	0.288	1.00

Table 3 Values of α, n, t at iteration 4 for different values of μ

μ	0	0.1	0.2	0.3	0.4	0.5
α	1.5	1.612	1.756	1.944	1.188	2.50
n	1	1.112	1.256	1.444	1.688	2.00
t	0	0.112	0.256	0.444	0.688	1.00

12.3 Iteration 3

Similarly to the previous iteration, Q_2 will remain constant. The collapse mechanism remains the same, having always three active rates $\dot{\theta}_1$, $\dot{\theta}_3$ and $\dot{u}$, and one has

$$\dot{w}_3^T = \dot{w}_2^T = (0.5, 0.5, 1, 0), \qquad \dot{q}_3^T = \dot{q}_2^T = (2, 0, -1, 0, 2, 0)$$

From the equilibrium equations, and putting $n = 1 + 2\,\mu$, one has

$$Q_3^T = (\mu\,(1+\mu), \mu\,(1+2\,\mu), -1, 0.5, 1, -\mu\,(1+2\,\mu)), \qquad \alpha_3 = 1.5 + \mu\,(1+2\,\mu)$$

Numerical values are given in Table 2. Once again, the sliding condition is violated and we continue the iterations.

12.4 Iteration 4

By following the same procedure as in the previous iterations, one has (Table 3)

$$\dot{w}_4^T = \dot{w}_3^T = (0.5, 0.5, 1, 0), \qquad \dot{q}_4^T = \dot{q}_3^T = (2, 0, -1, 0, 2, 0)$$

$$Q_4^T = (-\mu\,(1+\mu+2\,\mu^2), 1+\mu\,(1+\mu+2\,\mu^2), -1, 0.5, 1, -\mu\,(1+\mu)+2\,\mu^2)$$

$$\alpha_4 = 1.5 + \mu\,(1+\mu+2\,\mu^2).$$

An infinite number of iterations is required to reach the exact collapse mechanism, but according to the two previous iterations, the algorithm is convergent to an exact

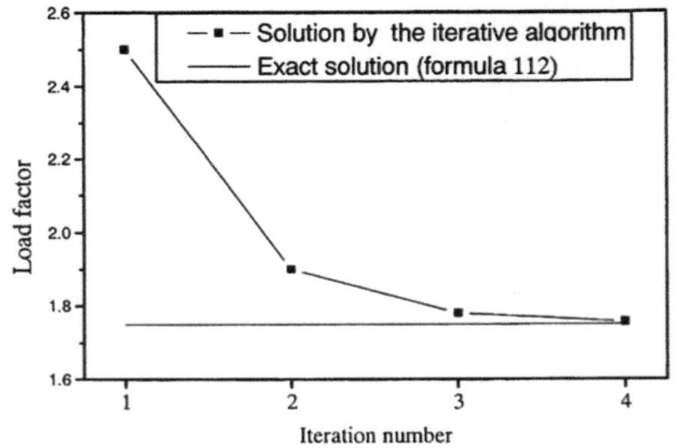

Fig. 14 Evolution of the limit factor during the iterations

solution (see Fig. 14). The last one can be determined easily in the present simple structure. For this, we rewrite all the equilibrium equations:

$$- m_1 + m_2 - m_4 = \alpha, \tag{49}$$

$$m_2 - 2 m_3 + m_4 = -\alpha, \tag{50}$$

$$m_4 + t = 0, \tag{51}$$

$$m_3 - m_4 - n = 0, \tag{52}$$

$$- m_1 + 2 m_3 - 2 m_4 = 2 \alpha, \tag{53}$$

$$- m_1 + 2 m_3 - 2 t = 2 \alpha. \tag{54}$$

The exact mechanism may be obtained by assuming $m_1 = -1$, $m_3 = 1$ and the sliding condition $\mid t \mid = \mu n$ occurs. Equations (51) and (52), combined with the sliding condition, provide

$$n = \frac{1}{1 - \mu}, \qquad t = \frac{\mu}{1 - \mu}$$

Equations (51)–(54) and (44) successively give us (Table 4)

$$\alpha^l = 1.5 + \frac{\mu}{1 - \mu}, \qquad m_4 = -\frac{\mu}{1 - \mu}, \qquad m_2 = 0.5.$$

Figures 15 and 16 show the variation of the exact limit load factor and contact and frictional reactions with respect to the friction coefficient.

We insist on the remarkable sensitivity of the limit state to the non-associativity caused by the presence of frictional contact in the right-hand column of the plane

Table 4 Values of α, n, t at iteration 5 for different values of μ

μ	0	0.1	0.2	0.3	0.4	0.5
α	1.5	1.611	1.750	1.928	2.166	2.50
n	1	1.111	1.250	1.428	1.666	2.00
t	0	0.111	0.250	0.428	0.666	1.00

Fig. 15 Variation of the contact and frictional reactions with respect to the friction coefficient

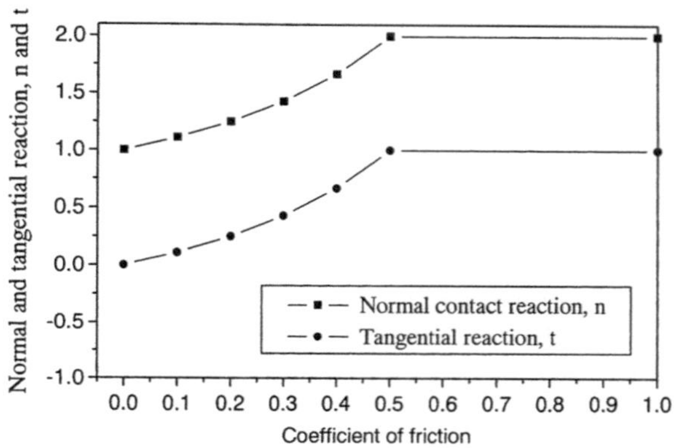

Fig. 16 Variation of the limit load factor with respect to the friction coefficient

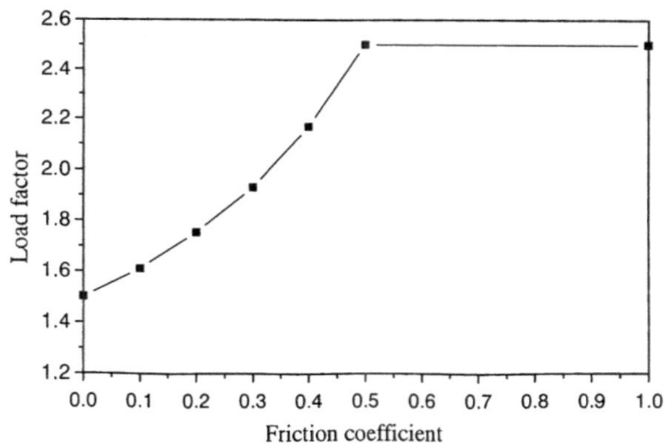

frame. For instance, the value 0.5 the of friction coefficient represents a boundary between two entirely different kinds of collapse mechanism. Indeed, for $\mu \geq 0.5$, the collapse mechanism is the one represented in Case 1 of Fig. 13, having three plastic hinges. On the other hand, if $\mu < 0.5$, the collapse mechanism is the one illustrated in Case 2 of Fig. 13 and has only two plastic hinges. A similar algorithm based on the statical Proposition 11.3 can be developed [3].

13 Conclusions

In the theoretical part, we showed that the bipotential of Coulomb's friction law is related to a specific bipotential convex cover with the property that any graph of the cover is non-maximal cyclically monotone. On this ground, we proposed a general algorithm explicitly for constructing the bipotential for the modelling of a given constitutive law.

In the numerical part, a robust algorithm, initially developed by de Saxcé and Feng [12] for the isotropic frictional contact law with associated sliding rule, has been adapted to handle non-associated sliding rules occurring when the friction is orthotropic. This algorithm has been successfully tested on examples, including cyclic loading for which a strong hysteretic behavior has been demonstrated.

In the last part, we introduced a variational version of the structural problem based on the concept of a bifunctional. A solution to the structural problem is simultaneously a solution to two coupled variational principles. On this ground, we generalized the limit analysis theorems for elastoplastic frames in the presence of frictional contact supports. The effect of the friction on the limit load and the collapse mechanism has been illustrated by considering a simple example with a single frictional contact support.

References

1. Bodovillé G (1999) On damage and implicit standard materials. CR Académie des Sciences de Paris, Série II, Fasc. b. Mécanique Physique Astrononomie 327:715–720
2. Bodovillé G, de Saxcé G (2001) Plasticity with non linear kinematic hardening: modelling and shakedown analysis by the bipotential approach. Eur J Mech A/Solids 20:99–112
3. Bousshine L, Chaaba A, de Saxcé G (2002) Plastic limit load of plane frames with frictional contact supports. Int J Mech Sci 44:2189–2216
4. Buliga M, de Saxcé G, Vallée C (2008) Existence and construction of bipotentials for graphs of multivalued laws. J Convex Anal 15(1):87–104
5. Buliga M, de Saxcé G, Vallée C (2010) Non maximal cyclically monotone graphs and construction of a bipotential for the Coulomb's dry friction law. J Convex Anal 17:081–094
6. Buliga M, de Saxcé G, Vallée C (2010) Bipotentials for non monotone multivalued operators: fundamental results and applications. Acta Applicandae Mathematicae 110:955–972
7. de Saxcé G (1992) Une généralisation de l'inégalité de Fenchel et ses applications aux lois constitutives. CR Académie des Sciences de Paris. Sér. II 314:125–129
8. de Saxcé G (1995) The bipotential method, a new variational and numerical treatment of the dissipative laws of materials. In: 10th international conference on mathematical and computer modelling and scientific computing, Boston, Massachusetts
9. de Saxcé G, Bousshine L (1993) On the extension of limit analysis theorems to the non associated flow rules in soils and to the contact with Coulomb's friction. In: Proceedings of XI Polish conference on computer methods in mechanics, Kielce (Poland), pp 815–822
10. de Saxcé G, Bousshine L (2002) Implicit standard materials. In: Weichert D, Maier G (eds) Inelastic behaviour of structures under variable repeated loads, vol 432. CISM courses and lectures. Springer, Wien
11. de Saxcé G, Feng ZQ (1991) New inequality and functional for contact friction: the implicit standard material approach. Mech Struct Mach 19:301–325

12. de Saxcé G, Feng ZQ (1998) The bi-potential method: a constructive approach to design the complete contact law with friction and improved numerical algorithms. Math Comput Model 6:225–245
13. Fenchel W (1949) On conjugate convex functions. Can J Math 1:73–77
14. Feng ZQ, Hjiaj M, de Saxcé G, Mróz Z (2006) Effect of frictional anisotropy on the quasistatic motion of a deformable solid sliding on a planar surface. Comput Mech 37:349–361
15. Feng ZQ, Hjiaj M, de Saxcé G, Mróz Z (2006) Influence of frictional anisotropy on contacting surfaces during loading/unloading cycles. Int J Non-Linear Mech 41:936–948
16. Fortin J, Hjiaj M, de Saxcé G (2002) An improved discrete element method based on a variational formulation of the frictional contact law. Comput Geotech 29:609–640
17. Halphen B, Nguyen Quoc S (1975) Sur les matériaux standard généralisés. CR Académie des Sciences de Paris 14:39–63
18. Hjiaj M, Bodovillé G, de Saxcé G (2002) Matériaux viscoplastiques et loi de normalité implicites. CR Académie des Sciences de Paris, Sér. II. Fasc. b 328:519–524
19. Hjiaj M, de Saxcé G, Mróz Z (2002) A variational-inequality based formulation of the frictional contact law with a non-associated sliding rule. Eur J Mech A/Solids 21:49–59
20. Hjiaj M, Feng ZQ, de Saxcé G, Mróz Z (2004) Three dimensional finite element computations for frictional contact problems with on-associated sliding rule. Int J Numer Methods Eng 60:2045–2076
21. Laborde P, Renard Y (2008) Fixed points strategies for elastostatic frictional contact problems. Math Meth Appl Sci 31:415–441
22. Michałowski R, Mróz Z (1978) Associated and non-associated sliding rules in contact friction problems. Arch Mech 11:259–276
23. Moreau JJ (2003) Fonctionnelles convexes. Istituto Poligrafico e zecca dello stato, Roma
24. Mróz Z, Stupkiewicz S (1994) An anisotropic fricition and wear model. Int J Solids Struct 31:1113–1131
25. Neals BG (1956) The plastic methods of structural analysis. Wiley, NewYork
26. Rockafellar RT (2979) Convex analysis, vol 238. Princeton University Press, Princeton
27. Save MA, Massonnet CE, de Saxcé G (1997) Plastic limit analysis of plates, shells and disks. Elsevier, New York
28. Symonds PS, Neal BG (1951) Recent progress in the plastic methods of structural analysis. J Franklin Inst 5–6
29. Vallée C, Lerintiu C, Fortuné D, Ban M, de Saxcé G (2005) Hill's bipotential. In: Mihailescu-Suliciu M (ed) New trends in continuum mechanics. Theta series in advanced mathematics. Theta Foundation, Bucarest, pp 339–351

Passive Control of Differential Algebraic Inclusions - General Method and a Simple Example

Claude-Henri Lamarque and Alireza Ture Savadkoohi

Abstract In this chapter, we consider a master system consisting of a nonlinear differential inclusion and an algebraic equation of constraint (resulting in a Differential Algebraic Inclusion (DAI) system). This system is coupled to a nonlinear energy sink (NES) corresponding to a one degree-of-freedom essentially nonlinear differential equation. We examine how a resonance capture can lead to a reduced order dynamical system. To obtain this reduced order model, we describe a multiple time scale analysis governed by the introduction of multi-timescales via a small parameter ε that is finite and strictly positive. The mass of the NES is small versus the mass of the master system, and it governs a mass ratio defining the small parameter ε. The first timescale is the fast scale. Introducing the Manevitch complexification leads to the definition of slow time envelope coordinates. These envelope coordinates either do not directly depend on the fast time scale or do not depend on this fast time scale via introduction of the so-called Slow Invariant Manifold (SIM). The slow time dynamics of the master system components is analyzed through introduction of equilibrium points, corresponding to periodic solutions, or singular points (governing bifurcations around the SIM), corresponding to quasi-periodic behaviors. We present a simple example of semi-implicit Differential Algebraic Equation (DAE), including a friction term coupled to a cubic NES. Analytical developments of a 1:1:1 resonance case permit us to predict passive control of a DAI by a NES.

C.-H. Lamarque (✉) · A. Ture Savadkoohi
University of Lyon, ENTPE, LTDS UMR CNRS 5513, 3, rue Maurice Audin, 69 518 Vaulx-en-Velin Cedex, France
e-mail: lamarque@entpe.fr

A. Ture Savadkoohi
e-mail: alireza.turesavadkoohi@entpe.fr

R. I. Leine et al. (eds.), *Advanced Topics in Nonsmooth Dynamics*,
https://doi.org/10.1007/978-3-319-75972-2_7

1 Introduction

Nonsmooth dynamical systems correspond to difficult problems, among which the theoretical resolution, the numerical approximation of the solutions and the control gave rise to numerous research achievements ([1–8], for example). We are going to focus here on a particular aspect: the passive control of dynamical systems of second order involving nonsmooth terms and an algebraic constraint. Passive control solutions in engineering were first explored a long time ago. A detailed study was carried out by Roberson [9] showing that the suppression band of an absorber is increased (with respect to traditional tuned mass dampers [10]) by including a cubic term in its restoring forcing function. Since then, several investigations have been performed on the passive control of main structural systems by nonlinear absorbers. Some "few" examples of extensive research results in regard to this domain that we can cite are: pendulum type [11] and autoparametric vibration absorber [12], buckled systems [13], impact dampers [14], nonlinear energy sink [15, 16], and nonsmooth [17] and nonlinear tuned vibration control systems [18]. Some works have been carried out to consider the passive control of main nonsmooth systems by nonlinear absorbers; for instance, we can mention the passive control process of a main system including a Dahl model or Bouc-Wen type hysteresis behavior by a nonlinear (smooth or nonsmooth) absorber [19, 20]. In these studies, the nonsmooth behaviors of main systems are included by means of internal variables in the form of differential equations. Vibratory energy control of main nonsmooth systems with Saint-Venant elements in which their behaviors are represented via "differential inclusions" have been studied by Schmidt and Lamarque [21], Weiss et al. [22] and Lamarque and Ture Savadkoohi [23]. In the current chapter, we consider a forced principal structure in which its behavior is modeled by nonlinear differential inclusion and an algebraic equation of constraint leading to a Differential Algebraic Inclusion (DAI) system. The system is treated via a time multi-scale method, which leads to the detection of its slow invariant manifold and characteristic points.

The chapter is structured as follows: In Sect. 2, previous research concerning the passive control of problems including a nonsmooth term are briefly recalled. In Sect. 3, the treated problems are presented and the method is described for the general case of a master system with a finite number of degrees-of-freedom submitted to an algebraic condition and with one NES coupled to the first mode of the master system. In Sect. 4, a simple academic example is treated so as to illustrate the method. Finally, the chapter is concluded in Sect. 5.

2 Presentation of the Problems

We consider mechanical dynamical problems of second order in time with a finite number of degrees-of-freedom including nonlinear terms that could be described by smooth nonlinear terms, piecewise linear models, or maximal monotone graphs.

These problems may involve internal variables described by additional problems of first order in time: these additional problems may correspond to differential inclusions or piecewise smooth nonlinear problems. In order to proceed to a passive control, the previous systems, potentially with additional ones, could be coupled to an essentially nonlinear system called a "Nonlinear Energy Sink" (NES). This second order problem contains nonlinear terms that are smooth or piecewise smooth (piecewise linear, for example). Typical situations and basic cases explaining our purpose have been treated: one degree-of-freedom with nonsmooth internal variable (Dahl or Bouc-Wen model) coupled to one NES [19, 20], one degree-of-freedom with nonsmooth term (Saint-Venant element modelling friction) coupled to one NES [21–23], for example. Here, we extend this study to the following case: semi-implicit Differential Algebraic Equation (DAE) [24] involving a nonsmooth term coupled to one NES.

3 General DAI Coupled to One NES

We consider a Differential Algebraic Inclusion (DAI) coupled to one NES depending on a small parameter ε. The ratio of masses of the NES and the master system governed by DAI introduces this small parameter. In this section, we describe a method for analyzing the resonant capture phenomenon between the NES and one mode of the DAI system, so that rough nonlinear behaviour can be split on two timescales and approximated by the main harmonic of resonance.

3.1 Model of the System

Let us consider the following model of the system without an NES written in modal form:

$$\ddot{X}(t) + DX(t) + \varepsilon\xi_0\dot{X}(t) + \varepsilon A_0(X(t), \dot{X}(t)) + \varepsilon h_0(Z(t)) - \varepsilon \sin(\omega t) f_0 \ni 0, \tag{1}$$

or in condensed form as

$$\ddot{X}(t) - \mathscr{F}(\dot{X}(t), X(t), t) \ni 0,$$

with an additional algebraic constraint

$$G(Z, X, \dot{X}) = 0, \tag{2}$$

with $1 >> \varepsilon > 0$ a small parameter, $X(t) \in \mathbb{R}^n$, $n \in \mathbb{N}^\star$, $f_0 \in \mathbb{R}^n$, $Z(t) \in \mathbb{R}^m$, $n \geq m \in \mathbb{N}^\star$, $h_0 : \mathbb{R}^m \to \mathbb{R}^n$ a smooth function, and A_0 is assumed to be a nonsmooth term defined via a maximal monotone operator so that existence and uniqueness questions are solved in the frame of the theoretical results given in [8]. Typically, $A_0(\dot{X}(t), X(t)) = A_0(\dot{X}(t))$ with A_0 maximal monotone graph on $\mathbb{R}^n$. $G : \mathbb{R}^m \times \mathbb{R}^n \times \mathbb{R}^n \to \mathbb{R}^n$ is a smooth (nonlinear) function verifying that $\nabla_Z G(Z(t), \dot{X}(t), X(t))$ is invertible, and either $\nabla_{\dot{X}} G(Z(t), \dot{X}(t), X(t)) = 0$ or, more generally, differentiation of the Eq. (2) leads to the following differential inclusion:

$$\begin{gathered}\Big(\nabla_Z G(Z(t), \dot{X}(t), X(t))\dot{Z}(t) + \nabla_X G(Z(t), \dot{X}(t), X(t))\dot{X}(t) + \\ \nabla_{\dot{X}} G(Z(t), \dot{X}(t), X(t))\mathscr{F}(\dot{X}(t), X(t), t)\Big) \ni 0,\end{gathered} \tag{3}$$

corresponding to a well-posed coupled problem for the nonsmooth terms occurring via $\mathscr{F}$. In order to simplify, we consider here the case $\nabla_{\dot{X}} G(Z(t), \dot{X}(t), X(t)) = 0$. We also assume $G(0, 0, 0) = 0$. D is a diagonal matrix with diagonal associated with the n frequencies $(\omega_1^2, \omega_2^2, \ldots, \omega_n^2)$ and ∇_U stands for the gradient operator versus the variable U.

For control purposes, we assume that a NES (with a very small mass) is coupled on one chosen mode of the main system (it counts as ω_1). So, the DAI (1)–(2) problem coupled to the NES can be written in the form

$$\begin{cases} \ddot{X}(t) + DX(t) + \varepsilon\xi_0\dot{X}(t) + \varepsilon A_0(X(t), \dot{X}(t)) + \varepsilon h_0(Z(t)) + \varepsilon(\lambda(\dot{x}_1 - \dot{y}) \\ \qquad + \Gamma(x_1 - y)) \begin{pmatrix} 1 \\ 0 \\ \vdots \\ 0 \end{pmatrix} - \varepsilon \sin(\omega t) f_0 \ni 0, \\ \nabla_Z G(Z(t), X(t), \dot{X}(t))\frac{dZ}{dt}(t) + \nabla_X G(Z(t), X(t), \dot{X}(t))\frac{dX}{dt}(t) = 0, \\ \varepsilon\ddot{y}(t) + \varepsilon\lambda(\dot{y}(t) - \dot{x}_1(t)) - \varepsilon\Gamma(x_1 - y) = 0, \end{cases} \tag{4}$$

where ξ_0 is a diagonal matrix of specific damping, λ governs the damping due to the coupling, Γ is an essentially nonlinear function (typically, $\Gamma(z) = \gamma z^3$), $y(t) \in \mathbb{R}$, and f_0 is a vector of $\mathbb{R}^n$. In spite of this, we assume that we do not have internal resonance in the system. We consider a detuning relation for all the frequencies:

$$\omega = \omega_j + \sigma_j \varepsilon,$$

which means that the frequency of the external excitation, i.e., ω, is varying around the jth frequency of the main system, i.e., ω_j. This variation of the frequency is controlled by the detuning parameter σ_j. Let us write

$$\begin{aligned} &\nabla_Z G(Z(t), X(t), \dot{X}(t))\frac{dZ}{dt}(t) + \nabla_X G(Z(t), X(t), \dot{X}(t))\frac{dX}{dt}(t) \\ &= P(Z(t), X(t), \dot{X}(t))\frac{dZ}{dt}(t) + Q(Z(t), X(t), \dot{X}(t))\frac{dX}{dt}(t), \end{aligned}$$

with P and Q matrices of size $m \times m$ and $m \times n$, respectively.

For such a system, we can directly start an analytical study of the evolution of the amplitudes of oscillations under principal resonance of the system. Another approach could be to introduce new coordinates derived from X, y. For example, center of unit mass of the first coordinate x_1 of X and of ε mass of the NES with coordinate y, and relative displacement $x_1 - y$, keeping the other coordinates $x_2, \ldots, x_n$, since the NES is coupled to the first mass. Here, we simplify the presentation of the method, and we keep the X, y coordinates.

3.2 Introduction of Complexification of Manevitch

We assume that, due to the resonance type, the response of the DAI coupled with the NES can be approached by modulation of the dynamical behavior around harmonics of a Fourier analysis according to the ω frequency. So, Manevitch complexification [25] could be introduced as

$$\begin{cases} \phi_1 \exp(i\omega t) = \dot{x}_1(t) + i\omega x(t), \ldots, \phi_n \exp(i\omega t) = \dot{x}_n(t) + i\omega x_n(t), \\ \phi_{n+2} \exp(i\omega t) = \dot{z}_1(t) + i\omega z_1(t), \ldots, \phi_{n+m+1} \exp(i\omega t) = \dot{z}_m(t) + i\omega z_m(t), \\ \phi_{n+1} \exp(i\omega t) = \dot{y}(t) + i\omega y(t) \end{cases} \tag{5}$$

with conjugate values

$$\begin{cases} \phi_1^\star \exp(-i\omega t) = \dot{x}_1(t) - i\omega x(t), \ldots, \phi_n^\star \exp(-i\omega t) = \dot{x}_n(t) - i\omega x_n(t), \\ \phi_{n+2}^\star \exp(-i\omega t) = \dot{z}_1(t) - i\omega z_1(t), \ldots, \phi_{n+m+1}^\star \exp(-i\omega t) = \dot{z}_m(t) - i\omega z_m(t), \\ \phi_{n+1}^\star \exp(-i\omega t) = \dot{y}(t) - i\omega y(t) \end{cases} \tag{6}$$

and classically, for $j = 1, \ldots, n$ and $l = 1, \ldots, m$,

$$\begin{cases} \dot{x}_j = \frac{1}{2}(\phi_j \exp(i\omega t) + \phi_j^\star \exp(-i\omega t)), \\ x_j = \frac{1}{2i\omega}(\phi_j \exp(i\omega t) - \phi_j^\star \exp(-i\omega t)), \\ \dot{z}_l = \frac{1}{2}(\phi_{l+n+1} \exp(i\omega t) + \phi_{l+n+1}^\star \exp(-i\omega t)), \\ z_l = \frac{1}{2i\omega}(\phi_{l+n+1} \exp(i\omega t) - \phi_{l+n+1}^\star \exp(-i\omega t)), \\ \dot{y} = \frac{1}{2}(\phi_{n+1} \exp(i\omega t) + \phi_{n+1}^\star \exp(-i\omega t)), \\ y = \frac{1}{2i\omega}(\phi_{n+1} \exp(i\omega t) - \phi_{n+1}^\star \exp(-i\omega t)), \end{cases} \tag{7}$$

The next step of the analysis is to describe the dynamics with coordinates $\phi_j, \phi_j^\star, j = 1, \ldots, n+m+1$ that are not constants, as in the usual study of stationary solutions (Harmonic Balance Method). The evolution will be written according to a multi-time scale analysis governed by the small parameter ε.

3.3 Multi-timescale Analysis. Slow Invariant Manifold

Let us set

$$\tau_j = \varepsilon^j t,\ j = 0, 1, 2, \ldots \tag{8}$$

As usual, the time derivative can be expanded at different scales of ε according to

$$\frac{d}{dt} = \frac{\partial}{d\tau_0} + \varepsilon\frac{\partial}{d\tau_1} + \ldots \tag{9}$$

We assume that all the $\phi_j, \phi_j^\star$, $j = 1, \ldots, n+m+1$ coordinates do not depend on the τ_0 timescale. We will see that either it will be verified a posteriori or it will be demanded a posteriori up to an asymptotic assumption at time scale τ_0. Because of this assumption, it is possible to calculate integrals versus τ_0 considering that all the ϕ_j are constant versus τ_0 (but not independent of time, since variation versus $\tau_1, \tau_2, \ldots$ could occur). The complex variables of Manevitch verify

$$\begin{cases} \dot{\phi}_j \exp(i\omega t) = \ddot{x}_j + \omega^2 x_j,\ j = 1, \ldots, n, \\ \dot{\phi}_{n+1} \exp(i\omega t) = \ddot{y} + \omega^2 y, \\ \dot{\phi}_{l+n+1} \exp(i\omega t) = \ddot{z}_l + \omega^2 z_l,\ l = 1, \ldots, m, \end{cases} \tag{10}$$

Let us consider a projection of Eq. (4) onto the first harmonic of the underlying Fourier analysis. We set $T = 2\pi/\omega$. Assuming ϕ_j, $j = 1, \ldots, n+m+1$ do not depend on τ_0 provides us with

$$\begin{aligned} &\frac{1}{T}\int_0^T \dot{\phi}_j \exp(i\omega t)\exp(-i\omega\tau_0)d\tau_0 = \\ &\frac{1}{T}\int_0^T \dot{\phi}_j(\tau_1, \tau_2, \ldots)\exp(i\omega\tau_0)\exp(-i\omega\tau_0)d\tau_0 = \dot{\phi}_j(\tau_1, \ldots). \end{aligned} \tag{11}$$

In the same way, we obtain, for $j = 1$,

$$\begin{aligned} &\dot{\phi}_1 + (\omega_1^2 - \omega^2)\tfrac{\phi_1}{2i\omega} + \varepsilon[\tfrac{\xi_{01}}{2}\phi_1 + F_1(\phi_1, \phi_1^\star, \ldots, \phi_n, \phi_n^\star) + H_{01}(\phi_{n+2}, \ldots, \phi_{n+m+1}^\star) \\ &+\tfrac{\lambda}{2}(\phi_1 - \phi_{n+1}) + \mathscr{C}(\phi_1, \phi_1^\star, \phi_{n+1}, \phi_{n+1}^\star) - \tfrac{f_{01}}{2}] = 0. \end{aligned} \tag{12}$$

Then, for $j = 2, \ldots, n$,

$$\begin{aligned} &\dot{\phi}_j + (\omega_j^2 - \omega^2)\tfrac{\phi_j}{2i\omega} + \varepsilon[\tfrac{\xi_{0j}}{2}\phi_j + F_j(\phi_1, \phi_1^\star, \ldots, \phi_n, \phi_n^\star) \\ &+H_{0j}(\phi_{n+2}, \ldots, \phi_{n+m+1}^\star) - \tfrac{f_{0j}}{2}] = 0. \end{aligned} \tag{13}$$

For $j = n+1$ corresponding to the equation associated with the NES behavior, we have

$$\varepsilon\left(\dot{\phi}_{n+1} + i\frac{1}{2\omega}\phi_{n+1} + \frac{\lambda}{2i\omega}(\phi_{n+1} - \phi_1) - \mathscr{C}(\phi_1, \phi_1^\star, \phi_{n+1}, \phi_{n+1}^\star)\right) = 0, \tag{14}$$

and it is possible to simplify by ε. For $j = n+2, \ldots, n+m+1$, we obtain algebraic relations from the calculations

$$\frac{1}{T}\int_0^T [\sum_{l=1}^{m} P_{jl}(Z(t), X(t), \dot{X}(t))\dot{z}_l(t) + \sum_{l=1}^{n} Q_{jl}\dot{x}_l(t)e^{-i\omega\tau_0}]d\tau_0 \tag{15}$$

denoted as

$$\mathscr{G}_j(\phi_1, \ldots, \phi_N^\star, \phi_{n+2}, \ldots, \phi_{n+m+1}^\star) = 0, \tag{16}$$

for $j = n+2, \ldots, n+m+1$, where we consider that each component of Z or X or $\dot{X}$ can be expressed versus the ϕ_j with

$$\begin{cases} \dot{x}_j = \frac{1}{2}(\phi_j(\tau_1, \ldots)e^{i\omega\tau_0} + \phi_j^\star(\tau_1, \ldots)e^{-i\omega\tau_0}), \\ x_j = \frac{1}{2i\omega}(\phi_j(\tau_1, \ldots)e^{i\omega\tau_0} - \phi_j^\star(\tau_1, \ldots)e^{-i\omega\tau_0}), \\ \dot{z}_l = \frac{1}{2}(\phi_{l+n+1}(\tau_1, \ldots)e^{i\omega\tau_0} + \phi_{l+n+1}^\star(\tau_1, \ldots)e^{-i\omega\tau_0}), \\ z_l = \frac{1}{2i\omega}(\phi_{l+n+1}(\tau_1, \ldots)e^{i\omega\tau_0} - \phi_{l+n+1}^\star(\tau_1, \ldots)e^{-i\omega\tau_0})), \\ \dot{y} = \frac{1}{2}(\phi_{n+1}(\tau_1, \ldots)e^{i\omega\tau_0} + \phi_{n+1}^\star(\tau_1, \ldots)e^{-i\omega\tau_0}), \\ y = \frac{1}{2i\omega}(\phi_{n+1}(\tau_1, \ldots)e^{i\omega\tau_0} - \phi_{n+1}^\star(\tau_1, \ldots)e^{-i\omega\tau_0}). \end{cases} \tag{17}$$

Functions F_j, H_{0j} and $\mathscr{C}$ are defined by

$$F_j(\phi_1, \phi_1^\star, \ldots, \phi_n, \phi_n^\star) = \frac{1}{T}\int_0^T A_{0j}(\tau_0, \Phi(\tau_1, \ldots), \Phi^\star(\tau_1, \ldots))\exp(-i\omega\tau_0)d\tau_0, \tag{18}$$

where $\Phi = (\phi_1, \ldots, \phi_n)^t$. Then,

$$\mathscr{C}(\phi_1, \phi_1^\star, \phi_{n+1}, \phi_{n+1}^\star) = \frac{1}{T}\int_0^T \Gamma(x_1 - y)e^{-i\omega\tau_0}d\tau_0, \tag{19}$$

where

$$x_1 - y = \frac{(\phi_1(\tau_1, \ldots) - \phi_{n+1}(\tau_1, \ldots))e^{i\omega\tau_0} - (\phi_1^\star(\tau_1, \ldots) - \phi_{n+1}^\star(\tau_1, \ldots))e^{-i\omega\tau_0}}{2i\omega}$$

$$H_{0j} = \frac{1}{T}\int_0^T h_{0j}(\phi_{n+2}(\tau_1, \ldots)e^{i\omega\tau_0}, \ldots, \phi_{n+m+1}^\star(\tau_1, \ldots)e^{-i\omega\tau_0})e^{-i\omega\tau_0}d\tau_0. \tag{20}$$

Now, let us start to organize all of these equations versus the orders of ε. First, we consider the ε^0 order. We have

$$\begin{aligned} \tfrac{\partial\phi_1}{\partial\tau_0} + o(\varepsilon) = 0 &\Rightarrow \phi_1 = \phi_1(\tau_1, \ldots) \\ &\vdots \\ \tfrac{\partial\phi_n}{\partial\tau_0} + o(\varepsilon) = 0 &\Rightarrow \phi_n = \phi_n(\tau_1, \ldots), \end{aligned} \tag{21}$$

since ε is factorized in front of all the terms except $(\omega_j^2 - \omega^2)\dfrac{\phi_j}{2i\omega} = I\sigma_j\varepsilon + o(\varepsilon^2)$. Let us simply denote

$$\mathscr{G}_{j0}(\phi_1, \ldots, \phi_N^\star, \phi_{n+2}, \ldots, \phi_{n+m+1}^\star) = 0, \tag{22}$$

for $j = n+2, \ldots, n+m+1$, the ε^0 order of equations $\mathscr{G} = 0$. Then, we have, for $j = n+1$,

$$\frac{\partial \phi_{n+1}}{\partial \tau_0} + \frac{\lambda}{2i\omega_1}(\phi_{n+1} - \phi_1) - \mathscr{C}_0(\phi_1, \phi_1^\star, \phi_{n+1}, \phi_{n+1}^\star) = 0, \tag{23}$$

where we consider $\mathscr{C}_0$ the ε^0 order of $\mathscr{C}$, and approximation $\omega = \omega_1 + o(\varepsilon)$, since we focus on the passive control of the first mode. Then, considering asymptotic behavior versus τ_0, we assume that when $\tau_0 \longrightarrow +\infty$, $\dfrac{\partial \phi_{n+1}}{\partial \tau_0} = 0$, i.e.,

$$\frac{\lambda}{2i\omega_1}(\phi_{n+1} - \phi_1) - \mathscr{C}_0(\phi_1, \phi_1^\star, \phi_{n+1}, \phi_{n+1}^\star) = 0. \tag{24}$$

This relation and Eqs. (22) define the so-called Slow Invariant Manifold (SIM). It can be seen that considering the dynamical behavior around the SIM leads to a reduced order n model, since we start from $n+m+1$ complex variables ϕ_j, but we add $m+1$ relations defining the SIM.

3.4 Analysis of the Dynamics of the Envelope at Fast Time Scale

Now, we analyse the nonlinear strongly modulated motion around the SIM. It is enough to consider equations at the ε^1 order for ϕ_j, $j = 1, \ldots, n$, written in the form

$$\frac{\partial \phi_1}{\partial \tau_1} + (i\sigma_j + \frac{\xi_{01}}{2})\phi_1 + F_1 + H_{01} + \frac{\lambda}{2}(\phi_1 - \phi_{n+1}) + \mathscr{C} = 0 \tag{25}$$

and

$$\frac{\partial \phi_j}{\partial \tau_1} + (i\sigma_j + \frac{\xi_{0j}}{2})\phi_j + F_j + H_{0j} = 0,\ j = 2, \ldots, n. \tag{26}$$

The analysis of the behavior can be done by looking for equilibrium points of the previous equations (corresponding to approximations of periodic solutions of the initial system, around the SIM) and by looking for potential "singular points" associated with the introduction of the $m+1$ relations defining the SIM into the previous equations. It means that if one see the variables $\phi_1, \ldots, \phi_N$ as functions of the

$\phi_{n+1}, \ldots, \phi_{n+m+1}$, the previous Eqs. (25) and (26) could be expressed versus these last variables as

$$M(\phi_{n+1}, \ldots, \phi^{\star}_{n+m+1}) \begin{pmatrix} \phi_{n+1} \\ \vdots \\ \phi_{m+n+1} \\ \phi^{\star}_{n+1} \\ \vdots \\ \phi^{\star}_{m+n+1} \end{pmatrix} = S(\phi_{n+1}, \ldots, \phi^{\star}_{n+m+1}), \tag{27}$$

which is obtained from differentiation of the SIM relations [28]. Singular points correspond to the occurrence of singularity of the $2(m+1) \times 2(m+1)$ matrix M when $S = 0$ corresponds to the equilibrium points. The application of the method for designing a passive controller of the initial system leads to answering of the following question: Is it possible to design the NES "parameters" $(\varepsilon, \lambda, \Gamma)$ so that there is no periodic equation with amplitudes (modulus of ϕ_j, $j = 1, \ldots, n$) higher than the given thresholds and singular points that appear governing quasi-periodic exchanges between the master system and the NES, limiting the same amplitudes around the same thresholds?

4 Example of a DAE Including Nonsmooth Terms Coupled to a NES

In order to illustrate the purpose, we consider a simple mathematical example of low dimension. We consider a semi-explicit DAE with friction coupled to a NES reduced to an essentially cubic nonlinearity. The model is governed by a semi-explicit Differential Algebraic Inclusion (DAI), such as

$$\begin{cases} \ddot{x} + \omega_1^2 x + \varepsilon a_0 \dot{x} + \alpha\rho(\dot{x}) + h(z) + \varepsilon\lambda(\dot{x} - \dot{y}) + \varepsilon\Gamma(x - y) \ni f(t), \\ g(\dot{x}, x, z) = 0, \\ \varepsilon\ddot{y} + \varepsilon\lambda(\dot{y} - \dot{x}) - \varepsilon\Gamma(x - y) = 0, \end{cases} \tag{28}$$

where $\omega_1, \lambda, \gamma, \alpha, a_0$ are constants. The parameter ε is small and positive, which is generally associated with a mass ratio (mass of the main system divided by the mass of the NES). So, ε is finite and is not a bookkeeping parameter that could tend towards 0^+. The graph of the "sign" is represented by ρ. The variable z governs the algebraic equation defined by g, while h is a function of z. We examine the principal resonance. We assume

$$\Gamma(x-y)=\gamma(x-y)^3,\ f(t)=\varepsilon f_0\sin(\omega t),\ \alpha=\varepsilon\alpha_0,\ h(z)=\varepsilon h_0(z),\ \omega=\omega_1+\sigma\varepsilon. \tag{29}$$

We consider the simple case with

$$g(\dot{x},x,z)=-z+\frac{x^3}{3},\ h_0(z)=\beta z, \tag{30}$$

where p and β are given parameters. Since we have

$$\frac{\partial g}{\partial \dot{x}}=0,\ \frac{\partial g}{\partial x}=x^2,\ \frac{\partial g}{\partial z}=-1, \tag{31}$$

the model of the semi-DAE system coupled with one NES can be written as

$$\begin{cases} \ddot{x}+\omega_1^2x+\varepsilon(a_0\dot{x}+\alpha_0\rho(\dot{x})+h_0(z)+\lambda(\dot{x}-\dot{y})+\gamma(x-y)^3-f_0\sin(\omega t))\ni 0,\\ \varepsilon(\ddot{y}+\lambda(\dot{y}-\dot{x})+\gamma(y-x)^3)=0,\\ \dot{z}=x^2\dot{x}. \end{cases} \tag{32}$$

4.1 Analytical Approximated Approach

Let us introduce new coordinates v and w (center of mass and relative displacement), so that

$$v=\frac{x+\varepsilon y}{1+\varepsilon},\ w=x-y. \tag{33}$$

We obtain

$$\begin{cases} \ddot{v}+\frac{\omega_1^2}{1+\varepsilon}(v+\frac{\varepsilon}{1+\varepsilon}w)+\\ \frac{\varepsilon}{1+\varepsilon}(a_0[\dot{v}+\frac{\varepsilon}{1+\varepsilon}\dot{w}]+\alpha_0\rho(\dot{v}+\frac{\varepsilon}{1+\varepsilon}\dot{w})+h_0(z)-f_0\sin(\omega t))\ni 0,\\ \ddot{w}+\omega_1^2(v+\frac{\varepsilon}{1+\varepsilon}w)+(1+\varepsilon)(\lambda\dot{w}+\gamma w^3)+\varepsilon(a_0[\dot{v}+\frac{\varepsilon}{1+\varepsilon}\dot{w}]+\\ \alpha_0\rho(\dot{v}+\frac{\varepsilon}{1+\varepsilon}\dot{w})+h_0(z)-f_0\sin(\omega t))\ni 0,\\ \dot{z}=(v+\frac{\varepsilon}{1+\varepsilon}w)^2[\dot{v}+\frac{\varepsilon}{1+\varepsilon}\dot{w}]=v^2\dot{v}+o(\varepsilon). \end{cases} \tag{34}$$

Multiple-time scales are introduced so that

$$\tau_j=\varepsilon^j t,\ j=0,1,2,\ldots \tag{35}$$

We use the complexification of Manevitch [25] with the main harmonic, by setting

$$\begin{cases} \phi_1 \exp(i\omega\tau_0) = \dot{v} + i\omega v, \phi_1^\star \exp(-i\omega\tau_0) = \dot{v} - i\omega v, \\ \phi_2 \exp(i\omega\tau_0) = \dot{w} + i\omega w, \phi_2^\star \exp(-i\omega\tau_0) = \dot{w} - i\omega w, \\ \phi_3 \exp(i\omega\tau_0) = \dot{z} + i\omega z, \phi_3^\star \exp(-i\omega\tau_0) = \dot{z} - i\omega z. \end{cases} \tag{36}$$

Following the previous method, we calculate the mean values of all equations to obtain the projection of each equation on the main harmonic $\exp(i\omega\tau_0)$. For the calculations, we assume (it will be verified or imposed later) that during the integrals versus τ_0, all functions $\phi_j = \phi_j(\tau_1, \tau_2, \ldots)$, $j = 1, 2, 3$ are seen as constants versus τ_0 (as well as their conjugates $\phi_j^\star$). In fact, we assume a very simple mono-harmonic form of the solution, so that analytical calculations can also be done for the nonsmooth terms (see [26, 27]). Cumbersome calculations provide

$$\begin{cases} \dot{\phi}_1 + (\frac{\omega_1^2}{1+\varepsilon} - \omega^2)\frac{\phi_1}{2i\omega} + \frac{\omega_1^2}{(1+\varepsilon)^2}\varepsilon\frac{\phi_2}{2i\omega} + \frac{\varepsilon\alpha_0}{1+\varepsilon}F(N_1, N_2, \delta_1, \delta_2) + \frac{\varepsilon}{1+\varepsilon}H_0(\phi_3, \phi_3^\star) - \\ \qquad \frac{\varepsilon f_0}{2i(1+\varepsilon)} = 0, \\ \dot{\phi}_2 - \frac{\omega}{2i}\phi_2 + \omega_1^2[\frac{\phi_1}{2i\omega} + \frac{\varepsilon\phi_2}{2i\omega(1+\varepsilon)}] + (1+\varepsilon)\lambda\frac{\phi_2}{2} + \varepsilon a_0(\phi_1 + \frac{\varepsilon\phi_2}{1+\varepsilon}) + \\ \varepsilon\alpha_0 F(N_1, N_2, \delta_1, \delta_2) + \varepsilon H_0(\phi_3, \phi_3^\star) + (1+\varepsilon)\gamma\frac{3i}{8\omega^3} \mid \phi_2 \mid^2 \phi_2 - \frac{\varepsilon f_0}{2i} = 0, \\ \phi_3 - \frac{1}{4\omega^2} \mid \phi_1 \mid^2 \phi_1 = 0, \end{cases} \tag{37}$$

where F and H_0 are given in Appendix 1, and by introducing the polar form of each ϕ_j, $j = 1, 2, 3$ in the form

$$\phi_j = N_j \exp(i\delta_j),\ N_j \text{ and } \delta_j \text{ real functions of } \tau_1, \tau_2, \ldots;\ N_j \geq 0,\ j = 1, 2, 3. \tag{38}$$

Taking into account the detuning relation $\omega = \omega_1 + \sigma\varepsilon$, and since

$$\dot{\phi}_j = \frac{d}{dt}\phi_j = \frac{\partial\phi_j}{\partial\tau_0} + \varepsilon\frac{\partial\phi_j}{\partial\tau_1} + \varepsilon^2\frac{\partial\phi_j}{\partial\tau_2} + \cdots = \frac{\partial\phi_j}{\partial\tau_0} + \varepsilon\frac{\partial\phi_j}{\partial\tau_1} + o(\varepsilon^2),\ j = 1, 2, 3, \tag{39}$$

these equations can be organized versus powers of ε.

$$\begin{cases} \frac{\partial\phi_1}{\partial\tau_0} + \varepsilon[\frac{\partial\phi_1}{\partial\tau_1} + \frac{i}{2}(\omega_1 + 2\sigma)\phi_1 + \alpha_0 F(N_1, N_2, \delta_1, \delta_2) + H_0(\phi_3, \phi_3^\star) - \frac{f_0}{2i}] + o(\varepsilon^2) = 0, \\ \frac{\partial\phi_2}{\partial\tau_0} + \frac{i\omega_1}{2}(\phi_2 - \phi_1) + \frac{\lambda}{2}\phi_2 - \frac{3i\gamma}{8\omega_1^3} \mid \phi_2 \mid^2 \phi_2 + o(\varepsilon) = 0, \\ \phi_3 - \frac{1}{4\omega_1^2} \mid \phi_1 \mid^2 \phi_1 + o(\varepsilon) = 0. \end{cases} \tag{40}$$

At the ε^0 order, we can see that $\frac{\partial\phi_1}{\partial\tau_0} = 0$, so the assumption $\phi_1 = \phi_1(\tau_1, \tau_2, \ldots)$ is verified. Moreover, the algebraic constraint becomes an algebraic equation, truncated at order ε^0 here. The slow invariant manifold (SIM) can be introduced by

$$\phi_1 = \phi_2 - \frac{i\lambda}{\omega_1}\phi_2 - \frac{3\gamma}{4\omega_1^4} \mid \phi_2 \mid^2 \phi_2. \tag{41}$$

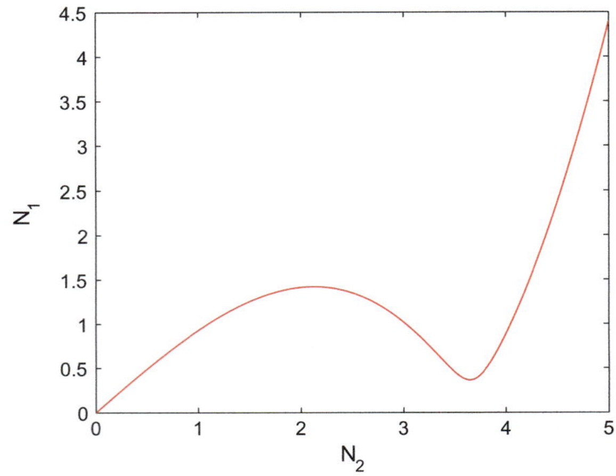

Fig. 1 The SIM of the system for following parameters: $\varepsilon = 10^{-3}$, $\omega_1 = 1, \lambda = 0.1, \gamma = 0.1$

By considering that if $\tau_0 \to +\infty$, the evolution of the motion is modulated around the SIM, so that one has $\frac{\partial \phi_2}{\partial \tau_0} = 0$, and in that case again, $\phi_2 = \phi_2(\tau_1, \tau_2, \dots)$. One can obtain the following expressions to define the SIM:

$$N_1 = N_2\sqrt{\frac{\lambda^2}{\omega_1^2} + (1 - \frac{3\gamma}{4\omega_1^4}N_2^2)^2} = H_1(N_2)$$

$$\delta_1 = \delta_2 - \arctan\frac{N_2}{N_1}(\frac{\lambda}{\omega_1(1 - \frac{3\gamma}{4\omega_1^4}N_2^2)}) = H_2(\delta_2, N_2) = \delta_2 + H_3(N_2). \quad (42)$$

An example of the SIM that is obtained by Eq. (42) is illustrated in Fig. 1.

At the ε^1 order, we keep the following equation governing the main complex coordinate:

$$\frac{\partial \phi_1}{\partial \tau_1} = \frac{1}{2i}[(\omega_1 + 2\sigma)\phi_1 - \omega_1\phi_2] - \alpha_0 F - H_0 + \frac{f_0}{2i} = RHS. \quad (43)$$

4.2 *Equilibrium Points of the Reduced Model*

From the evolution equation of ϕ_1 versus τ_1, equilibrium points are obtained by demanding that, at ε^1 order, $\frac{\partial \phi_1}{\partial \tau_1} = 0$. We obtain

$$RHS = \frac{1}{2i}[(\omega_1 + 2\sigma)\phi_1 - \omega_1\phi_2] - \alpha_0 F - H_0 + \frac{f_0}{2i} = 0, \quad (44)$$

so that two real relations can be obtained:

$$\frac{\omega_1}{2}N_2\sin(\delta_1-\delta_2)+\frac{2\alpha_0\eta}{\pi}\sin(2\delta_1)+\frac{f_0}{2}\sin(\delta_1)=0,$$
$$\frac{1}{2}(\omega_1+2\sigma)N_1-\frac{\omega_1}{2}N_2\cos(\delta_1-\delta_2)+\frac{2\alpha_0\eta}{\pi}\cos(2\delta_1)-\frac{\beta}{8\omega_1^3}N_1^3+\frac{f_0}{2}\sin(\delta_1)=0, \quad (45)$$

where η reads as

$$\eta=\mathbf{sgn}(N_1\cos(\delta_1)+\frac{\varepsilon}{1+\varepsilon}N_2\cos(\delta_2)). \quad (46)$$

sgn(...) stands for the sign function. By adding relations of the SIM (see Eq. (42)), we have real relations leading to the definition of the equilibrium points:

$$\begin{aligned} N_1 &= H_1(N_2),\\ \delta_1 &= H_2(N_2,\delta_2)=\delta_2+H_3(N_2),\\ \cos(\delta_1-\delta_2) &= \frac{N_2}{N_1}(1-\frac{3\gamma}{4\omega_1^4}N_2^2)\\ \sin(\delta_1-\delta_2) &= -\frac{N_2}{N_1}\frac{\lambda}{\omega_1}\\ N_1(\frac{4\alpha_0\eta}{\pi}\sin(2\delta_1)+f_0\sin(\delta_1)) &= \lambda N_2^2,\\ N_1(\frac{4\alpha_0\eta}{\pi}\cos(2\delta_1)+f_0\cos(\delta_1)) &= \omega_1 N_2^2(1-\frac{3\gamma}{4\omega_1^4}N_2^2)-(\omega_1+2\sigma)N_1^2 \end{aligned} \quad (47)$$

It is possible to express N_1, δ_1 and δ_2 versus N_2 so that one equation depending only on N_2 can be calculated. First, combining previous relations, we have

$$\sin(3\delta_1)=\pi\frac{\lambda^2N_2^4+(\omega_1N_2^2(1-\frac{3\gamma}{4\omega_1^4}N_2^2)-(\omega_1+2\sigma)H_1^2)^2-H_1^2((\frac{4\alpha_0}{\pi})^2+f_0^2)}{8\alpha_0\eta f_0H_1^2}, \quad (48)$$

so that we can express $\delta_1=H_4(N_2)$. Then, we obtain an equation depending only on N_2 with $\eta=\pm1$:

$$-\lambda N_2^2+\frac{4\alpha_0\eta}{\pi}H_1\sin(2H_4)+f_0H_1\cos(H_4)=0. \quad (49)$$

One has simply to check that a solution N_2 (giving corresponding δ_1, N_1, δ_2) obtained for a given value of η is compatible with the calculation of η (see Eq. (65) in Appendix 1).

4.3 Singular Points of the Reduced Model

Equation (43) can be rewritten in the real form

$$\begin{cases} \frac{\partial N_1}{\partial\tau_1}=\mathrm{Re}(RHS)\\ N_1\frac{\partial\delta_1}{\partial\tau_1}=\mathrm{Im}(RHS). \end{cases} \quad (50)$$

Since N_1 and δ_1 are functions of N_2 and δ_2, Eq. (43) can be written in the form

$$\begin{pmatrix} \frac{\partial H_1}{\partial N_2} & 0 \\ \frac{\partial H_2}{\partial N_2} & \frac{\partial H_2}{\partial \delta_2} \end{pmatrix} \begin{pmatrix} \frac{\partial N_2}{\partial \tau_1} \\ \frac{\partial \delta_2}{\partial \tau_1} \end{pmatrix} = \begin{pmatrix} \mathrm{Re}(RHS) \\ \mathrm{Im}(RHS) \end{pmatrix}, \tag{51}$$

which yields to

$$\begin{pmatrix} \frac{\partial H_1}{\partial N_2} & 0 \\ \frac{\partial H_2}{\partial N_2} & 1 \end{pmatrix} \begin{pmatrix} \frac{\partial N_2}{\partial \tau_1} \\ \frac{\partial \delta_2}{\partial \tau_1} \end{pmatrix} = \begin{pmatrix} \mathrm{Re}(RHS) \\ \mathrm{Im}(RHS) \end{pmatrix}. \tag{52}$$

Singular points of the system demand that

$$\mathbf{det} \begin{pmatrix} \frac{\partial H_1}{\partial N_2} & 0 \\ \frac{\partial H_2}{\partial N_2} & 1 \end{pmatrix} = 0 \tag{53}$$

or

$$\frac{\partial H_1}{\partial N_2} = 0. \tag{54}$$

It can be seen that Eq. (54) provides positions of local maxima of the SIM, which is defined in Eq. (42). As a result, singular points of the system under study are confined on local maxima of the SIM.

4.4 Numerical Scheme for the DAI

In order to compare the behaviors forecasted by the (rather rough) analytical approach, we need to process numerical simulations. Based on previous works (see, for example, [8]), we use an implicit Euler numerical scheme, which is correct mathematically and can guarantee the convergence of the approximated solution towards the exact solution (but with -quite small- order 1). Let us write the semi-explicit DAI (32) as a first-order differential inclusion. Let us set

$$X = \begin{pmatrix} \dot{x} \\ x \\ \dot{y} \\ y \\ z \end{pmatrix} = \begin{pmatrix} x_1 \\ x_2 \\ x_3 \\ x_4 \\ x_5 \end{pmatrix}. \tag{55}$$

Then, since we have $\frac{\partial g}{\partial x_1} = 0$ and $\frac{\partial g}{\partial x_5} = -1$, let us write

$$\dot{x}_5 = \frac{\partial g}{\partial x_2} \dot{x}_2 = x_2^2 x_1. \tag{56}$$

Finally, the equations can be expressed via

$$\dot{X} + \mathscr{L}X + G(X,t) + A(X) \ni \mathbf{0}, \tag{57}$$

where $\mathbf{0} = (0,0,0,0,0)^t$,

$$\mathscr{L} = \begin{pmatrix} \varepsilon(a_0+\lambda) & \omega_1^2 & -\varepsilon\lambda & 0 & 0 \\ -1 & 0 & 0 & 0 & 0 \\ -\lambda & 0 & \lambda & 0 & 0 \\ 0 & 0 & -1 & 0 & 0 \\ 0 & 0 & 0 & 0 & 0 \end{pmatrix} = \begin{pmatrix} L_1 \\ L_2 \\ L_3 \\ L_4 \\ L_5 \end{pmatrix} \tag{58}$$

and

$$G(X,t) = \begin{pmatrix} \varepsilon[h_0(x_5) + \gamma(x_2-x_4)^3 - f_0(t)] \\ 0 \\ \gamma(x_4-x_2)^3 \\ 0 \\ -\frac{\partial g}{\partial x_2}x_1 \end{pmatrix}, \quad \frac{\partial g}{\partial x_2} = x_2^2, \tag{59}$$

with

$$A(X) = \begin{pmatrix} \varepsilon\alpha_0\rho(x_1) \\ 0 \\ 0 \\ 0 \\ 0 \end{pmatrix}. \tag{60}$$

The DAI problem of Eq. (57) with given initial conditions in the phase space of X possesses a unique solution (see Ref. [8]). Discretization of the problem can easily be done. Let us choose a time step $\Delta t > 0$. For any integer n, let us set $t_n = n\Delta t$, $X(t_n) \simeq X_n$, $X(t_{n+1}) \simeq X_{n+1}$. An Euler implicit scheme can be built from

$$\begin{cases} \frac{1}{\Delta t}(X_{n+1} - X_n) + \mathscr{L}X_n + G(X_n, t_n) + A(X_{n+1}) = \mathbf{0}, \\ X_0 \text{ given.} \end{cases} \tag{61}$$

The Euler implicit scheme algorithm is given in Appendix 2.

4.5 An Example: SMR of a System

Here, the system (32) is integrated numerically via an implicit Euler scheme with the time step as $\Delta t = 10^{-4}$. System parameters are collected in Table 1. The following initial conditions are assumed for the system:

$$\big(\dot{x}(0), x(0), \dot{y}(0), y(0), z(0)\big) = \big(0, 2, 0, 2, \frac{8}{3}\big). \tag{62}$$

Table 1 System parameters relevant to Eq.(32)

ε	ω_1	a_0	α_0	β^*	σ	λ	γ
10^{-3}	1	0.1	0.1	0.1	0.1	0.1	0.1

P* $h_0(z) = \varepsilon\beta z$

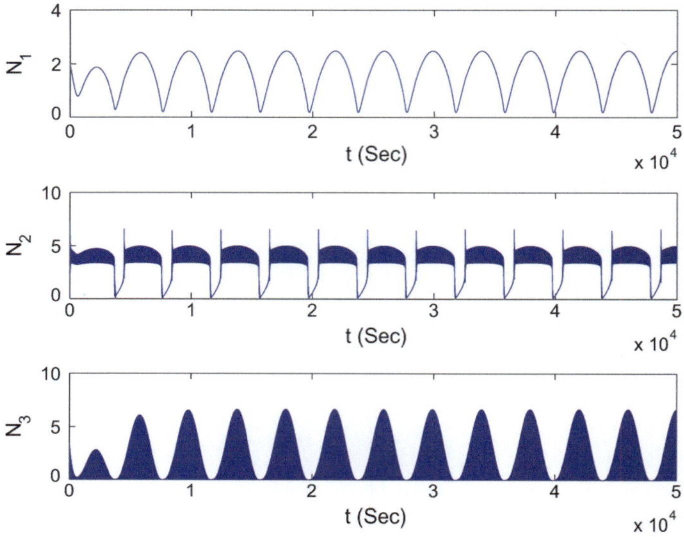

Fig. 2 Time histories of N_2 and N_1. Results are obtained by numerical integration of the system (32) via an implicit Euler scheme with the time step as $\Delta t = 10^{-4}$

The SIM of the mentioned system is depicted in Fig. 1. Local maxima of the SIM, which are also positions of singular points of the system, can be obtained via Eq. (54). They read as

$$N_2 = 2.124 \ , \ 3.642. \tag{63}$$

Considering that all parameters of the system are fixed, the necessary forcing amplitude for leading the system to a SMR can be obtained via Eq. (49). One should replace N_2 with values that are reported in Eq. (63). They correspond to

$$f_0 = 0.242 \ , \ 4.019. \tag{64}$$

Here, we take $f_0 = 4.019$. Time histories of system amplitudes in terms of N_j, $j = 1, 2, 3$ and system variables x, y and z are presented in Figs. 2 and 3, respectively. These figures show that the system experiences persisting bifurcations due to the existence of a singular point on $N_2 = 3.642$. The SIM of the system obtained via Eq. (42) and corresponding numerical values for N_2 and N_1 are collected in Fig. 4. It is seen that the system oscillates around the SIM and it bifurcates persistently as soon as it reaches its local maxima, which are positions of singular points. It should

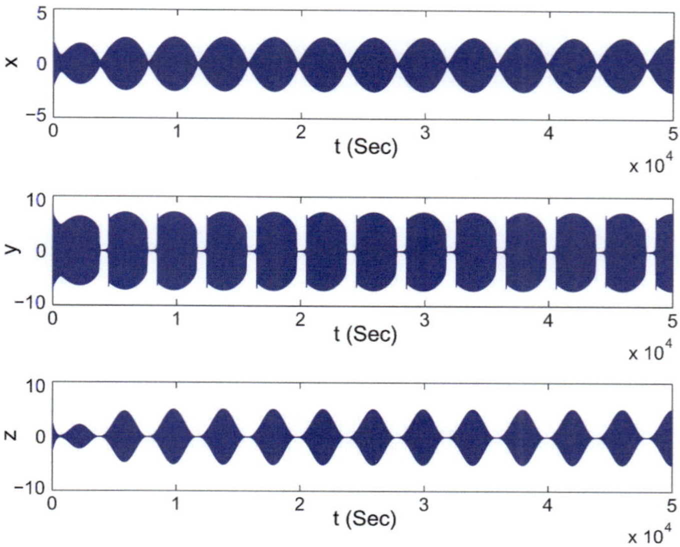

Fig. 3 Time histories of x, y and z. Results are obtained by numerical integration of the system (32) via an implicit Euler scheme with the time step as $\Delta t = 10^{-4}$

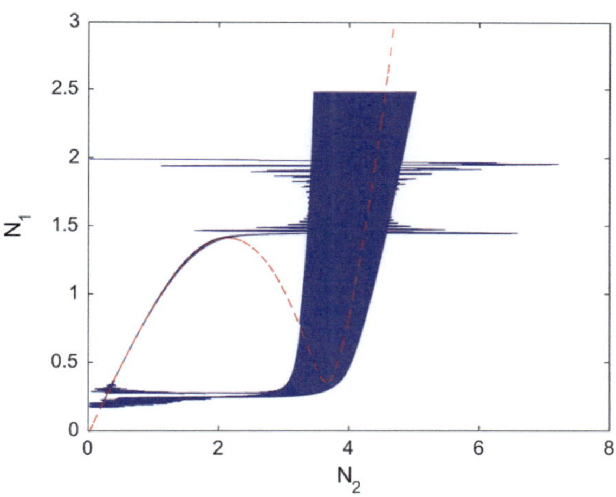

Fig. 4 The SIM of the system obtained through Eq. (42) (red dashed line) and corresponding numerical results (blue solid line) collected by numerical integration of the system (32) via an implicit Euler scheme with the time step as $\Delta t = 10^{-4}$

be mentioned that the SIM that is defined in Eq. (42) is obtained by keeping only the first harmonics of the system, while the numerical results contain all harmonics of the system. That is why the numerical results presented in Fig. 4 oscillate around the SIM. The effects of higher harmonics can be seen in Fig. 5 where the analytical relations between N_1 and N_3 (see the third equation of the system 40) are compared with those obtained from the numerical scheme. Finally, to get an idea about the tolerance of the error imposed by the implicit Euler scheme, the time histories of

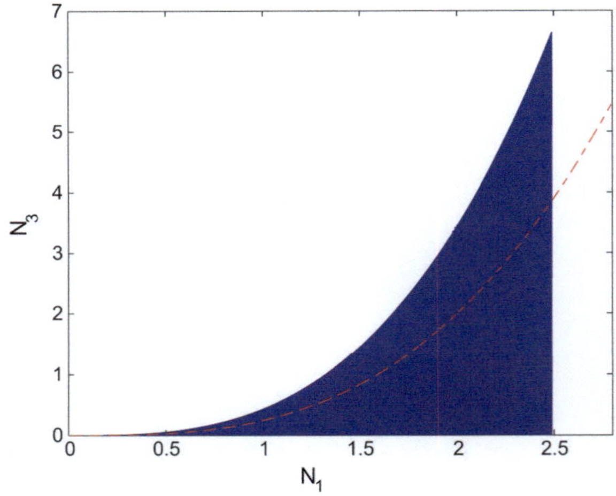

Fig. 5 Relation between N_3 and N_1: The analytical curve (red dashed line) obtained from the third equation of the system (40) and the numerical one (solid blue line) collected by numerical integration of the system (32) via an implicit Euler scheme with the time step as $\Delta t = 10^{-4}$

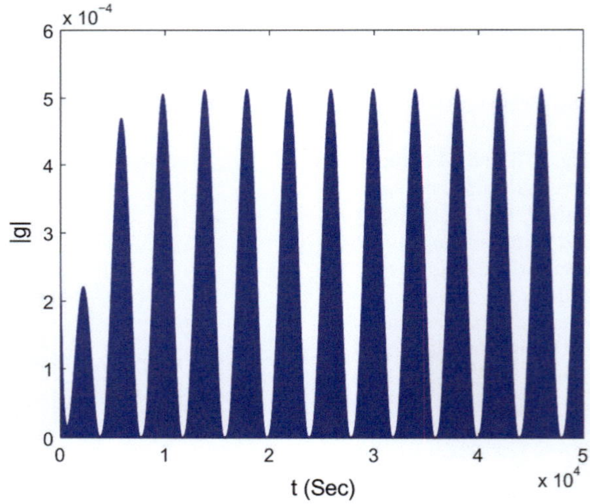

Fig. 6 The absolute value of the $g(\dot{x}, x, z)$ function (see Eq. (28)), obtained from an implicit Euler scheme

$g(\dot{x}, x, z)$ function (see Eq. (28)) are depicted in Fig. 6. This figure shows that the error stays at the $o(\Delta t)$ with $\Delta t = 10^{-4}$, which is coherent with the error tolerance of the Euler scheme.

5 Conclusion

This work can be extended according to the following perspectives: Manevitch complexification can be generalized so as to involve more than one harmonic and the two timescales approach can be adapted so as to define the SIM and study a reduced order model of the dynamics at scale τ_1. From the example, one can see that analytical developments are quite convenient for the design of a NES in engineering applications. To have a straightforward explicit relation will be quite impossible if one increases the dimension of the master system, the number of NES and the implicit character of the algebraic relations. But algebraic relations and the reduced order model can be studied numerically so as to proceed with the design of the NES. Such a generalization of the approach for the nonsmooth case is also an interesting perspective for the parametric investigation of the NES parameters and for applications in engineering. Generalization to systems with more than one NES, or to systems with mass, stiffness and damping matrices and smooth geometrical nonlinearities with algebraic constraints and nonsmooth terms, is possible if existence and uniqueness questions can be studied for the obtained model. For master systems, NES and algebraic relations without any nonsmooth terms, the research of singular points can be linked to the occurrence of Neimark-Sacker bifurcations.

Acknowledgements The authors want to thank LABEX CELYA (ANR-10-LABX-0060) of the "Université de Lyon," within the Investissement d'Avenir program (ANR-11-IDEX-0007), operated by the French National Research Agency(ANR).

APPENDIX 1 - Expressions of F and H_0

The function F is defined as

$$\int_0^{\frac{2\pi}{\omega}} \rho(\dot{v} + \frac{\varepsilon \dot{w}}{1+\varepsilon}) \exp(-i\omega\tau_0) d\tau_0.$$

We have

$$\dot{v} + \frac{\varepsilon \dot{w}}{1+\varepsilon} = \frac{1}{2}(\phi_1 + \frac{\varepsilon}{1+\varepsilon}\phi_2) exp(i\omega\tau_0) + c.c.,$$

and we assume that ϕ_1, ϕ_2 do not depend on τ_0. Using the polar form $\phi_j = N_j \exp(i\delta_j)$, $j = 1, 2$, we also have

$$\dot{v} + \frac{\varepsilon \dot{w}}{1+\varepsilon} = N_1 \cos(\omega\tau_0 + \delta_1) + \frac{\varepsilon}{1+\varepsilon} N_2 \cos(\omega\tau_0 + \delta_2).$$

So, we obtain

$$F \frac{2i\eta}{\pi} \exp(-i\omega t_1^\star),$$

where

$$N_1 \cos(\omega t_1^\star + \delta_1) + \frac{\varepsilon}{1+\varepsilon} N_2 \cos(\omega t_1^\star + \delta_2) = 0$$

and

$$t_1^\star \in [0, \frac{\pi}{\omega}],$$

$$\eta = \text{sign}(N_1 \cos(\delta_1) + \frac{\varepsilon}{1+\varepsilon} N_2 \cos(\delta_2)). \quad (65)$$

Finally, we derive

$$\tan(\omega t_1^\star) = \frac{N_1 \cos(\delta_1) + \frac{\varepsilon}{1+\varepsilon} N_2 \cos(\delta_2)}{N_1 \sin(\delta_1) + \frac{\varepsilon}{1+\varepsilon} N_2 \sin(\delta_2)},$$

and keeping the main orders of ε, we have

$$F(N_1, N_2, \delta_1, \delta_2) = \tfrac{2i}{\pi} \text{sign}(\cos(\delta_1)) \exp(-i\delta_1)[1 - \varepsilon i \tfrac{N_2}{N_1} \sin(\delta_1 - \delta_2) + o(\varepsilon^2)],$$
$$H_0(\phi_3, \phi_3^\star) = \tfrac{\beta \phi_3}{2i\omega_1} = \tfrac{\beta}{8i\omega_1^3} \mid \phi_1 \mid^2 \phi_1. \quad (66)$$

APPENDIX 2 - Euler Implicit Numerical Scheme

Let us note that $X_n = \begin{pmatrix} x_{1n} \\ x_{2n} \\ x_{3n} \\ x_{4n} \\ x_{5n} \end{pmatrix}$, $X_{n+1} = \begin{pmatrix} x_{1n+1} \\ x_{2n+1} \\ x_{3n+1} \\ x_{4n+1} \\ x_{5n} \end{pmatrix}$. Let X_0 be given. For $n \geq 0$, we have

$$aux_n = x_{1n} - \Delta t L_1 X_n - \Delta t \varepsilon (h_0(x_{5n}) + \gamma (x_{2n} - x_{4n})^3 - f_0(t_n)),$$
$$L_1 X_n = \varepsilon (a_0 + \lambda) x_{1n} + \omega_1^2 x_{2n} - \varepsilon \lambda x_{3n},$$
$$x_{1n+1} = \begin{cases} 0 \text{ if } \mid aux_n \mid \leq \varepsilon \alpha_0 \Delta t \\ aux_n - \varepsilon \alpha_0 \Delta t \text{ if } aux_n \geq \varepsilon \alpha_0 \Delta t \\ aux_n + \varepsilon \alpha_0 \Delta t \text{ if } aux_n \leq -\varepsilon \alpha_0 \Delta t \end{cases},$$
$$x_{2n+1} = x_{2n} + \Delta t x_{1n},$$
$$x_{3n+1} = x_{3n} - \Delta t (L_2 X_n + \gamma (x_{4n} - x_{2n})^3) = x_{3n} - \Delta t (x_{1n} + \gamma (x_{4n} - x_{2n})^3),$$
$$x_{4n+1} = x_{4n} + \Delta t x_{3n},$$
$$x_{5n+1} = x_{5n} + \Delta t x_{2n}^2 x_{1n}.$$

References

1. Moreau JJ (1988) Unilateral contact and dry friction in finite freedom dynamics. In: Nonsmooth mechanics and applications, CISM courses and lectures, vol 302. Springer, Wien
2. Brogliato B (1996) Nonsmooth impact mechanics: models, dynamics and control. Springer, London

3. Jean M (1999) The non-smooth contact dynamics method. Comput Methods Appl Mech Eng 177:235–257
4. Glocker C (2001) Set-valued force laws, dynamics of non-smooth systems. Springer, Berlin
5. Pfeiffer F, Foerg M, Ulbrich H (2006) Numerical aspects of non-smooth multibody dynamics. Comput Methods Appl Mech Eng. 195:6891–6908
6. Acary V, Brogliato B (2008) Numerical methods for nonsmooth dynamical dystems: applications in mechanics and electronics. Springer, London
7. Leine RI, van de Wouw N (2008) Stability and convergence of mechanical systems with unilateral constraints. Springer, Berlin
8. Bastien J, Bernardin F, Lamarque C-H (2013) Non smooth deterministic or stochastic discrete dynamical systems: applications to models with friction or impact. Wiley-ISTE, Surrey
9. Roberson RE (1952) Synthesis of a nonlinear dynamic vibration absorber. J Franklin Inst 254:205–220
10. Frahm H, (1911) Device for damping vibrations of bodies. US 989958
11. Sevin E (1961) On the parametric excitation of pendulum-type vibration absorber. J Appl Mech 28:330–334
12. Haxton RS, Barr ADS (1972) The autoparametric vibration absorber. J Eng Ind 94:119–125
13. Rice HJ, McCarith JR (1987) Practical non-linear vibration absorber design. J Sound Vib 116:545–559
14. Ema S, Marui E (1996) Damping characteristics of an impact damper and its applications. Int J Mach Tools Manuf 36:293–306
15. Vakakis AF, Gendelman OV (2000) Energy pumping in nonlinear mechanical oscillators: part II-resonance capture. J Appl Mech 68:42–48
16. Vakakis AF, Gendelman OV, Bergman LA, McFarland DM, Kerschen G, Lee YS (2009) Nonlinear targeted energy transfer in mechanical and structural systems, vol I & II. Springer, Netherlands
17. Lamarque C-H, Gendelman OV, Ture Savadkoohi A, Etcheverria E (2011) Targeted energy transfer in mechanical systems by means of non-smooth nonlinear energy sink. Acta Mech 221:175–200
18. Habib G, Kerschen G (2015) Suppression of limit cycle oscillations using the nonlinear tuned vibration absorber. Proc R Soc A-Math Phy 471:0140976
19. Ture Savadkoohi A, Lamarque C-H (2013) Dynamics of coupled Dahl type and non-smooth systems at different scales of time. Int J Bifurc Chaos 23:1350114
20. Lamarque C-H, Ture Savadkoohi A (2014) Dynamical behavior of a Bouc-Wen type oscillator coupled to a nonlinear energy sink. Meccanica 49:1917–1928
21. Schmidt F, Lamarque C-H (2010) Energy pumping for mechanical systems involving nonsmooth Saint-Venant terms. Int J Non Linear Mech 45:866–875
22. Weiss M, Ture Savadkoohi A, Gendelman OV, Lamarque C-H (2014) Dynamical behavior of a mechanical system including Saint-Venant component coupled to a nonlinear energy sink. Int J Non Linear Mech 63:10–18
23. Lamarque C-H (2015) Ture savadkoohi: targeted energy transfer between a system with a set of Saint-Venant elements and a nonlinear energy sink. Contin Mech Thermodyn 27:819–833
24. Hairer E, Wanner G (1996) Springer series in computational mathematics. In: Solving ordinary differential equations II, stiff and differential-algebraic problems. Springer, Berlin (GMbH)
25. Manevitch LI (2001) The description of localized normal modes in a chain of nonlinear coupled oscillators using complex variables. Nonlinear Dyn 25:95–109
26. Ture Savadkoohi A, Lamarque C-H, Dimitrijevic Z (2012) Vibratory energy exchange between a linear and a nonsmooth system in the presence of the gravity. Nonlinear Dyn 70:1473–1483
27. Weiss M, Chenia M, Ture Savadkoohi A, Lamarque C-H, Vaurigaud B, Hammouda A (2016) Multi-scale energy exchanges between an elasto-plastic oscillator and a light nonsmooth system with external pre-stress. Nonlinear Dyn 83:109–135
28. Ture Savadkoohi A, Lamarque C-H, Weiss M, Vaurigaud B, Charlemagne S (2016) Analysis of the 1:1 resonant energy exchanges between coupled oscillators with rheologies. Nonlinear Dyn 86:2145–2159

Experimental Validation of Torsional Controllers for Drilling Systems

N. van de Wouw, T. Vromen, M. J. M. van Helmond, P. Astrid, A. Doris and H. Nijmeijer

Abstract Torsional stick-slip vibrations decrease the performance, reliability and fail-safety of drilling systems used for the exploration and harvesting of oil, gas, minerals and geo-thermal energy. Current industrial controllers regularly fail to eliminate stick-slip vibrations, especially when multiple torsional flexibility modes in the drill-string dynamics play a role in the onset of stick-slip vibrations. This chapter presents the experimental validation of novel robust output-feedback controllers designed to eliminate stick-slip vibrations in the presence of multiple dominant torsional

---

N. van de Wouw (✉) · H. Nijmeijer
The Department of Mechanical Engineering, Eindhoven University of Technology, Eindhoven, The Netherlands
e-mail: n.v.d.wouw@tue.nl

H. Nijmeijer
e-mail: h.nijmeijer@tue.nl

N. van de Wouw
Delft Center for Systems and Control, Delft University of Technology, Delft, The Netherlands

N. van de Wouw
The Department of Civil, Environmental & Geo-Engineering, University of Minnesota, Minneapolis, USA

T. Vromen
Océ Technologies B.V., Venlo, The Netherlands
e-mail: Thijs.vromen@oce.com

M. J. M. van Helmond
YER, Amsterdam, The Netherlands
e-mail: m.j.m.v.helmond@alumnus.tue.nl

P. Astrid
Shell Global Solutions International B.V., Rijswijk, The Netherlands
e-mail: patricia.astrid@shell.com

A. Doris
GeoSea nv, Geotechnical and Offshore Solutions, Zwijndrecht, Belgium
e-mail: doris.apostolos@deme-group.com

R. I. Leine et al. (eds.), *Advanced Topics in Nonsmooth Dynamics*,
https://doi.org/10.1007/978-3-319-75972-2_8

flexibility modes. For this purpose, a representative experimental test setup is designed, using a model of a real-life drilling rig as a basis. The model of the dynamics of the experimental setup can be cast in Lur'e-type form with set-valued nonlinearities representing an (uncertain) model for the complex bit-rock interaction and the interaction between the drill-string and the borehole. The proposed controller design strategy is based on skewed-μ-DK-iteration and aims at optimizing the robustness with respect to uncertainty in the non-smooth bit-rock interaction. Moreover, a closed-loop stability analysis for the non-smooth drill-string model is provided. Experimental results confirm that stick-slip vibrations are indeed eliminated using the designed controller in realistic drilling scenarios in which state-of-practice controllers have failed to achieve the same.

1 Introduction

Efficiency, reliability and safety are important aspects in the drilling of deep wells for the exploration and production of oil, gas, mineral resources and geo-thermal energy. Drill-strings several kilometers in length are used to transmit the axial force and torque necessary to drill the rock formations. These drill-string systems are known to exhibit different types of self-excited vibration, which decrease the drilling efficiency, accelerate bit wear, may cause sudden failure of expensive Measure-While-Drilling (MWD) tools, and may cause drill-string failure due to fatigue. This chapter focuses on the controlled mitigation of *torsional* stick-slip vibrations.

Modelling of the torsional dynamics of the drill-string is an important step towards the control of torsional vibrations. Most controller designs presented in literature rely on one- or two degree-of-freedom (DOF) models for the torsional dynamics only, see e.g., [4, 14, 31, 35]. The resisting torque-on-bit (TOB) is typically modelled as a frictional contact with a velocity weakening effect. Although experiments using single cutters to identify the bit-rock interaction law, see [5], do not reveal such a velocity weakening effect, analysis of models that take the coupled axial and torsional dynamics into account shows that such coupling effectively leads to a velocity weakening effect in the TOB [30]. This motivates a modelling-for-control approach that only involves the torsional dynamics and a set-valued, velocity weakening bit-rock interaction law. In contrast to other studies, however, we use a multi-modal model of the torsional dynamics, as field observations have revealed that multiple torsional resonance modes play a role in the onset of stick-slip oscillations.

Controllers for drilling systems aim to achieve drill-string rotation at a constant velocity and the mitigation of stick-slip vibrations. Moreover, the following control specifications are important. First, only surface measurements can be used for feedback. Second, the controller should be able to cope with dynamics related to multiple torsional flexibility modes. Third, robustness with respect to uncertainty in the non-smooth bit-rock interaction has to be guaranteed and, fourth control performance specifications, related to, e.g., measurement noise sensitivity and actuator constraints, need to be taken into account in the controller design.

A well-known control method, which aims at damping the first torsional mode, is the *Soft Torque Rotary system*, see [12]. The same objective is set in [14], which uses a PI-controller based on the top drive velocity. Other control methods have been developed, including torsional rectification [35], observer-based output-feedback [4, 6, 39], impedance matching [8], adaptive output-feedback for infinite dimensional drill-string models [1], weight-on-bit control [2] and robust control [15, 31]. Although important steps forward have been take in these works, an approach that satisfies all mentioned requirements has not yet been developed. A robust control approach, as proposed in [15, 31], is particularly suitable for this problem, since both robustness with respect to uncertainty of the system dynamics and control performance specifications can be taken into account in the control design. In [31], an $\mathscr{H}_\infty$ controller synthesis method is applied to a 2-DOF drill-string model and the twist in the drill-string is used as measurement, i.e., knowledge of the angular position of the bit is assumed. [15] uses the μ-synthesis technique through the DK-iteration procedure for the purpose of obtaining less conservative bounds on the uncertainty to obtain robustness with respect to the nonlinear bit-rock interaction. The model used is a similar 2-DOF model, and down-hole measurements (for assessing the twist of the drill-string) are also used in this case. Moreover, the employed 2-DOF models only take the first flexibility mode into account. In this chapter, we present experimental results of a robust control approach for the control of torsional drill-string vibrations, of which preliminary model-based results have been presented in [38] and which can cope with *multiple* torsional resonance modes.

Because of the high costs involved in testing on a real drilling rig, experimental lab-scale setups representing the drilling dynamics used for multiple purposes can be found in the literature, some examples of which are mentioned here (see [25] for a more comprehensive overview). In [22], an experimental 2-DOF drill-string system is used for the analysis of friction-induced stick-slip limit cycles. The same setup is used in [4] for experimental validation of an observer-based output-feedback controller. In [17], an experimental setup is developed that can emulate various excitation mechanisms of the drill-string, including stick-slip, well-borehole contact, and drilling fluid interaction. The aforementioned test setups both use brake systems to implement bit-rock interaction laws. A different approach is taken in [18], in which an experimental setup for exploring stick-slip phenomena is used that involves real cutting using a bit. In [36], an experimental setup is used to investigate whirling effects in drilling systems, involving both torsional and lateral dynamics. Another example of the experimental validation of a controller design approach to torsional vibrations in drilling systems can be found in [20]. Also for the testing of down-hole tools, experimental setups are used as a stepping stone towards implementation of the technique. For example, experimental results of Resonance Enhanced Drilling (RED) technology are presented in [40], and in [29], an experimental setup for investigating the Anti Stick-slip Tool (AST) is shown.

The need for a new experimental setup design stems from the fact that the controllers proposed in this thesis focus on the robustness with respect to multiple dominant torsional flexibility modes in the drill-string dynamics. To investigate this robustness, it is important that the experimental setup represents such a drilling system with multiple dominant flexibility modes (in contrast to, e.g., [22, 36], in which setups with a single flexibility mode are considered).

The main contributions of this chapter are, firstly, the model-based design of a representative (lab-scale) experimental drill-string setup, secondly, the design of a robust output-feedback controller methodology for eliminating stick-slip vibrations and, thirdly, experimental results showing the merit of the proposed control approach.

In Sect. 2, the design of the experimental setup is motivated and detailed. This design is based on a non-smooth model of a real drilling rig. Section 3 deals with the controller design strategy aiming to eliminate the torsional vibrations. In Sect. 4, the proposed control strategy is validated experimentally. The chapter closes with concluding remarks in Sect. 5.

2 Design of the Experimental Drill-String Setup

2.1 Model-Based Design of the Experimental Setup

Consider a drilling system, as schematically shown in Fig. 1. The investigated system is a realistic drill-string model of an offshore jack-up drilling rig, and the reservoir sections of the wells are drilled with a 6" PDC bit to reach depths of more than 6000 m along-hole and with an inclination angle up to 60°, resulting in significant resistive torques along the drill-string due to frictional borehole drill-string interaction. The rig is equipped with an AC top drive and fitted with a modern SoftTorque system [14, 19]. However, for this depth and hole size, stick-slip vibrations have been observed in

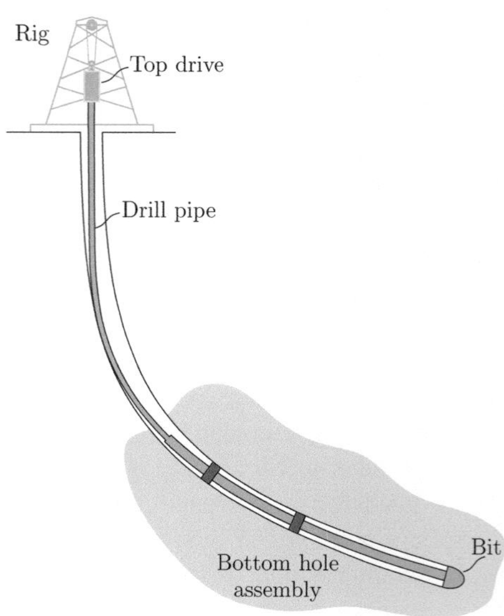

Fig. 1 Schematic drilling system, not to scale (adapted from [27])

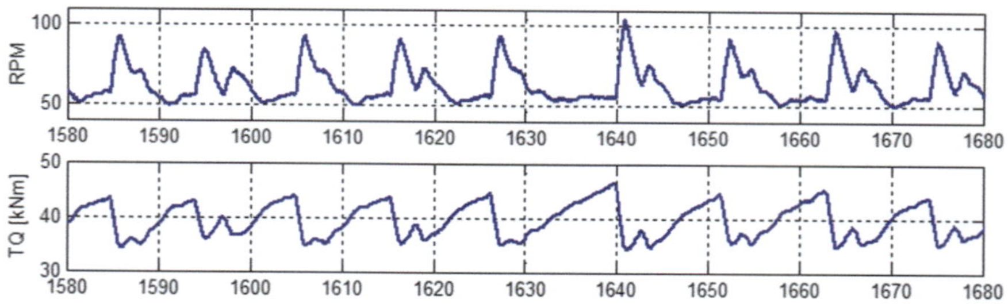

Fig. 2 Field data of the drilling rig under investigation, indicating severe stick-slip oscillations, see [11] (desired angular velocity is approximately 50 rpm). Top plot: top drive angular velocity in RPM; bottom plot: top drive torque

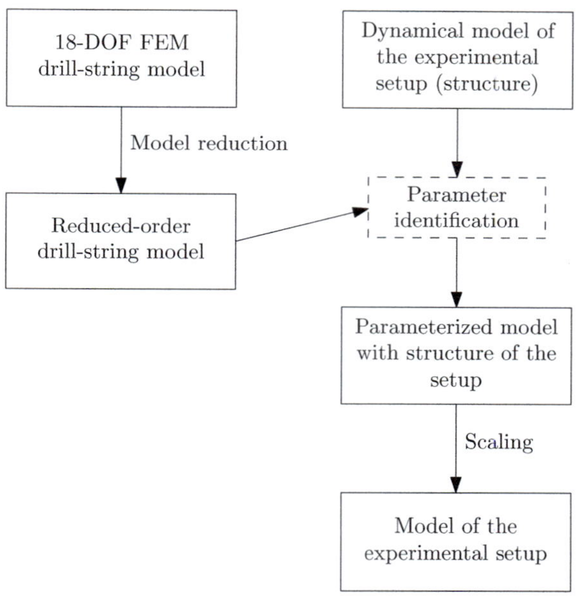

Fig. 3 Step-wise development of a model to be used as a basis for the design of the experimental setup

the field for this rig (see [11]), as shown in Fig. 2. In this figure, measurement data of the real rig is shown. The top drive angular velocity (RPM) and top drive torque (TQ) show severe oscillations, indicating stick-slip oscillations at the bit. The fact that a control strategy, which only damps the first flexibility mode of the torsional drill-string dynamics, fails to eliminate stick-slip vibrations shows that multiple resonance modes play a role and motivates construction of multi-modal drill-string models and development of a controller based on these models.

A finite-element method (FEM) model of this real-life drilling system is used as a basis for the design of the experimental drill-string setup. A detailed description of the FEM model is given in Sect. 2.1.1. Here, the focus is on the steps that are taken to develop a model of the experimental setup based on this 18-DOF FEM model. These steps are summarized in Fig. 3 and are discussed in more detail in the following sections. In Sect. 2.1.2, the model reduction strategy that is used for obtaining a reduced-order drill-string model is discussed. Next, the model of the experimental setup is explained in more detail in Sect. 2.1.3 and the identification approach for obtaining the parameters for this model based on the reduced-order model is given in Sect. 2.1.4. Since it is impossible to scale down an oil-field drill-string to a lab-scale setup that still exhibits the main (torsional) dynamics we aim to study, we propose a model with four rotating discs, coupled with (steel) strings, as shown in Fig. 4. It is important to mention that the proposed model of the experimental setup has a specific structure, due to the mechanical elements (i.e., inertias and springs) that are used in the setup resulting in a lumped-parameter model, while on the other hand, the reduced-order drill-string model does not have such a specific structure. The identified parameters of the obtained model are still of the same order of magnitude as the original drill-string model (e.g., inertia and stiffness properties of the system as a whole are still of the same order of magnitude, and are hence not (yet) scaled). As a consequence, the representative torsional velocity and torque levels of the setup match those of a real drill-string system. Therefore, scaling is used to obtain suitable torque levels and velocities for a lab-scale drill-string setup, but also to obtain feasible inertias and stiffnesses for the lab-scale experimental system design. This scaling procedure is discussed in Sect. 2.1.5.

2.1.1 Finite-Element Model

A finite-element model of this drilling system, which represents a drill-string 6249 m in length, has been developed, and the simulation results of this model have been validated with field data for a range of operational conditions (such as weight-on-bit (WOB) and angular velocity). The 18-DOF finite-element model is obtained by representing the drill-string with a number of equivalent pipe sections in order to accurately describe the torsional dynamics relevant to stick-slip vibrations. The model is validated by comparing the simulations of the non-smooth model (i.e., including bit-rock and borehole-drillstring interaction torques) with field measurements of the drill-string system. Figures 5 and 6 show two cases of this validation study, i.e., the simulation results of the finite-element model are compared with the field data under two different operating conditions; in both cases, the drill-string system exhibited stick-slip vibrations at the bit. As can be seen from these figures, the simulation results match the field data, both in terms of the amplitude and the frequency of the oscillations. The latter observations further motivate the usage of the developed model as a basis for controller design in this thesis.

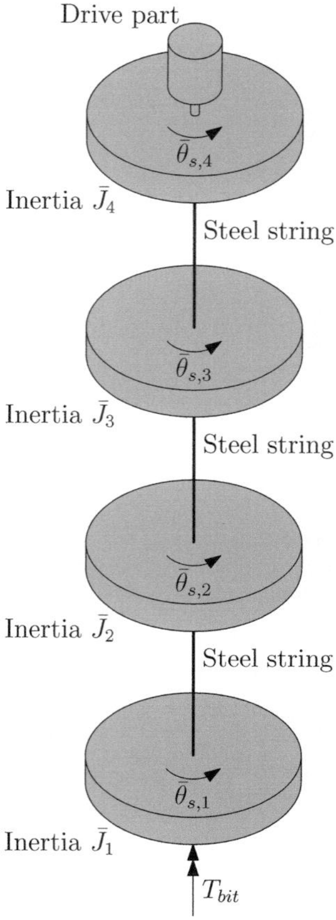

Fig. 4 Schematic representation of a model with four discs

The finite-element method (FEM) representation of the drill-string is a model with 18 elements. The element at the top is a rotational inertia to model the top drive inertia, and the subsequent elements are equivalent pipe sections based on the dimensions and material properties of the drill-string (see [37] for more details). The resulting model can be written as a second-order differential equation of the following form:

$$M\ddot{\theta} + D\dot{\theta} + K_t\theta_d = S_w T_w(\dot{\theta}) + S_b T_{bit}(\dot{\theta}_1) + S_t T_{td} \tag{1}$$

with the rotational displacement coordinates $\theta \in \mathbb{R}^m$ with $m = 18$, the top drive motor torque input $T_{td} \in \mathbb{R}$ being the control input, the bit-rock interaction torque $T_{bit} \in \mathbb{R}$ and the interaction torques $T_w \in \mathbb{R}^{m-1}$ between the borehole and the drill-string acting on the nodes of the FEM model. The coordinates $\theta = \begin{bmatrix}\theta_1 \cdots \theta_m\end{bmatrix}^\top$ represent the angular displacements of the nodes of the finite-element representation. Next, we define the difference in angular position between adjacent nodes as

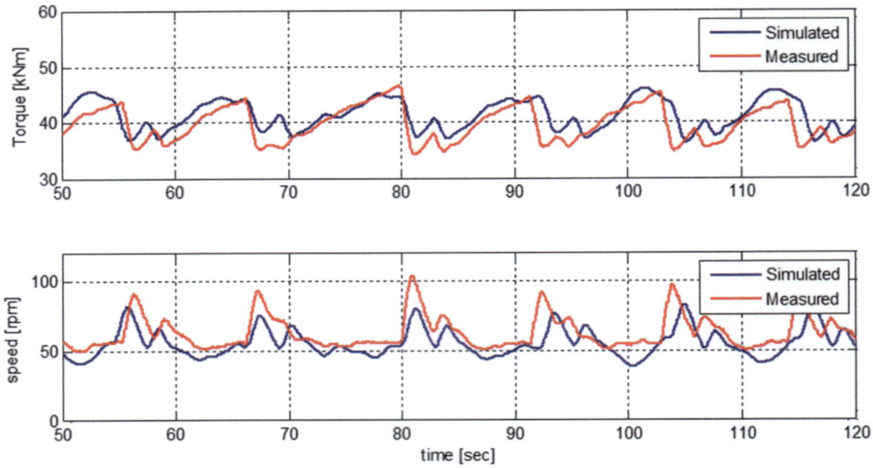

Fig. 5 Comparison between a simulation result of the FEM model and actual field data from the rig (top plot: top drive torque; bottom plot: top drive velocity); the desired angular velocity is approximately 50 rpm [11]

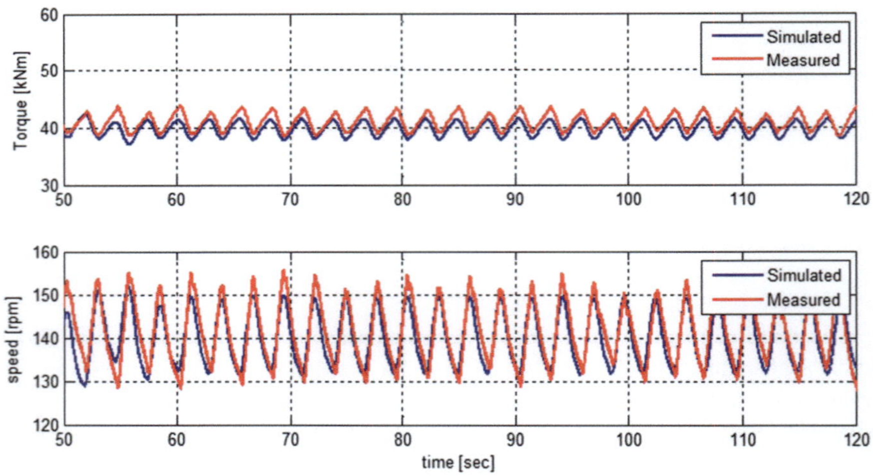

Fig. 6 Comparison between a simulation result of the FEM model and actual field data from the rig (top plot: top drive torque; bottom plot: top drive velocity); the desired angular velocity is approximately 140 rpm, [11]

follows: $\theta_d := \left[\theta_1 - \theta_2\ \theta_2 - \theta_3\ \cdots\ \theta_{m-1} - \theta_m\right]^\top$. In (1), the mass, damping and stiffness matrices are, respectively, given by $M \in \mathbb{R}^{m\times m}$, $D \in \mathbb{R}^{m\times m}$ and $K_t \in \mathbb{R}^{m\times m-1}$, and the matrices $S_w \in \mathbb{R}^{m\times m-1}$, $S_b \in \mathbb{R}^{m\times 1}$ and $S_t \in \mathbb{R}^{m\times 1}$ represent the generalized force directions of the interaction torques, the bit torque and the input torque, respectively. The coordinates θ are chosen such that the first element (θ_1) describes

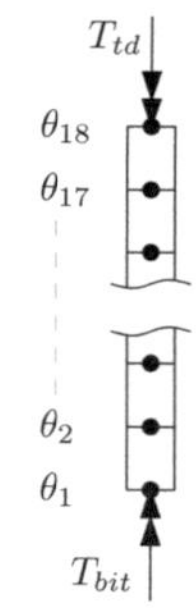

Fig. 7 Schematic representation of the 18-DOF finite-element model

the rotation of the bit and the last element (θ_{18}) the rotation of the top drive at the surface, as illustrated in Fig. 7. The interaction between the borehole and the drill-string is modelled as (set-valued) Coulomb friction, that is,

$$T_{w,i} \in T_i \ \mathrm{Sign}\left(\dot{\theta}_i\right), \quad \text{for } i = 2, \dots, m, \tag{2}$$

with T_i representing the amount of friction at each element and the set-valued sign function defined as

$$\mathrm{Sign}\,(y) := \begin{cases} -1, & y < 0 \\ [-1, 1]\,, & y = 0 \\ 1, & y > 0. \end{cases} \tag{3}$$

Note that possible viscous effects between the drill-string and the borehole are captured in the damping matrix D, which motivates only Coulomb effects being taken into account in the interaction torques T_w. The set-valued bit-rock interaction model is given by

$$T_{bit}(\dot{\theta}_1) \in \mathrm{Sign}\left(\dot{\theta}_1\right)\left(T_d + (T_s - T_d)\, e^{-v_d|\dot{\theta}_1|}\right), \tag{4}$$

where T_s is the static torque, T_d the dynamic torque and $v_d := \frac{30}{N_d \pi}$ s/rad indicates the decrease from static to dynamic torque. A schematic representation of the bit-rock interaction is shown in Fig. 8. For typical parameter settings, the ratio between T_s and T_d is within the range 2–5, i.e., the static torque is 2 to 5 times higher than the dynamic torque. Moreover, typical parameter settings for N_d are such that the decrease from static to dynamic torque is mainly between 0 and 20–30 rpm, which results in a severe velocity-weakening effect in the bit-rock interaction for low angular velocities.

2.1.2 Reduced-Order Model

The FEM model presented above has 18 degrees of freedom. For the design of the setup, we rely on a reduced-order model. The purpose of this reduced-order model is to approximate the higher-order FEM model with a reduced number of states, while

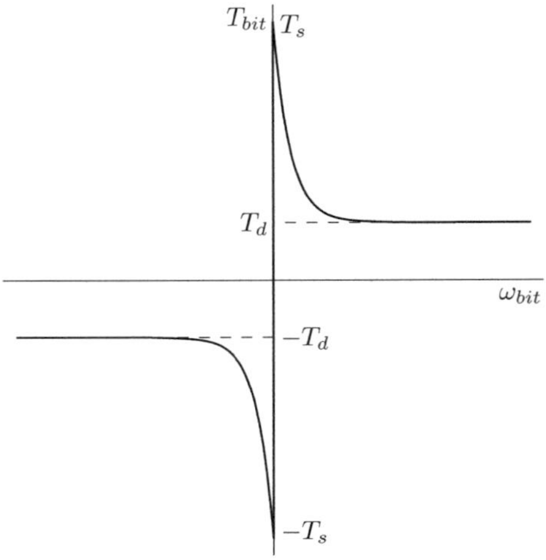

Fig. 8 Schematic representation of the bit-rock interaction T_{bit} in (4); $\omega_{bit} := \dot{\theta}_1$

still preserving the key dynamic system properties. As mentioned before, models with multiple flexibility modes are considered, because field observations have revealed that higher flexibility modes of the drill-string also play a role in the onset of stick-slip vibrations (see [23]). As mentioned in [37], the first three resonance modes, with resonance frequencies at $f_1 \approx 0.15$, $f_2 \approx 0.38$ and $f_3 \approx 0.53$ Hz, are dominant in the drill-string dynamics (see Figs. 9, 10 and 11). Therefore, a drill-string model with at least four degrees of freedom is considered capable of enabling the accurate capture of those first three flexibility modes and the rigid body mode by the reduced-order model.

For the design of the experimental setup, we aim to accurately approximate the torsional flexibility modes of the drill-string system associated with the lowest resonance frequencies. Therefore, an eigenmode-based reduction strategy is used, also known as the mode displacement method [10]. Now, let us consider the undamped (and unforced) drill-string system and, in addition, the stiffness matrix K related to the absolute angular positions $\theta = \begin{bmatrix}\theta_1 \cdots \theta_m\end{bmatrix}^\top$, instead of the stiffness matrix K_t, related to the difference in angular position θ_d as in (1), hence $M\ddot{\theta} + K\theta = 0$. Then, the mode displacement method is based on the free vibration modes of these structural dynamics. This leads to the following generalized eigenvalue problem: $\begin{bmatrix}K - \lambda_i^2 M\end{bmatrix} v_i = 0$, where v_i is the mode shape vector corresponding to the eigenfrequency λ_i, with $i \in [1, \ldots, m]$. The resulting eigenfrequencies are grouped in ascending order, i.e., $\lambda_1 \leq \lambda_2 \leq \cdots \leq \lambda_m$, and the corresponding eigenmodes $v_1, v_2, \ldots, v_m$ are collected in the square $(m \times m)$ modal matrix $V = \begin{bmatrix}v_1 \; v_2 \cdots v_m\end{bmatrix}$. Using this matrix, we employ the following coordinate transformation to modal coordinates η:

$$\theta = V\eta. \tag{5}$$

The general idea of the reduction approach is to keep the first $m_r < m$ eigenvectors, which correspond to the lowest eigenfrequencies in the reduced-order model. Hereto, consider the following transformation matrix $T = \begin{bmatrix} v_1 \ v_2 \ \cdots \ v_{m_r} \end{bmatrix}$. Using this transformation matrix, (5) can be rewritten as

$$\theta = \begin{bmatrix} T \ U \end{bmatrix} \begin{bmatrix} \theta_r \\ \eta_2 \end{bmatrix} = T\theta_r + U\eta_2, \tag{6}$$

where U contains the truncated eigenmodes, that is, the eigencolumns $m_r + 1$ to m, and η_2 contains the states that correspond to these modes; the coordinates preserved in the reduced-order model are denoted by θ_r. Using (1) and (6) and projecting the resulting equations of motion on the expansion basis T results in the following reduced-order dynamics:

$$M_r\ddot{\theta}_r + D_r\dot{\theta}_r + K_r\theta_r = T^\top S_w T_w(\dot{\check{\theta}}) + T^\top S_b T_{bit}(\dot{\check{\theta}}_1) + T^\top S_t T_{td} \tag{7}$$

with $M_r = T^\top M T \in \mathbb{R}^{m_r \times m_r}$, $D_r = T^\top D T \in \mathbb{R}^{m_r \times m_r}$, $K_r = T^\top K T \in \mathbb{R}^{m_r \times m_r}$ and $\check{\theta} := T\theta_r \in \mathbb{R}^m$ being the estimated (full-order) angular displacements based on the reduced-order estimates.

In this work, the case in which $m_r = 4$ is considered, that is, we take the rigid body mode and three torsional flexibility modes into account. The relevant frequency response functions of the (linear) drill-string dynamics are shown in Figs. 9, 10 and 11. These frequency response functions describe the (linear) drill-string dynam-

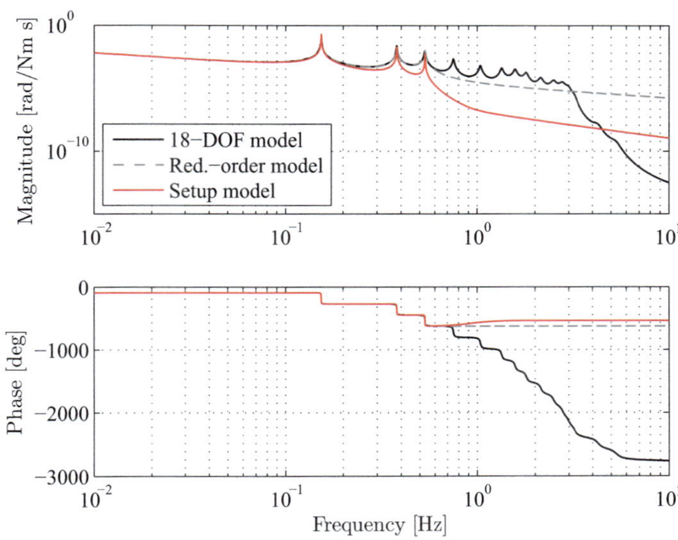

Fig. 9 Frequency response function of the 18-DOF model, the reduced-order model and the setup model with the identified parameters from input torque T_{td} to bit velocity ω_{bit}

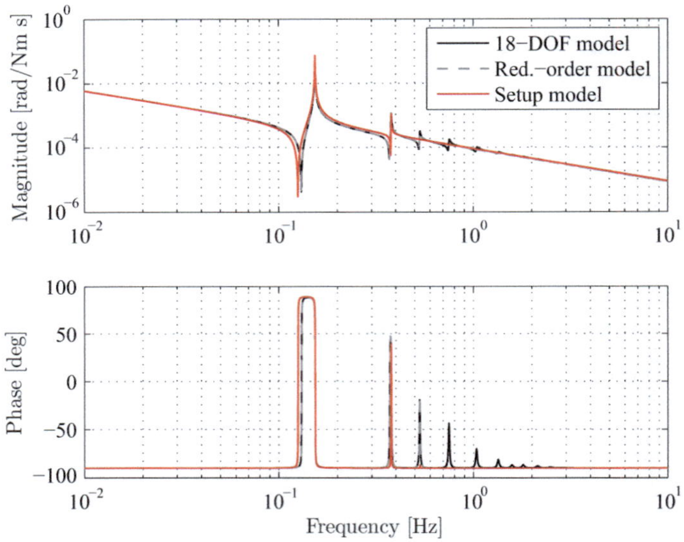

Fig. 10 Frequency response function of the 18-DOF, the reduced-order model and the setup model with the identified parameters from input torque T_{td} to top drive velocity ω_{td}

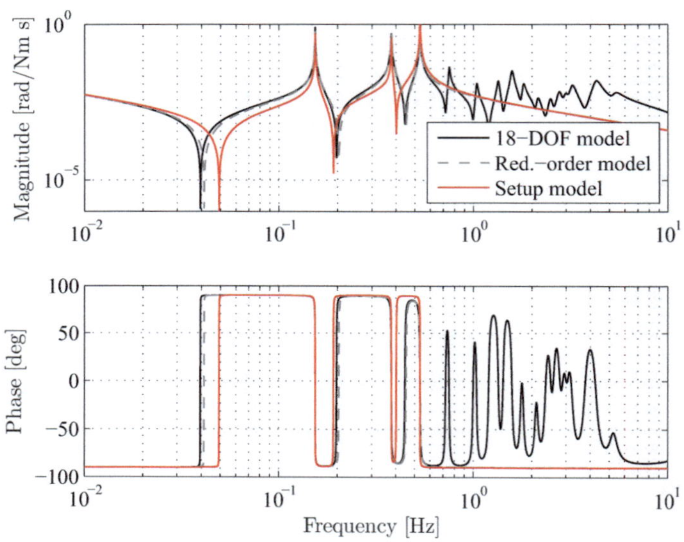

Fig. 11 Frequency response function of the 18-DOF, the reduced-order model and the setup model with the identified parameters from bit torque T_{bit} to bit velocity ω_{bit}, i.e., bit-mobility

ics from the relevant inputs (top drive torque and bit-rock interaction torque) to the angular velocity outputs at the top drive and bit, i.e., respectively, $\omega_{td} := \dot{\theta}_{18}$ and $\omega_{bit} := \dot{\theta}_1$. As can be observed, the first three eigenmodes are indeed accurately matched by the reduced-order model.

2.1.3 Dynamical Model of the Experimental Setup

In this section, the model that is used for the design of the experimental setup, as shown in Fig. 4, is discussed in more detail. For the model, we will not restrict ourselves to connections between adjacent discs, but will also take potential connections between all the discs into account. The coordinates $\bar{\theta}_s = \left[\bar{\theta}_{s,1} \cdots \bar{\theta}_{s,4}\right]^\top$ represent the angular displacements of the discs. The equations of motion of the system are given by:

$$\bar{M}_s\ddot{\bar{\theta}}_s + \bar{D}_s\dot{\bar{\theta}}_s + \bar{K}_s\bar{\theta}_s = S_{ws}T_{ws}(\dot{\bar{\theta}}_s) + S_{bs}T_{bit}(\dot{\bar{\theta}}_{s,1}) + S_{ts}T_{td}, \tag{8}$$

with

$$\bar{M}_s = \begin{bmatrix} \bar{J}_1 & 0 & 0 & 0 \\ 0 & \bar{J}_2 & 0 & 0 \\ 0 & 0 & \bar{J}_3 & 0 \\ 0 & 0 & 0 & \bar{J}_4 \end{bmatrix} \tag{9}$$

$$\bar{D}_s = \begin{bmatrix} \bar{d}_{12}+\bar{d}_{13}+\bar{d}_{14} & -\bar{d}_{12} & -\bar{d}_{13} & -\bar{d}_{14} \\ -\bar{d}_{12} & \bar{d}_{12}+\bar{d}_{23}+\bar{d}_{24} & -\bar{d}_{23} & -\bar{d}_{24} \\ -\bar{d}_{13} & -\bar{d}_{23} & \bar{d}_{13}+\bar{d}_{23}+\bar{d}_{34} & -\bar{d}_{34} \\ -\bar{d}_{14} & -\bar{d}_{24} & -\bar{d}_{34} & \bar{d}_{14}+\bar{d}_{24}+\bar{d}_{34} \end{bmatrix}, \tag{10}$$

$$\bar{K}_s = \begin{bmatrix} \bar{k}_{12}+\bar{k}_{13}+\bar{k}_{14} & -\bar{k}_{12} & -\bar{k}_{13} & -\bar{k}_{14} \\ -\bar{k}_{12} & \bar{k}_{12}+\bar{k}_{23}+\bar{k}_{24} & -\bar{k}_{23} & -\bar{k}_{24} \\ -\bar{k}_{13} & -\bar{k}_{23} & \bar{k}_{13}+\bar{k}_{23}+\bar{k}_{34} & -\bar{k}_{34} \\ -\bar{k}_{14} & -\bar{k}_{24} & -\bar{k}_{34} & \bar{k}_{14}+\bar{k}_{24}+\bar{k}_{34} \end{bmatrix}, \tag{11}$$

$$S_{ws} = \begin{bmatrix} 0&0&0 \\ 1&0&0 \\ 0&1&0 \\ 0&0&1 \end{bmatrix}, \quad S_{bs} = \begin{bmatrix} 1 \\ 0 \\ 0 \\ 0 \end{bmatrix}, \quad S_{ts} = \begin{bmatrix} 0 \\ 0 \\ 0 \\ 1 \end{bmatrix}, \tag{12}$$

and the resistive torques at discs 2, 3 and 4 for modeling the borehole drill-string interaction are given by T_{ws}. Recall that T_{td} denotes the top drive motor torque and T_{bit} denotes the bit-rock interaction torque.

2.1.4 Parameter Identification for the Setup Model

The next step is to determine the parameters of the 4-DOF model of the experimental setup based on the reduced-order model presented in Sect. 2.1.2. First, the inertias of the four discs are determined. The total inertia of the 4-disc setup is chosen to be equal to the total inertia of the original 18-DOF model. In addition, we require the inertia of the upper disc ($\bar{J}_4$) to be equal to the inertia of the top drive, such that the upper disc actually represents the top drive. Doing so, the torque in the drill-string below disc 4 comes to represent the pipe torque that is used as measurement in the linear robust controller approach (presented in Sect. 3). The inertia of the bottom disc ($\bar{J}_1$) is determined based on the "high"-frequency behavior (i.e., above the eigenfrequencies) of the reduced-order model. The remaining part of the total inertia is equally distributed over the two remaining discs. The remaining damping and stiffness parameters, are determined using an optimization-based identification approach. The objective of the optimization procedure is to find the model parameters such that the difference in the complex plane between the frequency response function of the reduced-order model and the model of the setup is minimized over all frequencies within the frequency range of interest. Hence, we seek to solve the following optimization problem:

$$\min_{p\in\left[\underline{p},\, \overline{p}\right]} J(p), \tag{13}$$

where $p := \left[\bar{k}_{12}\ \bar{k}_{23}\ \bar{k}_{34}\ \bar{k}_{13}\ \bar{k}_{14}\ \bar{k}_{24}\ \bar{d}_{12}\ \bar{d}_{23}\ \bar{d}_{34}\ \bar{d}_{13}\ \bar{d}_{14}\ \bar{d}_{24}\right]$ are the parameters of the setup to be determined, $\underline{p}$ and $\overline{p}$, represent a lower and upper bound for the parameters and the objective function $J(p)$ is given by

$$J(p) = \sum_{\omega_l} w(j\omega)\left(\left|W(j\omega)H_r^{T_{td}\omega_{bit}}(j\omega) - W(j\omega)H_s^{T_{td}\omega_{bit}}(j\omega)\right|^2\right) \tag{14}$$

with $H_r^{T_{td}\omega_{bit}}$ and $H_s^{T_{td}\omega_{bit}}$ the frequency response functions from top drive torque input to bit velocity output of the reduced-order model and the setup model, respectively. The frequency response function from top drive torque input to bit velocity output is chosen for the parameter identification because it captures the relevant dynamics of the drilling system that should be represented in the setup. Note that $H_s^{T_{td}\omega_{bit}}$ depends on the parameters p. The frequency grid ω_l is a discrete grid of frequencies between 0.05 and 6 Hz, because that is the relevant frequency range of the reduced-order drill-string dynamics (see Fig. 9). The frequency-dependent weighting filter $W(j\omega)$ is chosen to be $W(j\omega) = J_{tot}j\omega$, to compensate for the negative slope of the frequency response function from top drive torque input to bit velocity output. The (scalar) multiplication factor $w(j\omega)$ in (14) is used to give extra weighting in specific frequency ranges. This multiplication factor is equal to 1.5 for $0.14 < f < 0.165$ (i.e., around the first resonance frequency), equal to 2 for $0.5 < f < 0.57$ (i.e., around the third resonance frequency) and equal to 1 for all other frequencies.

The results of the fitting procedure are shown in Figs. 9, 10 and 11. Note that the identified parameters are still of the same order of magnitude as for the original drill-string model. For example, the inertia of the upper disc is equal to the inertia of a real top drive (approximately 1800 kgm^2) and a driving torque at the top drive is typically on the order of 40 kNm. These settings are infeasible for a lab-scale setup. Therefore, scaling of the parameters in order to obtain feasible dimensions for the lab-scale setup is discussed in Sect. 2.1.5. It turned out that it is not possible to fit the reduced-order model of the original drill-string and the model of the setup. This is mainly caused by the fact that in the finite-element model (and therefore also in the reduced-order model), the drill-string's properties are distributed along it, whereas the model of the setup is based on a lumped mass approach (multiple discs representing discrete inertias). This is particularly visible in Fig. 9: due to the lumped inertias of the setup model, the slope of the magnitude of the FRF decreases by 2 (on a loglog-scale) after each resonance peak and the phase decreases by 180 degrees (due to the 2 poles associated with the resonance). However, the FRFs of the 18-DOF and reduced-order model do not show this behavior; this is caused by zeros of these models that are in the right-half-plane of the complex plane (i.e., non-minimum phase). Nevertheless, a satisfactory match of the dominant resonances is achieved and, moreover, simulation results of the non-smooth setup model (see Fig. 12) confirm that the response of the setup model is in good correspondence with the response of the reduced-order model and the original 18-DOF FEM model. In Fig. 12, the response of the 18-DOF drill-string model is compared with the response of the 4-DOF setup model. In both simulations, the system is controlled with a SoftTorque controller (see (37) in Sect. 4.2) and the parameters $c_t = 1829$ and $k_t = 1177$), and the desired angular velocity is equal to 50 rpm. Clearly, the response of the setup model is similar to the response of the original FEM model. This illustrates that the dominant dynamics of the original 18-DOF model is captured by the 4-DOF setup model, also in the scope of the non-smooth dynamics leading to stick-slip oscillations.

2.1.5 Scaling of the Drill-String Model

An identified set of parameters for the experimental setup has been obtained in the previous section. However, these parameters are based on a full-scale drilling rig and, as mentioned before, such parameter values are infeasible for a lab-scale experimental setup. To obtain feasible parameter values for the experimental setup, a scaling of the variables and parameters is in order, while retaining the resonance frequencies of the drill-string system. Therefore, two scaling factors are introduced: c_1 is used to scale the torque level and c_2 to scale the states of the system. The states are scaled according to $\theta_s = \frac{1}{c_2}\bar{\theta}_s$ and the equations of motion are pre-multiplied with a factor $\frac{1}{c_1}$ to scale the torque level. This results in the (scaled) equations of motion given by

$$M_s\ddot{\theta}_s + D_s\dot{\theta}_s + K_s\theta_s = S_{ws}\hat{T}_{ws}(\dot{\theta}_s) + S_{bs}\hat{T}_{bit}(\dot{\theta}_{s,1}) + S_{ts}\hat{T}_{td} \tag{15}$$

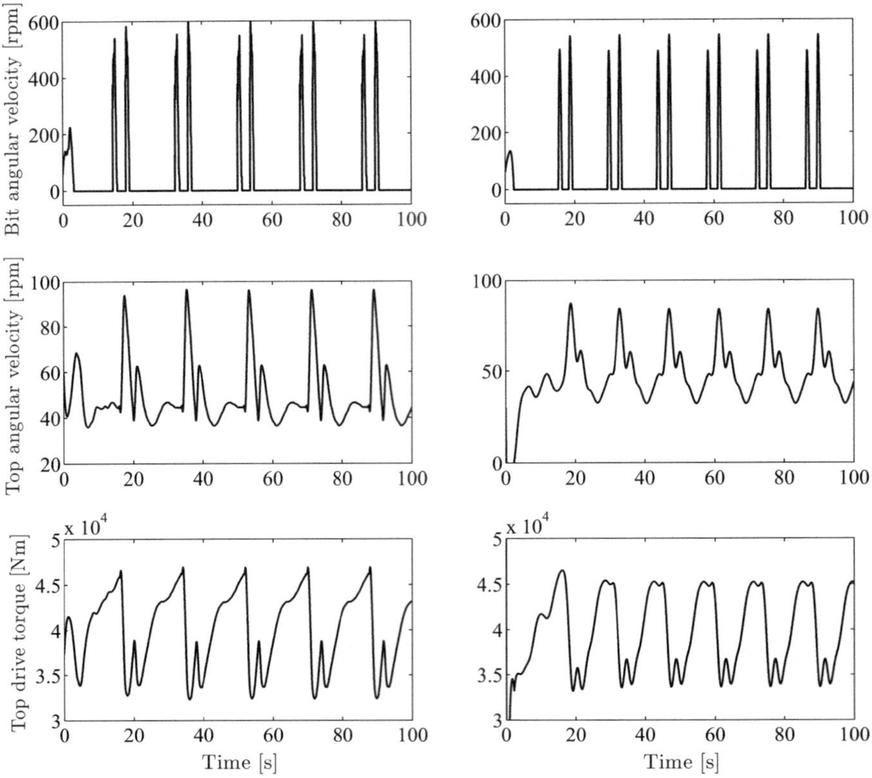

Fig. 12 Simulation result of the 18-DOF drill-string model (left-hand side) compared with a simulation result of the 4-DOF model of the experimental setup (right-hand side)

with $M_s := \frac{c_2}{c_1}\bar{M}_s$, $D_s := \frac{c_2}{c_1}\bar{D}_s$, $K_s := \frac{c_2}{c_1}\bar{K}_s$, $\hat{T}_{ws} := \frac{1}{c_1}T_{ws}$, $\hat{T}_{bit} := \frac{1}{c_1}T_{bit}$ and $\hat{T}_{td} := \frac{1}{c_1}T_{td}$. The scaled bit-rock interaction torque $\hat{T}_{bit}$ is given by the following scaled law:

$$\hat{T}_{bit}(\dot{\theta}_{s,1}) \in \text{Sign}\left(\dot{\theta}_{s,1}\right)\left(\hat{T}_d + \left(\hat{T}_s - \hat{T}_d\right)e^{\left(-30\,|\dot{\theta}_{s,1}|\right)/\left(\hat{N}_d\pi\right)}\right) \tag{16}$$

with $\hat{T}_d = \frac{1}{c_1}T_d$, $\hat{T}_s = \frac{1}{c_1}T_s$ and $\hat{N}_d = \frac{1}{c_2}N_d$, and the scaled drill-string borehole interaction torques can be written as

$$\hat{T}_{ws,i} \in \hat{T}_{s,i}\ \text{Sign}\left(\dot{\theta}_{s,i}\right), \quad \text{for } i = 2, \ldots, m, \tag{17}$$

where $\hat{T}_{s,i} = \frac{1}{c_1}T_{s,i}$. The scaling factors are determined to be $c_1 = 6250$ and $c_2 = 10$. This scaling is chosen first, to obtain feasible torque levels for typical motors that can be used in lab-scale systems (mainly influenced by c_1) and, second, to achieve angular position differences between adjacent discs that are sufficiently small so as to avoid plastic deformation of the steel strings between those discs. The latter aspect, of course, also depends on the length and diameter of the strings, which need to have feasible dimensions. The scaled parameters are summarized in Table 1, the scaled parameters regarding the interaction torques are given in Table 2. The top drive torque is on the order of 40 kNm for the full scale system, whereas this is scaled to approximately 6.4 Nm for the setup, and since the states are scaled with a factor 10, a desired angular velocity of 50 rpm in practice is equal to a desired angular velocity of 5 rpm in the setup. Note that no time-scaling applied, which implies that the resonance frequencies of the system have not been changed.

By applying the described scaling, the model of the experimental drill-string setup is scaled to feasible dimensions for designing a lab-scale setup. With the method described in this section, a set of prescribed model parameters is obtained for the design of the setup. The setup design is discussed in more detail in the next section.

2.2 *The Experimental Drill-String Setup*

The experimental setup is designed to be adjustable and modular. In particular, it is designed such that it should be possible to change the inertia of the discs and the stiffness of the strings, and, by using a hardware-in-the-loop approach, other parameters such as damping can also be adjusted. With this hardware-in-the-loop approach, additional dynamics is emulated in software and implemented using motors driving all the individual discs. In addition, the setup is designed such that it is possible to

Table 1 Parameters of the setup model

Symbol	Value (kgm^2)	Symbol	Value (Nms/rad)	Symbol	Value (Nm/rad)
J_1	0.064	k_{12}	0.630	d_{12}	0
J_2	0.708	k_{23}	1.799	d_{23}	0.0018
J_3	0.708	k_{34}	1.097	d_{34}	0.0024
J_4	2.845	k_{13}	0	d_{13}	0
		k_{14}	0	d_{14}	0.0005
		k_{24}	0	d_{24}	0

Table 2 Parameters of the scaled bit-rock interaction model and drill-string borehole interaction torques

Symbol	Value	Symbol	Value (Nm)
$\hat{T}_s$	1.232 Nm	$\hat{T}_{s,2}$	2.297
$\hat{T}_d$	0.272 Nm	$\hat{T}_{s,3}$	3.038
$\hat{N}_d$	0.5 rad/s	$\hat{T}_{s,4}$	0.662

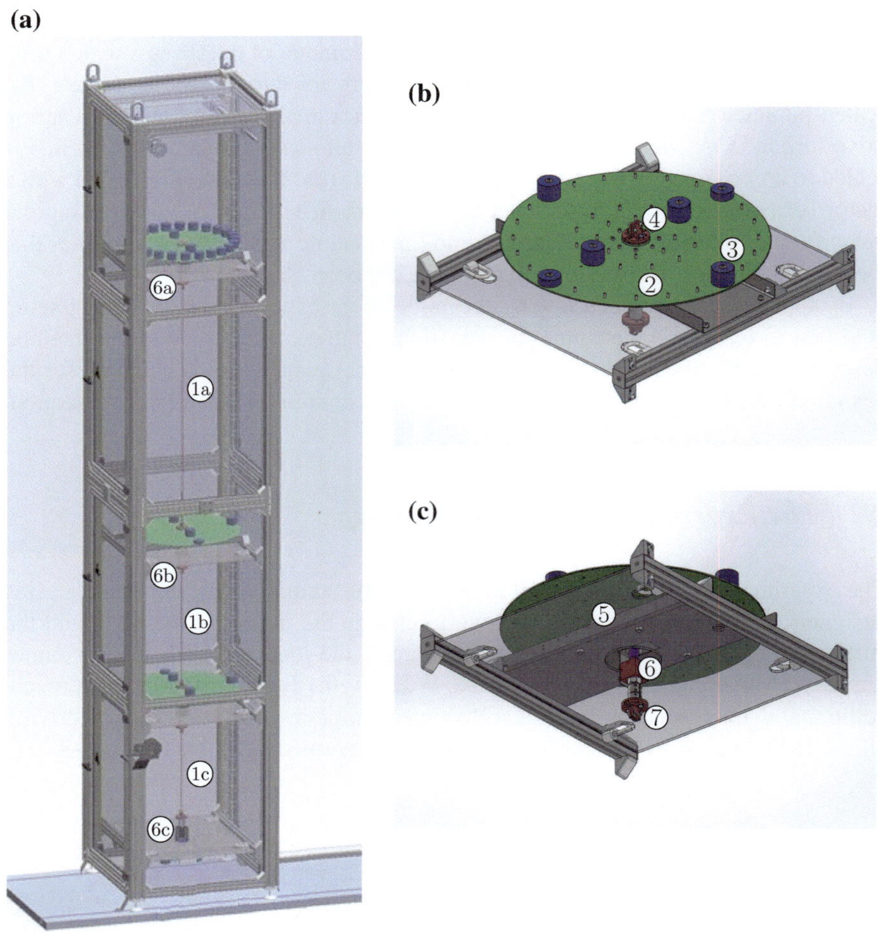

Fig. 13 Schematic representation of the experimental drill-string setup. **a** is an overview of the setup, **b** is a top view on one of the disc platforms, **c** is a bottom view of one of the disc platforms. Different parts are numbered as follows: 1- (steel) strings between the different discs; 2- disc (representing inertia); 3- additional masses to change the inertia of the disc; 4- upward connection for the string; 5- flat hollow shaft torque motor (embedded in the frame); 6- torque sensors; 7- downward connection for the string

investigate a system with additional flexibility modes by adding an extra disc to the setup. A schematic overview of the setup is shown in Fig. 13.

Let us now discuss the design of the setup in more detail. The total setup is 5 m tall and has a footprint of 1×1 m. As can be seen in Fig. 13a, the setup has 4 disc platforms to support the 4 discs of the model (see Fig. 13b, c; note that the bottom disc platform is slightly different, which is explained in more detail later). These discs are interconnected by steel strings to represent the torsional stiffness of the drill-string system and each disc is equipped with a motor. For the top disc, this motor is used to drive the system and to apply the desired control action. At the bottom disc, the motor is used to emulate the desired bit-rock interaction, and at the intermediate discs, the drill-string borehole interaction torques are implemented using these motors. In addition, these motors are used to emulate the hardware-in-the-loop components, such as damping torque associated with the damping constant d_{14}, and to compensate for undesired effects, such as friction and cogging in the motors. Each of the motors is equipped with an encoder, and the setup contains three torque sensors for measuring the torques in the interconnecting strings. Furthermore, a DS1103 controller board from dSPACE [7] is used as a real-time control and data acquisition platform. A photo of the lab-scale drill-string system is shown in Fig. 14.

The three upper disc platforms are identical and equipped with Georgii Kobold KTY-F torque motors ([9]). These are flat direct-drive brushless DC motors with a maximum torque of 26 Nm and a maximum angular velocity of 250 rpm. To actuate and control these motors, Siep & Meyer SD2S motor amplifiers ([33]) are used, and to measure the angular position of the discs, built-in 19-bit Heidenhain ECI 119 inductive encoders ([13]) are used. The 19-bit encoder signal is converted, in the motor amplifiers, into a 15-bit quadrature signal that is used by the dSPACE system to determine the angular position of the discs. The angular velocities of the discs are determined by numerical differentiation of the angular positions measured by the encoders. The discs have an inertia of approximately 0.350 kgm^2, including the inertia of the motor. By adding additional masses at a certain radius on the discs, the inertia of the discs can be adjusted (in steps of approximately 0.05 kgm^2) to obtain the desired inertia, as specified in Table 1.

The bottom disc platform is shown in Fig. 15 and is different from the other platforms. This difference has two main reasons: first, the specified inertia of the bottom disc is much lower compared to the inertias of the other discs and, second, in order to accurately implement the desired bit-rock interaction law, it is important that this disc has a low static friction. To realize these two aspects, a disc with a smaller diameter and a different type of motor is used. The installed motor is a brushed DC motor from Printed Motor Works (type: GN16RE), see [28], with a maximum torque of 2.55 Nm and a maximum angular velocity of 3000 rpm. The static friction in this motor is approximately 0.05 Nm, which is sufficiently lower than the dynamic torque level $\hat{T}_d$ to be implemented (see Table 2). In addition, a 16-bit Sick DFS60A incremental encoder ([32]) is used together with a Copley Controls Xenus Plus motor amplifier (type: XTL-230-40), see [3]. The bottom disc has an inertia of approximately 0.03 kgm^2 and can be adjusted in steps of approximately 0.01 kgm^2 to achieve the prescribed inertia.

To represent the torsional stiffness of the drill-string model, steel strings with a specific length and diameter are used. The length and diameter are chosen such that the prescribed stiffnesses (see Table 1) are achieved. The specified damping factors are obtained by implementing the damping using the motors (i.e., in a hardware-in-the-loop fashion) based on the measured difference in angular velocity of the discs, while compensating for the material damping that is already present in the strings.

Fig. 14 The experimental drill-string setup

The setup is also equipped with three PCM TQ-RT2A-25NM torque sensors [26]. These sensors can measure up to 25 Nm with an accuracy of ±0.2%. The torque sensors are placed below the upper two discs and just above the bottom disc, as indicated in Figure 13a with 6a-c. The torque sensor below the top drive disc will be used for the pipe torque measurement, to be used in the scope of feedback control.

The foregoing description of the experimental setup shows that the setup is equipped with multiple sensors: encoders in all the discs to measure the angular

Fig. 15 Bottom disc platform, with 1: the disc with additional weights; 2: the motor; 3: the encoder; 4: the torque sensor

position (and determine the angular velocity) and three torque sensors to measure the torques in the steel strings between the discs. However, the control design strategies to be presented in later sections will only require surface (top-side) measurements. The extra sensors, which are not required for the proposed control strategies, are used for parameter identification and validation of the setup dynamics and for analyses of the obtained experimental results.

2.3 Summary

In this section, the design of the experimental setup is discussed. First, a model of the experimental setup, based on the 18-DOF FEM drill-string model, is presented. The 4-DOF setup model is designed such that it represents the dominant dynamics of the dynamics of an oil-field drill-string system that exhibits torsional vibrations. The non-smooth model of the experimental drill-string setup is scaled to feasible dimensions in support of the design of the lab-scale setup. Finally, the mechanical and electrical design of the designed setup is presented in detail.

3 Output-Feedback Controller Design

In this section, a design approach for torsional controllers, aiming to eliminate torsional stick-slip oscillations, is described. In Sect. 3.1, a model reformulation is presented rendering the model suitable in the scope of controller synthesis. Section 3.2

details the control problem formulation in system-theoretic terms. Next, Sect. 3.3 describes the proposed output-feedack control strategy inducing robustness with respect to uncertainties in the bit-rock interaction torque.

3.1 Non-smooth Modelling for Control

The dynamic model of the setup in second -order form, as given in (15)–(17), can be cast into a first-order Lur'e-type system form as follows:

$$
\begin{aligned}
\dot{x} &= Ax + Gv + G_2 v_2 + Bu_t \\
q &= Hx \\
q_2 &= H_2 x \\
y &= Cx \\
v &\in -\varphi(q) \\
v_2 &\in -\phi(q_2).
\end{aligned}
\tag{18}
$$

Herein, $x = \left[\theta_{s,1} - \theta_{s,2}, \dot{\theta}_{s,1}, \dot{\theta}_{s,2}, \theta_{s,2} - \theta_{s,3}, \dot{\theta}_{s,3}, \theta_{s,3} - \theta_{s,4}, \dot{\theta}_{s,4}\right]^\top \in \mathbb{R}^7$ is the state, where $\theta_{s,i}$, $i = 1, 2, 3, 4$, describes the rotational displacement of the inertias of the setup, and the bit velocity is defined as $q := \dot{\theta}_{s,1}$. Furthermore, $q_2 := \left[\dot{\theta}_{s,2}\ \dot{\theta}_{s,3}\ \dot{\theta}_{s,4}\right]^\top$. Note that only relative angular positions are taken into account, such that the 4-DOF system is described with only 7 state variables. Moreover, the bit-rock interaction torque $\hat{T}_{bit}$ is denoted by $v \in \mathbb{R}$ and the drill-string-borehole interaction torques $\hat{T}_{ws}$ are denoted by $v_2 \in \mathbb{R}^3$. As a consequence, the nonlinearities $\varphi(\cdot)$ and $\phi(\cdot)$ are defined by the set-valued nonlinearities in the right-hand sides of (16) and (17), respectively. In addition, $u_t := \hat{T}_{td} \in \mathbb{R}$ is the (top drive torque) control input and $y := \left[\ \omega_{td}\ T_{pipe}\right]^\top \in \mathbb{R}^2$ is the measured output, where $\omega_{td} := \dot{\theta}_{s,4}$ is the top drive angular velocity. The so-called pipe torque T_{pipe} is the torque in the drill-string directly below the top drive (sometimes also referred to as the saver sub torque). In the experimental setup, this torque is measured using a torque sensor directly below the top-most inertia.

3.2 Control Problem Formulation

The desired operation of the drill-string system is a constant angular velocity ω_{eq} for all four inertias. So, the objective is to regulate this set-point of the non-smooth drill-string system by means of an output-feedback controller. The available output measurements for the controller are the top drive angular velocity ω_{td} and the pipe torque T_{pipe}. The system can be controlled by the top drive torque u_t. The controller should:

1. locally stabilize the desired velocity of the drill-string, therewith eliminating torsional (stick-slip) vibrations;
2. ensure robustness with respect to uncertainty in the non-smooth bit-rock interaction φ;
3. guarantee the satisfaction of closed-loop performance specifications, in particular, on measurement noise sensitivity, i.e., limitation of the amplification of measurement noise, and limitation of the control action such that top drive limitations can be satisfied;
4. guarantee robust stability and performance in the presence of multiple flexibility modes dominating the torsional dynamics.

To facilitate controller synthesis, the drill-string dynamics (18) are reformulated. The desired constant angular velocity ω_{eq} for all discs can be associated with a desired equilibrium x_{eq} for the state of the system. To ensure that x_{eq} is an equilibrium of the closed-loop system, the control input $u_t = u_c + \tilde{u}$ is decomposed in a constant feedforward torque u_c (inducing x_{eq}) and the feedback torque $\tilde{u}$. For the feedforward design, we assume that $\dot{\theta}_{s,i} > 0$, for $i = 2, 3, 4$, hence ϕ is constant and can be compensated for by constant u_c, and we determine x_{eq} and u_c using the equilibrium equation of system (18), i.e., $Ax_{eq} - G\varphi(Hx_{eq}) - G_2\phi(H_2x_{eq}) + Bu_c \ni 0$. Next, we define $\xi := x - x_{eq}$ and apply a linear loop transformation such that the slope of a transformed nonlinearity $\tilde{\varphi}(q)$ (associated with $\varphi(q)$ through the loop transformation) is equal to zero at the equilibrium velocity, i.e., $\partial\tilde{\varphi}/\partial q|_{q=\omega_{eq}} = 0$. This results in a state-space representation of the transformed drill-string dynamics in perturbation coordinates:

$$\dot{\xi} = A_t\xi + B\tilde{u} + G\tilde{v} \tag{19a}$$

$$\tilde{y} = C\xi \tag{19b}$$

$$\tilde{q} = H\xi \tag{19c}$$

$$\tilde{v} \in -\tilde{\varphi}(\tilde{q}) \tag{19d}$$

with $A_t := A + \delta GH$, $\delta = -\left.\partial\varphi/\partial q\right|_{q=\omega_{eq}} > 0$, $\tilde{y} := y - Cx_{eq}$, $\tilde{q} := q - Hx_{eq}$, $\tilde{\varphi}(\tilde{q}) := \varphi(\tilde{q} + Hx_{eq}) - \varphi(Hx_{eq}) + \delta\tilde{q}$ and $\tilde{v} := v - v_{eq} - \delta\tilde{q}$. The dynamics in (19) represents a Lur'e-type system, with the linear dynamics (19a)–(19c), with transfer function G_{ol}, and having inputs $\tilde{u}$ and $\tilde{v}$ and outputs $\tilde{y}$ and $\tilde{q}$, and the nonlinearity $\tilde{\varphi}$ in the feedback loop. The open-loop transfer function $G_{ol}(s)$ is defined as

$$\begin{bmatrix}\tilde{q}(s)\\ \tilde{y}(s)\end{bmatrix} := G_{ol}(s)\begin{bmatrix}\tilde{v}(s)\\ \tilde{u}(s)\end{bmatrix} = \begin{bmatrix}g_{11}(s) & g_{12}(s)\\ g_{21}(s) & g_{22}(s)\end{bmatrix}\begin{bmatrix}\tilde{v}(s)\\ \tilde{u}(s)\end{bmatrix}. \tag{20}$$

In the context of the second controller objective above, we model the nonlinearity $\tilde{\varphi}$ (Fig. 16a) by an uncertainty Δ (Fig. 16b). This model formulation is used in the controller design approach developed in Sect. 3.3. Note that $\tilde{\varphi}$ describes a nonlinear (set-valued) mapping from $\tilde{q}$ to $\tilde{v}$, while the uncertainty Δ is assumed to be a (complex) LTI uncertainty (with output $\check{v}$). This means that, for example, stability of

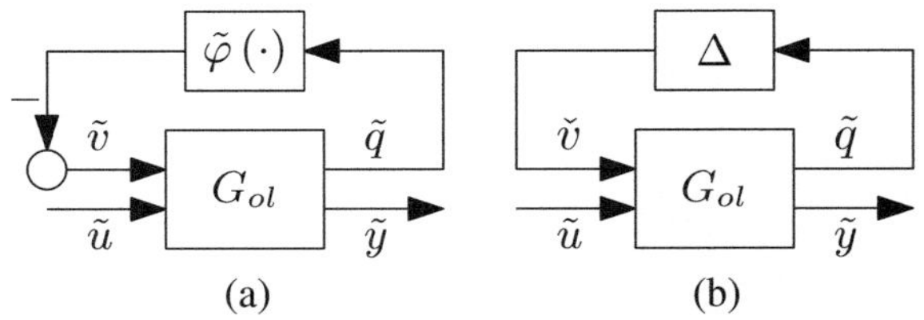

Fig. 16 Block diagram of the system dynamics (19) in Lur'e type form (**a**) and the linear dynamics G_{ol} with (complex) model uncertainty Δ (**b**)

the closed-loop system with uncertainty Δ does not directly imply stability for the closed-loop system with nonlinearity $\tilde{\varphi}$. Nevertheless, the model in Fig. 16b is used as a basis for controller synthesis in the next section. Subsequently, the stability of the nonlinear (non-smooth) closed-loop system is analyzed in detail in Sect. 3.3.3.

3.3 Design of a Robust Output-Feedback Controller

In this section, we present a robust control design approach based on skewed-μ DK-iteration ([38]).

First, we formulate the general control configuration that is used in such a robust control context. Next, in Sect. 3.3.1, we analyze nominal performance of the (linear) system, i.e., without uncertainty. This is extended to robust performance for the (linear) system with uncertainty taken into account in Sect. 3.3.2. The stability of the closed-loop nonlinear system is investigated in Sect. 3.3.3.

This robust control technique combines several concepts from robust control theory to design a controller that achieves robust stability and performance of a system with model uncertainties [34].

Robust control methods focus on the design of controllers, while system uncertainties are explicitly taken into account in the design. The general control configuration for a (LTI) plant P with an uncertainty Δ and (LTI) controller K is shown in Fig. 17, where e is the error in the measured output, u the control output and w and z represent the (weighted) exogenous inputs and outputs. This structure is similar to the block diagram in Fig. 16b with $\bar{v}$ and $\bar{q}$ as weighted representations of $\check{v}$ and $\tilde{q}$ (see Sect. 3.3.2) and, in addition, includes the controller K. The system P, in Fig. 17, is described by

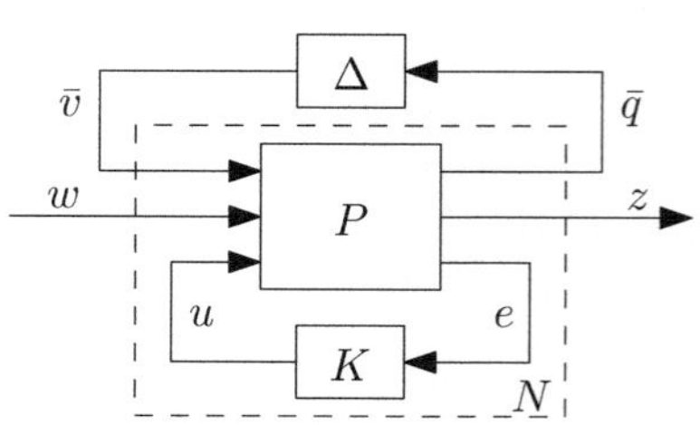

Fig. 17 General control configuration with uncertainty block Δ

$$\begin{bmatrix} \tilde{q} \\ z \\ e \end{bmatrix} = \begin{bmatrix} P_{11} & P_{12} & P_{13} \\ P_{21} & P_{22} & P_{23} \\ P_{31} & P_{32} & P_{33} \end{bmatrix} \begin{bmatrix} \tilde{v} \\ w \\ u \end{bmatrix}. \tag{21}$$

The system $N := F_l(P, K)$ is defined as the lower linear fractional transformation (LFT) of the plant P with the controller K, that is:

$$N = \begin{bmatrix} P_{11} & P_{12} \\ P_{21} & P_{22} \end{bmatrix} + \begin{bmatrix} P_{13} \\ P_{23} \end{bmatrix} K (I - P_{33}K)^{-1} \begin{bmatrix} P_{31} & P_{32} \end{bmatrix}.$$

With the introduction of the controller K, we can also introduce the closed-loop bit-mobility function. The closed-loop bit-mobility transfer function G_{cl} from the input $\tilde{v}$ to the output $\tilde{q}$, of system (19) with controller K, is defined by

$$G_{cl} := g_{11} - g_{12}K(I + g_{22}K)^{-1}g_{21}. \tag{22}$$

This bit-mobility plays an important role in the stability of the closed-loop system (see Sect. 3.3.3 for the role of G_{cl} in the scope of a nonlinear stability analysis), and is therefore important in the controller design methodology.

3.3.1 Nominal Stability and Nominal Performance

As mentioned above, the controller design aims at stability, performance, and robustness for the uncertainty Δ. In this section, the focus is on the first two aspects. Robustness is considered in the next section. Based on the system representation in Fig. 16b, the closed-loop system of the linear drill-string dynamics G_{ol} in feedback with the linear, dynamic controller K to be designed is shown in Fig. 18. In this representation, additional inputs n and d are introduced, representing measurement noise and actuator noise, respectively.

Consider the system without uncertainty given by

$$\begin{bmatrix} z \\ e \end{bmatrix} := \underline{P} \begin{bmatrix} w \\ u \end{bmatrix} = \begin{bmatrix} P_{22} & P_{23} \\ P_{32} & P_{33} \end{bmatrix} \begin{bmatrix} w \\ u \end{bmatrix} \tag{23}$$

with w and z weighted versions of $\underline{w} := \begin{bmatrix} n & d \end{bmatrix}^\top$ and $\underline{z} := \begin{bmatrix} e & u \end{bmatrix}^\top$, respectively. The weighted inputs and outputs are discussed in more detail in Sect. 3.3.2. Moreover,

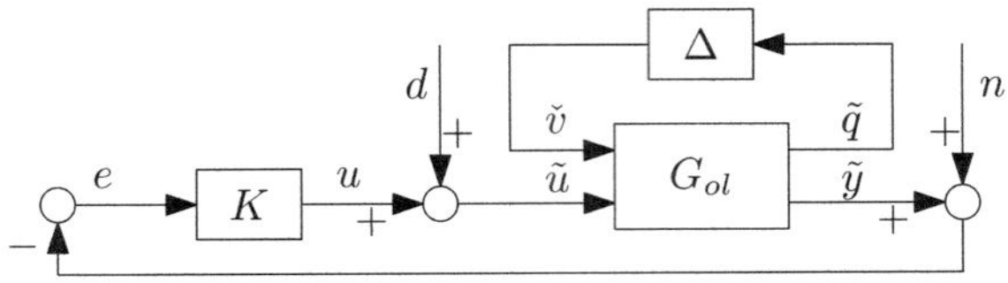

Fig. 18 Linear drill-string dynamics G_{ol} in closed loop with the controller K and including model uncertainty Δ with disturbances d and n

define the lower LFT of $\underline{P}$ with the controller K, that is, $N_{22} := F_l\left(\underline{P}, K\right)$. Next, the concept of nominal performance is defined as follows: for a system without uncertainty Δ, the closed-loop system $N_{22} = F_l(\underline{P}, K)$ is internally stable and the $\mathcal{H}_\infty$-norm of this system (from w to z) is smaller than 1, that is,

$$\|N_{22}\|_\infty = \sup_\omega \bar{\sigma}\left(F_l(\underline{P}, K)\right) < 1, \tag{24}$$

where we used the definition of the $\mathcal{H}_\infty$-norm $\|H(s)\|_\infty := \text{ess}\sup_{\omega\in\mathbb{R}} \bar{\sigma}\left(H\left(j\omega\right)\right)$ and $\bar{\sigma}$ indicates the maximum singular value. This means that nominal performance can be achieved by solving the "standard" $\mathcal{H}_\infty$ optimal control problem, in which the aim is to find the internally stabilizing controller K that minimizes $\|F_l(\underline{P}, K)\|_\infty$ (see [34] for details). Internal stability of the closed-loop can be guaranteed by a proper choice of the inputs w and outputs z. As proved in [41, Sect. 5.3], by choosing w and z as defined earlier, the $\mathcal{H}_\infty$ controller synthesis guarantees internal stability of the closed-loop system. Specification of the weighting filters is treated in more detail in Sect. 3.3.2. Moreover, the system *with* uncertainty is addressed in the next section, leading to the concept of (alternative) robust performance.

3.3.2 Alternative Robust Performance

Robust performance means that the stability and performance objective, addressed in Sect. 3.3.1, is achieved for all possible models in the uncertainty set **D** [34], i.e., for all $\Delta \in \mathbf{D}$. Standard robust performance techniques typically aim at optimizing the performance for all possible plants induced by the uncertainty set. In contrast, we aim to optimize the robustness with respect to the uncertainty while still guaranteeing internal stability and satisfaction of given performance objectives. This is what we call *alternative robust performance*. In the drilling context, this means that, for example, a (fixed) bound on the control action should be satisfied (see controller objective 3 in Sect. 3.2), while the robustness with respect to the nonlinear bit-rock interaction is optimized (as specified in the second controller objective).

Consider the system P in Fig. 17, including the uncertainty block, Δ. The input-output pair $\bar{v}$, $\bar{q}$ is related to this uncertainty block and the (weighted) closed-loop transfer function $N(s) = F_l\left(P, K\right)$ is given by

$$\begin{bmatrix} \bar{q} \\ w \end{bmatrix} = N \begin{bmatrix} \bar{v} \\ z \end{bmatrix} = F_l\left(P, K\right) \begin{bmatrix} \bar{v} \\ z \end{bmatrix}. \tag{25}$$

Robust stability is obtained by designing a controller K such that the system N is internally stable and the upper LFT, $F := F_u(N, \Delta)$, is stable for all $\Delta \in \mathbf{D}$. Herein, the uncertainty set $\mathbf{D}$ is a norm-bounded subset of $\mathcal{H}_\infty$,[1] i.e., $\mathbf{D} =$

[1] $\mathcal{H}_\infty$ is a (closed) Banach space of matrix-valued functions that are analytic in the open right-half plane and bounded on the imaginary axis. The real rational subspace of $\mathcal{H}_\infty$ is denoted by $\mathcal{RH}_\infty$, which consists of all proper and real rational stable transfer matrices [41, Sect. 4.3].

$\{\Delta \in \mathscr{RH}_\infty | \|\Delta\|_\infty < 1\}$. The aim is to find a stabilizing controller that also meets certain performance specifications. Therefore, we use a similar approach as in [34, Sect. 8.10] and consider the fictitious 'uncertainty' Δ_P. The uncertainty Δ_P is a complex unstructured uncertainty block which represents the $\mathscr{H}_\infty$ performance specifications. Moreover, note that $\Delta_P \in \mathbf{D}_P$, with $\mathbf{D}_P = \{\Delta_P \in \mathscr{RH}_\infty \,|\, \|\Delta_P\|_\infty \leq 1\}$. The result given in [41, Theorem 11.8] states that a robust performance problem is equivalent to a robust stability problem with the augmented uncertainty

$$\hat{\Delta} = \begin{bmatrix} \Delta & 0 & 0 \\ 0 & & \\ 0 & \Delta_P & \end{bmatrix} \tag{26}$$

with $\hat{\Delta}$ a block-diagonal matrix. In other words, both the performance specifications and uncertainty are taken into account in a similar fashion. Moreover, $\hat{\mathbf{D}}$ is the uncertainty set with a structure as given in (26) and any $\Delta \in \mathbf{D}$ and $\Delta_P \in \mathbf{D}_P$. The robust performance condition can now be formulated as follows:

$$\mu_{\hat{\mathbf{D}}}\left(N(j\omega)\right) \leq 1, \quad \forall\omega, \tag{27}$$

where $\mu_{\hat{\mathbf{D}}}$ is the structured singular value with respect to $\hat{\mathbf{D}}$. The structured singular value is defined as the real non-negative function

$$\mu_{\hat{\mathbf{D}}}(N) = \frac{1}{\bar{k}_m}, \; \bar{k}_m = \min\left\{k_m \,\middle|\, \det\left(I - k_m N \hat{\Delta}\right) = 0\right\} \tag{28}$$

with complex matrix N and block-diagonal uncertainty $\hat{\Delta}$.

To optimize the robustness with respect to the uncertainty Δ (i.e., part of $\hat{\Delta}$ in (26)), the skewed structured singular value μ^s can be used. The skewed structured singular value is used if some uncertainty blocks in $\hat{\Delta}$ are kept fixed (Δ_P in this case) to investigate how large another source of uncertainty (Δ in this case) can be, before robust stability/performance can no longer be guaranteed. In this case, we aim to optimize the robustness of the closed-loop system with respect to uncertainty Δ in the bit-rock interaction. Thus, we aim to obtain the largest uncertainty set Δ, given a fixed Δ_P (i.e., fixed performance specifications). Therefore, we introduce the matrix $K_m^s := \operatorname{diag}\left(k_m^s,\; I\right)$, and the skewed structured singular value $\mu^s_{\hat{\Delta}}(N)$ is defined as

$$\mu^s_{\hat{\mathbf{D}}}(N) = \frac{1}{\bar{k}^s_m}, \; \bar{k}^s_m = \min\left\{k^s_m \,\middle|\, \det\left(I - K^s_m N \hat{\Delta}\right) = 0\right\}. \tag{29}$$

Thus, the robust performance condition (27), with additional scaling (through K_m^s) in terms of the skewed structured singular value, is written as the *alternative* robust performance condition

$$\mu^s_{\hat{\mathbf{D}}}\left(N(j\omega)\right) \leq 1, \quad \forall\omega. \tag{30}$$

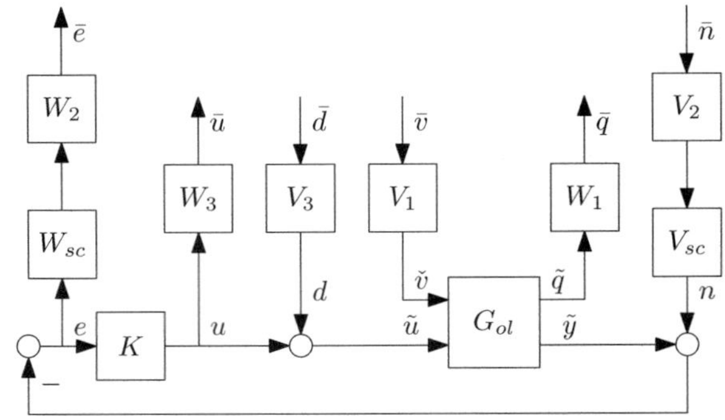

Fig. 19 Closed-loop system with weighting filters and scaling matrices

To support controller design satisfying particular performance specifications, weighting filters and scaling matrices are introduced in the loop in Fig. 18, as shown in Fig. 19. Those frequency-domain weighting filters allow us to specify the (inverse) maximum allowed magnitudes of the closed-loop transfer functions. Moreover, the scaling matrices are introduced to improve the numerical conditioning of the problem and to tune the desired bandwidth. The (weighted) generalized plant P with input weighting filters $V_i(s)$ and output weighting filters $W_i(s)$, with $i \in \{1, 2, 3\}$, and scaling matrices W_{sc} and V_{sc}, is specified by

$$\begin{bmatrix} \bar{q} \\ \bar{e} \\ \bar{u} \\ e \end{bmatrix} = \underbrace{\begin{bmatrix} W_1 & 0 & 0 & 0 \\ 0 & W_2 W_{sc} & 0 & 0 \\ 0 & 0 & W_3 & 0 \\ 0 & 0 & 0 & I_2 \end{bmatrix} \bar{P} \begin{bmatrix} V_1 & 0 & 0 & 0 \\ 0 & V_{sc} V_2 & 0 & 0 \\ 0 & 0 & V_3 & 0 \\ 0 & 0 & 0 & 1 \end{bmatrix}}_{P} \begin{bmatrix} \bar{v} \\ \bar{n} \\ \bar{d} \\ u \end{bmatrix}.$$

Herein, $\bar{P}(s)$ is the MIMO transfer function of the unweighted system $\bar{P}$ with inputs $\begin{bmatrix} \tilde{v} & n & d & u \end{bmatrix}^\top$ and outputs $\begin{bmatrix} \tilde{q} & e & u & e \end{bmatrix}^\top$ with its state-space realization given by

$$\bar{P} \stackrel{s}{=} \left[\begin{array}{c|cccc} A_t & G & 0 & B & B \\ \hline H & 0 & 0 & 0 & 0 \\ -C & 0 & -I & 0 & 0 \\ 0 & 0 & 0 & 0 & I \\ -C & 0 & -I & 0 & 0 \end{array} \right]. \tag{31}$$

In this section, we have introduced an alternative robust performance framework. To design a controller that minimizes the skewed structured singular value $\mu^s_{\hat{\mathbf{D}}}$, for the purpose of obtaining robust performance, a procedure for synthesizing such a controller, known as the DK-iteration procedure [34, Sect. 8.12], is treated concisely below.

The first step in such a DK-iteration procedure is the introduction of D-scaling matrices. This scaling uses the fact that $\hat{\Delta}$ is structured, hence, the inputs and out-

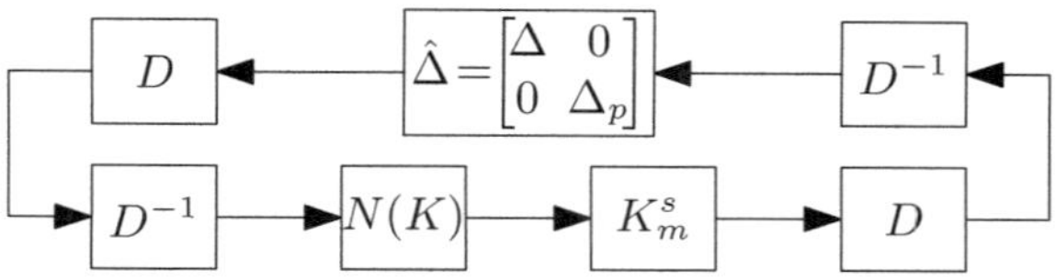

Fig. 20 Block diagram of the implementation for the skewed-μ DK-iteration procedure

puts to $\hat{\Delta}$ and N are scaled by inserting the matrices D and D^{-1}, as shown in Fig. 20. Using such scaling generally enables one to find potentially tighter robust stability/performance conditions. For further details on the procedure, the reader is referred to [24, 34].

The skewed-μ DK-iteration procedure aims at designing a controller that minimizes the peak value over frequency of the upper bound on the skewed structured singular value, i.e., a controller K should be designed by solving the following optimization problem:

$$\min_{K}\left(\min_{D}\|DK_m^sN(K)D^{-1}\|_\infty\right). \tag{32}$$

Here, the original scaling matrix $D(\omega)$ is replaced by a stable minimum-phase transfer function fit $D(s)$ of $D(\omega)$. The dependency of the closed-loop transfer function N on the controller K is indicated by $N(K)$. In DK-iterations, a μ-analysis (D-step) and $\mathscr{H}_\infty$-optimization (K-step) are solved alternately (see [24]). In other words, the skewed-μ DK-iteration procedure alternates between minimizing (32) with respect to either K or D (while holding the other fixed) and recursively updating k_m^s (which characterizes K_m^s) during the D-step.

3.3.3 Closed-Loop Stability Analysis

The main purpose of the controller is to stabilize the equilibrium $\xi = 0$ of the nonlinear system (19). Let us that assume that a controller K has been designed that meets the performance specifications and is robust with respect to the uncertainty Δ. Hence, the designed controller guarantees stability for the *linear* closed-loop system $N(s)$ and achieves robustness with respect to the uncertainty Δ. In this section, the stability of the *nonlinear* closed-loop system is considered. Therefore, we define a symmetric sector condition on the nonlinearity $\tilde{\varphi}$ such that, for any (locally Lipschitz) nonlinearity which (locally) satisfies this sector condition, (local) asymptotic stability of the origin of the closed-loop system can be guaranteed.

We use the circle criterion [16, Theorem 7.1] to determine a (symmetric) sector on the nonlinearity $\tilde{\varphi}$ for which robust stability can be guaranteed. Consider the closed-loop bit-mobility (22) and a symmetric sector condition on the nonlinearity which is satisfied for all $\tilde{q} \in \mathscr{S}$ with $\mathscr{S} := \{\tilde{q} \in \mathbb{R} | \tilde{q}_l < \tilde{q} < \tilde{q}_u\}$ and $\tilde{q}_l < 0 < \tilde{q}_u$, i.e. $\tilde{\varphi}(\tilde{q}) \in \left[-\gamma, \gamma\right]$ $\forall\, \tilde{q} \in \mathscr{S}$ and $\gamma > 0$. We note that, although $\tilde{\varphi}$ is a set-valued nonlinearity, we have that, for $\omega_{eq} > 0$ (i.e., for a nominal velocity away from the discontuinity in the bit-rock interaction at zero velocity), there indeed exist $\tilde{q}_l$ and $\tilde{q}_u$ such that the latter symmetric sector condition is satisfied. The nonlinear system is locally absolutely stable (i.e., $\xi = 0$ is locally asymptotically stable for any $\tilde{\varphi}(\tilde{q}) \in [-\gamma, \gamma]$ with $\tilde{q} \in \mathscr{S}$) if

$$H(s) = (1 + \gamma G_{cl}(s))\,(1 - \gamma G_{cl}(s))^{-1}, \tag{33}$$

is strictly positive real. Applying Lemma 6.1 in [16], a scalar transfer function $H(s)$ is strictly positive real if the following conditions are satisfied:

1. $H(s)$ is Hurwitz;
2. $\mathrm{Re}\,[H(j\omega)] = \mathrm{Re}\left[\frac{1+\gamma G_{cl}(j\omega)}{1-\gamma G_{cl}(j\omega)}\right] > 0, \quad \forall \omega \in \mathbb{R}$;
3. $H(\infty) > 0$.

For the symmetric sector, the condition on $H(s)$ being Hurwitz is equivalent to $G_{cl}(s)$ being Hurwitz. The closed-loop transfer function $G_{cl}(s)$ of the feedback interconnection is Hurwitz by the design of the stabilizing controller K. Moreover, G_{cl} is strictly proper, and therefore $H(\infty) = 1$, such that the third condition is satisfied. The second condition is equivalent to the condition:

$$\|G_{cl}(j\omega)\|_\infty < \frac{1}{\gamma}. \tag{34}$$

Hence, the $\mathscr{H}_\infty$-norm of the closed-loop bit-mobility G_{cl} gives an upper bound on the sector that the nonlinearity $\tilde{\varphi}$ should comply with, for the system to be absolutely stable. With the DK-iteration procedure, presented above, a controller K can be designed such that $\|G_{cl}\|_\infty$ is indeed minimized. In other words, the robustness with respect to uncertainty in the bit-rock interaction is optimized. This shows the benefit of employing the alternative robust performance technique (see Sect. 3.3.2) in terms of optimizing the robustness of the closed-loop drill-string dynamics with respect to the uncertainty in the bit-rock interaction, also in the nonlinear context.

In the following section, design guidelines for the tuning of the weighting filters tailored to the drilling context are given, and the designed controller is presented and validated through experiments.

4 Experimental Controller Validation

In this section, the implementation and experimental results obtained with the controller design strategy of the previous section are presented. First, a startup scenario for the experiments on the drill-string setup is discussed in Sect. 4.1. Next, in Sect. 4.2, the implementation of the SoftTorque controller, being the industrial standard, is discussed, and an experimental result of this industrial controller having been applied to the setup is shown. Finally, in Sect. 4.3, the implementation and experimental results are discussed for the linear robust output-feedback controller, as presented in Sect. 3.

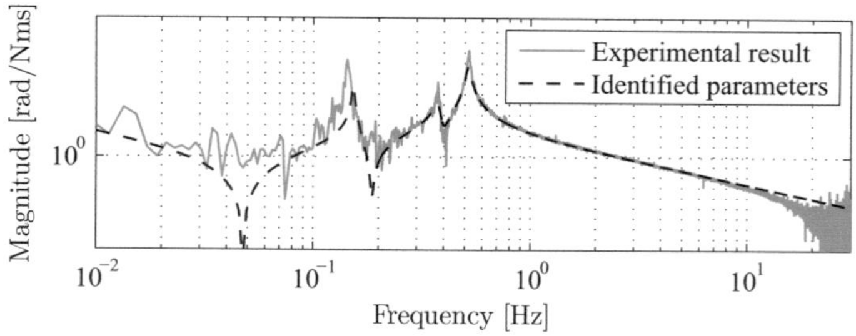

Fig. 21 Open-loop bit-mobility of the setup, i.e., the frequency response function from bit torque T_{bit} to bit velocity ω_{bit}

Before going into detail about the implementation of the controllers and the experimental results, let us consider the (open-loop) bit-mobility of the experimental drill-string setup. As advocated earlier, the bit-mobility plays an important role in the onset of stick-slip vibrations and the proposed control strategy aims to minimize its $\mathcal{H}_\infty$-norm. The measured open-loop bit-mobility of the setup is shown in Fig. 21. In the same figure, the bit-mobility of the setup model based on experimentally identified parameters is shown. For details on the performed parametric model identification, we refer to [37]. Clearly, the third resonance mode is well captured by the model and the second flexibility mode is more damped in the model compared to the actual bit-mobility of the setup. Moreover, the first resonance mode exhibits some discrepancy between the model and the experiments. Here, we opt for an identified model that aims to capture, in particular, the third resonance mode accurately, as it is precisely this dominant mode (in the bit-mobility) that is responsible for the occurrence of stick-slip oscillations. Moreover, it is well-known that it is relatively easy to design a controller that robustly damps the first mode despite such uncertainties (already guaranteed by a Soft-Torque controller).

4.1 Experimental Startup Scenario Description

For the experiments, we introduce a so-called startup scenario, which is based on practical startup procedures for drilling rigs. Herein, the drill-string is first accelerated to a low constant rotational velocity with the bit above the formation (off bottom) and, subsequently, the angular velocity and weight-on-bit (WOB) are gradually increased to the desired operating conditions. The increase in WOB is modelled as a scaling of the bit-rock interaction torque (TOB).

The startup scenario for the experiments is visualized in Fig. 22. The reference angular velocity of the upper discs is shown in the upper plot and the scaling of the TOB, indicated by α, is shown in the bottom plot. The timing of the transitions in the startup scenario can be summarized as follows:

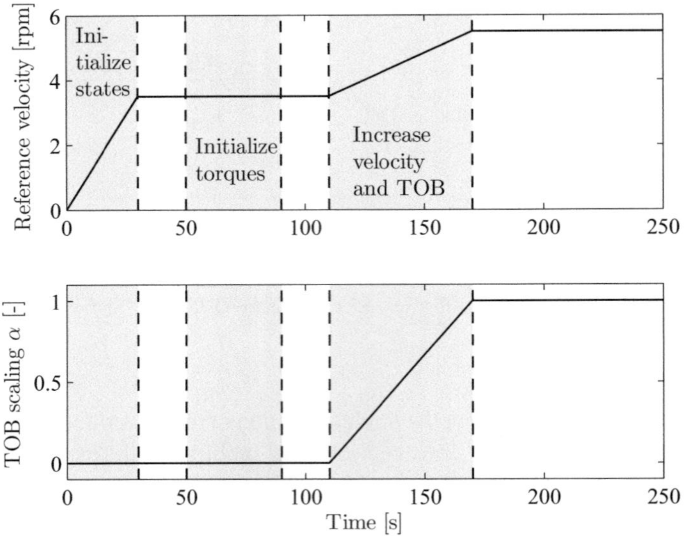

Fig. 22 Reference velocity and TOB scaling of the startup scenario for the experimental setup

1. Start with zero initial velocities and linearly increase the reference angular velocity from zero to 3.5 rpm[2] in the period between $t = 0$ and $t = 30$ s. At the same time, increase the feedforward torque (u_c) to its nominal value;
2. Between $t = 50$ and $t = 90$ s, adapt the drill-string borehole interaction torques T_w to obtain the desired values, based on the torque sensor readings, in order to compensate for possibly changed friction characteristics (in the bearings supporting the discs);
3. Gradually increase the reference angular velocity until the desired operating velocity (ω_{eq} being 5.5 rpm) is reached (in the time window $110 \leq t < 170$ s). At the same time, gradually change the TOB to emulate that the bit bites the formation, and finally, obtain the nominal operating condition in both the angular velocity and the TOB. Adapting the torque on bit is done as follows. The bit-rock interaction model is scaled by using the scaling factor $\alpha(t)$ according to:

$$\hat{T}_{bit}(t) = \text{Sign}(\omega_{bit}) \left(T_{ini} + \alpha(t) \left(\hat{T}_d - T_{ini} + \left(\hat{T}_s - \hat{T}_d \right) e^{-\frac{30}{\hat{N}_d \pi} |\omega_{bit}|} \right) \right), \quad (35)$$

where T_{ini} is the amount of resisting torque that is still present at the bit-rock interface, even when the bit is off bottom (e.g., due to drilling mud and interactions with the borehole). For WOB $= 0$ (off bottom; characterized by $\alpha = 0$), there is no velocity-weakening in the TOB. The scaling factor $\alpha(t)$ in (35) is given by

[2]Note that due to scaling, this corresponds to 35 rpm on a real drilling rig.

$$\alpha(t) = \begin{cases} 0, & t_0 \leq t \leq t_1 \\ \frac{t-t_1}{t_2-t_1}, & t_1 < t < t_2 \\ 1, & t \geq t_2 \end{cases} \tag{36}$$

with $t_1 = 110$ and $t_2 = 170$ in this case.

4.2 SoftTorque Controller

The SoftTorque controller ([14]) is a controller for drill-string systems, widely used in industry. This controller aims at damping of the first torsional flexibility mode of the drill-string system only. This active damping system is a PI-controller, based only on the velocity error e_y between the measured top drive velocity $y = \omega_{td}$ and the reference angular velocity $\omega_{td,ref}$, i.e., $e_y := \omega_{td,ref} - \omega_{td}$. The controller is given by the transfer function

$$T_{fb}(s) = \left(c_t + \frac{k_t}{s}\right) e_y(s), \; s \in \mathbb{C}, \tag{37}$$

with $c_t = 2.93$ and $k_t = 1.87$ tuned such that damping of the first torsional flexibility mode of the setup is obtained (note that these controller parameter settings corresponds to unscaled system parameters, as mentioned in Sect. 2.1.4). In Fig. 23, the measured closed-loop bit-mobility of the drill-string setup with the SoftTorque controller is shown. It is clearly visible that the first torsional mode is damped using the SoftTorque controller, but the amplitude of the second and third modes are similar in the open-loop and closed-loop cases, illustrating a key deficiency of the SoftTorque controller.

An experimental result of the closed-loop drill-string system with SoftTorque controller (with the same constant feedforward active as for the controller proposed in Sect. 3) is shown in Fig. 24. In the response of the bit angular velocity, stick-slip oscillations can be observed. The onset of these oscillations starts when the reference

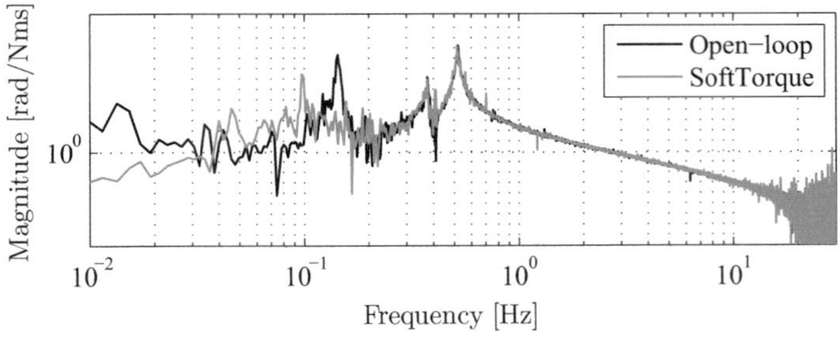

Fig. 23 Bit-mobility of the setup with SoftTorque controller

angular velocity and scaling factor α (for emulating an increase of the WOB) start to increase at $t = 110$ s. This experimentally shows that the SoftTorque controller is indeed unable to avoid stick-slip oscillations for the setup.

In Fig. 24, the filtered and unfiltered responses of the system are shown. The filtered response of the system is compared with a simulation result of the model of the setup with the identified parameters. The results are shown side-by-side in Fig. 25. To allow for a clear comparison, a shift of the time axis has been applied for

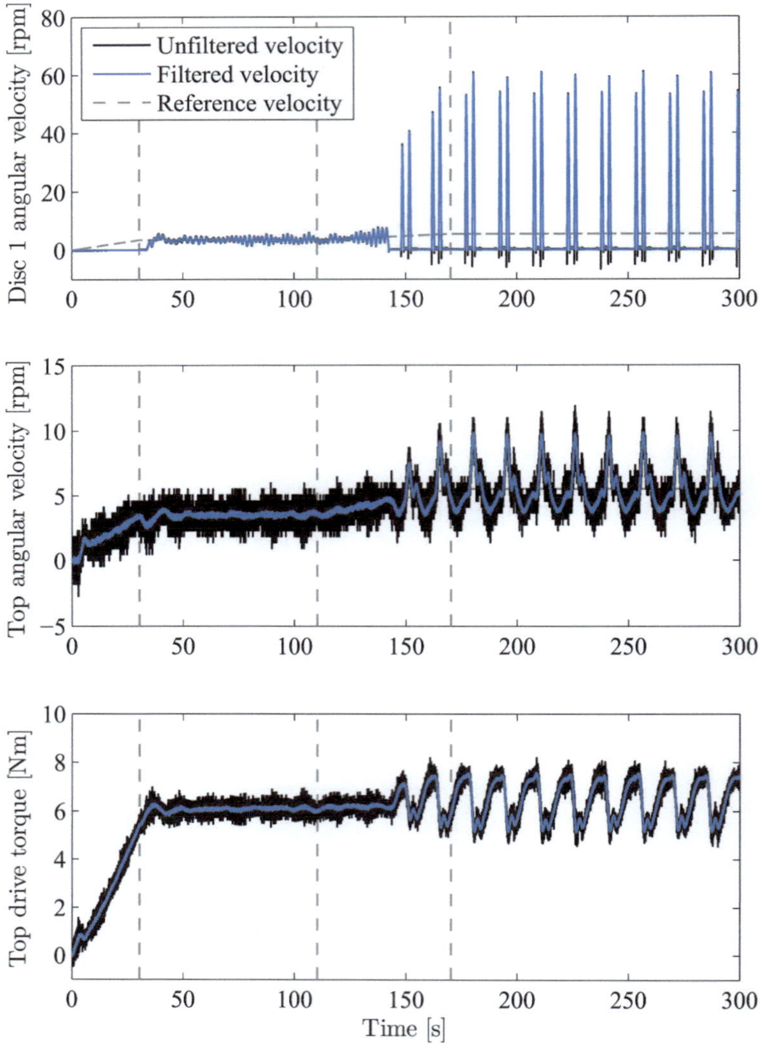

Fig. 24 Experimental result of the drill-string setup with the SoftTorque controller in the startup scenario

the experimental results. As can be seen from this figure, the closed-loop response of the experimental setup is very similar to the response of the simulation results. The only difference is the somewhat shorter sticking period in the simulation results between two successive groups of two slipping periods (i.e., the long sticking period). This result further illustrates that the setup is capable of accurately emulating the non-smooth drill-string dynamics to be investigated, also in closed-loop operation.

4.3 $\mathscr{H}_\infty$-Based Output-Feedback Controller

The linear robust output-feedback controller design methodology, presented in Sect. 3.3, is also used to design a controller for the experimental drill-string setup. The results of the drill-string setup in closed loop with the $\mathscr{H}_\infty$-based controller are presented in this section.

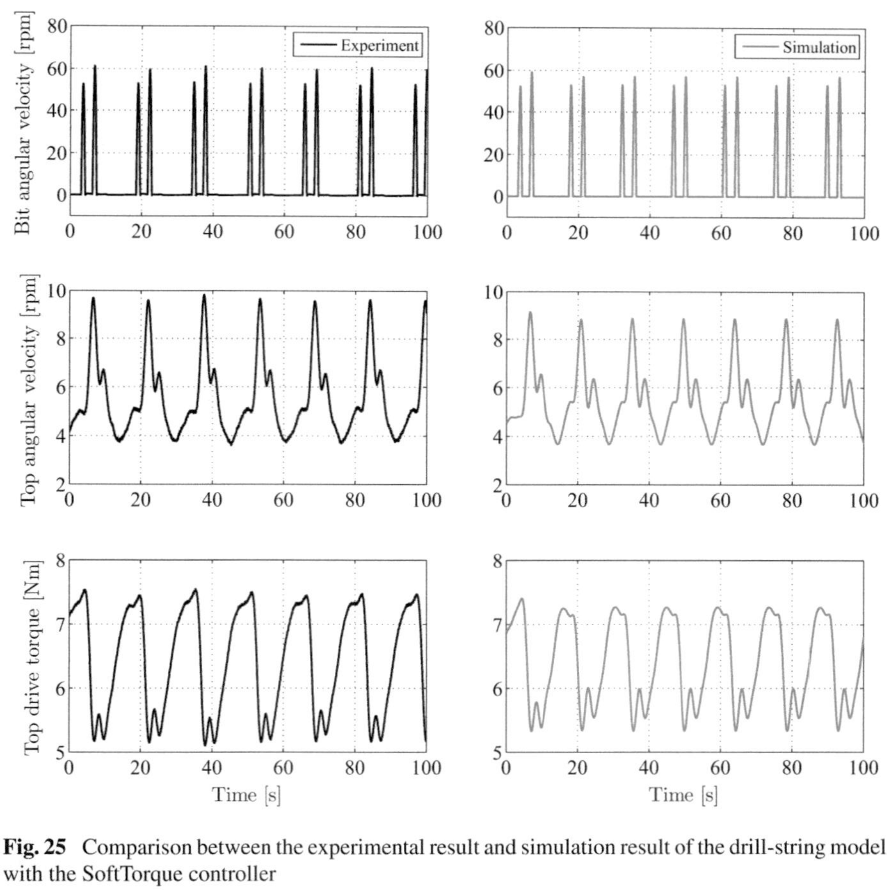

Fig. 25 Comparison between the experimental result and simulation result of the drill-string model with the SoftTorque controller

Weighting filter design is key to satisfying the performance specifications related to, e.g., measurement noise sensitivity and actuator limitations. Moreover, achieving specific design targets such as the inclusion of integral action and high-frequency roll-off can be achieved by absorbing these filters into the loop see [21]. High-frequency roll-off reduces measurement noise amplification. Also, integral action is desired from a practical point of view, e.g., in case of a mismatch between the (model-based) feedforward torque u_c and the actual required feedforward torque due to uncertainty in the respective models for the bit-rock interaction and the drill-string borehole interaction. In that case, integral action will compensate for this mismatch so as to obtain convergence to the desired setpoint.

For the design of a controller for the drill-string model (18), the following objectives are set:

- Integral action for low-frequencies;
- Second-order roll-off for high freqencies
- Cross-over frequency of the open-loop transfer function KG_{ol} (at the plant input) at 0.6 Hz, i.e., just above the third eigenfrequency of the drill-string system (see Fig. 11);
- Plant output scaling, i.e., scale the plant output $y = \begin{bmatrix}\omega_{td} & T_{pipe}\end{bmatrix}^{\top}$ such that the components of the weighted plant output are of the same order of magnitude.

These objectives are obtained through specific choices for several settings of the weighting filters, as displayed in Fig. 19.

First, we apply plant scaling by using the scaling matrices W_{sc} and V_{sc}. This scaling is applied to compensate for the different order of magnitude of the two plant outputs ω_{td} and T_{pipe}. This is important for a system with multiple outputs in a norm-based controller synthesis method such as skewed-μ DK-iteration. When the plant outputs are not scaled and the outputs differ in order of magnitude, one off-diagonal term in the closed-loop sensitivity function will be large and the other small. In the synthesis, it is then possible that the emphasis is on reducing the large off-diagonal element at the expense of other elements. The plant scaling matrices W_{sc} and V_{sc} are tuned to compensate for this effect. The matrices are given by

$$W_{sc} = \begin{bmatrix} w_{sc1} & 0 \\ 0 & w_{sc2} \end{bmatrix}, \quad V_{sc} = \begin{bmatrix} 1 & 0 \\ 0 & 1 \end{bmatrix}.$$

The filters $V_i(s)$ and $W_i(s)$, $i = 1, 2, 3$, are so-called performance filters and are used to tune the performance-related properties of the closed-loop system. The filters $V_1(s)$ and $W_1(s)$ can be used to tune the closed-loop bit-mobility (G_{cl}). Ideally, the bit-mobility should be damped as much as possible (as follows from the stability analysis in Sect. 3.3.3). However, this typically results in high control action. To deal with this trade-off, the weighting filter $V_1(s)$ has a notch filter and is defined as follows:

$$\begin{aligned} V_1 &= v_1 V_{notch} \\ &= v_1 \frac{\frac{1}{(2\pi f_1)^2} s^2 + \frac{2b_1}{2\pi f_1} s + 1}{\frac{1}{(2\pi f_2)^2} s^2 + \frac{2b_2}{2\pi f_2} s + 1}, \end{aligned} \tag{38}$$

where f_j $(j = 1, 2)$ is the frequencies of the notch filter $V_{notch}(s)$ and b_1 and b_2 the parameters for tuning the depth of the notch filter. The output weighting filter $W_1(s)$ is set to a constant w_1.

The remaining weighting filters are the filters for tuning the closed-loop performance transfer functions. Let us first focus on the input weighting filters $V_2(s)$ and $V_3(s)$. The filter $V_2(s)$ is given by

$$V_2 = \begin{bmatrix} v_{21} & 0 \\ 0 & v_{22} \end{bmatrix}, \tag{39}$$

where v_{21} and v_{22} are static gains. These gains, as well as static gains in other weighting filters, are used to scale those filters. Scaling is necessary to obtain a feasible controller design with respect to the performance uncertainty $\Delta_P(s)$ and changing the gains allows for the synthesis of different controllers. The input weighting filter $V_3(s)$ is set as

$$V_3(s) = v_3 \|g_{co}\|^{-1} \frac{1}{w_{sc1}}, \tag{40}$$

where v_3 is a static gain and $g_{co} := g_{22,1}(j2\pi f_{co})$, i.e., the sub plant gain, related to input $\tilde{u}$ and output $\tilde{y}_1 = \omega_{td} - \omega_{eq}$, at the target cross-over frequency f_{co}. This gain is chosen to obtain a cross-over frequency of the open-loop transfer function KG_{ol} at 0.6 Hz, as specified. This cross-over frequency is chosen to achieve damping of the dominant resonance modes.

The output weighting filters $W_2(s)$ and $W_3(s)$ are also used to tune the closed-loop transfer functions, as well as to meet the first two controller objectives, i.e., to include integral action and first-order roll-off. The controller $K_t(s)$ to be designed has two inputs and a single output (due to the two measured signals of the plant), i.e., $K_t(s) = \left[K_{\omega_{td}}(s)\ K_{T_{pipe}}(s)\right]$. The controller aims at stabilizing the desired angular velocity setpoint. Hence, an integrator should be specified in the top drive angular velocity control loop. Note that it is not possible (and not necessary) to include an integrator in both control loops $K_{\omega_{td}}(s)$ and $K_{T_{pipe}}(s)$. An integrator would force the sensitivity function to zero for $s = 0$; however, this is not possible for both sensitivity functions, due to the fact that we are dealing with a non-square plant. In other words, there is only one control signal that can eliminate the steady-state error for one of the two measurements. However, forcing ω_{td} to its equilibrium value also results in T_{pipe} converging to its equilibrium. So, by only requiring integral action in the control loop related to ω_{td}, the output weighting filter $W_2(s)$ is given by

$$W_2(s) = \begin{bmatrix} W_I(s) & 0 \\ 0 & w_{22} \end{bmatrix} = \begin{bmatrix} P_I \frac{s+2\pi f_I}{s} & 0 \\ 0 & w_{22} \end{bmatrix}, \tag{41}$$

using $W_I(s)$ to obtain an integral action in $K_{\omega_{td}}(s)$ and w_{22} a static gain. To obtain high-frequency roll-off, a roll-off filter is included in the output filter $W_3(s)$, hence

$$W_3(s) = w_3 w_{sc1} \|g_{co}\| W_R^{-1}, \tag{42}$$

where w_3 is a static gain, and $W_R = \frac{(2\pi f_R)^2}{s^2+4\pi\beta f_R s+(2\pi f_R)^2}$ the second-order roll-off filter with roll-off frequency f_R.

The weighting filters $W_2(s)$ and $W_3(s)$ are unstable and non-proper weighting filters, respectively. Therefore, these filters are not applicable in the $\mathscr{H}_\infty$-controller synthesis. To circumvent this limitation and still obtain a controller that includes integral action and high-frequency roll-off, we add filters in the loop [21]. We require high-frequency roll-off on both input signals (top drive velocity and pipe torque) of the controller and integral action on the top drive velocity. To acheive this, the actual plant that is used in the controller synthesis algorithm is given by

$$G_t(s) = \text{diag}\left(1,\ W_I(s),\ 1\right) G_{ol}(s) \text{diag}\left(1,\ W_R(s)\right), \tag{43}$$

where $W_R(s)$ and $W_I(s)$ are the roll-off and integrator filters, respectively. The resulting controller $K(s)$ from the DK-iteration procedure, treated in Sect. 3.3.2, for this plant G_t, has no integrator and roll-off properties. However, the actual controller (for the plant G_{ol}) can be calculated as follows:

$$K_t(s) = W_R(s) K(s) \text{diag}\left(W_I(s),\ 1\right), \tag{44}$$

which does include the desirable integrator and roll-off properties.

Now, two different controllers will be synthesized based on the skewed-μ DK-iteration procedure and the proposed weighting filters from the previous section. Of course, it is possible to change all weighting filters so as to obtain a different controller; however, the weighting filters have been chosen such that the controller objectives can be met, and tuning of the parameters already allows us to synthesize different controllers. The two controllers mainly differ in the allowed control action and will be referred to as a *high-gain* (hg) controller and a *low-gain* (lg) controller. The extra allowed control action for the high-gain controller is used for even greater suppression of the bit-mobility compared to the low-gain controller. In Table 3, the parameters of the weighting filters are given for both controllers. The notch filter in $V_1(s)$ is used to allow for a higher bit-mobility within specific frequency ranges.

Performing the DK-iteration procedure for the drill-string system with the weighting filters as specified above, results in the controller $K_t(s) = \left[K_{\omega_{td}}(s),\ \ K_{T_{pipe}}(s)\right]$, as shown in Fig. 26 for both the high-gain and the low-gain controller. These controllers only use the measured top drive angular velocity ω_{td} and the pipe torque measurement T_{pipe}. In the experimental setup, the pipe torque measurement is based on the torque sensor reading just below the upper disc, compensated for the additional damping term between disc 1 and 4. From this figure, the integral action in the controller, $K_{\omega_{td}}(s)$, which uses the top drive angular velocity, can be clearly recognized. This feature is also present (single-input-single-output) in the SoftTorque controller, also depicted in Fig. 26. This figure shows that both the high-gain and the low-gain controller have a second-order roll-off filter. It can also be observed that the designed

Table 3 Parameter settings for the performance weighting filters for the designed high-gain and low-gain controller

Filter/setting	Parameters	
	Low-gain controller	High-gain controller
W_{sc}	$w_{sc1} = 1, w_{sc2} = 10$	$w_{sc1} = 1, w_{sc2} = 10$
V_1	$v_1 = 0.1$	$v_1 = 0.7$
	$f_1 = f_2 = 0.517\,\text{Hz}$	$f_1 = f_2 = 0.518\,\text{Hz}$
	$b_1 = 0.125$	$b_1 = 0.033$
	$b_2 = 0.91$	$b_2 = 0.8$
W_1	$w_1 = 1.2$	$w_1 = 1$
V_2	$v_{21} = 4, v_{22} = 0.125$	$v_{21} = 5, v_{22} = 0.167$
V_3	$v_3 = 1.286$	$v_3 = 1.135$
W_2	$P_I = 0.1$	$P_I = 0.01$
	$f_I = 0.134$	$f_I = 1$
	$w_{22} = 0.5$	$w_{22} = 0.01$
W_3	$w_3 = 0.243$	$w_3 = 0.0044$
	$f_R = 0.469$	$f_R = 1$
	$\beta = 0.1$	$\beta = 0.1$

controllers have distinct frequency-dependent characteristics within the frequency range of the torsional resonance modes of the drill-string system (see Fig. 21), which is not the case for the SoftTorque controller. This industrial controller, which only uses top drive velocity measurements, is a properly tuned active damping system (i.e., PI-control of the angular velocity), which aims at damping only the first torsional mode of the drill-string dynamics.

The resulting measured bit-mobilities are shown in Fig. 27. It can be seen that the designed controllers suppress the first and second flexibility mode in the bit-mobility. However, the third mode is only slightly damped using these controllers. Clearly, the high-gain controller ($\mathscr{H}_\infty$ (hg)) achieves more damping of the third mode than the low-gain controller ($\mathscr{H}_\infty$ (lg)). The limited amount of damping of this mode is caused by the fact that it is difficult to synthesize a controller that suppresses the third flexibility mode and at the same time satisfies the performance specifications regarding measurement noise sensitivity. The sensitivity with respect to measurement noise plays an important role in the design of controllers for the experimental setup, because the level of noise (especially on the top drive angular velocity) is relatively high. In addition, the third mode is almost unobservable in, e.g., the frequency response function from top-drive torque to top-drive velocity, see Fig. 10. Therefore, it is difficult to suppress the third torsional flexibility mode.

Remark 1 We conjecture that (e.g., torque) sensors in the drill-string can significantly improve the observability properties of such essential flexibility modes, and can hence potentially be used in a feedback strategy to improve the damping of such modes that are poorly observable in surface measurements.

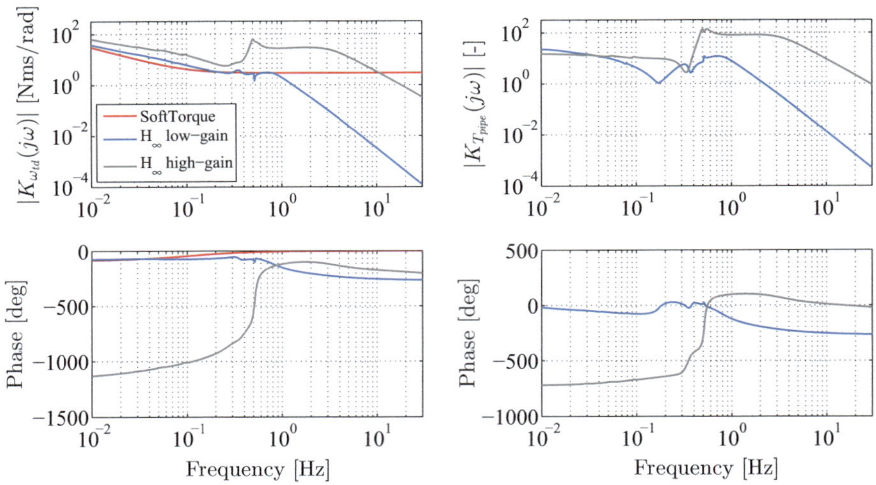

Fig. 26 Designed linear dynamic controllers (and Soft-Torque controller) for the experimental drill-string setup. Left plot is the controller that uses the top drive angular velocity, while the controller in the right plot is based on the pipe torque measurement

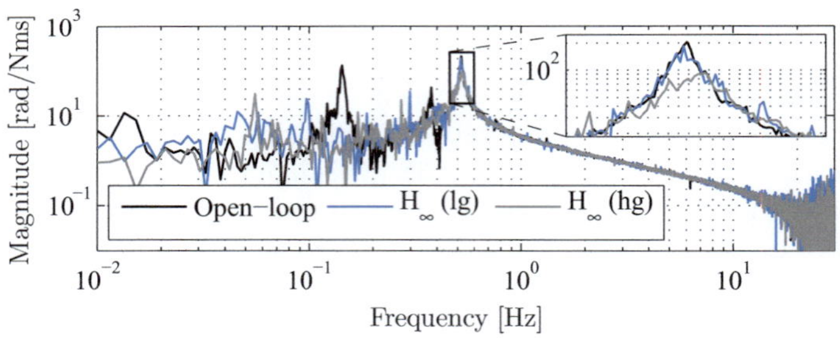

Fig. 27 Bit-mobility of the setup with two different $\mathscr{H}_\infty$-controllers

The measured response is shown in Fig. 28. First, the low-gain $\mathscr{H}_\infty$-controller is used, and after approximately 210 s, we switch to the high-gain controller. This switch is not necessary, and the desired setpoint can also be stabilized using the low-gain controller only. However, the high-gain controller has improved robustness properties (due to the improved damping of the third mode), which can be beneficial. By only using the high-gain controller in the startup scenario, it is not possible to stabilize the desired setpoint. A closer look at the experimental results with the $\mathscr{H}_\infty$-controllers shows that the low-gain controller is able to stabilize the desired setpoint of 5.5 rpm with limited control action (i.e., at least the controller acts less aggressively compared to the high-gain $\mathscr{H}_\infty$-controller). The oscillations in the bit angular velocity are still relatively large in amplitude; however, the oscillations are sufficiently damped to mitigate stick-slip vibrations. In addition, it has to be noted that, due to the presence of the roll-off filters in the controller, high-frequency (measurement) noise is not amplified

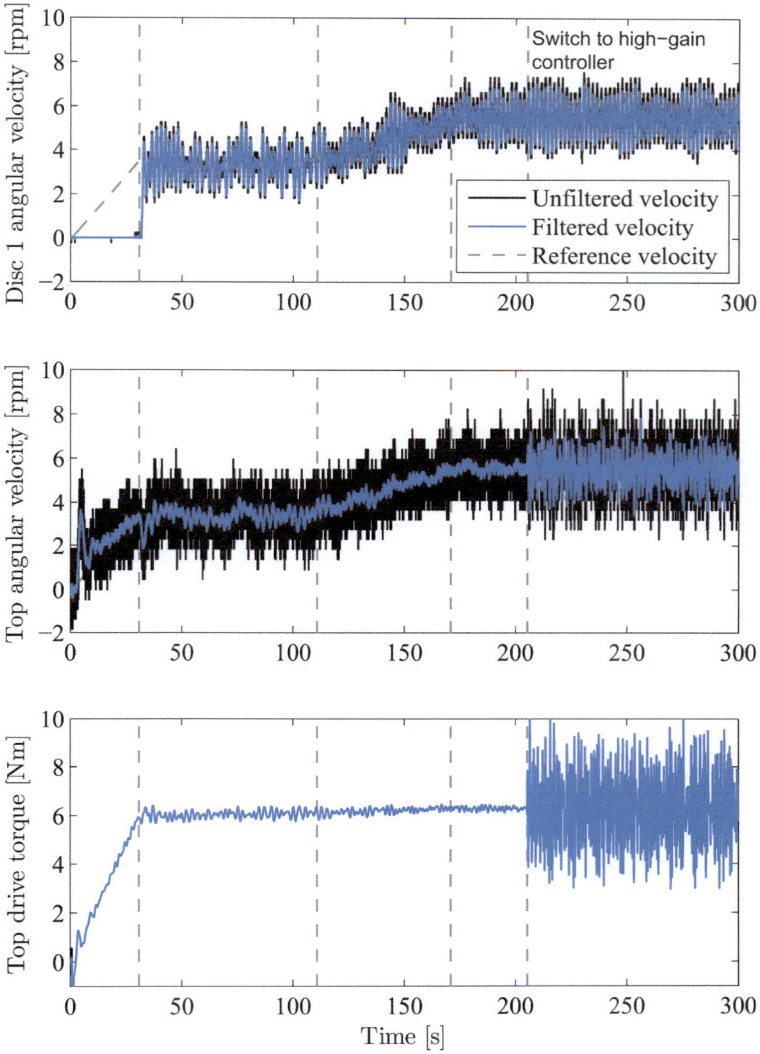

Fig. 28 Experimental result of the drill-string setup with the designed $\mathscr{H}_\infty$-controllers in the startup scenario, after approximately 210 s, the controller is switched to a high-gain $\mathscr{H}_\infty$-controller

by the controller, such that possible oscillations caused by such disturbances are avoided. The high-gain controller clearly uses more control action, which also results in more oscillations in the top drive angular velocity. The high-gain controller induces slightly larger oscillations in the bit angular velocity than those induced by the low-gain controller. The latter effect is related to the (measurement) noise sensitivity of these controllers. Still, the high-gain controller ensures a higher robustness against uncertainties in the bit-rock-interaction, as evidenced by an improved attenuation of the (third) resonance in the bit mobility, see Fig. 27.

Summarizing, with the designed $\mathscr{H}_\infty$-controllers, it is possible to stabilize a desired angular velocity of 5.5 rpm and to avoid stick-slip oscillations in a realistic scenario in which the SoftTorque controller could not avoid such oscillations.

5 Concluding Remarks

In this chapter, we have presented the design of an experimental, lab-scale drill-string setup based on a non-smooth model of a real-life drilling rig. The setup was designed to reflect multiple dominant torsional flexibility modes of the system dynamics, as field tests have shown that multiple modes can be associated with the occurrence of stick-slip oscillations. Next, we have proposed a robust control design strategy that can be used to design controllers that (1) stabilize a constant velocity setpoint, and hence avoid such stick-slip limit cycling, (2) guarantee robust stability in the presence of uncertainties in the bit-rock interaction, (3) take into account practically relevant performance specifications, and (4) guarantee robust stability and performance in the presence of multi-modal torsional drill-string dynamics. Finally, such controllers have been implemented and tested on the experimental setup, and it has been shown that these can eliminate stick-slip oscillations in realistic startup scenarios in which an industrial SoftTorque controller fails to do so.

References

1. Bresch-Pietri D, Krstic M (2014) Adaptive output-feedback for wave PDE with anti-damping—application to surface-based control of oil drilling stick-slip instability. In: Proceedings of the 53rd IEEE conference on decision and control, Los Angeles, California, U.S.A., pp 1295–1300
2. Canudas-de Wit C, Rubio F, Corchero M (2008) D-OSKIL: a new mechanism for controlling stick-slip oscillations in oil well drillstrings. IEEE Trans Control Syst Technol 16(6):1177–1191
3. Copley Controls (2015) Xenus Plus XTL-230-40. Website: http://www.copleycontrols.com/Motion/Products/Drives/Digital/xenusPlus.html
4. De Bruin JCA, Doris A, Van de Wouw N, Heemels WPMH, Nijmeijer H (2009) Control of mechanical motion systems with non-collocation of actuation and friction: a Popov criterion approach for input-to-state stability and set-valued nonlinearities. Automatica 45(2):405–415
5. Detournay E, Defourny P (1992) A phenomenological model for the drilling action of drag bits. Int J Rock Mech Mining Sci Geomech Abstracts 29(1):13–23
6. Doris A (2013) Method and system for controlling vibrations in a drilling system. Patent WO 2013(076184):A2
7. dSPACE (2015) DS1103 ppc controller board, https://www.dspace.com/en/inc/home/products/hw/singbord/ppcconbo.cfm
8. Dwars S (2015) Recent advances in soft torque rotary systems. In: SPE/IADC Drilling conference and exhibition, SPE/IADC 173037, London, United Kingdom
9. Georgii Kobold (2015) KTY-F torque motors, http://www.georgiikobold.de/en/products/Torque-Motors-KTY.b7059.php
10. Géradin M, Rixen D (1997) Mechanical vibrations: theory and application to structural dynamics, 2nd edn. Wiley, Chichester, England

11. Grauwmans R, Stulemeijer I (2009) Drilling efficiency optimization: vibration study, presentation shell exploration and production, Rijswijk, the Netherlands
12. Halsey GW, Kyllingstad A, Kylling A (1988) Torque feedback used to cure slip-stick motion. In: SPE annual technical conference and exhibition, Houston, Texas, U.S.A., SPE 18049, pp 277–282
13. Heidenhain (2015) Encoders for servo drives, pp 70–71. http://www.heidenhain.com/
14. Jansen JD, Van den Steen L (1995) Active damping of self-excited torsional vibrations in oil-well drillstrings. J Sound Vib 179(4):647–668
15. Karkoub M, Zribi M, Elchaar L, Lamont L (2010) Robust μ-synthesis controllers for suppressing stick-slip induced vibrations in oil well drill strings. Multibody Syst Dyn 23(2):191–207
16. Khalil H (2002) Nonlinear systems, 3rd edn. Prentice-Hall, Upper Saddle River
17. Khulief YA, Al-Sulaiman FA (2009) Laboratory investigation of drillstring vibrations. Proc Inst Mech Eng Part C: J Mech Eng Sci 223:2249–2262
18. Kovalyshen Y (2014) Experiments on stick-slip vibrations in drilling with drag bits. In: Proceedings of the 48th US rock mechanics geomechanics symposium, Minneapolis, Minnesota, USA
19. Kyllingstad A, Nessjøen PJ (2009) A new stick-slip prevention system. In: SPE/IADC drilling conference and exhibition, Amsterdam, Netherlands, SPE/IADC 199660
20. Lu H, Dumon J, Canudas-de Wit C (2009) Experimental study of the d-oskil mechanism for controlling the stick-slip oscillations in a drilling laboratory testbed. In: Proceedings of the IEEE multi-conference on systems and control, Saint Petersburg, Russia, pp 1551–1556
21. Meinsma G (1995) Unstable and nonproper weights in $\mathscr{H}_\infty$ control. Automatica 31(11):1655–1658
22. Mihajlović N, van Veggel AA, van de Wouw N, Nijmeijer H (2004) Analysis of friction-induced limit cycling in an experimental drill-string system. ASME J Dyn Syst Meas Control 126(4):709–720
23. Nessjøen PJ, Kyllingstad A, D'Ambrosio P, Fonseca IS, Garcia A, Levy B (2011) Field experience with an active stick-slip prevention system. In: SPE/IADC drilling conference and exhibition, SPE/IADC 139956, Amsterdam, Netherlands
24. Oomen T, van Herpen R, Quist S, van de Wal M, Bosgra O, Steinbuch M (2014) Connecting system identification and robust control for next-generation motion control of a wafer stage. IEEE Trans Control Syst Technol 22(1):102–118
25. Patil P, Teodoriu C (2013) A comparative review of modelling and controlling torsional vibrations and experimentation using laboratory setups. J Petrol Sci Eng 112:227–238
26. PCM (2015) RT2-rotating torque transducer. http://www.pcm-uk.com/transducer-torque-rt2.html
27. Perneder L, Detournay E (2013) Steady-state solutions of a propagating borehole. Int J Solids Struct 50:1226–1240
28. Printed Motor Works (2015) GN series. http://www.printedmotorworks.com/wp-content/uploads/GN-Series-Overview.pdf
29. Reimers N (2014) Mitigation of torsional vibrations and stick-slip induced by aggressive, energy efficient drill-bits. In: Detournay E, Denoël V, van de Wouw N, Zhou Y (eds) Third international colloquium on nonlinear dynamics and control of deep drilling systems, Minneapolis, Minnesota, USA, pp 9–14
30. Richard T, Germay C, Detournay E (2007) A simplified model to explore the root cause of stick-slip vibrations in drilling systems with drag bits. J Sound Vib 305(3):432–456
31. Serrarens AFA, Van De Molengraft MJG, Kok JJ, Van Den Steen L (1998) $\mathscr{H}_\infty$ control for suppressing stick-slip in oil well drillstrings. IEEE Control Syst Mag 18(2):19–30
32. Sick (2015) DFS60A incremental encoder. http://www.sick.com/
33. Sieb & Meyer (2015) SD2S drive amplifier. http://www.sieb-meyer.de/
34. Skogestad S, Postlethwaite I (2005) Multivariable feedback control, 2nd edn. Wiley
35. Tucker RW, Wang C (1999) On the effective control of torsional vibrations in drilling systems. J Sound Vib 224(1):101–122

36. Vlajic N, Balachandran B (2014) Nonlinear and delay effects in drilling mechanics. In: Detournay E, Denoël V, van de Wouw N, Zhou Y (eds) Third international colloquium on nonlinear dynamics and control of deep drilling systems, Minneapolis, Minnesota, USA, pp 51–66
37. Vromen T (2015) Control of stick-slip vibrations in drilling systems. PhD thesis, Eindhoven University of Technology, Eindhoven, The Netherlands
38. Vromen T, Dai CH, van de Wouw N, Oomen T, Astrid P, Nijmeijer H (2015) Robust output-feedback control to eliminate stick-slip oscillations in drill-string systems. In: 2nd IFAC workshop on automatic control in offshore oil and gas production, Florianopolis, Brazil, pp 272–277
39. Vromen T, van de Wouw N, Doris A, Astrid P, Nijmeijer H (2017) Nonlinear output-feedback control of torsional vibrations in drilling systems. Int J Robust Nonlinear Control. https://doi.org/10.1002/rnc.3759
40. Wiercigroch M, Pavlovskaia E, Vaziri V, Kapitaniak M (2014) Resonance enhanced drilling: modelling, experiments and design. In: Detournay E, Denoël V, van de Wouw N, Zhou Y (eds) Third international colloquium on nonlinear dynamics and control of deep drilling systems, Minneapolis, Minnesota, USA, pp 1–7
41. Zhou K, Doyle J, Glover K (1996) Robust and optimal control. Prentice Hall, Upper Saddle River

On the Constraints Formulation in the Nonsmooth Generalized-α Method

Olivier Brüls, Vincent Acary and Alberto Cardona

Abstract The simulation of flexible multibody systems with unilateral contact conditions and impacts requires advanced numerical methods. The nonsmooth generalized-α method was developed in order to combine an accurate and second-order time discretization of the smoother part of the dynamics and a consistent but first-order time discretization of the impulsive contributions. Compared to the Moreau-Jean scheme, this approach improves the quality of the numerical solution, especially for the representation of the vibrating response of flexible bodies. It relies on the formal definition of a so-called smooth motion that captures a non-impulsive part of the total nonsmooth motion. This definition may account for some contributions of the bilateral constraints and/or of the active unilateral constraints at the velocity or at the acceleration level. This chapter shows that the formulation of the constraints strongly influences the numerical stability and the computational cost of the method. A strategy for enforcing the bilateral and unilateral constraints simultaneously at the position, velocity and acceleration levels is also established, with a careful formulation of the activation criteria based on augmented Lagrange multipliers. In the special case of smooth systems, a comparison is made with more standard solvers for differential-algebraic equations. The properties of this method are demonstrated using illustrative numerical examples of smooth and nonsmooth mechanical systems.

O. Brüls (✉)
University of Liège, Allée de la Découverte 9, 4000 Liège, Belgium
e-mail: o.bruls@uliege.be

V. Acary
INRIA Rhône-Alpes, Grenoble, France
e-mail: vincent.acary@inria.fr

A. Cardona
CIMEC, Universidad Nacional del Litoral-CONICET, Santa Fe, Argentina
e-mail: acardona@unl.edu.ar

R. I. Leine et al. (eds.), *Advanced Topics in Nonsmooth Dynamics*,
https://doi.org/10.1007/978-3-319-75972-2_9

1 Introduction

This chapter addresses the numerical simulation of mechanical systems composed of rigid and flexible bodies interconnected by kinematic joints and subject to frictionless contact conditions. These models are intended for the analysis of the dynamic interactions between motion, impacts and vibrations in various industrial applications, such as in automotive, wind turbine, and robotic systems. The kinematic joints impose restrictions on the relative motions of the bodies and are modelled as bilateral constraints, whereas the non-penetration conditions at the contact points are modelled as unilateral constraints. These unilateral constraints may cause impact phenomena, so that the dynamic response becomes nonsmooth, involving velocity jumps and impulsive reaction forces.

In many practical situations, the nonsmooth behaviours are nevertheless localized in space and/or in time. After spatial and time discretization, this implies that velocity jumps and impulsive forces are only observed for a limited number of coordinates and/or during a limited number of time steps. Even though the correct description of these velocity jumps and impulsive forces is of the utmost importance for the global consistency of the simulation, the quality of the results within the smooth parts of the motion is also essential.

The most popular time-stepping methods for nonsmooth systems, such as the Moreau-Jean scheme [25, 27] or the Schatzman-Paoli scheme [29, 30], are robust with respect to the treatment of nonsmooth phenomena. An interesting overview of related mathematical results can be found in Chap. 4 of this book. However, these time-stepping methods lead to rather poor first-order approximations of the smooth parts of the motion and to high levels of numerical dissipation, which is particularly penalizing for the accurate representation of vibration phenomena in flexible systems. In the Moreau-Jean scheme, the constraints are imposed at the velocity level, so that a constraint drift generally appears at the position level. Alternatively, event-driven techniques, which adapt their time steps to the impact instants, can be used in combination with a higher order scheme during the free flight phases [20]. However, their performance decreases if the frequency of impacts increases, and they cannot be used if accumulation phenomena, involving an infinite series of impact in a finite time interval, are present. A more detailed description of numerical methods for the simulation of nonsmooth systems can be found in [2].

These observations motivated the recent development of more sophisticated time-stepping algorithms for nonsmooth systems, which involve improved approximations of the smoother parts of the motion [12, 14, 34, 35, 37]. Several authors [1, 12, 36] have also investigated the development of algorithms that simultaneously enforce the bilateral and unilateral constraints at the velocity and position levels, so that any drift-off phenomenon is avoided. In this chapter, we revisit the nonsmooth generalized-α method introduced in [12, 14]. It relies on a splitting of the motion into smooth (non-impulsive) and nonsmooth (impulsive) contributions. The smooth contributions are integrated using the second-order generalized-α method, whereas the nonsmooth contributions are integrated using a first-order backward Euler scheme. This method leads to qualitatively better solutions than the Moreau-Jean method, both for rigid and flexible systems.

If the splitting of the dynamics into smooth and nonsmooth contributions leads to algorithms with improved performance, some freedom remains in the precise definition of the smooth motion, especially regarding the contributions of the bilateral and unilateral constraints. This question has a significant influence on the numerical stability of the solution in the presence of impacts and velocity jumps. In [14], the smooth motion was defined as an unconstrained motion, whereas the bilateral constraints at the velocity level were imposed in [12]. Here, we propose a definition of the smooth motion that involves the bilateral constraints and the active unilateral constraints at the acceleration level.

After a description of the equations of motion in Sect. 2 and of the nonsmooth generalized-α method in Sect. 3, the special case of a smooth mechanical system without impact is addressed in Sect. 4 and a comparison with more standard solvers for differential-algebraic equations (DAEs), which are commonly used for the analysis of smooth multibody systems, is performed. We show that the proposed algorithm can be interpreted as an index-1 formulation that simultaneously enforces the constraints at the position, velocity and acceleration levels. In Sect. 5, the behaviour of the algorithm in the smooth case is studied based on the numerical example of a pendulum modelled as a DAE. In this example, a post-impact numerical solution is also reproduced by considering disturbed initial conditions at the acceleration level. This analysis reveals the high robustness and stability of the proposed algorithm.

Three examples of nonsmooth dynamic systems are studied in Sect. 6: a bouncing rigid pendulum, a bouncing flexible pendulum and the horizontal impact of an elastic bar. These examples intend to reveal the good properties of the algorithm for systems with bilateral constraints, impacts, accumulation phenomena, flexible bodies, finite contact duration, dynamic activation and deactivation of unilateral constraints. Also, it is shown that the numerical damping of the generalized-α is no longer necessary for the stabilization of the constraints, but is only useful for the stabilization of the spurious high frequency modes resulting from the finite element discretization of flexible bodies. The conclusions of the study are finally summarized in Sect. 7.

2 Nonsmooth Dynamics

2.1 Mechanical Systems with Unilateral Constraints

Let us consider a mechanical system with bilateral and unilateral constraints. For example, the bilateral constraints may represent the restrictions imposed by a kinematic joint that connects two bodies of the system, whereas a unilateral constraint may represent a non-penetration condition when two bodies are in contact. In a first step, we assume that no impact occurs in the system, but that detachment phenomena may occur during the motion. The equations of motion are then expressed as

$$\dot{\mathbf{q}} = \mathbf{v}, \tag{1a}$$

$$\mathbf{M}(\mathbf{q})\,\dot{\mathbf{v}} - \mathbf{g}_{\mathbf{q}}^{T}(\mathbf{q})\,\boldsymbol{\lambda} = \mathbf{f}(\mathbf{q}, \mathbf{v}, t), \tag{1b}$$

$$\mathbf{g}^{\overline{\mathscr{U}}}(\mathbf{q}) = \mathbf{0}, \tag{1c}$$

$$\mathbf{0} \leq \mathbf{g}^{\mathscr{U}}(\mathbf{q}) \perp \boldsymbol{\lambda}^{\mathscr{U}} \geq \mathbf{0}, \tag{1d}$$

where t is the time, $\mathbf{q}$ is the vector of coordinates, e.g., the nodal coordinates of a finite element mesh, $\mathbf{v}$ is the vector of velocities, $\mathbf{M}(\mathbf{q})$ is the mass matrix, $\mathbf{f}(\mathbf{q}, \mathbf{v}, t) = \mathbf{f}^{\text{ext}}(t) - \mathbf{f}^{\text{damp}}(\mathbf{q}, \mathbf{v}) - \mathbf{f}^{\text{int}}(\mathbf{q})$ collects the external, damping and internal forces, $\mathbf{g}$ is the combined set of bilateral and unilateral constraints, $\mathbf{g}_{\mathbf{q}}(\mathbf{q})$ is the matrix of constraint gradients, $\boldsymbol{\lambda}$ is the vector of Lagrange multipliers that represents the unilateral and bilateral reaction forces, $\mathscr{U}$ is the set of indices of the unilateral constraints, $\overline{\mathscr{U}}$ is its complementarity set, i.e., the set of bilateral constraints, and $\mathscr{T} = \mathscr{U} \cup \overline{\mathscr{U}}$ is the total set of constraints, and we have

$$\mathbf{g} = \begin{bmatrix} \mathbf{g}^{\mathscr{U}} \\ \mathbf{g}^{\overline{\mathscr{U}}} \end{bmatrix}, \qquad \boldsymbol{\lambda} = \begin{bmatrix} \boldsymbol{\lambda}^{\mathscr{U}} \\ \boldsymbol{\lambda}^{\overline{\mathscr{U}}} \end{bmatrix}. \tag{2}$$

Equation (1d) takes the form of a complementarity condition known as the Signorini condition. For one contact $j \in \mathscr{U}$, the function $g^j(\mathbf{q})$ represents the signed gap distance, which can be obtained from the contact kinematics. The contact condition imposes $g^j(\mathbf{q})\,\lambda^j = 0$ with both $g^j(\mathbf{q})$ and λ^j being non-negative, i.e., we do not authorize penetration and the reaction force can only be compressive.

The equations of motion (1) can be solved through time integration from given initial conditions $\mathbf{q}(0) = \mathbf{q}_0$ and $\mathbf{v}(0) = \mathbf{v}_0$ in order to obtain the trajectory $\mathbf{q}(t)$, $\mathbf{v}(t)$ and the Lagrange multipliers $\boldsymbol{\lambda}(t)$ on a given time interval $[0, T]$. However, the equations of motion also hide a purely algebraic relationship between $\mathbf{q}(t)$, $\mathbf{v}(t)$ and $\boldsymbol{\lambda}(t)$. Indeed, at a given time t, the constraint reaction forces $\boldsymbol{\lambda}(t)$ can be evaluated as an algebraic function of the current position $\mathbf{q}(t)$ and velocity $\mathbf{v}(t)$. As described below, the expression of this function is obtained by constraint differentiation.

If the bilateral constraints are satisfied at the position level, then their first and second time-derivatives also vanish, leading to the expression of the bilateral constraints at the velocity level

$$\frac{\mathrm{d}\mathbf{g}(\mathbf{q}(t))}{\mathrm{d}t} = \mathbf{g}_{\mathbf{q}}^{\overline{U}}(\mathbf{q})\,\mathbf{v} = \mathbf{0} \tag{3}$$

and at the acceleration level

$$\frac{\mathrm{d}^2\mathbf{g}(\mathbf{q}(t))}{\mathrm{d}t^2} = \mathbf{g}_{\mathbf{q}}^{\overline{U}}(\mathbf{q})\,\dot{\mathbf{v}} + \mathbf{h}^{\overline{U}}(\mathbf{q}, \mathbf{v}) = \mathbf{0}, \tag{4}$$

where $\mathbf{h}(\mathbf{q}, \mathbf{v})$ is a quadratic operator with respect to its second argument. This operator is defined as

$$\mathbf{h}(\mathbf{q}, \mathbf{v}) = \frac{\partial \mathbf{s}(\mathbf{q}, \mathbf{v})}{\partial \mathbf{q}} \mathbf{v}, \tag{5}$$

with $\mathbf{s}(\mathbf{q}, \mathbf{v}) = \mathbf{g_q}(\mathbf{q})\, \mathbf{v}$.

The unilateral constraint $j \in \mathscr{U}$ is active at the position level at time t_i if $g^j(\mathbf{q}(t_i)) = 0$. As $\lambda^j \geq 0$, this constraint is such that $\lambda^j(t_i) - r\, g^j(\mathbf{q}(t_i)) \geq 0$, where $r > 0$ is a strictly positive, yet arbitrary, real number. The variable $\lambda^j(t) - r\, g^j(\mathbf{q}(t))$ is an augmented Lagrange multiplier, as encountered in augmented Lagrangian formulations [3, 31, 32]. The set of active unilateral constraints at the position level is thus defined as

$$\mathscr{U}_A(t) = \{j \in \mathscr{U} : \lambda^j(t) - r\, g^j(\mathbf{q}(t)) \geq 0\}. \tag{6}$$

In order to avoid penetration right after t_i, any constraint j in $\mathscr{U}_A(t_i)$ needs to be increasing so the gap velocity $\dot{g}^j = g^j_{\mathbf{q}}(\mathbf{q}(t_i))\, \mathbf{v}(t_i)$ can only be non-negative. Hence, the unilateral constraint is transferred at the velocity level as [33]

$$0 \leq g^j_{\mathbf{q}}(\mathbf{q}(t))\, \mathbf{v}(t) \perp \lambda^j \geq 0, \quad \forall j \in \mathscr{U}_A(t). \tag{7}$$

The unilateral constraint $j \in \mathscr{U}$ is active at the velocity level at time t_i if $g^j(\mathbf{q}(t_i)) = 0$ and $g^j_{\mathbf{q}}(\mathbf{q}(t_i))\, \mathbf{v}(t_i) = 0$. As $\lambda^j \geq 0$, this constraint satisfies $\lambda^j(t_i) - r\, g^j_{\mathbf{q}}(\mathbf{q}(t_i))\, \mathbf{v}(t_i) \geq 0$ for $r > 0$. The set of active unilateral constraints at the velocity level is thus defined as

$$\mathscr{U}_B(t) = \{j \in \mathscr{U}_A(t) : \lambda^j(t) - r\, g^j_{\mathbf{q}}(\mathbf{q}(t))\, \mathbf{v}(t) \geq 0\}. \tag{8}$$

In order to avoid penetration right after t_i, the gap acceleration $\ddot{g}^j = g^j_{\mathbf{q}}(\mathbf{q}(t_i))\, \dot{\mathbf{v}}(t_i) + h^j(\mathbf{q}(t_i), \mathbf{v}(t_i))$ needs to be non-negative for any constraint $j \in \mathscr{U}_B(t_i)$. The unilateral constraint is thus further transferred at the acceleration level as [33]

$$0 \leq g^j_{\mathbf{q}}(\mathbf{q}(t))\, \dot{\mathbf{v}}(t) + h^j(\mathbf{q}(t), \mathbf{v}(t)) \perp \lambda^j(t) \geq 0, \quad \forall j \in \mathscr{U}_B(t). \tag{9}$$

The unilateral constraint $j \in \mathscr{U}$ is active at the acceleration level at time t_i if $g^j(\mathbf{q}(t_i)) = 0$, $g^j_{\mathbf{q}}(\mathbf{q}(t_i))\, \mathbf{v}(t_i) = 0$ and $g^j_{\mathbf{q}}(\mathbf{q}(t_i))\, \dot{\mathbf{v}}(t_i) + h^j(\mathbf{q}(t_i), \mathbf{v}(t_i)) = 0$. Following a similar argument as above, the set of active unilateral constraints at the acceleration level is thus defined as

$$\mathscr{U}_C(t) = \{j \in \mathscr{U}_B(t) : \lambda^j(t) - r\, (g^j_{\mathbf{q}}(\mathbf{q}(t))\, \dot{\mathbf{v}}(t) + h^j(\mathbf{q}(t), \mathbf{v}(t))) \geq 0\}. \tag{10}$$

For convenience, we also introduce the active sets $\mathscr{A}(t) = \overline{\mathscr{U}} \cup \mathscr{U}_A(t)$, $\mathscr{B}(t) = \overline{\mathscr{U}} \cup \mathscr{U}_B(t)$, $\mathscr{C}(t) = \overline{\mathscr{U}} \cup \mathscr{U}_C(t)$ and the inactive sets $\overline{\mathscr{A}}(t) = \mathscr{T} \setminus \mathscr{A}(t)$, $\overline{\mathscr{B}}(t) = \mathscr{T} \setminus \mathscr{B}(t)$ and $\overline{\mathscr{C}}(t) = \mathscr{T} \setminus \mathscr{C}(t)$.

Using these definitions of the active sets $\mathscr{A}$, $\mathscr{B}$ and $\mathscr{C}$, which implicitly depend on $\mathbf{q}$, $\mathbf{v}$, $\dot{\mathbf{v}}$ and $\boldsymbol{\lambda}$, the equations of motion can be represented in three equivalent ways as:

- the formulation with the constraints at the position level:

$$\dot{\mathbf{q}} = \mathbf{v}, \tag{11a}$$
$$\mathbf{M}(\mathbf{q})\,\dot{\mathbf{v}} - \mathbf{g}_{\mathbf{q}}^T(\mathbf{q})\,\boldsymbol{\lambda} = \mathbf{f}(\mathbf{q}, \mathbf{v}, t), \tag{11b}$$
$$\mathbf{g}^{\mathscr{A}}(\mathbf{q}) = \mathbf{0}, \tag{11c}$$
$$\boldsymbol{\lambda}^{\overline{\mathscr{A}}} = \mathbf{0}, \tag{11d}$$

- the formulation with the constraints at the velocity level:

$$\dot{\mathbf{q}} = \mathbf{v}, \tag{12a}$$
$$\mathbf{M}(\mathbf{q})\,\dot{\mathbf{v}} - \mathbf{g}_{\mathbf{q}}^T(\mathbf{q})\,\boldsymbol{\lambda} = \mathbf{f}(\mathbf{q}, \mathbf{v}, t), \tag{12b}$$
$$\mathbf{g}_{\mathbf{q}}^{\mathscr{B}}(\mathbf{q})\,\mathbf{v} = \mathbf{0}, \tag{12c}$$
$$\boldsymbol{\lambda}^{\overline{\mathscr{B}}} = \mathbf{0}, \tag{12d}$$

- the formulation with the constraints at the acceleration level:

$$\dot{\mathbf{q}} = \mathbf{v}, \tag{13a}$$
$$\mathbf{M}(\mathbf{q})\,\dot{\mathbf{v}} - \mathbf{g}_{\mathbf{q}}^T(\mathbf{q})\,\boldsymbol{\lambda} = \mathbf{f}(\mathbf{q}, \mathbf{v}, t), \tag{13b}$$
$$\mathbf{g}_{\mathbf{q}}^{\mathscr{C}}(\mathbf{q})\,\dot{\mathbf{v}} + \mathbf{h}^{\mathscr{C}}(\mathbf{q}, \mathbf{v}) = \mathbf{0}, \tag{13c}$$
$$\boldsymbol{\lambda}^{\overline{\mathscr{C}}} = \mathbf{0}, \tag{13d}$$

The expression of the Lagrange multipliers can now be obtained from the formulation with the constraints at the acceleration level. Indeed, if the mass matrix is nonsingular, the acceleration can be evaluated from Eq. (13b) as

$$\dot{\mathbf{v}} = \mathbf{M}^{-1}(\mathbf{q})\,(\mathbf{f}(\mathbf{q}, \mathbf{v}, t) + \mathbf{g}_{\mathbf{q}}^T(\mathbf{q})\,\boldsymbol{\lambda}), \tag{14}$$

so that Eqs. (13c) and (13d) give the equation for the Lagrange multipliers as

$$\mathbf{g}_{\mathbf{q}}^{\mathscr{C}}(\mathbf{q})\,\mathbf{M}^{-1}(\mathbf{q})\,(\mathbf{f}(\mathbf{q}, \mathbf{v}, t) + \mathbf{g}_{\mathbf{q}}^T(\mathbf{q})\,\boldsymbol{\lambda}) + \mathbf{h}^{\mathscr{C}}(\mathbf{q}, \mathbf{v}) = \mathbf{0}, \tag{15a}$$
$$\boldsymbol{\lambda}^{\overline{\mathscr{C}}} = \mathbf{0}, \tag{15b}$$

If the position $\mathbf{q}(t)$ and velocity $\mathbf{v}(t)$ are known at a given time t and if all constraints in $\mathscr{C}$ are independent, the Lagrange multipliers $\boldsymbol{\lambda}(t)$ can be evaluated by solving this linear set of algebraic equations. Actually, this problem includes a linear complementarity condition, as the active set $\mathscr{C}$ implicitly depends on the unknown value of $\boldsymbol{\lambda}(t)$.

This constraint differentiation process revealed the existence of hidden bilateral and unilateral constraints at the position, velocity and acceleration levels, which are satisfied by the exact solution. Clearly, the initial conditions $\mathbf{q}_0$ and $\mathbf{v}_0$ should be consistent with the constraints at the position and velocity levels. In the context of DAE (i.e., systems without unilateral constraint), these hidden constraints are at the core of so-called index reduction methods, which have been proposed to improve the numerical stability of time integration schemes [4]. In the context of unilaterally constrained systems, these hidden constraints can also be exploited to formulate efficient numerical algorithms, as will be discussed later.

2.2 Mechanical Systems with Impacts

2.2.1 Equations of Motion

Now, the formulation is extended to deal with impact phenomena, which means that impulsive reaction forces and jumps in the velocity field may arise, though the position field remains absolutely continuous in time. Assuming that the velocity is a function of bounded variation, the right and left limits are introduced:

$$\dot{\mathbf{q}}^+(t) = \lim_{\tau\to t, \tau>t} \dot{\mathbf{q}}(\tau), \tag{16}$$

$$\dot{\mathbf{q}}^-(t) = \lim_{\tau\to t, \tau<t} \dot{\mathbf{q}}(\tau), \tag{17}$$

$$\mathbf{v}^+(t) = \lim_{\tau\to t, \tau>t} \mathbf{v}(\tau), \tag{18}$$

$$\mathbf{v}^-(t) = \lim_{\tau\to t, \tau<t} \mathbf{v}(\tau), \tag{19}$$

For the sake of notation simplicity, the convention $\mathbf{v}(t) = \mathbf{v}^+(t)$ and $\dot{\mathbf{q}}(t) = \dot{\mathbf{q}}^+(t)$ shall be used in the remaining part of this chapter.

When an impact occurs, the velocity is discontinuous and the acceleration is not well-defined in the usual sense. This motivates the representation of the dynamics in terms of the measure associated with the velocity $\mathrm{d}\mathbf{v}$ [26]. This measure satisfies the property

$$\mathbf{v}(t_2) - \mathbf{v}(t_1) = \int_{(t_1,t_2]} \mathrm{d}\mathbf{v} \tag{20}$$

and, if the singular continuous part of the measure is neglected, it admits the decomposition

$$\mathrm{d}\mathbf{v} = \dot{\mathbf{v}}\,\mathrm{d}t + \sum_i (\mathbf{v}(t_i) - \mathbf{v}^-(t_i))\,\delta_{t_i}, \tag{21}$$

where $\mathrm{d}t$ is the standard Lebesgue measure, the summation is performed over all impacts, and δ_{t_i} is the Dirac delta supported at t_i. Similarly, a measure $\mathrm{d}\mathbf{i}$ is introduced to represent the reaction forces with possible impulsive contributions. This measure is such that the integral

$$\boldsymbol{\Lambda}^*(t_1; t_2) = \int_{(t_1,t_2]} \mathrm{d}\mathbf{i} \tag{22}$$

represents the total impulse of the reaction forces over the time interval $(t_1, t_2]$, and it admits the decomposition

$$\mathrm{d}\mathbf{i} = \boldsymbol{\lambda}\, \mathrm{d}t + \sum_i \mathbf{p}_i\, \delta_{t_i}, \tag{23}$$

where $\boldsymbol{\lambda}$ is the vector of nonimpulsive Lagrange multipliers associated with the Lebesgue measurable constraint forces and $\mathbf{p}_i$ is the impulse producing the jump at the instant t_i.

Then, the equations of motion can be expressed in the following form:

$$\dot{\mathbf{q}} = \mathbf{v}, \tag{24a}$$

$$\mathbf{M}(\mathbf{q})\, \mathrm{d}\mathbf{v} - \mathbf{g}_{\mathbf{q}}^T(\mathbf{q})\, \mathrm{d}\mathbf{i} = \mathbf{f}(\mathbf{q}, \mathbf{v}, t)\, \mathrm{d}t, \tag{24b}$$

$$\mathbf{g}^{\overline{\mathscr{U}}}(\mathbf{q}) = \mathbf{0}, \tag{24c}$$

$$\mathbf{0} \le \mathbf{g}^{\mathscr{U}}(\mathbf{q}(t)) \perp \mathrm{d}\mathbf{i}^{\mathscr{U}} \ge \mathbf{0}, \tag{24d}$$

2.2.2 Impact Equation

For almost every time t, when there is no impact, the equations of motion given in Eq. (11) with the definition of the active unilateral constraint $\mathscr{U}_A$ in Eq. (6) are still valid. At each impact time t_i, Eq. (24d) leads to

$$\mathbf{0} \le \mathbf{g}^{\mathscr{U}}(\mathbf{q}(t_i)) \perp \mathbf{p}_i^{\mathscr{U}} \ge \mathbf{0}, \tag{25}$$

so that the definition of the set of active unilateral constraints $\mathscr{U}_A$ at the impact time t_i is adapted as

$$\mathscr{U}_A(t_i) = \{j \in \mathscr{U} : p_i^j - r_p\, g^j(\mathbf{q}(t_i)) \ge 0\}, \tag{26}$$

with the strictly positive scalar number $r_p > 0$. The equations of motion at the impact time become

$$\mathbf{M}(\mathbf{q}(t_i))\, (\mathbf{v}(t_i) - \mathbf{v}^-(t_i)) - \mathbf{g}_{\mathbf{q}}^T(\mathbf{q}(t_i))\, \mathbf{p}_i = \mathbf{0}, \tag{27a}$$

$$\mathbf{g}^{\mathscr{A}}(\mathbf{q}(t_i)) = \mathbf{0}, \tag{27b}$$

$$\mathbf{p}_i^{\overline{\mathscr{A}}} = \mathbf{0}, \tag{27c}$$

An impact law is then needed to specify the post-impact velocity. The Newton impact law defines the normal velocity jump in the case of an impact for the constraint $j \in \mathscr{U}_A(t_i)$ as

$$g^j_{\mathbf{q}}(\mathbf{q}(t_i))\,\mathbf{v}(t_i) = -e^j\, g^j_{\mathbf{q}}(\mathbf{q}(t_i))\,\mathbf{v}^-(t_i), \tag{28}$$

where $e^j \in [0, 1]$ is the coefficient of restitution. The present formalism is developed for the analysis of contact conditions between rigid or flexible bodies. For rigid bodies, the coefficient of restitution defines the amount of energy dissipated during an impact. For flexible bodies, the physical meaning of a coefficient of restitution is not clear. The spatial discretization of a flexible body using the finite element method leads to a finite dimensional system with finite masses. An impact law with a coefficient of restitution is thus needed to describe contact conditions. In practice, for flexible bodies, a value $e^j = 0$ may often be used so that the condition $g^j_{\mathbf{q}}\,\mathbf{v}(t_i) = 0$ is imposed when the constraint is active. Based on this impact law, the contact condition at the impact time is expressed at the velocity level as

$$\mathbf{0} \le \mathbf{g}^{\mathscr{U}_A}_{\mathbf{q}}(\mathbf{q}(t_i))\,\mathbf{v}(t_i) + \mathbf{E}^{\mathscr{U}_A}\,\mathbf{g}^{\mathscr{U}_A}_{\mathbf{q}}(\mathbf{q}(t_i))\,\mathbf{v}^-(t_i) \perp \mathbf{p}^{\mathscr{U}_A}_i \ge \mathbf{0}, \tag{29}$$

where $\mathbf{E}^{\mathscr{U}}$ is a diagonal matrix formed with the coefficients of restitutions of all contact points. At the impact time t_i, the set of active unilateral constraints at the velocity level $\mathscr{U}_B$ is adapted as

$$\mathscr{U}_B(t_i) = \left\{ j \in \mathscr{U}_A(t_i) : p^j_i - r_p\,(g^j_{\mathbf{q}}(\mathbf{q}(t_i))\,\mathbf{v}(t_i) + e^j\, g^j_{\mathbf{q}}(\mathbf{q}(t_i))\,\mathbf{v}^-(t_i)) \ge 0 \right\}. \tag{30}$$

The equation for evaluating the velocity jump and the impact at time t_i is obtained as

$$\mathbf{M}(\mathbf{q}(t_i))\,(\mathbf{v}(t_i) - \mathbf{v}^-(t_i)) - \mathbf{g}^T_{\mathbf{q}}(\mathbf{q}(t_i))\,\mathbf{p}_i = \mathbf{0}, \tag{31a}$$

$$\mathbf{g}^{\mathscr{B}}_{\mathbf{q}}(\mathbf{q}(t_i))\,\mathbf{v}(t_i) + \mathbf{E}^{\mathscr{B}}\,\mathbf{g}^{\mathscr{B}}_{\mathbf{q}}(\mathbf{q}(t_i))\,\mathbf{v}^-(t_i) = \mathbf{0}, \tag{31b}$$

$$\mathbf{p}^{\overline{\mathscr{B}}}_i = \mathbf{0}. \tag{31c}$$

Equation (31b) accounts for the bilateral and active unilateral constraints. The size of the matrix of restitution coefficients $\mathbf{E}$ is thus adapted to include the bilateral constraints with artificial restitution coefficients fixed to zero.

2.2.3 Active Set Formulations

The definitions of $\mathscr{U}_A$ in Eqs. (6) and (26) can be merged into a single definition valid for every time as

$$\mathscr{U}_A = \{ j \in \mathscr{U} : \mathrm{d}i^j - g^j(\mathbf{q})\,\mathrm{d}\rho \ge 0 \}, \tag{32}$$

where $\mathrm{d}\rho > 0$ is a measure defined from the strictly positive and constant scalar numbers r and r_p as

$$\mathrm{d}\rho = r\,\mathrm{d}t + r_p \sum_i \delta_{t_i}. \tag{33}$$

Then, the combination of Eqs. (11) and (27) leads to a formulation in terms of measures

$$\dot{\mathbf{q}} = \mathbf{v}, \tag{34a}$$
$$\mathbf{M}(\mathbf{q}(t))\,\mathrm{d}\mathbf{v} - \mathbf{g}_{\mathbf{q}}^T(\mathbf{q})\,\mathrm{d}\mathbf{i} = \mathbf{f}(\mathbf{q}, \mathbf{v}, t)\,\mathrm{d}t, \tag{34b}$$
$$\mathbf{g}^{\mathscr{A}}(\mathbf{q}(t)) = \mathbf{0}, \tag{34c}$$
$$\mathrm{d}\mathbf{i}^{\overline{\mathscr{A}}} = \mathbf{0}, \tag{34d}$$

which is valid for every time and in which the constraints are expressed at the position level. Notice that Eq. (34) should be combined with the impact law to obtain a complete set of equations.

Similarly, the definitions of $\mathscr{U}_B$ in Eq. (8) and (30) can be merged into a single definition for every time as

$$\mathscr{U}_B = \left\{ j \in \mathscr{U}_A : \mathrm{d}i^j - (g_{\mathbf{q}}^j(\mathbf{q})\,\mathbf{v} + e^j\, g_{\mathbf{q}}^j(\mathbf{q})\,\mathbf{v}^-)\,\mathrm{d}\rho \geq 0 \right\}. \tag{35}$$

Then, the formulation of the equations of motion in terms of measures is obtained from Eqs. (12) and (31) as

$$\dot{\mathbf{q}} = \mathbf{v}, \tag{36a}$$
$$\mathbf{M}(\mathbf{q}(t))\,\mathrm{d}\mathbf{v} - \mathbf{g}_{\mathbf{q}}^T(\mathbf{q})\,\mathrm{d}\mathbf{i} = \mathbf{f}(\mathbf{q}, \mathbf{v}, t)\,\mathrm{d}t, \tag{36b}$$
$$\mathbf{g}_{\mathbf{q}}^{\mathscr{B}}(\mathbf{q})\,\mathbf{v} + \mathbf{E}^{\mathscr{B}}\,\mathbf{g}_{\mathbf{q}}^{\mathscr{B}}(\mathbf{q})\,\mathbf{v}^- = \mathbf{0}, \tag{36c}$$
$$\mathrm{d}\mathbf{i}^{\overline{\mathscr{B}}} = \mathbf{0}, \tag{36d}$$

which is valid for every time and in which the constraints are expressed at the velocity level.

As in the Moreau-Jean method, the formulation in Eq. (36) embeds the impact law in the expression of the constraints at the velocity level. However, the activation criterion defined by Eqs. (32) and (35) involves the augmented Lagrange multipliers $\mathrm{d}i^j - g^j\,\mathrm{d}\rho$, and thereby differs from the activation strategy initially proposed by Moreau, which only involves the gap distance g^j. In our notations, the set of active unilateral constraints in the original Moreau-Jean method would be defined as

$$\mathscr{U}_A^{\mathrm{Moreau}}(t) = \{ j \in \mathscr{U} : g^j(\mathbf{q}(t)) \leq 0 \}. \tag{37}$$

After time discretization, the set $\mathscr{U}_A^{\mathrm{Moreau}}$ at time step t_{n+1} is evaluated based on a prediction of the displacement $\mathbf{q}^*(t_{n+1})$, whose definition affects the numerical solution. In practice, it turns out that, in the Moreau-Jean method, $\mathbf{q}^*(t_{n+1})$ cannot be merely chosen as the actual displacement $\mathbf{q}(t_{n+1})$. In contrast, we will show that the proposed activation criterion based on augmented Lagrange multipliers according to Eqs. (6) and (26) leads to a simpler and more implicit discrete activation strategy.

Equation (36) can be discretized in time using the Moreau-Jean θ-method [25, 27]. This method is known for its robustness and its ability to deal consistently with unilateral constraints and impacts in mechanical systems. However, as the constraints are only imposed at the velocity level, the numerical integration error will induce a drift of the constraints at the position level that will accumulate as time goes by. Also, for standard applications, the numerical parameter θ is selected in the interval $(1/2, 1]$. This implies that the equations of motion are integrated with only first-order accuracy and that the overall solution is affected by a rather large level of numerical dissipation.

For nonsmooth systems, it is not possible to formulate the equations of motion in terms of measures with the constraints at the acceleration level, because, while the acceleration variable is defined for almost every time, it is not so at the impact instants.

3 Nonsmooth Generalized-α Method

3.1 Splitting Method

Following [12, 14], the motion is split at one time step into a smooth trajectory with continuous positions and velocities and nonsmooth contributions representing impulsive forces, velocity jumps and position corrections. The smooth trajectory is constructed by integration of an acceleration variable $\tilde{\dot{\mathbf{v}}}$ that shall be defined below. The advantage of this approach comes from the possibility of using a second-order scheme to integrate $\tilde{\dot{\mathbf{v}}}$ instead of a first-order θ-method.

Let us introduce the set of constraints $\mathscr{S}(t)$ that shall be included in the definition of the smooth motion. It can be selected in several different manners, which shall be studied later in Sect. 3.2. At a given time t and for given values of $\mathbf{q}(t)$ and $\mathbf{v}(t)$, the smooth acceleration $\tilde{\dot{\mathbf{v}}}(t)$ and the smooth Lagrange multiplier $\tilde{\boldsymbol{\lambda}}(t)$ are defined as the solution to the well-posed algebraic system

$$\mathbf{M}(\mathbf{q})\,\tilde{\dot{\mathbf{v}}} - \mathbf{g}_{\mathbf{q}}^T(\mathbf{q})\,\tilde{\boldsymbol{\lambda}} = \mathbf{f}(\mathbf{q}, \mathbf{v}, t), \tag{38a}$$

$$\mathbf{g}_{\mathbf{q}}^{\mathscr{S}}(\mathbf{q})\,\tilde{\dot{\mathbf{v}}} + \mathbf{h}^{\mathscr{S}}(\mathbf{q}, \mathbf{v}) = \mathbf{0}, \tag{38b}$$

$$\tilde{\boldsymbol{\lambda}}^{\overline{\mathscr{S}}} = \mathbf{0}. \tag{38c}$$

An important point is that the resulting acceleration $\tilde{\dot{\mathbf{v}}}(t)$ is defined for every time, including the impact instants. The values of $\tilde{\dot{\mathbf{v}}}$ and $\tilde{\boldsymbol{\lambda}}$ at time t only depend on the values of $\mathbf{q}$, $\mathbf{v}$ and $\mathscr{S}$ at time t. In general, $\mathscr{S}(t)$ implicitly depends on $\tilde{\dot{\mathbf{v}}}(t)$ and

$\tilde{\boldsymbol{\lambda}}(t)$. As $\mathbf{q}(t)$ is a continuous function and $\mathbf{v}(t)$ is a function of bounded variations, the acceleration $\dot{\tilde{\mathbf{v}}}(t)$ and the multiplier $\tilde{\boldsymbol{\lambda}}(t)$ are also functions of bounded variations and, by construction, they are free from any impulsive contribution. Also, we use the conventions $\dot{\tilde{\mathbf{v}}}(t) = \dot{\tilde{\mathbf{v}}}^+(t)$ and $\tilde{\boldsymbol{\lambda}}(t) = \tilde{\boldsymbol{\lambda}}^+(t)$. Notice that a discontinuity of $\dot{\tilde{\mathbf{v}}}(t)$ can be either caused by a jump in the velocity $\mathbf{v}(t)$ or by a constraint activation or deactivation in the set $\mathscr{S}(t)$. The velocity field $\tilde{\mathbf{v}}(t)$ and the position field $\tilde{\mathbf{q}}(t)$ of the smooth trajectory, which are obtained by time integration of $\dot{\tilde{\mathbf{v}}}(t)$ over the time step, are absolutely continuous functions of time.

The nonsmooth contributions to the total motion are then represented by the differential measure $\mathrm{d}\mathbf{w}$, which is defined such that

$$\mathrm{d}\mathbf{v} = \dot{\tilde{\mathbf{v}}}\,\mathrm{d}t + \mathrm{d}\mathbf{w}. \tag{39}$$

We obtain, using Eqs. (36b), (38a) and (39),

$$\mathbf{M}(\mathbf{q})\,\mathrm{d}\mathbf{w} - \mathbf{g}_{\mathbf{q}}^T(\mathbf{q})\,(\mathrm{d}\mathbf{i} - \tilde{\boldsymbol{\lambda}}\,\mathrm{d}t) = \mathbf{0}. \tag{40}$$

We insist on the fact that the smooth trajectory is a mere artificial construction, which is only intended for the formulation of an appropriate time integration procedure. The physical response is represented by the total motion $\mathbf{q}(t)$ and $\mathbf{v}(t)$ and the total impulse $\mathrm{d}\mathbf{i}$.

The formulation of the constraints at the acceleration level in Eq. (38b) departs from the definition of the smooth motion based on the velocity constraints that was proposed in [12], but leads to several advantages that will be investigated throughout the paper. Firstly, it is not necessary to evaluate explicitly the smooth trajectory at the position or velocity levels, which simplifies the initialization of these variables. Secondly, the sensitivity of this formulation to disturbances induced by the coupling with nonsmooth phenomena, such as velocity jumps or constraint activation and deactivation, is reduced. Thirdly, this formulation can tolerate the dynamic activation and deactivation of unilateral constraints in the set $\mathscr{S}(t)$.

In summary, the dynamics is now represented by the following set of equations:

$$\dot{\mathbf{q}} = \mathbf{v}, \tag{41a}$$

$$\mathrm{d}\mathbf{v} = \dot{\tilde{\mathbf{v}}}\,\mathrm{d}t + \mathrm{d}\mathbf{w}, \tag{41b}$$

$$\mathbf{M}(\mathbf{q})\,\dot{\tilde{\mathbf{v}}} - \mathbf{g}_{\mathbf{q}}^T(\mathbf{q})\,\tilde{\boldsymbol{\lambda}} = \mathbf{f}(\mathbf{q}, \mathbf{v}, t), \tag{41c}$$

$$\mathbf{g}_{\mathbf{q}}^{\mathscr{S}}(\mathbf{q})\,\dot{\tilde{\mathbf{v}}} + \mathbf{h}^{\mathscr{S}}(\mathbf{q}, \mathbf{v}) = \mathbf{0}, \tag{41d}$$

$$\tilde{\boldsymbol{\lambda}}^{\overline{\mathscr{S}}} = \mathbf{0}, \tag{41e}$$

$$\mathbf{M}(\mathbf{q})\,\mathrm{d}\mathbf{w} - \mathbf{g}_{\mathbf{q}}^T(\mathbf{q})\,(\mathrm{d}\mathbf{i} - \tilde{\boldsymbol{\lambda}}\,\mathrm{d}t) = \mathbf{0}, \tag{41f}$$

$$\mathbf{g}_{\mathbf{q}}^{\mathscr{B}}(\mathbf{q})\,\mathbf{v} + \mathbf{E}^{\mathscr{B}}\,\mathbf{g}_{\mathbf{q}}^{\mathscr{B}}(\mathbf{q}^-)\,\mathbf{v}^- = \mathbf{0}, \tag{41g}$$

$$\mathrm{d}\mathbf{i}^{\overline{\mathscr{B}}} = \mathbf{0}. \tag{41h}$$

3.2 *Activation Strategy for the Constraints on the Smooth Motion*

This section addresses the possible contribution $\tilde{\boldsymbol{\lambda}}$ of the constraint reaction forces in the definition of the smooth motion. The choice to include such contributions or not bears some arbitrariness. Indeed, the value of $\tilde{\boldsymbol{\lambda}}$ has no physical meaning; only the total impulse represented by $\mathrm{d}\mathbf{i}$ can receive a physical interpretation. Even though some contributions of the reaction forces are disregarded in the definition of $\tilde{\boldsymbol{\lambda}}$, they will be consistently incorporated into the total impulse $\mathrm{d}\mathbf{i}$ that satisfies the discrete complementarity condition.

However, it is appealing to define the smooth motion so that it evolves as closely as possible to the physical motion for at least two reasons. Firstly, the smooth motion is integrated using a higher-order scheme, so we can expect a higher accuracy if the smooth motion is closer to the total (physical) one. Secondly, when the nonsmooth corrections are reduced, the convergence of the iterative procedure at each time step, which is at the core of the implicit integration procedure, is accelerated.

In the proposed method, the acceleration $\dot{\tilde{\mathbf{v}}}$ and the multipliers $\tilde{\boldsymbol{\lambda}}$ are well-defined at any time (though they can be discontinuous) by Eq. (38), so that, by construction, no impulsive term can appear. This observation remains valid when some constraints on the smooth motion are activated and deactivated. This means that we are relatively free to dynamically activate and deactivate some unilateral constraints in $\mathscr{S}$ as we feel appropriate without inducing inconsistent impulsive excitations on the smooth motion.

Three different activation strategies for the smooth constraints are now considered:

- Strategy 1: $\mathscr{S} = \emptyset$, i.e., no bilateral constraint and no unilateral constraint is taken into account, as proposed in [14]. This means that the smooth motion is considered as a constraint-free motion.
- Strategy 2: $\mathscr{S} = \overline{\mathscr{U}}$, i.e., only the bilateral constraints are taken into account, but all unilateral constraints are excluded, as proposed in [12]. This means that the smooth motion satisfies the bilateral constraints, but does not account for the contact forces.
- Strategy 3: $\mathscr{S} = \overline{\mathscr{U}} \cup \tilde{\mathscr{U}}_C$, with $\tilde{\mathscr{U}}_C$ the time-dependent set of active unilateral constraints at the acceleration level defined according to

$$\tilde{\mathscr{U}}_C = \{j \in \mathscr{U}_B : \tilde{\lambda}^j - r\,(g^j_{\mathbf{q}}(\mathbf{q})\,\dot{\tilde{\mathbf{v}}} + h^j(\mathbf{q},\mathbf{v})) \geq 0\}. \tag{42}$$

  This strategy is a new approach considered in this chapter. Notice that the definition of $\tilde{\mathscr{U}}_C$ relies on the acceleration $\dot{\tilde{\mathbf{v}}}$, which is well-posed for every time (including the impact times), and thus slightly differs from the definition of $\mathscr{U}_C$, which is not defined at the impact time. With this strategy, for almost every time (when there is no impact), Eqs. (13) and (38) are strictly equivalent, so that $\dot{\tilde{\mathbf{v}}} = \dot{\mathbf{v}}$ and $\tilde{\boldsymbol{\lambda}} = \boldsymbol{\lambda}$. This means that, for almost every time, $\dot{\tilde{\mathbf{v}}}$ and $\tilde{\boldsymbol{\lambda}}$ represent the standard accelerations and reaction forces, but that they exclude impulsive contributions at the impact instants.

Compared to strategy 1, we clearly expect that strategy 2 brings the smooth motion closer to the physical motion, as it satisfies the bilateral constraints. For this reason, strategy 2 should be preferred to strategy 1.

When all active unilateral constraints remain closed, the physical motion becomes smooth and satisfies the active constraints at the acceleration level. In this case, for the exact solution, the smooth motion defined in strategy 3 is equal to the total motion, i.e., $\dot{\tilde{\mathbf{v}}} = \dot{\mathbf{v}}$ and $\tilde{\boldsymbol{\lambda}} = \boldsymbol{\lambda}$. This means that the total motion is integrated with second-order accuracy. In the numerical scheme, numerical errors may lead to small differences between the smooth motion and the total motion, but we expect that these differences are much smaller compared to the position corrections and velocity jumps in strategy 2. Compared to strategy 2, strategy 3 should thus be preferred when the constraints remain closed.

When some unilateral constraints are active but some impulsive phenomena are present in the system, the acceleration is not well-defined and the physical interpretation of the constraint at the acceleration level becomes irrelevant. In this case, it is not clear whether strategy 2 or strategy 3 should be preferred. This question will be investigated through numerical tests in Sect. 6.

3.3 Gear-Gupta-Leimkuhler Formulation

In Eq. (41g), the constraints on the total (physical) motion are imposed at the velocity level. Due to numerical integration errors, a drift of the constraints is expected at the position level. In order to remedy this situation, an adaptation of the Gear-Gupta-Leimkuhler formulation [18] to nonsmooth systems was considered by several authors [1, 12, 36]. The algorithm discussed here is built upon the formulation proposed in [12]. An additional Lagrange multiplier $\boldsymbol{\mu}$ is thus introduced in Eq. (41a), leading to

$$\mathrm{d}\mathbf{v} = \dot{\tilde{\mathbf{v}}}\,\mathrm{d}t + \mathrm{d}\mathbf{w}, \tag{43a}$$

$$\mathbf{M}(\mathbf{q})\,\dot{\tilde{\mathbf{v}}} - \mathbf{g}_{\mathbf{q}}^{T}(\mathbf{q})\,\tilde{\boldsymbol{\lambda}} = \mathbf{f}(\mathbf{q},\mathbf{v},t), \tag{43b}$$

$$\mathbf{g}_{\mathbf{q}}^{\mathscr{S}}(\mathbf{q})\,\dot{\tilde{\mathbf{v}}} + \mathbf{h}^{\mathscr{S}}(\mathbf{q},\mathbf{v}) = \mathbf{0}, \tag{43c}$$

$$\tilde{\boldsymbol{\lambda}}^{\overline{\mathscr{S}}} = \mathbf{0}, \tag{43d}$$

$$\mathbf{M}(\mathbf{q})(\dot{\mathbf{q}} - \mathbf{v}) - \mathbf{g}_{\mathbf{q}}^{T}(\mathbf{q})\boldsymbol{\mu} = \mathbf{0}, \tag{43e}$$

$$\mathbf{g}^{\mathscr{A}}(\mathbf{q}) = \mathbf{0}, \tag{43f}$$

$$\boldsymbol{\mu}^{\overline{\mathscr{A}}} = \mathbf{0}, \tag{43g}$$

$$\mathbf{M}(\mathbf{q})\,\mathrm{d}\mathbf{w} - \mathbf{g}_{\mathbf{q}}^{T}(\mathbf{q})\,(\mathrm{d}\mathbf{i} - \tilde{\boldsymbol{\lambda}}\,\mathrm{d}t) = \mathbf{0}, \tag{43h}$$

$$\mathbf{g}_{\mathbf{q}}^{\mathscr{B}}(\mathbf{q})\,\mathbf{v} + \mathbf{E}^{\mathscr{B}}\,\mathbf{g}_{\mathbf{q}}^{\mathscr{B}}(\mathbf{q}^{-})\,\mathbf{v}^{-} = \mathbf{0}, \tag{43i}$$

$$\mathrm{d}\mathbf{i}^{\overline{\mathscr{B}}} = \mathbf{0}. \tag{43j}$$

One can easily check that the solution to Eq. (41) also satisfies Eq. (43) with $\boldsymbol{\mu} = \mathbf{0}$. So, the introduction of the new Lagrange multiplier preserves the original solution to the problem.

3.4 Discrete Smooth and Nonsmooth Variables

In order to prepare the time discretization procedure, several global variables that represent the total jumps and total impulses over the time step $(t_n, t_{n+1}]$ are introduced. Over the current time step, the smooth motion is first constructed by integration of the smooth acceleration $\dot{\tilde{\mathbf{v}}}(t)$ from the physical initial conditions $\mathbf{q}(t_n)$ and $\mathbf{v}(t_n)$ to the end of the time step

$$\tilde{\mathbf{v}}(t) = \mathbf{v}(t_n) + \int_{t_n}^{t} \dot{\tilde{\mathbf{v}}}(\tau)\, \mathrm{d}\tau, \tag{44}$$

$$\tilde{\mathbf{q}}(t_{n+1}) = \mathbf{q}(t_n) + h\, \mathbf{v}(t_n) + \int_{t_n}^{t_{n+1}} \int_{t_n}^{t} \dot{\tilde{\mathbf{v}}}(\tau)\, \mathrm{d}\tau\, \mathrm{d}t, \tag{45}$$

where $h = t_{n+1} - t_n$ is the the time-step size. Even if the total velocity $\mathbf{v}(t)$ undergoes a discontinuity, $\tilde{\mathbf{v}}(t)$ is, by construction, a continuous function of time in $(t_n, t_{n+1}]$.

The velocity jump is defined as

$$\mathbf{W}(t_n; t_{n+1}) = \int_{(t_n, t_{n+1}]} \mathrm{d}\mathbf{w} \tag{46}$$

Using Eqs. (39) and (44), we get

$$\mathbf{W}(t_n; t_{n+1}) = \mathbf{v}(t_{n+1}) - \tilde{\mathbf{v}}(t_{n+1}). \tag{47}$$

Similarly, the position correction is defined as

$$\mathbf{U}(t_n; t_{n+1}) = \int_{t_n}^{t_{n+1}} (\dot{\mathbf{q}}(t) - \tilde{\mathbf{v}}(t))\, \mathrm{d}t, \tag{48}$$

so that, using Eqs. (39), (44) and (45),

$$\mathbf{U}(t_n; t_{n+1}) = \mathbf{q}(t_{n+1}) - \tilde{\mathbf{q}}(t_{n+1}). \tag{49}$$

Then, the relative impulse variable

$$\boldsymbol{\Lambda}(t_n; t_{n+1}) = \int_{(t_n, t_{n+1}]} (\mathrm{d}\mathbf{i} - \tilde{\boldsymbol{\lambda}}(t)\, \mathrm{d}t) \tag{50}$$

and the relative double integral variable

$$\boldsymbol{\nu}(t_n; t_{n+1}) = \int_{t_n}^{t_{n+1}} \left(\boldsymbol{\mu}(t) + \int_{(t_n,t]} (\mathrm{d}\mathbf{i} - \tilde{\boldsymbol{\lambda}}(\tau)\,\mathrm{d}\tau) \right) \mathrm{d}t \tag{51}$$

are introduced so that, according to Theorem 1 in [12],

$$\mathbf{M}(\mathbf{q}(t_{n+1}))\,\mathbf{W}(t_n; t_{n+1}) - \mathbf{g}_{\mathbf{q}}^T(\mathbf{q}(t_{n+1}))\,\boldsymbol{\Lambda}(t_n; t_{n+1}) = \mathscr{O}(h), \tag{52a}$$

$$\mathbf{M}(\mathbf{q}(t_{n+1}))\,\mathbf{U}(t_n; t_{n+1}) - \mathbf{g}_{\mathbf{q}}^T(\mathbf{q}(t_{n+1}))\,\boldsymbol{\nu}(t_n; t_{n+1}) = \mathscr{O}(h^2). \tag{52b}$$

It is important to observe that $\boldsymbol{\Lambda}(t_n; t_{n+1})$ does not represent the total impulse of the reaction forces, but only a part of it as the contribution of the non-impulsive reaction forces $\tilde{\boldsymbol{\lambda}}$ is excluded in the definition (50). The total (physical) impulse, denoted as $\boldsymbol{\Lambda}^*(t_n; t_{n+1})$, is evaluated by time integration of the measure of the reaction forces $\mathrm{d}\mathbf{i}$

$$\boldsymbol{\Lambda}^*(t_n; t_{n+1}) = \int_{(t_n,t_{n+1}]} \mathrm{d}\mathbf{i} = \boldsymbol{\Lambda}(t_{n+1}) + \int_{t_n}^{t_{n+1}} \tilde{\boldsymbol{\lambda}}(\tau)\,\mathrm{d}t. \tag{53}$$

Similarly, the total double integral $\boldsymbol{\nu}^*(t_n; t_{n+1})$ is defined as

$$\boldsymbol{\nu}^*(t_n; t_{n+1}) = \int_{t_n}^{t_{n+1}} \left(\boldsymbol{\mu}(t) + \int_{(t_n,t]} \mathrm{d}\mathbf{i} \right) \mathrm{d}t = \boldsymbol{\nu}(t_{n+1}) + \int_{t_n}^{t_{n+1}} \int_{t_n}^{t} \tilde{\boldsymbol{\lambda}}(t)\,\mathrm{d}\tau\,\mathrm{d}t. \tag{54}$$

The contribution of $\boldsymbol{\mu}$ is introduced in Eq. (54) so that $\boldsymbol{\nu}^*(t_n; t_{n+1})$ is conveniently expressed in terms of the variables $\boldsymbol{\nu}(t_n; t_{n+1})$ and $\tilde{\boldsymbol{\lambda}}$.

3.5 Active Sets in the Discrete Time System

Following a similar argumentation as developed in [12], the set of active unilateral constraints at the position level over the time step $(t_n, t_{n+1}]$ is defined as

$$\mathscr{U}_A(t_n; t_{n+1}) = \left\{ j \in \mathscr{U} : \nu^{*j}(t_n; t_{n+1}) - r\,g^j(\mathbf{q}(t_{n+1})) \geq 0 \right\}. \tag{55}$$

This activation rule based on the augmented Lagrange multiplier fixes the problem of the spurious oscillations reported in [1] in a simple way.

The active unilateral constraints at the velocity level over the time step $(t_n, t_{n+1}]$ are defined as

$$\mathscr{U}_B(t_n; t_{n+1}) = \left\{ j \in \mathscr{U}_A(t_n; t_{n+1}) : \Lambda^{*j}(t_n; t_{n+1}) - r(g_{\mathbf{q}}^j \mathbf{v}(t_{n+1}) + e^j\, g_{\mathbf{q}}^j \mathbf{v}(t_n)) \geq 0 \right\}. \tag{56}$$

Finally, if the third strategy is used for the activation of the constraints on the smooth motion (see Sect. 3.2), the active unilateral constraints at the acceleration level over the time step $(t_n, t_{n+1}]$ are defined as

$$\begin{aligned} \mathscr{U}_C(t_n; t_{n+1}) = \{ & j \in \mathscr{U}_B(t_n; t_{n+1}) : \\ & \tilde{\lambda}^j(t_{n+1}) - r(g^j_{\mathbf{q}}\dot{\tilde{\mathbf{v}}}(t_{n+1}) + h^j(\mathbf{q}(t_{n+1}), \mathbf{v}(t_{n+1}))) \geq 0 \Big\} . \end{aligned} \tag{57}$$

In [12], the unilateral constraints were never activated in the smooth equation, so that $\tilde{\boldsymbol{\lambda}}^{\mathscr{U}} = \mathbf{0}$, $\boldsymbol{\nu}^{*\mathscr{U}} = \boldsymbol{\nu}^{\mathscr{U}}$ and $\boldsymbol{\Lambda}^{*\mathscr{U}} = \boldsymbol{\Lambda}^{\mathscr{U}}$. But if $\tilde{\boldsymbol{\lambda}}^{\mathscr{U}}$ differs from $\mathbf{0}$, it contributes directly to the physical contact forces. This is the reason why the activation criteria in Eqs. (55) and (56) need to be established based on the total impulse and total double integral represented by $\boldsymbol{\nu}^*$ and $\boldsymbol{\Lambda}^*$ (and not $\boldsymbol{\nu}$ and $\boldsymbol{\Lambda}$).

The definition of $\mathscr{U}_B$ also differs from [12] in the following way. Here, the definition of $\mathscr{U}_B$ involves the augmented Lagrange multipliers at the position level (as it is a subset of $\mathscr{U}_A$) and a criterion on the augmented Lagrange multiplier at the velocity level. In [12], the criterion on the augmented Lagrange multiplier at the position level is replaced by a criterion on the penetration of the smooth motion $g^j(\tilde{\mathbf{q}}(t_{n+1})) \leq 0$. This modification allows us to completely eliminate the variable $\tilde{\mathbf{q}}$ from the algorithm and to simplify the formulation.

As discussed in [12], in this scheme, the variables $\boldsymbol{\nu}$ and $\boldsymbol{\nu}^*$ do not have a clear physical meaning, but are only useful for the exact enforcement of all active constraints at the position level at the end of the time step. So, the physical contact impulse is solely represented by the variable $\boldsymbol{\Lambda}^*$.

3.6 Generalized-α Time Integration

The integrals in Eqs. (44) and (45) can be approximated according to the generalized-α method as

$$\int_{(t_n, t_{n+1}]} \dot{\tilde{\mathbf{v}}}\, \mathrm{d}t = h(1-\gamma)\mathbf{a}_n + h\gamma\mathbf{a}_{n+1}, \tag{58}$$

$$\int_{t_n}^{t_{n+1}} \int_{t_n}^{t} \dot{\tilde{\mathbf{v}}}(\tau)\, \mathrm{d}\tau\, \mathrm{d}t = h^2(0.5-\beta)\mathbf{a}_n + h^2\beta\mathbf{a}_{n+1}, \tag{59}$$

$$(1-\alpha_m)\mathbf{a}_{n+1} + \alpha_m\mathbf{a}_n = (1-\alpha_f)\dot{\tilde{\mathbf{v}}}_{n+1} + \alpha_f\dot{\tilde{\mathbf{v}}}_n, \tag{60}$$

where $\mathbf{a}_{n+1}$ can be interpreted as a shifted approximation of the acceleration at time $t_{n+1} + (\alpha_m - \alpha_f)h$. In the initialization procedure, the value of $\mathbf{a}_0$ at time $t = 0$ can be approximated (i) by $\mathbf{a}_0 = \dot{\mathbf{v}}((\alpha_m - \alpha_f)h)$ by solving Eq. (38) at $t = (\alpha_m - \alpha_f)h$ or (ii) by the order h approximation $\mathbf{a}_0 = \dot{\mathbf{v}}(0)$. This second and simpler option is retained in this work. The numerical parameters β, γ, α_m, α_f can be selected according to the methods of Newmark [28], Hilber-Hughes-Taylor [21] or Chung and Hulbert [15]. This last option is considered here. The Chung-Hulbert method is a second-order scheme with an adjustable level of numerical dissipation in the high-frequency range. More precisely, based on the user-prescribed value of the spectral radius at infinite frequencies $\rho_\infty \in [0, 1]$, which is an image of the level of numerical dissipation in the high-frequency range ($\rho_\infty = 1$ means no dissipation, $\rho_\infty = 0$ means maximal dissipation such that any high-frequency disturbance is eliminated in one time step), the coefficients of the Chung-Hulbert method are determined as

$$\alpha_m = \frac{2\rho_\infty - 1}{\rho_\infty + 1}, \; \alpha_f = \frac{\rho_\infty}{\rho_\infty + 1}, \; \gamma = 0.5 + \alpha_f - \alpha_m, \; \beta = 0.25(\gamma + 0.5)^2. \tag{61}$$

Finally, the integrals of the multipliers $\tilde{\boldsymbol{\lambda}}(t)$ that appear in the definition of the active sets $\mathscr{A}$ and $\mathscr{B}$ are evaluated using a similar strategy as

$$\int_{t_n}^{t_{n+1}} \tilde{\boldsymbol{\lambda}}(t)\,\mathrm{d}t = h(1-\gamma)\boldsymbol{\eta}_n + h\gamma\boldsymbol{\eta}_{n+1}, \tag{62}$$

$$\int_{t_n}^{t_{n+1}} \int_{t_n}^{t} \tilde{\boldsymbol{\lambda}}(\tau)\,\mathrm{d}\tau\,\mathrm{d}t = h^2(0.5-\beta)\boldsymbol{\eta}_n + h^2\beta\boldsymbol{\eta}_{n+1}, \tag{63}$$

$$(1-\alpha_m)\boldsymbol{\eta}_{n+1} + \alpha_m\boldsymbol{\eta}_n = (1-\alpha_f)\tilde{\boldsymbol{\lambda}}_{n+1} + \alpha_f\tilde{\boldsymbol{\lambda}}_n, \tag{64}$$

where $\boldsymbol{\eta}_{n+1}$ is a shifted approximation of the multiplier $\tilde{\boldsymbol{\lambda}}$ at time $t_{n+1} + (\alpha_f - \alpha_m)h$, which is initialized as $\boldsymbol{\eta}_0 = \tilde{\boldsymbol{\lambda}}_0$.

3.7 Summary of the Time Stepping Scheme

Based on the definitions and results presented in the previous sections, the discrete system of equations is finally obtained as

$$\mathbf{M}(\mathbf{q}_{n+1})\,\dot{\tilde{\mathbf{v}}}_{n+1} - \mathbf{g}_{\mathbf{q}}^T(\mathbf{q}_{n+1})\,\tilde{\boldsymbol{\lambda}}_{n+1} = \mathbf{f}(\mathbf{q}_{n+1}, \mathbf{v}_{n+1}, t_{n+1}), \tag{65a}$$

$$\mathbf{g}_{\mathbf{q}}^{\mathscr{S}}(\mathbf{q}_{n+1})\,\dot{\tilde{\mathbf{v}}}_{n+1} + \mathbf{h}^{\mathscr{S}}(\mathbf{q}_{n+1}, \mathbf{v}_{n+1}) = \mathbf{0}, \tag{65b}$$

$$\tilde{\boldsymbol{\lambda}}_{n+1}^{\overline{\mathscr{S}}} = \mathbf{0}, \tag{65c}$$

$$\mathbf{M}(\mathbf{q}_{n+1})\mathbf{U}_{n+1} - \mathbf{g}_{\mathbf{q}}^T(\mathbf{q}_{n+1})\boldsymbol{\nu}_{n+1} = \mathbf{0}, \tag{65d}$$

$$\mathbf{g}^{\mathscr{A}}(\mathbf{q}_{n+1}) = \mathbf{0}, \tag{65e}$$

$$\boldsymbol{\nu}_{n+1}^{\overline{\mathscr{A}}} = \mathbf{0}, \tag{65f}$$

$$\mathbf{M}(\mathbf{q}_{n+1})\,\mathbf{W}_{n+1} - \mathbf{g}_{\mathbf{q}}^T(\mathbf{q}_{n+1})\,\boldsymbol{\Lambda}_{n+1} = \mathbf{0}, \tag{65g}$$

$$\mathbf{g}_{\mathbf{q}}^{\mathscr{B}}(\mathbf{q}_{n+1})\,\mathbf{v}_{n+1} + \mathbf{E}^{\mathscr{B}}\,\mathbf{g}_{\mathbf{q}}^{\mathscr{B}}(\mathbf{q}_n)\,\mathbf{v}_n = \mathbf{0}, \tag{65h}$$

$$\boldsymbol{\Lambda}_{n+1}^{\overline{\mathscr{B}}} = \mathbf{0}, \tag{65i}$$

combined with the time integration formulae

$$\mathbf{q}_{n+1} - \mathbf{q}_n = h\mathbf{v}_n + h^2(0.5 - \beta)\mathbf{a}_n + h^2\beta\mathbf{a}_{n+1} + \mathbf{U}_{n+1}, \tag{65j}$$

$$\mathbf{v}_{n+1} - \mathbf{v}_n = h(1 - \gamma)\mathbf{a}_n + h\gamma\mathbf{a}_{n+1} + \mathbf{W}_{n+1}, \tag{65k}$$

$$(1 - \alpha_m)\mathbf{a}_{n+1} + \alpha_m\mathbf{a}_n = (1 - \alpha_f)\dot{\tilde{\mathbf{v}}}_{n+1} + \alpha_f\dot{\tilde{\mathbf{v}}}_n. \tag{65l}$$

The active sets $\mathscr{A}$, $\mathscr{B}$ and $\mathscr{S}$ are evaluated as described in Sect. 3.5 based on the discrete variables at time step $n+1$, in particular, based on the variables $\boldsymbol{\Lambda}^*_{n+1}$ and $\boldsymbol{\nu}^*_{n+1}$ defined as

$$\boldsymbol{\Lambda}^*_{n+1} = \boldsymbol{\Lambda}(t_{n+1}) + h(1 - \gamma)\boldsymbol{\eta}_n + h\gamma\boldsymbol{\eta}_{n+1}, \tag{65m}$$

$$\boldsymbol{\nu}^*_{n+1} = \boldsymbol{\nu}(t_{n+1}) + h^2(0.5 - \beta)\boldsymbol{\eta}_n + h^2\beta\boldsymbol{\eta}_{n+1}, \tag{65n}$$

$$(1 - \alpha_m)\boldsymbol{\eta}_{n+1} + \alpha_m\boldsymbol{\eta}_n = (1 - \alpha_f)\tilde{\boldsymbol{\lambda}}_{n+1} + \alpha_f\tilde{\boldsymbol{\lambda}}_n. \tag{65o}$$

The sets $\mathscr{A}$, $\mathscr{B}$ and $\mathscr{S}$ thus implicitly depend on the solution at step t_{n+1}. Let us remark that the variables $\boldsymbol{\Lambda}^*_{n+1}$ and $\boldsymbol{\nu}^*_{n+1}$ do not explicitly appear in the equations of motion, but are necessary for the definition of the active sets $\mathscr{A}$ and $\mathscr{B}$.

Initial conditions should be specified for the variables $\mathbf{q}_0$, $\mathbf{v}_0$, which should be compatible with the constraints at the position and velocity levels. Based on these initial conditions, the initial values of $\dot{\tilde{\mathbf{v}}}_0$ and $\tilde{\boldsymbol{\lambda}}_0$ are obtained by solving the algebraic system (65a–65c). Finally, one can initialize $\mathbf{a}_0 = \dot{\tilde{\mathbf{v}}}_0$ and $\boldsymbol{\eta}_0 = \tilde{\boldsymbol{\lambda}}_0$.

One also observes that the smooth positions $\tilde{\mathbf{q}}_{n+1}$ and velocities $\tilde{\mathbf{v}}_{n+1}$ do not appear in this scheme, which is a difference compared to the algorithm presented in [12].

3.8 Solution of the Discretized Problem

At each time step, the system of nonlinear equations represented by Eq. (65) should be solved for the different variables at time t_{n+1}. As the activation status of the constraints depends on the unknowns of the problem, the problem implicitly includes complementarity conditions.

For the sake of numerical efficiency, Eq. (65) can be condensed by elimination of the linear equations that represent the time integration formulae (65j–65o). This elimination relies on a distinction between the independent variables selected as $\dot{\tilde{\mathbf{v}}}_{n+1}$, $\tilde{\boldsymbol{\lambda}}_{n+1}$, $\mathbf{U}_{n+1}$, $\boldsymbol{\nu}_{n+1}$, $\mathbf{W}_{n+1}$, $\boldsymbol{\Lambda}_{n+1}$, and the remaining dependent variables $\mathbf{q}_{n+1}$, $\mathbf{v}_{n+1}$, $\mathbf{a}_{n+1}$, $\boldsymbol{\eta}_{n+1}$, $\boldsymbol{\nu}^*_{n+1}$ and $\boldsymbol{\Lambda}^*_{n+1}$. For a system with n_q coordinates in $\mathbf{q}$ and n_g constraints in $\mathbf{g}$, the problem is represented by a system of $3(n_q + n_g)$ nonlinear equations with complementarity conditions for the $3(n_q + n_g)$ independent variables.

This nonlinear system can be solved using a semi-smooth Newton process, which can also be interpreted as an active set method [10, 22–24]. This method relies on iterations based on the linearized system with an update of the activation status at each iteration.

Algorithm 1 Nonsmooth generalized-α time integration scheme

Inputs: initial values $\mathbf{q}_0$ and $\mathbf{v}_0$
Compute the consistent value of $\tilde{\dot{\mathbf{v}}}_0$ and $\tilde{\boldsymbol{\lambda}}_0$ and initialize $\mathbf{a}_0 := \tilde{\dot{\mathbf{v}}}_0$ and $\boldsymbol{\eta}_0 = \tilde{\boldsymbol{\lambda}}_0$
for $n = 0$ to $n_{\text{final}} - 1$ **do**
  Predict the variables $\mathbf{q}_{n+1}$, $\mathbf{v}_{n+1}$, $\tilde{\dot{\mathbf{v}}}_{n+1}$, $\boldsymbol{\nu}_{n+1}$, $\boldsymbol{\Lambda}_{n+1}$, $\tilde{\boldsymbol{\lambda}}_{n+1}$, $\mathbf{a}_{n+1}$,$\boldsymbol{\nu}^*_{n+1}$, $\boldsymbol{\eta}_{n+1}$ $\boldsymbol{\Lambda}^*_{n+1}$
  for $i = 1$ to i_{max} **do**
    Evaluate the sets $\mathscr{A}$, $\mathscr{B}$ and $\mathscr{S}$ at time t_{n+1}
    Evaluate the residuals of the equations of motion given by Eqs. (65a–65i)
    if all residuals are below the tolerance **then**
      break
    end if
    Evaluate the iteration matrix of Eqs. (65a–65c) with respect to $\tilde{\dot{\mathbf{v}}}_{n+1}$ and $\tilde{\boldsymbol{\lambda}}_{n+1}$
    Solve the resulting linearized problem and evaluate the corrections of $\tilde{\dot{\mathbf{v}}}_{n+1}$ and $\tilde{\boldsymbol{\lambda}}_{n+1}$
    Update the dependent variables $\mathbf{q}_{n+1}$, $\mathbf{v}_{n+1}$, $\mathbf{a}_{n+1}$, $\boldsymbol{\eta}_{n+1}$, $\boldsymbol{\nu}^*_{n+1}$ and $\boldsymbol{\Lambda}^*_{n+1}$
    Evaluate the residuals of Eqs. (65d–65f)
    Evaluate the iteration matrix of Eqs. (65d–65f) with respect to $\mathbf{U}_{n+1}$ and $\boldsymbol{\nu}_{n+1}$
    Solve the resulting linearized problem and evaluate the corrections of $\mathbf{U}_{n+1}$ and $\boldsymbol{\nu}_{n+1}$
    Update the dependent variables $\mathbf{q}_{n+1}$ and $\boldsymbol{\nu}^*_{n+1}$
    Evaluate the residuals of Eqs. (65g–65i)
    Evaluate the iteration matrix of Eqs. (65g–65i) with respect to $\mathbf{W}_{n+1}$ and $\boldsymbol{\Lambda}_{n+1}$
    Solve the resulting linearized problem and evaluate the corrections of $\mathbf{W}_{n+1}$ and $\boldsymbol{\Lambda}_{n+1}$
    Update the dependent variables $\mathbf{v}_{n+1}$ and $\boldsymbol{\Lambda}^*_{n+1}$
  end for
end for

A simplification of the linearized system can be obtained if some coupling terms between equations are neglected in the iteration matrix that appears in the linearized problem. In this case, the solution to the full linearized problem within each iteration can be approximated by a sequence of three subproblems of size $n_q + n_g$, as described in Algorithm 1. A similar procedure was used in [12], and more implementation details can be found in that paper. In many practical cases, it turns out that this approximation of the iteration matrix does not significantly penalize the convergence of the process, but significantly reduces the computational cost.

During the inner semismooth Newton iterations, the activation criteria are evaluated in non-converged states for which the equilibrium is not reached. The definition of these criteria based on the augmented Lagrange multipliers is essential for ensuring the robustness of the activation strategy and the convergence of the iterations towards the equilibrium state. Another important detail is that, even though the dependent variables are updated between the treatments of the different subsytems, the sets $\mathscr{A}$, $\mathscr{B}$ and $\mathscr{S}$ are evaluated only once at the begining of the global Newton iteration, but are not updated between the treatments of the different subsystems.

4 Special Case: Smooth Motion

The above algorithm is general and can deal with rigid and flexible multibody systems with bilateral constraints, unilateral contact conditions and impacts, involving velocity jumps and impulsive reaction forces. As a special case, it is also applicable to systems without unilateral constraints or with strictly closed unilateral constraints. In this case, no impact occurs and the dynamics evolves smoothly without velocity jumps or impulsive phenomena.

Even though we are interested in nonsmooth systems, the numerical performances of the method should also be investigated in the smooth phases of motion between impact phenomena. In this section, the equations of motion and the time integration algorithm are first particularized to smooth systems. Then, more usual DAE solvers for smooth systems will be reviewed and compared to the proposed algorithm.

If no impulsive contribution is present in Eqs. (21) and (23), we can write

$$\mathrm{d}\mathbf{v} = \dot{\mathbf{v}}\,\mathrm{d}t, \tag{66}$$

$$\mathrm{d}\mathbf{i} = \boldsymbol{\lambda}\,\mathrm{d}t, \tag{67}$$

and, if all active constraints remain closed, the dynamics can be represented by

$$\dot{\mathbf{q}} = \mathbf{v}, \tag{68a}$$

$$\mathbf{M}(\mathbf{q})\,\dot{\mathbf{v}} - \mathbf{g}_{\mathbf{q}}^T(\mathbf{q})\,\boldsymbol{\lambda} = \mathbf{f}(\mathbf{q},\mathbf{v},t), \tag{68b}$$

$$\mathbf{g}(\mathbf{q}) = \mathbf{0}. \tag{68c}$$

4.1 Special Form of the Proposed Algorithm

For a smooth dynamic system without impact, Eq. (43) becomes

$$\mathbf{M}(\mathbf{q})\,\dot{\tilde{\mathbf{v}}} - \mathbf{g}_{\mathbf{q}}^T(\mathbf{q})\,\tilde{\boldsymbol{\lambda}} = \mathbf{f}(\mathbf{q},\mathbf{v},t), \tag{69a}$$

$$\mathbf{g}_{\mathbf{q}}(\mathbf{q})\,\dot{\tilde{\mathbf{v}}} + \mathbf{h}(\mathbf{q},\mathbf{v}) = \mathbf{0}, \tag{69b}$$

$$\mathbf{M}(\mathbf{q})(\dot{\mathbf{q}} - \mathbf{v}) - \mathbf{g}_{\mathbf{q}}^T(\mathbf{q})\boldsymbol{\mu} = \mathbf{0}, \tag{69c}$$

$$\mathbf{g}(\mathbf{q}) = \mathbf{0}, \tag{69d}$$

$$\mathbf{M}(\mathbf{q})\,(\dot{\mathbf{v}} - \dot{\tilde{\mathbf{v}}}) - \mathbf{g}_{\mathbf{q}}^T(\mathbf{q})\,\boldsymbol{\xi} = \mathbf{0}, \tag{69e}$$

$$\mathbf{g}_{\mathbf{q}}(\mathbf{q})\,\mathbf{v} = \mathbf{0}, \tag{69f}$$

with $\boldsymbol{\xi} = \boldsymbol{\lambda} - \tilde{\boldsymbol{\lambda}}$. This equation has the structure of a stabilized index-1 DAE, which combines the constraints at the position, velocity and acceleration levels. One can check that any solution to Eq. (68) satisfies this formulation with $\boldsymbol{\mu} = \mathbf{0}, \boldsymbol{\xi} = \mathbf{0}, \boldsymbol{\lambda} = \tilde{\boldsymbol{\lambda}}$ and $\dot{\mathbf{v}} = \dot{\tilde{\mathbf{v}}}$. To the best of our knowledge, this form is not known in the multibody dynamics community. Nevertheless, it can be used in combination with various time integration schemes, as index-1 DAEs are known to be less numerically sensitive than higher index systems.

The discrete form of Eq. (69) becomes

$$\mathbf{M}(\mathbf{q}_{n+1})\,\dot{\tilde{\mathbf{v}}}_{n+1} - \mathbf{g}_{\mathbf{q}}^T(\mathbf{q}_{n+1})\,\tilde{\boldsymbol{\lambda}}_{n+1} = \mathbf{f}(\mathbf{q}_{n+1}, \mathbf{v}_{n+1}, t_{n+1}), \tag{70a}$$

$$\mathbf{g}_{\mathbf{q}}(\mathbf{q}_{n+1})\,\dot{\tilde{\mathbf{v}}}_{n+1} + \mathbf{h}(\mathbf{q}_{n+1}, \mathbf{v}_{n+1}) = \mathbf{0}, \tag{70b}$$

$$\mathbf{M}(\mathbf{q}_{n+1})\mathbf{U}_{n+1} - \mathbf{g}_{\mathbf{q}}^T(\mathbf{q}_{n+1})\,\boldsymbol{\nu}_{n+1} = \mathbf{0}, \tag{70c}$$

$$\mathbf{g}(\mathbf{q}_{n+1}) = \mathbf{0}, \tag{70d}$$

$$\mathbf{M}(\mathbf{q}_{n+1})\,\mathbf{W}_{n+1} - \mathbf{g}_{\mathbf{q}}^T(\mathbf{q}_{n+1})\,\boldsymbol{\Lambda}_{n+1} = \mathbf{0}, \tag{70e}$$

$$\mathbf{g}_{\mathbf{q}}(\mathbf{q}_{n+1})\,\mathbf{v}_{n+1} = \mathbf{0}, \tag{70f}$$

which needs to be combined with the time integration formulae in Eqs. (65j–65l). In this case, the position correction $\mathbf{U}_{n+1}$ and the velocity jump $\mathbf{W}_{n+1}$ are only needed to compensate for the drift of the constraints at the position and velocity levels that results from the time integration of the acceleration constraint at every time step. These corrections are thus expected to be small.

4.2 Other Formulations for Smooth Systems with Constraints at a Single Level

In multibody dynamics, one generally combines the kinematic equation and the dynamic equilibrium

$$\dot{\mathbf{q}} = \mathbf{v}, \tag{71a}$$

$$\mathbf{M}(\mathbf{q})\,\dot{\mathbf{v}} - \mathbf{g}_{\mathbf{q}}^T\,\boldsymbol{\lambda} = \mathbf{f}(\mathbf{q}, \mathbf{v}, t) \tag{71b}$$

with the constraints either expressed at the position level (index-3 formulation), velocity level (index-2 formulation) or acceleration level (index-1 formulation), or based on a linear combination according to the index-1 Baumgarte stabilization method as follows:

$$\left\{\begin{array}{rl} \mathbf{g}(\mathbf{q}) = \mathbf{0} & \text{if position constraint} \\ \mathbf{g}_{\mathbf{q}}(\mathbf{q})\,\mathbf{v} = \mathbf{0} & \text{if velocity constraint} \\ \mathbf{g}_{\mathbf{q}}(\mathbf{q})\,\dot{\mathbf{v}} + \mathbf{h}(\mathbf{q}, \mathbf{v}) = \mathbf{0} & \text{if acceleration constraint} \\ \mathbf{g}_{\mathbf{q}}(\mathbf{q})\,\dot{\mathbf{v}} + \mathbf{h}(\mathbf{q}, \mathbf{v}) + 2\alpha\,\mathbf{g}_{\mathbf{q}}(\mathbf{q})\,\mathbf{v} + \beta^2\,\mathbf{g}(\mathbf{q}) = \mathbf{0} & \text{if Baumgarte form.} \end{array}\right. \tag{71c}$$

These equations can be solved for given initial conditions $\mathbf{q}(0) = \mathbf{q}_0$ and $\mathbf{v}(0) = \mathbf{v}_0$. For the sake of consistency, these initial conditions need to verify the constraints at the position and velocity levels.

The index-3 formulation is widely used for the simulation of multibody systems [8, 19]. Numerous theoretical results are available for implicit time integration schemes based on this formulation. For example, using the generalized-α time integration scheme, all solution components (position, velocities, accelerations and Lagrange multipliers) converge to the exact solution with second-order accuracy on finite time intervals. This result was first obtained for mechanical systems modelled as DAEs on a vector space [5] and later extended to systems with finite rotation variables and modelled as DAEs on a Lie group [7, 13]. In order to reduce the influence of numerical disturbances, a careful scaling strategy is recommended for the different equations and variables of the discrete system [11]. The hidden constraints at the velocity and acceleration levels are not exactly satisfied, but the constraint violation error stays in certain limits and decreases with the time step as fast as $\mathscr{O}(h^2)$ on finite time intervals. However, order reduction phenomena were pointed out in [7], which may affect the initial phase of a simulation by spurious transient numerical oscillations in the accelerations and Lagrange multipliers with $\mathscr{O}(h)$ amplitude. Also, the index-3 formulation cannot be directly extended to build time-stepping schemes for systems with unilateral constraints, as it does not lend itself to the incorporation of the impact law.

The index-2 formulation based on the expression of the constraint at the velocity level is equivalent to Eq. (36) in the special case of a smooth system without impact. It is thus particularly relevant for nonsmooth systems, as the impact law may be incorporated into the velocity constraint according to Moreau's sweeping process. In nonsmooth dynamics, the problem is usually integrated in time using a θ-method [2, 25, 27]. In this approach, the numerical solution is not forced to satisfy the constraint at the position level so that drift-off phenomena can occur as a result of the accumulation of numerical integration errors.

The index-1 formulation based on the constraint at the acceleration level is even less sensitive from a numerical point of view and can be solved using non-stiff time integration methods. However, it suffers from important drift-off phenomena at the velocity and position levels [4]. These drift-off phenomena can be eliminated by the implementation of projection methods that bring the numerical solution back to the constraint manifold. The Baumgarte stabilization also enforces a single constraint, but is formed as a weighted linear combination of the constraints at the position, velocity and acceleration levels [9, 17]. In a strict sense, the resulting numerical solution does not satisfy any of these constraints individually. To the best of our knowledge, these index-1 formulations have not been used in time-stepping schemes for unilaterally constrained systems with impacts and velocity jumps because the acceleration variable is not properly defined at the impact time. One of the original contributions of this chapter is to exploit the acceleration variable that results from the splitting procedure and is well-defined at any time for the formulation of the active constraints at the acceleration level for nonsmooth mechanical systems.

4.3 Gear-Gupta-Leimkuhler Formulation

The Gear-Gupta-Leimkuhler (GGL) formulation is another index reduction method that was initially developed for smooth DAEs and that simultaneously enforces the constraints at the position and velocity levels [18]. It is based on the reformulation of the initial set of equations in index-2 form as

$$\dot{\mathbf{q}} - \mathbf{g}_{\mathbf{q}}^T \boldsymbol{\mu} = \mathbf{v}, \tag{72a}$$

$$\mathbf{M}(\mathbf{q})\,\dot{\mathbf{v}} - \mathbf{g}_{\mathbf{q}}^T \boldsymbol{\lambda} = \mathbf{f}(\mathbf{q}, \mathbf{v}, t), \tag{72b}$$

$$\mathbf{g}(\mathbf{q}) = \mathbf{0}, \tag{72c}$$

$$\mathbf{g}_{\mathbf{q}}(\mathbf{q})\,\mathbf{v} = \mathbf{0}. \tag{72d}$$

One can check that any exact solution to the initial DAE (68) is also a solution to this set of equations with $\boldsymbol{\mu} = \mathbf{0}$.

As shown in [6, 7], this index-2 problem can be solved using the generalized-α method. In this chapter, the notations from these references are slightly adapted to match our previous developments. At time step $n+1$, the unknown variables $\mathbf{q}_{n+1}$, $\mathbf{v}_{n+1}$, $\dot{\mathbf{v}}_{n+1}$, $\boldsymbol{\lambda}_{n+1}$, $\mathbf{U}_n = h(\dot{\mathbf{q}}_n - \mathbf{v}_n)$ and $\boldsymbol{\nu}_n = h\boldsymbol{\mu}_n$ should thus satisfy

$$\mathbf{U}_n - \mathbf{g}_{\mathbf{q}}^T(\mathbf{q}_n)\,\boldsymbol{\nu}_n = \mathbf{0}, \tag{73a}$$

$$\mathbf{M}(\mathbf{q}_{n+1})\,\dot{\mathbf{v}}_{n+1} - \mathbf{g}_{\mathbf{q}}^T(\mathbf{q}_{n+1})\,\boldsymbol{\lambda}_{n+1} = \mathbf{f}(\mathbf{q}_{n+1}, \mathbf{v}_{n+1}, t_{n+1}), \tag{73b}$$

$$\mathbf{g}(\mathbf{q}_{n+1}) = \mathbf{0}, \tag{73c}$$

$$\mathbf{g}_{\mathbf{q}}(\mathbf{q}_{n+1})\,\mathbf{v}_{n+1} = \mathbf{0}, \tag{73d}$$

together with the integration formula

$$\mathbf{q}_{n+1} = \mathbf{q}_n + h\mathbf{v}_n + h^2(0.5-\beta)\mathbf{a}_n + h^2\beta\mathbf{a}_{n+1} + \mathbf{U}_n, \tag{73e}$$

$$\mathbf{v}_{n+1} = \mathbf{v}_n + h(1-\gamma)\mathbf{a}_n + h\gamma\mathbf{a}_{n+1}, \tag{73f}$$

$$(1-\alpha_m)\mathbf{a}_{n+1} + \alpha_m\mathbf{a}_n = (1-\alpha_f)\dot{\mathbf{v}}_{n+1} + \alpha_f\dot{\mathbf{v}}_n. \tag{73g}$$

This method leads to a numerical solution that simultaneously satisfies the constraints at the position and velocity levels. Unlike in the analytical solution, the multiplier $\boldsymbol{\nu}_n$ of the numerical solution is not exactly $\mathbf{0}$, with the consequence that $\mathbf{U}_n \neq \mathbf{0}$, i.e., $\dot{\mathbf{q}}_n \neq \mathbf{v}_n$. Compared to the index-3 formulation, this method is less numerically sensitive and is not prone to the order reduction phenomenon mentioned in the previous section [7].

In order to highlight the connection with the nonsmooth algorithm discussed in this chapter and in [12], the method can be slightly adapted as

$$\mathbf{M}(\mathbf{q}_{n+1})\,\mathbf{U}_{n+1} - \mathbf{g}_{\mathbf{q}}^T(\mathbf{q}_{n+1})\,\boldsymbol{\nu}_{n+1} = \mathbf{0}, \tag{74a}$$

$$\mathbf{M}(\mathbf{q}_{n+1})\,\dot{\mathbf{v}}_{n+1} - \mathbf{g}_{\mathbf{q}}^T(\mathbf{q}_{n+1})\,\boldsymbol{\lambda}_{n+1} = \mathbf{f}(\mathbf{q}_{n+1}, \mathbf{v}_{n+1}, t_{n+1}), \tag{74b}$$

$$\mathbf{g}(\mathbf{q}_{n+1}) = \mathbf{0}, \tag{74c}$$

$$\mathbf{g}_{\mathbf{q}}(\mathbf{q}_{n+1})\,\mathbf{v}_{n+1} = \mathbf{0}, \tag{74d}$$

with the time integration formulae

$$\mathbf{q}_{n+1} = \mathbf{q}_n + h\mathbf{v}_n + h^2(0.5-\beta)\mathbf{a}_n + h^2\beta\mathbf{a}_{n+1} + \mathbf{U}_{n+1}, \tag{74e}$$

$$\mathbf{v}_{n+1} = \mathbf{v}_n + h(1-\gamma)\mathbf{a}_n + h\gamma\mathbf{a}_{n+1}, \tag{74f}$$

$$(1-\alpha_m)\mathbf{a}_{n+1} + \alpha_m\mathbf{a}_n = (1-\alpha_f)\dot{\mathbf{v}}_{n+1} + \alpha_f\dot{\mathbf{v}}_n. \tag{74g}$$

Two changes can be observed between Eqs. (73) and (74). Firstly, the mass matrix $\mathbf{M}$ now appears in Eq. (74a). Secondly, the position correction $\mathbf{U}_{n+1}$ that appears in the position update Eq. (74e) is evaluated at time step $n+1$ (and not at time step n, as in Eq. (73e).

Various investigations have addressed the extension of the GGL formulation for nonsmooth systems [1, 12, 36]. Also, the formulation presented in Sect. 4.1 can be interpreted as a recursive application of the GGL method, so that the constraints at the acceleration level are also incorporated.

4.4 *Emulation of Post-impact Conditions*

If an impact is followed by a free-flight phase on a finite time interval, the post-impact numerical solution will be affected by disturbances that will propagate dynamically in the free-flight phase. An important question is thus how to characterize the behaviour of the algorithm for smooth mechanical systems with a particular focus on the sensitivity to disturbances induced by impulsive phenomena and constraint activations. This section shows that the behaviour of the nonsmooth generalized-α method in the post-impact phase can be investigated based on the underlying smooth system with disturbed initial conditions.

Let us consider a nonsmooth system and imagine that an isolated impact occurs in the time interval $[t_{n-1}, t_n)$, but that no other nonsmooth phenomenon arises for $t > t_n$. Over the time interval $[t_{n-1}, t_n)$, the velocity is discontinuous, but the displacement remains continuous in time. If the system is simulated either using the method described in [12] or the method proposed in this paper, the numerical solution at t_{n+1} only depends on $\mathbf{q}_n$, $\mathbf{v}_n$, $\dot{\tilde{\mathbf{v}}}_n$ and $\mathbf{a}_n$ (we do not need to evaluate $\boldsymbol{\eta}_{n+1}$, $\boldsymbol{\Lambda}^*_{n+1}$ and $\boldsymbol{\nu}^*_{n+1}$, as the constraint status is assumed to be known for $t > t_n$). Let us analyze the consistency of these variables $(\mathbf{q}_n, \mathbf{v}_n, \dot{\tilde{\mathbf{v}}}_n, \mathbf{a}_n)$ with respect to the bilateral constraints in the post-impact phase.

The positions $\mathbf{q}_n$ and velocities $\mathbf{v}_n$ are, by construction, consistent with the bilateral constraints at the position and velocity levels. Therefore, at those levels, the discontinuity leads to new and consistent initial conditions and erases the pre-impact time history.

As the velocity is discontinuous, the acceleration $\dot{\tilde{\mathbf{v}}}$ defined according to our splitting method also undergoes an $\mathcal{O}(1)$ discontinuity over the time interval $[t_{n-1}, t_n)$. At t_n, consistent values of the acceleration $\dot{\tilde{\mathbf{v}}}_n$ and of the shifted value $\mathbf{a}_n$ could be computed from Eq. (38) based on the value of $\mathbf{q}_n$ and $\mathbf{v}_n$, using a similar technique as for the definition of the initial conditions. The results would thus be consistent and completely independent of the values of the pre-impact solution. This strategy would be interpreted as a reinitialization of the time integration procedure after the impact.

However, the method described in [12] and the method proposed here do not rely on a reinitialization procedure, as we do not want to perform specific treatments every time an impact occurs. Instead, the smooth acceleration is integrated over the impact according to the generalized-α method as if no discontinuity were present. Therefore, the pre-impact acceleration history influences the post-impact numerical solution as follows:

- In the method described in [12], for given values of $\mathbf{q}_n$ and $\mathbf{v}_n$, the values of $\dot{\tilde{\mathbf{v}}}_n$ and $\mathbf{a}_n$ still depend on the pre-impact values $\dot{\tilde{\mathbf{v}}}_{n-1}$, $\mathbf{a}_{n-1}$ and $\mathbf{v}_{n-1}$.
- In the algorithm proposed here, the value of $\dot{\tilde{\mathbf{v}}}_n$ is defined as an algebraic function of $\mathbf{q}_n$ and $\mathbf{v}_n$, and is thus independent of the pre-impact solution, but the value of $\mathbf{a}_n$ still depends on the pre-impact values $\dot{\tilde{\mathbf{v}}}_{n-1}$ and $\mathbf{a}_{n-1}$ (see Eq. (65l)).

Compared to a correct reinitialization of the acceleration variables solely based on the post-impact state, the pre-impact solution influences the values $\mathbf{a}_n$ and possibly $\dot{\tilde{\mathbf{v}}}_n$ in both algorithms, leading to $\mathcal{O}(1)$ disturbances. As a consequence, $\mathbf{a}_n$ and possibly $\dot{\tilde{\mathbf{v}}}_n$ may violate the constraint at the acceleration level with $\mathcal{O}(1)$ errors. Thus, the post-impact numerical solution can be emulated by a simulation of the underlying smooth system for $t > t_n$ if the initial accelerations $\dot{\tilde{\mathbf{v}}}_n$ and $\mathbf{a}_n$ are modified with $\mathcal{O}(1)$ disturbances.

This situation is also representative of the transition of a unilateral constraint from an open to a closed status in $\mathcal{S}$ over the time interval $[t_{n-1}, t_n)$. Indeed, in this case, the position $\mathbf{q}_n$ and velocity $\mathbf{v}_n$ satisfy the new constraint at the position and velocity levels, but the acceleration $\dot{\tilde{\mathbf{v}}}_n$ and the shifted variable $\mathbf{a}_n$ do not necessarily satisfy the new constraint at the acceleration level.

5 Application to a Smooth System

The properties of the proposed method are first investigated in the context of the numerical solution to smooth DAEs. The classical example of a pendulum modelled as a constrained mechanical system serves for the comparison. Numerical methods derived from the generalized-α method using four different formulations of the equations of motion are compared:

- the index-3 formulation with the constraints at the position level only, referred to as the "P-constrained" method;

- the index-2 formulation with the constraints at the velocity level only, referred to as the "V-constrained" method;
- the index-2 Gear-Gupta-Leimkuhler formulation with the constraints at the position and velocity levels, referred to as the "PV-constrained" method;
- the proposed index-1 formulation with the constraints imposed simultaneously at the position, velocity and acceleration levels, referred to as the "PVA-constrained" method.

5.1 *Problem Description*

Let us analyse the transient response of the pendulum depicted in Fig. 1. In order to study the behaviour of the algorithm in the presence of constraints, a set of 3 absolute but redundant coordinates is chosen $\mathbf{q} = [x \ y \ \theta]^T$, where x and y are the coordinates of the center of mass and θ is the angle of the pendulum. These coordinates have to satisfy 2 bilateral constraints

$$g^1(\mathbf{q}) \equiv x - L\cos\theta = 0, \tag{75}$$

$$g^2(\mathbf{q}) \equiv y - L\sin\theta = 0. \tag{76}$$

The physical parameters of the system are selected as: length of the pendulum $L = 1\,\text{m}$, mass $m = 1\,\text{kg}$, moment of inertia $J = 0.1\,\text{kg}\,\text{m}^2$, and gravity acceleration along the y-axis $a_g = 10\,\text{rad/s}^2$. The initial conditions at the position and velocity levels are defined as $\theta_0 = \pi/6\,\text{rad}$ and $\dot{\theta}_0 = 10\,\text{rad/s}$. The numerical parameters of the numerical solvers are selected as $h = 2.\,10^{-3}\,\text{s}$, $\rho_\infty = 0.9$.

5.2 *Results Based on Consistent Initial Conditions*

Consistent initial positions $\mathbf{q}$ and velocities $\mathbf{v}$ are established from the initial values θ_0 and $\dot{\theta}_0$. The initial acceleration $\dot{\mathbf{v}}$ is obtained by solving Eq. (38) at time t_0 and the shifted acceleration is initialized as $\mathbf{a}_0 = \dot{\mathbf{v}}_0$. The results are presented in Fig. 2. On certain graphs, high numerical oscillations are observed at the frequency of the step size, which means that the variable under study jumps between a low to a high value

Fig. 1 Pendulum

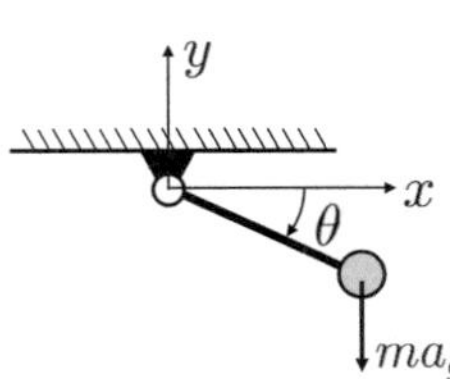

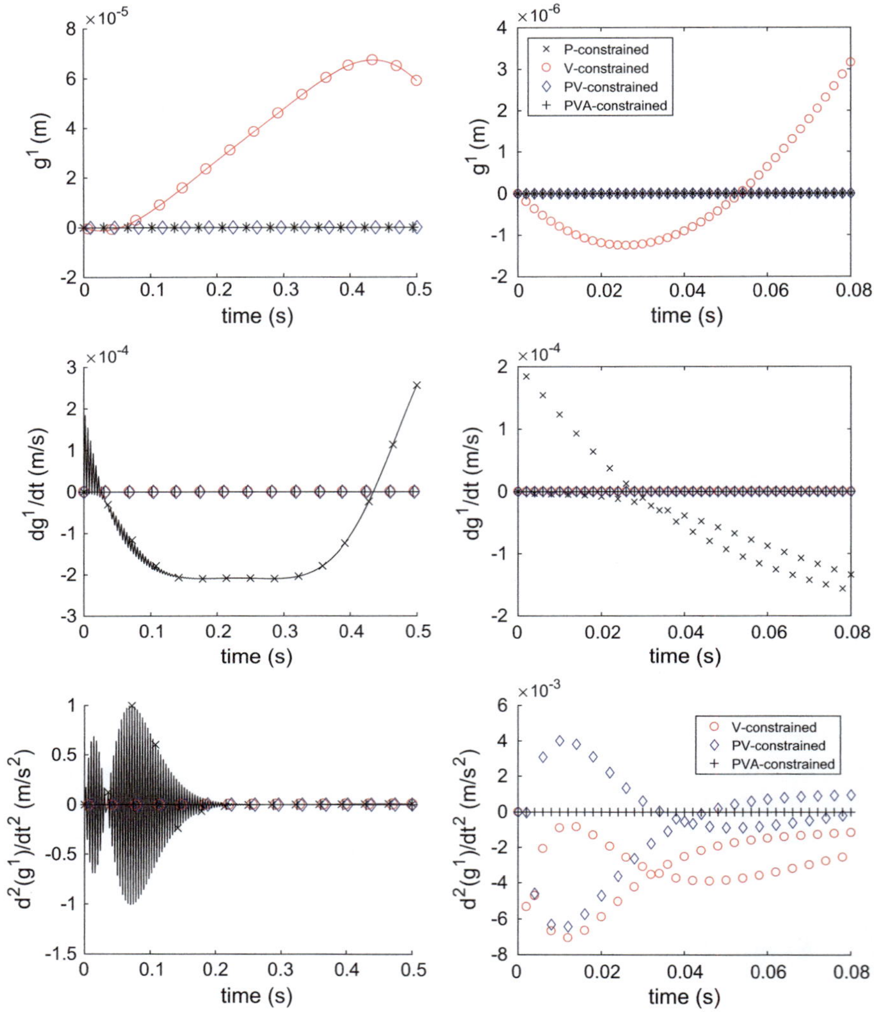

Fig. 2 Position (top), velocity (middle) and acceleration (bottom) constraints in the pendulum example - left: full time interval, right: zoom on the initial phase. In the bottom-right plot, the solution to the index-3 problem with the position constraint is not represented for the sake of readibility

at each step. For the sake of readability, when zooming on these phenomena, only the values at the successive time steps are represented by markers, but the interpolating line between the time steps is not necessarily displayed.

In the index-3 solution based on the sole position constraint, spurious high frequency oscillations of the constraint at the velocity and acceleration levels are observed in the initial phase. After a transient phase, these high-frequency oscillations are damped out and the hidden constraints do not converge to zero, but rather evolve in a continuous manner. The amplitude of the transient high-frequency oscillations of the acceleration constraint is particularly large, and it can be shown that it decreases only as $\mathcal{O}(h)$ when the time step is decreased, which reflects the presence of an order reduction phenomenon, as discussed in Sect. 4.2.

In the index-2 solution based on the sole velocity constraint, a constraint drift is observed at the position level, which increases as time goes by. Spurious high-frequency oscillations are observed at the acceleration level, but it can be shown that their amplitude is quite limited and decreases as fast as $\mathcal{O}(h^2)$ when the time step decreases, i.e., there is no order reduction phenomenon in this case. After a transient phase, the spurious oscillations disappear and the acceleration constraint evolves in a continuous manner.

In the index-2 GGL solution, which enforces the constraints at the position and velocity levels, the constraints are indeed satisfied up to machine precision at those levels. At the acceleration level, the behaviour is similar as for the other index-2 solution discussed in the previous paragraph.

In the proposed index-1 solution, the results confirm that the constraints are satisfied up to machine precision at the three levels (position, velocity and acceleration).

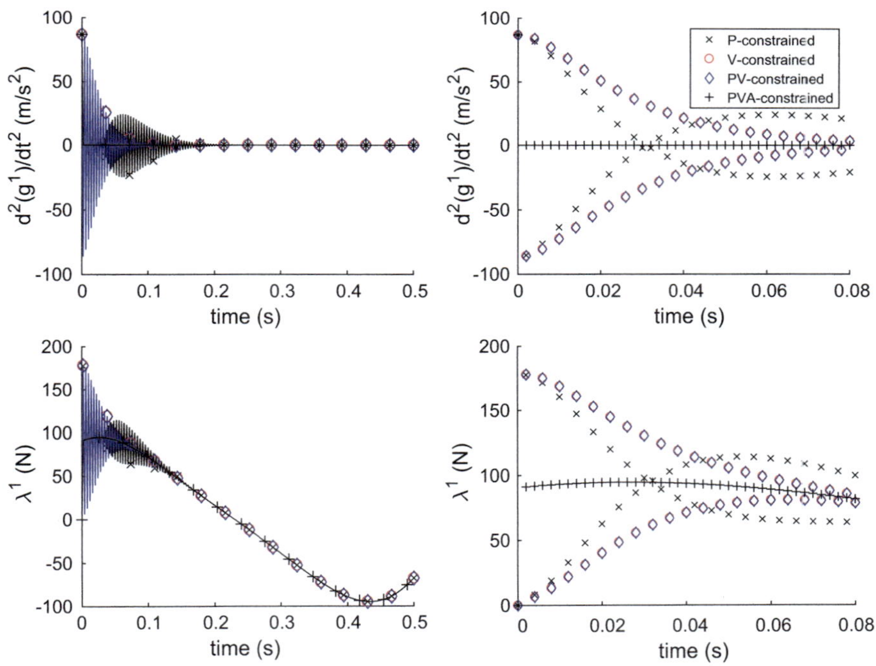

Fig. 3 Acceleration constraint (top) and Lagrange multiplier (bottom) in the pendulum example with post-impact initial conditions (left: full time interval, right: zoom on the initial phase)

5.3 Results Based on Post-impact Initial Conditions

In order to emulate the disturbances induced by an impact on the post-impact numerical solution, the simulation of the rigid pendulum is run using disturbed initial accelerations such that the constraint is not satisfied at the acceleration level.

Figure 3 presents the simulation results for the pendulum when the acceleration and shifted acceleration are initialized as $\dot{\mathbf{v}}_0 = \mathbf{a}_0 = \mathbf{0}$. Large spurious oscillations of the acceleration constraint and Lagrange multiplier are observed for all algorithms, except for the proposed method that enforces the constraints at the position, velocity and acceleration levels. Thus, the proposed method appears much less sensitive to the disturbances induced by impact phenomena.

6 Application to Nonsmooth Systems

In this section, three numerical examples are used to compare two algorithms for nonsmooth dynamic systems:

- The algorithm described in [12], in which the constraint on the smooth motion only includes the bilateral constraints that are imposed at the velocity level;
- The algorithm proposed here, in which the constraint on the smooth motion includes the bilateral constraints, as well as the active unilateral constraints, both imposed at the acceleration level.

These two algorithms will be respectively called the “PVV-constrained” method and the “PVA-constrained” method in the following. In both algorithms, the smooth motion is integrated using the generalized-α time integration formula.

The first example is a bouncing rigid pendulum, the second example is a bouncing elastic pendulum modelled as a geometrically exact beam, and the last example is the horizontal impact of an elastic bar. These three examples also served as a support for the analysis of several algorithms for nonsmooth systems in [12, 14]. Here, these examples are exploited to explore the properties of the PVA-constrained algorithm, which is a novel contribution of this chapter.

6.1 Bouncing Rigid Pendulum

We consider the same pendulum as described in Sect. 5.1, but, as shown in Fig. 4, a unilateral constraint restricts the motion of its center of mass as

$$g^3(\mathbf{q}) \equiv x - x_{\min} \geq 0, \tag{77}$$

Fig. 4 Bouncing pendulum

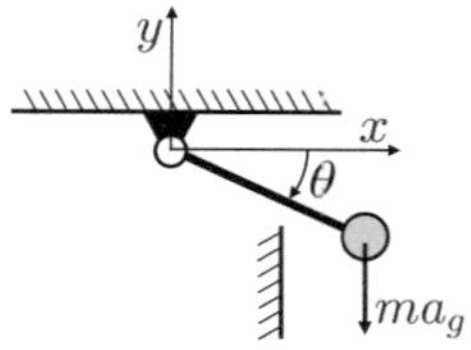

with $x_{\min} = \sqrt{2}/2$ m. The initial conditions are $\theta_0 = \pi/12$ rad and $\dot{\theta}_0 = 0$ rad/s. Consistent initial conditions are then defined for $\mathbf{q}$, $\mathbf{v}$, $\dot{\tilde{\mathbf{v}}}$ and $\mathbf{a}$. The time step and the spectral radius are selected as $h = 1.\,10^{-3}$ s and $\rho_\infty = 0.9$.

The evolution of the gap distance $g^3(\mathbf{q})$ during the motion is shown in Fig. 5. The pendulum bounces several times against the hurdle and, at the end of the trajectory, the system gets stabilized in the closed contact configuration after an accumulation phenomenon.

The evolution of the bilateral constraint at the acceleration level (Fig. 6) reveals significant numerical oscillations after each impact in the PVV-constrained algorithm. In contrast, the solution obtained using the PVA-constrained method exactly satisfies the acceleration constraints without any such oscillations. In the same figure, similar oscillations are observed in the smooth bilateral multiplier $\tilde{\lambda}^1$ evaluated using the PVV-constrained method. In the PVA-constrained method, a discontinuity occurs at each impact, but no oscillation is visible.

At the end of the trajectory, the nonsmooth phenomena disappear and the total horizontal reaction force in the rigid body becomes constant and can simply be estimated as $\tilde{\lambda}^1 + \Lambda^1/h$. Considering Figs. 6 (bottom-left) and 7 (right), the same total reaction force is obtained in the two methods at the end of the trajectory, but the value of the relative impulse Λ^1 is equal to zero in the proposed algorithm. Indeed, in this smooth part of the trajectory, the smooth equation captures the total motion and, in this case, the corrections at the position and velocity levels $\mathbf{W}$ and $\mathbf{U}$ tend to zero for the PVA-constrained algorithm.

In the transient phase before the unilateral constraint gets closed, the relative impulse Λ^1 can take negative values in the PVA-constrained method, as the complementarity condition is not applied to Λ^1, but rather to Λ^{*1}.

The PVA-constrained method generally brings less numerical dissipation, since the reaction forces are better integrated. This is in agreement with the observation of a later stabilization of the system in the closed contact state in Fig. 7.

Finally, the results in Fig. 8 were obtained using a spectral radius $\rho_\infty = 1$, i.e., without any numerical dissipation. The PVA-constrained method still gives the expected results without any spurious numerical oscillation, whereas the Lagrange multiplier obtained from the PVV-constrained method undergoes strong oscillations after the first impact that never disappear from the solution.

In summary, this example has shown that both algorithms give a satisfactory numerical solution that exactly satisfies the bilateral and unilateral constraints at the position and velocity levels. Their comparison reveals (i) that imposing the constraints at the acceleration level improves the handling of the bilateral constraints

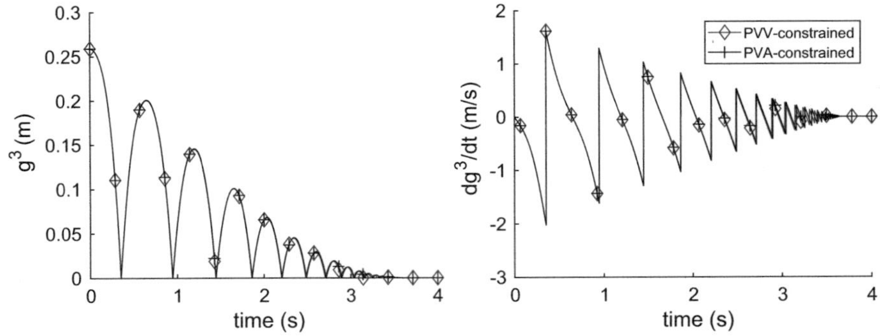

Fig. 5 Unilateral constraint in the bouncing rigid pendulum example (left: position level, right: velocity level)

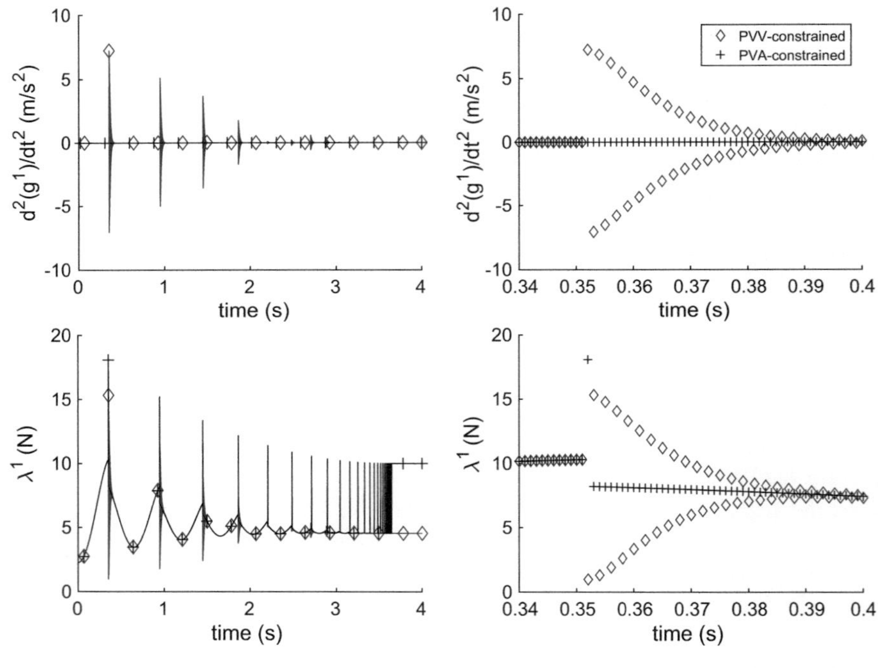

Fig. 6 Bouncing rigid pendulum: bilateral constraint at the acceleration level (top) and Lagrange multipliers $\tilde{\boldsymbol{\lambda}}$ (bottom) - left: full time interval, right: zoom on the first impact

after the impact phenomena and alleviates the need to introduce numerical dissipation in the time integration scheme in this example, and (ii) that the unilateral constraint can be activated at the acceleration level in the smooth motion. During the free flight mode or the closed constraint mode, the smooth motion then captures the full motion, which is thus integrated with second-order accuracy without any spurious oscillations.

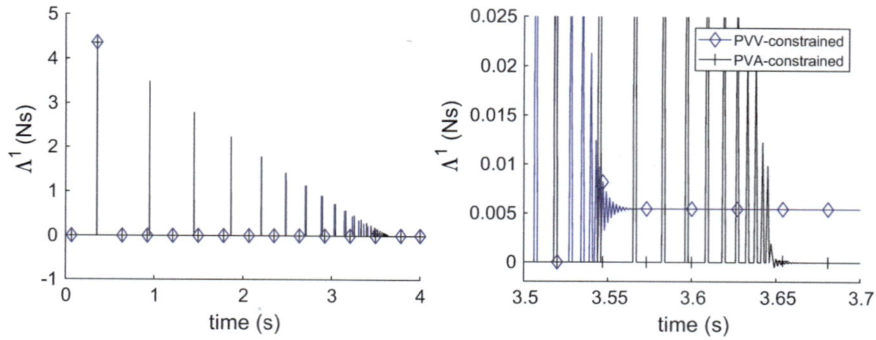

Fig. 7 Lagrange multiplier $\mathbf{\Lambda}$ of the bilateral constraint in the bouncing pendulum example (left: full time interval, right: zoom on the end phase)

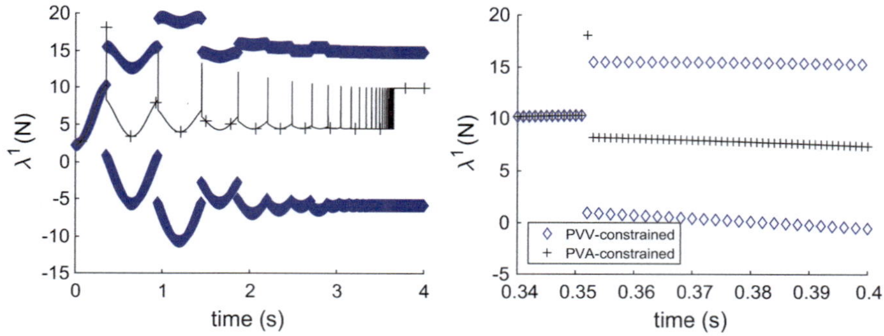

Fig. 8 No numerical damping - Lagrange multiplier $\tilde{\boldsymbol{\lambda}}$ of the bilateral constraint in the bouncing pendulum example (left: full time interval, right: zoom on the first impact)

6.2 *Bouncing Flexible Pendulum*

In this example, shown in Fig. 9, a flexible pendulum modelled as an elastic beam hits an obstacle. The beam is modelled according to the geometrically exact beam theory and discretized into nonlinear finite elements [19]. Thus, this example highlights nonlinear interactions between the beam and the non-penetration constraint at the contact point.

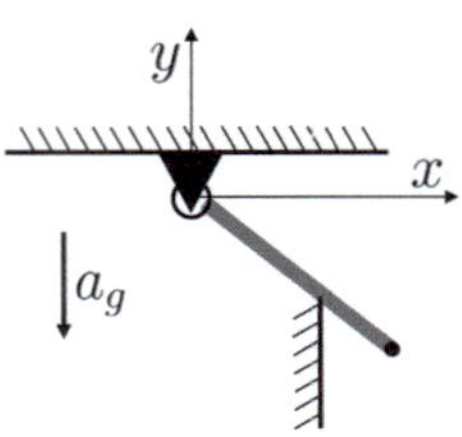

Fig. 9 Bouncing elastic pendulum

The contact condition is modelled as a unilateral constraint applied at the tip node of the beam mesh

$$g^1(\mathbf{q}) \equiv x_{\text{tip}} - x_{\text{min}} \geq 0. \tag{78}$$

There is no bilateral constraint in this example. The properties of the beam are: undeformed length $L = 1$ m, cross-section area $A = 10^{-4}$ m^2, cross-section inertia $I = 8.33\ 10^{-10}$ m^4, shear section area $A_s = (5/6)\ A$, Young modulus $E = 2.1\ 10^{11}$ N/m^2, density $\rho = 7800$ kg/m^3, and Poisson coefficient $\nu = 0.3$. At the initial time, the beam is horizontal with zero velocity. The unilateral constraint is defined as $x_{\text{min}} = L\sqrt{2}/2$.

The beam is modelled using four finite elements. The time step is $h = 5.10^{-6}$ s and the spectral radius is $\rho_\infty = 0.8$. A restitution coefficient is included in the formulation of the impact law and its value is defined as $e = 0$.

In the PVV-constrained method, the unilateral constraint is never activated at the acceleration level in the definition of the smooth motion. As there is no bilateral constraint in this case, the smooth motion is thus fully unconstrained. In the PVA-constrained method, the unilateral constraint at the acceleration level gets activated and deactivated in a dynamic manner, so that the constraint reaction force brings some stronger disturbances on the smooth motion.

As the step-size h is quite small, the mean number of Newton iterations at each time step is very close to one for both algorithms.

The constraints at the position, velocity and acceleration levels are depicted in Fig. 10. The numerical response is characterized by rather complex dynamic phenomena. The first contact phase is characterized by a finite duration on the interval [0.327,0.348] s. However, the contacts at the position, velocity and acceleration levels do not stay permanently activated over this time interval, but rather enter and leave the system in an intermittent manner. The zooms on the initial contact phase indicate a good agreement between the two algorithms at the position, velocity and acceleration levels. The solutions tend to diverge later on, as the problem is particularly sensitive. One also observes the activation of the constraint at the acceleration level for some time intervals in the PVA-constrained method, whereas this constraint is never activated in the PVV-constrained method.

The reaction forces at the contact point are represented in Fig. 11. During the first contact phase, one observes a collection of rather close impulses.

In Fig. 12, the energy decays monotonously during the motion. During the first contact phase, the energy decays progressively according to a kind of staircase function. One also observes the faster energy decay of the PVV-constrained method, which can be attributed to the higher level of numerical dissipation in this scheme.

In summary, the bouncing elastic pendulum example shows the ability of both algorithms to study the dynamics of a geometrically nonlinear beam with a unilateral constraint. Both methods show similar numerical performances in this case, which involves high frequency activation and deactivation phenomena during the contact phases.

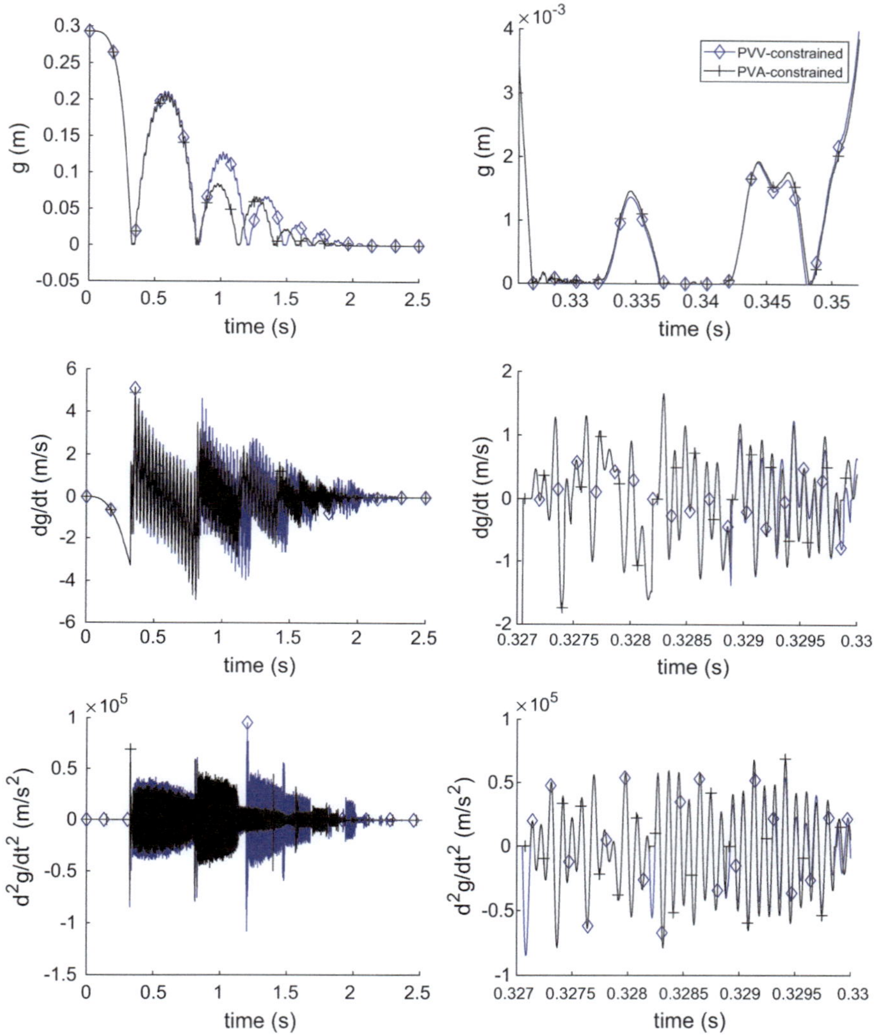

Fig. 10 Bouncing flexible pendulum: unilateral constraint at the position (top), velocity (middle) and acceleration levels (bottom) - left: full time interval, right: zoom on the first contact phase (the zoom interval is different for the position constraint)

6.3 *Horizontal Impact of an Elastic Bar*

The horizontal impact of an elastic bar, as shown in Fig. 13 is now considered. The problem was described in [16] and has an analytical solution. According to this analytical solution, the contact stays closed for a period of $\Delta t = 2L\sqrt{\rho/E}$ and the energy is conserved. The contact force remains finite, so there is no impact, even if the velocity undergoes a discontinuity at the contact point when the contact closes.

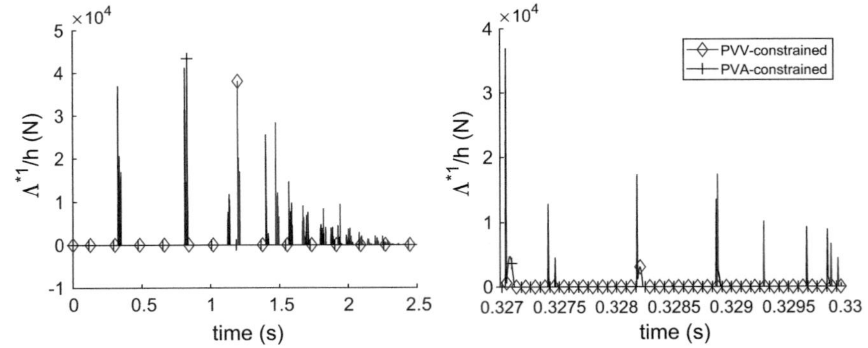

Fig. 11 Reaction force in the bouncing pendulum example (left: full time interval, right: zoom on the first contact phase)

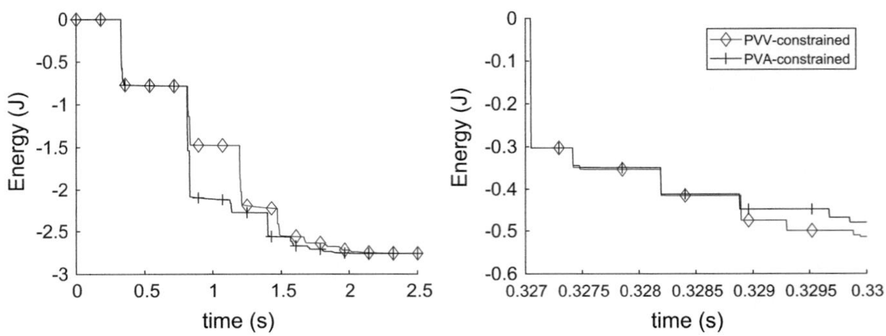

Fig. 12 Constraint at the velocity level and energy in the bouncing pendulum example (left: full time interval, right: zoom on the first contact phase)

Fig. 13 Horizontal impact of an elastic bar

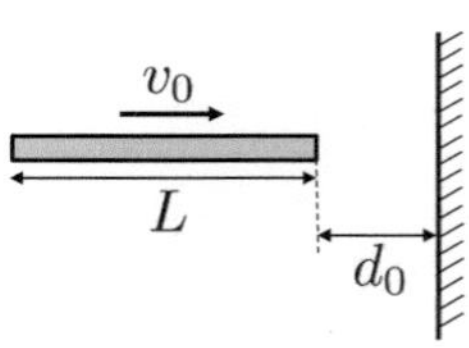

In our finite element model, a restitution coefficient e is needed at the level of the impact law. This coefficient has no physical meaning and simply represents the energy dissipation in the last element of the mesh. In order to be able to represent the instantaneous closing of the unilateral constraint, we propose choosing $e = 0$.

The physical parameters are defined as in [16]: Young modulus $E = 900\,\text{N/m}^2$, density $\rho = 1\,\text{kg/m}^3$, undeformed length $L = 10\,\text{m}$, initial distance from the obstacle $d_0 = 5\,\text{m}$, initial velocity $v_0 = 10\,\text{m/s}$. With these data, the closed contact period is $\Delta t = 2/3\,\text{s}$. The bar is discretized using 200 finite elements, the time step is taken as

$h = 2.10^{-3}$ s, and the spectral radius of the generalized-α time integrator is chosen as $\rho_\infty = 0.8$.

The results are presented in Figs. 14, 15 and 16. The two algorithms give very close results. The main difference is found in the mean number of Newton iterations at each time step. In the PVV-constrained method, we have 2.94 iterations per time step (about 10 iterations per step during the contact phase), whereas in the PVA-constrained method, only 0.80 iterations are needed on average. The explanation is that the PVV-constrained method completely disregards the unilateral constraint when evaluating the smooth motion. Therefore, the physical solution is rather far from the smooth solution, the position and velocity corrections $\mathbf{U}$ and $\mathbf{W}$ are quite significant, and more iterations are needed to solve the coupled problem.

This example shows that the two methods provide relevant numerical solutions to a unilaterally constrained structure with closed contacts. Once again, the constraints at the position and velocity levels are exactly satisfied by the numerical solution. This study also reveals the superiority of the PVA-constrained algorithm for flexible systems when some unilateral constraints stay closed during rather long time intervals.

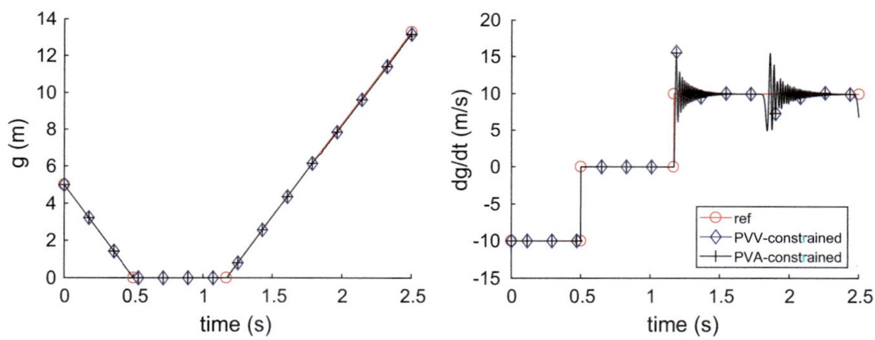

Fig. 14 Unilateral constraint in the bar impact example (left: position level, right: velocity level)

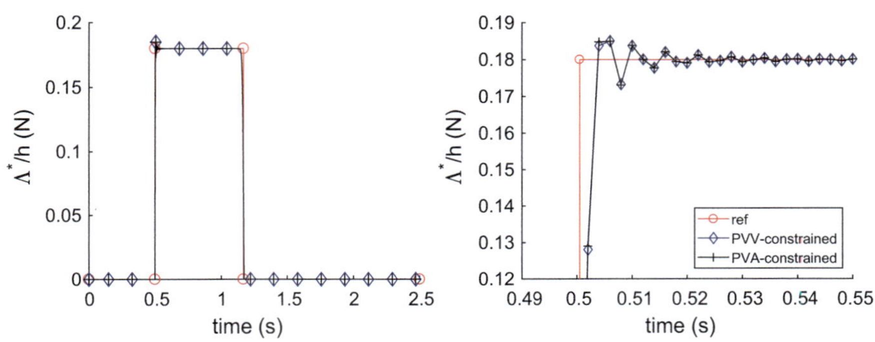

Fig. 15 Reaction force Λ^*/h in the bar impact example (left: full time interval, right: zoom on the post-impact phase)

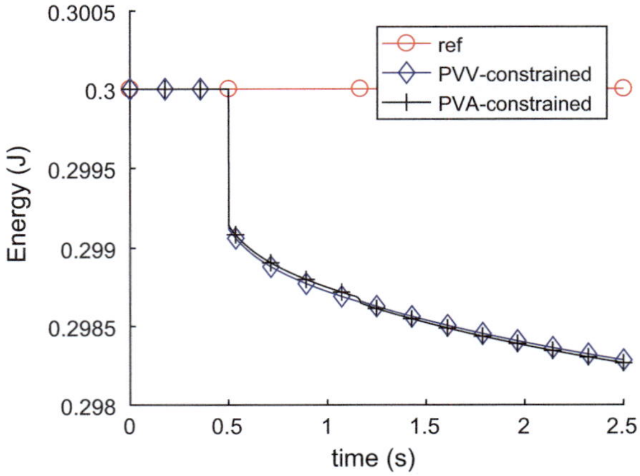

Fig. 16 Energy in the bar impact example

7 Conclusion

The nonsmooth generalized-α method was developed for the analysis of flexible multibody systems with contact conditions and impact phenomena. It relies on a splitting of the total motion into smooth (non-impulsive) and nonsmooth (impulsive) contributions. A second-order time integration scheme is then used for the smooth contributions, whereas a first-order scheme is used for the consistent integration of impulsive contributions. Compared to the classical Moreau-Jean method, this method leads to qualitatively better numerical solutions with less numerical dissipation.

This chapter addresses the formulation of the constraints that appear in the definition of the smooth motion and which can have a deep impact on the numerical properties of the scheme. We propose imposing all active constraints at the acceleration levels on the smooth part of the motion, while the total motion simultaneously satisfies the constraints at the position and velocity levels. Some advantages of this formulation are the elimination of spurious numerical oscillations of the constraints that generally occur after an impact and the possibility of accounting for the contributions of the unilateral constraints to the smooth motion. When the contact remains closed, the integration of the contact forces is performed with a higher accuracy, which comes with a reduced level of numerical dissipation, and the convergence of the approximated Newton iterations is accelerated as the amplitudes of the nonsmooth corrections are reduced. These properties were demonstrated in several numerical examples of smooth and nonsmooth mechanical systems. It is remarkable that, in rigid-body examples, the constraints and the overall numerical solution are inherently stabilized (in the sense that no spurious numerical oscillation is observed), even if no numerical dissipation is introduced at the level of the generalized-α time integrator.

Some key elements of the method can also be summarized. Firstly, the proposed splitting strategy leads to a definition of the acceleration variable $\dot{\tilde{\mathbf{v}}}$ as an algebraic function of the physical position and velocity at the current time, which permits the dynamic activation and deactivation of unilateral constraints in a very simple manner.

The acceleration $\dot{\tilde{\mathbf{v}}}$ represents the standard acceleration for almost every time, but it excludes impulsive contributions at the impact instants. Even though this acceleration is discontinuous, the position and velocity of the smooth trajectory are continuous, even in the presence of impacts. The definition of the activation criteria for the unilateral constraints at the position, velocity and acceleration levels is particularly critical for the robustness of the algorithm. The proposed criteria rely on the definition of augmented Lagrange multipliers at the position, velocity and acceleration levels, and can thus be used in a reliable way within the Newton semi-smooth iterations, even if the solution is not yet converged.

As a perspective, the present algorithm could be tested for more complex examples with a larger number of bodies and contact conditions. The extension to frictional contact conditions could also be investigated.

References

1. Acary V (2013) Projected event-capturing time-stepping schemes for nonsmooth mechanical systems with unilateral contact and Coulomb's friction. Comput Methods Appl Mech Eng 256(10):224–250
2. Acary V (2008) Numerical methods for nonsmooth dynamical systems - applications in mechanics and electronics. Lecture notes in applied and computational mechanics. Springer, Berlin, vol. 35
3. Alart P, Curnier A (1991) A mixed formulation for frictional contact problems prone to Newton like solution methods. Comput Methods Appl Mech Eng 92:353–375
4. Arnold M (2008) Numerical methods for simulation in applied dynamics. In: Schiehlen W, Arnold M (eds) Simulation techniques in applied dynamics - CISM Lecture notes. Springer, Wien, pp 191–246
5. Arnold M, Brüls O (2007) Convergence of the generalized-α scheme for constrained mechanical systems. Multibody Syst Dyn 18(2):185–202
6. Arnold M, Brüls O, Cardona A (2011) Convergence analysis of generalized-α Lie group integrators for constrained systems. In: Proceedings of multibody dynamics ECCOMAS thematic conference. Brussels
7. Arnold M, Brüls O, Cardona A (2015) Error analysis of generalized-α Lie group time integration methods for constrained mechanical systems. Numerische Mathematik 129:149–179
8. Bauchau O (2011) Flexible multibody dynamics. Springer, Dordrecht
9. Baumgarte J (1972) Stabilization of constraints and integrals of motion in dynamical systems. Comput Methods Appl Mech Eng 1:1–16
10. Ben Gharbia I, Gilbert J (2012) Nonconvergence of the plain Newton-min algorithm for linear complementarity problems with a P-matrix. Math Program Ser A 134:349–364
11. Bottasso C, Bauchau O, Cardona A (2007) Time-step-size-independent conditioning and sensitivity to perturbations in the numerical solution of index three differential algebraic equations. SIAM J Sci Comput 29(1):397–414
12. Brüls O, Acary V, Cardona A (2014) Simultaneous enforcement of constraints at position and velocity levels in the nonsmooth generalized-α scheme. Comput Methods Appl Mech Eng 281:131–161
13. Brüls O, Cardona A, Arnold M (2012) Lie group generalized-α time integration of constrained flexible multibody systems. Mech Mach Theory 48:121–137
14. Chen Qz, Acary V, Virlez G, Brüls O (2013) A nonsmooth generalized-α scheme for flexible multibody systems with unilateral constraints. Int J Numer Methods Eng 96:487–511

15. Chung J, Hulbert G (1993) A time integration algorithm for structural dynamics with improved numerical dissipation: the generalized-α method. ASME J Appl Mech 60:371–375
16. Doyen D, Ern A, Piperno S (2011) Time-integration schemes for the finite element dynamic Signorini problem. SIAM J Sci Comput 33(1):223–249
17. Flores P, Machado M, Seabra E, Tavares da Silva M (2010) A parametric study on the Baumgarte stabilization method for forward dynamics of constrained multibody systems. ASME J Comput Nonlinear Dyn 6:011,019–011,019–9
18. Gear C, Leimkuhler B, Gupta G (1985) Automatic integration of Euler-Lagrange equations with constraints. J Comput Appl Math 12–13:77–90
19. Géradin M, Cardona A (2001) Flexible multibody dynamics: a finite element approach. Wiley, Chichester
20. Haddouni M, Acary V, Garreau S, Beley JD, Brogliato B (2017) Comparison of several formulations and integration methods for the resolution of DAEs formulations in event-driven simulation of nonsmooth frictionless multibody dynamics. Multibody Syst Dyn 41:201–231
21. Hilber H, Hughes T, Taylor R (1977) Improved numerical dissipation for time integration algorithms in structural dynamics. Earthq Eng Struct Dyn 5:283–292
22. Hintermüller M, Ito K, Kunish K (2003) The primal-dual active set strategy as a semismooth Newton method. SIAM J Optim 13(3):865–888
23. Hüeber S, Stadler G, Wohlmuth BI (2008) A primal-dual active set algorithm for three-dimensional contact problems with Coulomb friction. SIAM J Sci Comput 30(2):572–596
24. Hüeber S, Wohlmuth B (2005) A primal-dual active set strategy for non-linear multibody contact problems. Comput Methods Appl Mech Eng 194(2729):3147–3166
25. Jean M (1999) The non-smooth contact dynamics method. Comput Methods Appl Mech Eng 177:235–257
26. Moreau J (1988) Bounded variation in time. In: Moreau J, Panagiotopoulos P, Strang G (eds) Topics in nonsmooth mechanics. Birkhäuser, Basel, pp 1–74
27. Moreau JJ (1988) Unilateral contact and dry friction in finite freedom dynamics. In: Moreau JJ, Panagiotopoulos P (eds) Non-smooth mechanics and applications, vol. 302. Springer, New York, pp 1–82
28. Newmark N (1959) A method of computation for structural dynamics. ASCE J Eng Mech Div 85:67–94
29. Paoli L, Schatzman M (2002) A numerical scheme for impact problems I: the one-dimensional case. SIAM J Numer Anal 40:702–733
30. Paoli L, Schatzman M (2002) A numerical scheme for impact problems II: the multi-dimensional case. SIAM J Numer Anal 40:734–768
31. Pfeiffer F (2008) On non-smooth dynamics. Meccanica 43:533–554
32. Pfeiffer F, Foerg M, Ulbrich H (2006) Numerical aspects of non-smooth multibody dynamics. Comput Methods Appl Mech Eng 195:6891–6908
33. Pfeiffer F, Glocker C (2004) Multibody dynamics with unilateral contacts. Wiley series in nonlinear science. Wiley-VCH, Weinheim
34. Schindler T, Acary V (2014) Timestepping schemes for nonsmooth dynamics based on discontinuous galerkin methods: definition and outlook. Math Comput Simul 95:180–199
35. Schindler T, Rezaei S, Kursawe J, Acary V (2015) Half-explicit timestepping schemes on velocity level based on time-discontinuous Galerkin methods. Comput Methods Appl Mech Eng 290:250–276
36. Schoeder S, Ulbrich H, Schindler T (2013) Discussion on the Gear-Gupta-Leimkuhler method for impacting mechanical systems. Multibody Syst Dyn 31:477–495
37. Studer C, Leine RI, Glocker C (2008) Step size adjustment and extrapolation for time-stepping schemes in non-smooth dynamics. Int J Numer Methods Eng 76(11):1747–1781

On Solving Contact Problems with Coulomb Friction: Formulations and Numerical Comparisons

Vincent Acary, Maurice Brémond and Olivier Huber

Abstract In this chapter, we review several formulations of the discrete frictional contact problem that arises in space and time discretized mechanical systems with unilateral contact and three-dimensional Coulomb's friction. Most of these formulations are well–known concepts in the optimization community, or more generally, in the mathematical programming community. To cite a few, the discrete frictional contact problem can be formulated as variational inequalities, generalized or semi–smooth equations, second–order cone complementarity problems, or optimization problems, such as quadratic programming problems over second-order cones. Thanks to these multiple formulations, various numerical methods emerge naturally for solving the problem. We review the main numerical techniques that are well-known in the literature, and we also propose new applications of methods such as the fixed point and extra-gradient methods with self-adaptive step rules for variational inequalities or the proximal point algorithm for generalized equations. All these numerical techniques are compared over a large set of test examples using performance profiles. One of the main conclusions is that there is no universal solver. Nevertheless, we are able to give some hints for choosing a solver with respect to the main characteristics of the set of tests.

1 Introduction

More than thirty years after the pioneering work of [21, 29, 30, 48–50, 61, 64, 83, 91, 93] on numerically solving mechanical problems with contact and friction,

---

V. Acary (✉) · M. Brémond
University Grenoble Alpes, Inria, CNRS, Grenoble INP (Institute of Engineering University of Grenoble Alpes), LJK, 38000 Grenoble, France
e-mail: vincent.acary@inria.fr

M. Brémond
e-mail: maurice.bremond@inria.fr

O. Huber
Wisconsin Institute for Discovery, University of Wisconsin-Madison, 330 N. Orchard St., Madison, WI 53715, USA
e-mail: ohuber2@wisc.edu

R. I. Leine et al. (eds.), *Advanced Topics in Nonsmooth Dynamics*,
https://doi.org/10.1007/978-3-319-75972-2_10

there are still active researches on this subject in the computational mechanics and applied mathematics communities. This can be explained by the fact that problems from mechanical systems with unilateral contact and Coulomb friction are difficult to numerically solve and the mathematical results of convergence of the numerical algorithms are rare, most of these requiring rather strong assumptions. In this chapter, we want to give some insight into the advantages and weaknesses of standard solvers found in the literature by comparing them on large sets of examples coming from the simulation of a wide range of mechanical systems. Some new numerical schemes are also introduced, mainly based on general solvers for variational inequalities and the proximal point algorithms.

1.1 Problem Statement

In this section, we formulate an abstract, algebraic finite–dimensional frictional contact problem. We cast this problem as a complementarity problem over cones, and discuss the properties of the latter. We end by presenting some instances with contact and friction phenomena that fit our problem description.

Abstract Problem

We want to discuss possible numerical solution procedures for the following three–dimensional finite–dimensional frictional contact problem and some of its variants. Let $n_c \in \mathbb{N}$ be the number of contact points and $n \in \mathbb{N}$ the number of degrees of freedom of a discrete mechanical system.

The problem data are: a positive definite matrix $M \in \mathbb{R}^{n\times n}$, a vector $f \in \mathbb{R}^n$, a matrix $H \in \mathbb{R}^{n\times m}$ with $m = 3n_c$, a vector $w \in \mathbb{R}^m$ and a vector of coefficients of friction $\mu \in \mathbb{R}^{n_c}$. The unknowns are two vectors $v \in \mathbb{R}^n$, a velocity–like vector and $r \in \mathbb{R}^m$, a contact reaction or impulse, solution to

$$\begin{cases} Mv = Hr + f \\ K^\star \ni \hat{u} \perp r \in K \end{cases} \quad \text{with} \quad \begin{array}{l} u := H^\top v + w \\ \hat{u} := u + g(u), \end{array} \tag{1}$$

where the set K is the Cartesian product of Coulomb's friction cone at each contact, that is,

$$K = \prod_{\alpha=1\ldots n_c} K^\alpha = \prod_{\alpha=1\ldots n_c} \{r^\alpha, \|r^\alpha_{\mathrm{T}}\| \leqslant \mu^\alpha |r^\alpha_{\mathrm{N}}|\}, \tag{2}$$

and $K^\star$ is the dual cone of K. The function $g\colon \mathbb{R}^m \to \mathbb{R}^m$ is a nonsmooth function defined as

$$g(u) = [[\mu^\alpha \|u^\alpha_{\mathrm{T}}\|, 0, 0]^\top, \alpha = 1\ldots n_c]^\top. \tag{3}$$

Note that the variables u and $\hat{u}$ do not appear as unknowns, since they can be directly obtained from v.

A Second Order Cone Complementarity Problem (SOCCP)

From the mathematical programming point of view, the problem appears to be a Second Order Cone Complementarity Problem (SOCCP) [39], which can be generically defined as

$$\begin{cases} y = f(x) \\ K^\star \ni y \perp x \in K, \end{cases} \tag{4}$$

where K is a second-order cone. If the nonlinear part of the problem (1) is neglected ($g(u) = 0$), the problem is an associated friction problem with dilatation and, by the way, is also gentle Second-Order Cone Linear Complementarity Problem (SOCLCP) with a positive definite matrix $W = H^\top M^{-1} H$ (possibly semi–definite). The assumption of an associated frictional law, i.e., a friction law in which the local sliding velocity is normal to the friction cone differs dramatically from the standard Coulomb friction, since it generates a non–vanishing normal velocity when the system slides. In other words, the sliding motion implies the separation of the bodies. When the non-associated character of the friction is taken into account through $g(u)$, the problem is non-monotone and nonsmooth, and therefore is very hard to solve efficiently. For a given numerical algorithm, it is not so difficult to design mechanical examples for which the algorithm runs into trouble [18].

Proofs of convergence of the numerical algorithms are rare, and most of these require strong assumptions, including the following: (a) small values of the friction coefficients, (b) full rank assumptions and the symmetry of the Delassus matrix W or (c) the assumption that the problem is two-dimensional. Among these results, we can cite the Czech school, where the coefficient of friction is assumed to be bounded and small. This assumption allows us to use fixed point methods on the convex sub–problems of Tresca friction (friction threshold that does depend on the normal reaction, and then transforms the cone into a semi-cylinder). We can also mention the results from [11, 94, 110], in which the friction cone is polyhedral (in 2D or by a faceting process). In that case, if $w = 0$ or $w \in \mathrm{im}(H^\top)$, Lemke's algorithm is able to solve the problem. The question of the existence of solutions has also been treated in [3, 68], recalled in Sect. 2.3, under similar assumptions but with different techniques. The question of uniqueness remains a difficult problem in the general case.

Range of Applicability

We clearly choose to greatly simplify the general problems of formulating the contact problems with friction by avoiding the inclusion of too many side effects that are themselves interesting but render the study too difficult to carry out in a single chapter. We choose finite dimensional systems in which the time dependency does not appear explicitly. Nevertheless, we believe that there is a strong interest in studying this problem, since it appears to be relatively generic in numerous simulations of systems with contact and friction. This problem is indeed at the heart of the simulation of mechanical systems with 3D Coulomb's friction and unilateral constraints in the following cases:

- It might be the result of the time–discretization by event–capturing time–stepping methods or event–detecting (event–driven) techniques of dynamical systems with friction; the variables are homogeneous to pairs of velocities/impulses or accelerations/forces.
- It might also be the result of space–discretization (by FEM, for instance) of the elastic quasi-static problems of frictional contact mechanics; in that case, the variables are homogenous to displacements/forces of displacement rates/forces.
- If the system is a dynamical mechanical system composed of flexible solids, the problem is again obtained through a space and time discretization.
- If the material follows a nonlinear mechanical bulk behavior, we can use this model after a standard Newton linearization procedure.

For a description of the derivation of such problems in various practical situations we refer to [1, 2, 76, 119].

1.2 Objectives and Outline of the Chapter

In this chapter, after stating the problem in more detail in Sect. 2, we recall the existence result of [3] for the problem (1) in Sect. 2.3. In this framework, we briefly present, in Sect. 3, a few alternative formulations of the problem that enable the design of numerical solution procedures: (a) finite–dimensional Variational Inequalities (VI) and Quasi-Variational Inequalities (QVI), (b) nonsmooth equations and (c) optimization-based formulations.

Right after these formulations, we list some of the most standard algorithms dedicated to one of the previous formulations:

1. the fixed point and projection numerical methods for solving VI are reviewed, with a focus on self-adaptive step rules (Sect. 4),
2. the nonsmooth (semi-smooth) Newton methods are described based on the various nonsmooth equations formulations (Sect. 5),
3. Section 6 is devoted to the presentation of splitting and proximal point techniques,
4. and finally, in Sect. 7, the Panagiotopoulos alternating optimization technique, the successive approximation technique and the SOCLCP approach are outlined.

Since it is difficult to be exhaustive on the approaches developed in the literature for solving frictional contact problems, we decided to leave out the following approaches, which we felt were outside the scope of the chapter:

- the approaches that alter the fundamental assumptions of the 3D Coulomb friction model by faceting the cone, as in the pioneering work of [67] and followed by [7, 11, 53, 94, 110], or by convexifying the Coulomb law (associated friction law with normal dilatancy) [12, 59, 73, 112–114], or finally, by regularizing the friction law [66].
- the recent developments of methods for the frictionless case [84, 85, 115].

- the approaches that are based on domain decomposition and parallel computing [17, 38, 58, 72, 100, 118]. We choose in this chapter to focus on single domain computation and to skip the discussion about distributed computing, mainly for the sake of the length of the chapter.

Finally, some possibly interesting approaches have not been reported. We are thinking mainly of the interior point methods approach [22, 69, 84]. Some basic implementations of such methods do not give satisfactory results. One of the reasons for this is the fact that we were not able to get robustness and efficiency over a large class of problems. As reported in [69, 73], it seems that it is necessary to alter the friction Coulomb's law by adding regularization or dilatancy into the model. In the same spirit, we also skip the comparison of the possibly very promising methods developed in [58, 59] that are based on Krylov subspace and spectral methods. It could be very interesting to bench these methods against the actual Coulomb friction model as well, that is to say, in the non-monotone case. Finally, our preliminary results on the use of direct general SOCP or SOCLCP solvers off the shelf were not convincing. Indeed, the structure of contact problems (product of a large number of small second-order cones) has to be taken into account to obtain efficiency, and unfortunately, these solvers are difficult to adapt to this structure.

Other comparisons have already been published in the literature. Some of the first comparison studies were done in [20, 99]. In this work, several formulations are detailed in the bidimensional case (variational inequality, linear complementarity problem (LCP) and augmented Lagrangian formulation) and comparisons of fixed point methods with projection, splitting methods and Lemke's method for solving LCP. Other comparisons have been done on 2D systems in [78–81]. In [23], a very interesting comparison in the three-dimensional case has been carried out showing the superiority of the semi-smooth Newton methods over the interior point methods. Comparisons on simple multi-body systems composed of kinematic chains can be found in [90].

As a difference with the previous publications, the comparisons are performed over a large set of examples using performance profiles in this chapter. Let us summarize the main conclusion from Sect. 8: on one hand, the algorithms based on Newton methods for nonsmooth equations solve the problem quickly when they succeed, but suffer from robustness issues, particularly if the matrix H is not full rank. On the other hand, the iterative methods dedicated to solving variational inequalities are quite robust but with an extremely slow rate of convergence. To sum up, as far as we know, there is no option that combines time efficiency and robustness. The set of problems used here are from the FCLIB collection.[1] In this work, this collection is solved with the software Siconos and its component Siconos/Numerics[2] [5].

[1]https://frictionalcontactlibrary.github.io/index.html, which aims to provide many problems to compare algorithms on a fair basis.

[2]http://siconos.gforge.inria.fr.

1.3 Notation

The following notation is used throughout the chapter: the 2-norm for a function g is denoted by $\|g\|$ and for a vector $x \in \mathbb{R}^n$ by $\|x\|$. The index $\alpha \in \mathbb{N}$ is used to identify the variable pertaining to a single contact. A multivalued mapping $T : \mathbb{R}^n \rightrightarrows \mathbb{R}^n$ is an operator whose images are sets. The second-order cone, also known as the Lorentz or ice–cream cone, is defined as $K_\mu := \{(x,t) \in \mathbb{R} \times \mathbb{R}_+ \mid \|x\| \leqslant \mu t\}$, $\mu \geqslant 0$. By polarity, the dual convex cone to a convex cone K is defined by

$$K^\star = \{x \in \mathbb{R}^n \mid y^\top x \geqslant 0, \quad \text{for all } y \in K\}. \tag{5}$$

The normal cone $N_K : \mathbb{R}^n \rightrightarrows \mathbb{R}^n$ to a closed convex set X is the set

$$N_K(x) = \{d \in \mathbb{R}^n \mid d^\top (y - x) \leqslant 0\}. \tag{6}$$

The notation $0 \leqslant x \perp y \geqslant 0$ denotes that $x \geqslant 0$, $y \geqslant 0$ and $x^\top y = 0$. A complementarity problem associated with a function $F : \mathbb{R}^n \to \mathbb{R}^n$ is to find $x \in \mathbb{R}^n$ such that $0 \leqslant F(x) \perp x \geqslant 0$. The generalized complementarity problem is given by $K^\star \ni F(x) \perp x \in K$, where K is a closed convex cone. Finite-dimensional Variational Inequality (VI) problems subsume complementarity problems and the system of equations. Solving a VI(X, F) is to find $x \in X$ such that

$$F(x)^\top (y - x) \geqslant 0 \qquad \text{for all } y \in X. \tag{7}$$

It is easy to see this problem is equivalent to solving a *generalized equation*

$$0 \in F(X) + N_X(x). \tag{8}$$

The Euclidean projector on a set X is denoted by P_X.

2 Description of the 3D Frictional Contact Problems

2.1 Signorini's Condition and Coulomb's Friction

Let us consider the contact between two bodies $A \subset \mathbb{R}^3$ and $B \subset \mathbb{R}^3$ with sufficiently smooth boundaries, as depicted in Fig. 1.

From the body A "perspective", the point $C_A \in \partial A$ is called a *master point to contact*. The choice of this master point C_A for writing the contact condition is crucial in practice and amounts to consistently discretizing the contact surface. The vector N defines an outward unit normal vector to A at the point C_A. With T_1, T_2 two vectors in the plane orthogonal to N, we can build an orthonormal frame $(C_A, \mathsf{N}, \mathsf{T}_1, \mathsf{T}_2)$ called the *local frame at contact*. The slave contact point $C_B \in \partial B$ is defined as the

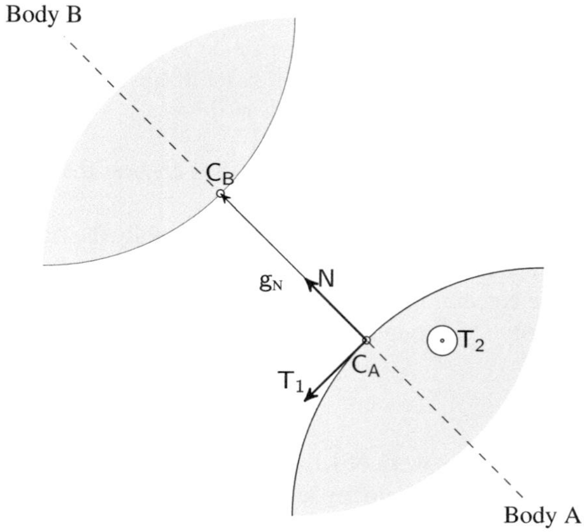

Fig. 1 Contact kinematic

projection of the point C_A on ∂B in the direction given by N. Note that we assume that such a point exists. The gap function is defined as the signed distance between C_A and C_B

$$g_{\mathrm{N}} = (C_B - C_A)^\top \mathsf{N}. \tag{9}$$

Consider two strictly convex bodies, which are non-penetrating, i.e., $A \cap B = \emptyset$; the master and slave contact points can be chosen as the proximal points of each body, and the normal vector N can be written as

$$\mathsf{N} = \frac{C_B - C_A}{\|C_B - C_A\|}. \tag{10}$$

The contact force exerted by A on B is denoted by $r \in \mathbb{R}^3$ and is decomposed in the local frame as

$$r := r_{\mathrm{N}}\mathsf{N} + r_{\mathrm{T}_1}\mathsf{T}_1 + r_{\mathrm{T}_2}\mathsf{T}_2, \quad \text{with } r_{\mathrm{N}} \in \mathbb{R} \text{ and } r_{\mathrm{T}} := [r_{\mathrm{T}_1}, r_{\mathrm{T}_2}]^\top \in \mathbb{R}^2. \tag{11}$$

The *Signorini condition* states that

$$0 \leqslant g_{\mathrm{N}} \perp r_{\mathrm{N}} \geqslant 0, \tag{12}$$

and models the unilateral contact. The condition (12), written at the *position level*, can also be defined at the *velocity level*. To this end, the relative velocity $u \in \mathbb{R}^3$ of the point C_B with respect to C_A is also decomposed in the local frame as

$$u := u_{\mathrm{N}}\mathsf{N} + u_{\mathrm{T}_1}\mathsf{T}_1 + u_{\mathrm{T}_2}\mathsf{T}_2 \quad \text{with } u_{\mathrm{N}} \in \mathbb{R} \text{ and } u_{\mathrm{T}} = [u_{\mathrm{T}_1}, u_{\mathrm{T}_2}]^\top \in \mathbb{R}^2. \tag{13}$$

At the velocity level, the Signorini condition is written as

$$\begin{cases} 0 \leqslant u_{\rm N} \perp r_{\rm N} \geqslant 0 & \text{if } g_{\rm N} \leqslant 0 \\ r_{\rm N} = 0 & \text{otherwise.} \end{cases} \tag{14}$$

The Moreau's viability Lemma [89] ensures that (14) implies (12) if $g_{\rm N} \geqslant 0$ holds in the initial configuration.

In some mechanical problems, especially the rigid multi-body systems dynamics, an impact law has to be introduced to complete the dynamics. The most simple law is the Newton impact law that relates the post impact velocity $u_{\rm N}$ to the pre-impact velocity $u_{\rm N}^-$ through a coefficient of restitution $e \geqslant 0$ as

$$u_{\rm N} = -e u_{\rm N}^- . \tag{15}$$

Following the work of J.J. Moreau [89], the impact law is embedded in the Signorini condition at the velocity level as

$$\begin{cases} 0 \leqslant u_{\rm N} + e u_{\rm N}^- \perp r_{\rm N} \geqslant 0 & \text{if } g_{\rm N} \leqslant 0 \\ r_{\rm N} = 0 & \text{otherwise,} \end{cases} \tag{16}$$

where $r_{\rm N}$ plays the role of an impulse. The pre-impact velocity is a known value, and thus can be treated as a constant term in w of Eq. (1). For the sake of simplicity, we will consider in the sequel that $-e u_{\rm N}^-$ is included in the vector w.

Coulomb's friction models the frictional behavior of the contact force law in the tangent plane spanned by (T_1, T_2). Let us define the Coulomb friction cone K, which is the isotropic second-order cone (Lorentz or ice–cream cone)

$$K = \{ r \in {\rm I\!R}^3 \mid \|r_{\rm T}\| \leqslant \mu r_{\rm N} \}, \tag{17}$$

where μ is the coefficient of friction. The Coulomb friction states for the sticking case that

$$u_{\rm T} = 0, \quad r \in K, \tag{18}$$

and for the sliding case that

$$u_{\rm T} \neq 0, \quad r \in \partial K, \quad \text{and } \; \exists \alpha > 0 \text{ such that } r_{\rm T} = -\alpha u_{\rm T}. \tag{19}$$

With the Coulomb friction model, there are two relations between $u_{\rm T}$ and $r_{\rm T}$. The distinction is based on the value of the relative velocity $u_{\rm T}$ between the two bodies. If $u_{\rm T} = 0$ (sticking case), we have $\|r_{\rm T}\| \leqslant \mu r_{\rm N}$. Otherwise, we get the sliding case.

Disjunctive Formulation of the Signorini–Coulomb Model

If we consider the velocity-level Signorini condition (14) together with the Coulomb friction (18)–(19), which is naturally expressed in terms of velocity, we obtain a disjunctive formulation of the frictional contact behavior as

$$\begin{cases} r = 0 & \text{if } g_{\mathrm{N}} > 0 \text{ (no contact)} \\ r = 0,\, u_{\mathrm{N}} \geqslant 0 & \text{if } g_{\mathrm{N}} \leqslant 0 \quad \text{(take–off)} \\ r \in K,\, u = 0 & \text{if } g_{\mathrm{N}} \leqslant 0 \quad \text{(sticking)} \\ r \in \partial K,\, u_{\mathrm{N}} = 0,\, \exists \alpha > 0,\, u_{\mathrm{T}} = -\alpha r_{\mathrm{T}} & \text{if } g_{\mathrm{N}} \leqslant 0 \quad \text{(sliding).} \end{cases} \tag{20}$$

In the computational practice, the disjunctive formulation is not suitable for solving the Coulomb problem, as it suggests the use of enumerative solvers, with an exponential complexity. In the sequel, alternative formulations of the Signorini–Coulomb model suitable for numerical applications are delineated. The core idea is to translate the cases in (20) into complementarity relations.

Inclusion into Normal Cones

The Signorini condition (12) and (14), in their complementarity forms, can be equivalently written as an inclusion into a normal cone to $\mathbb{R}_+$

$$-\, g_{\mathrm{N}} \in N_{\mathbb{R}_+}(r_{\mathrm{N}}) \qquad \text{and} \qquad -\, u_{\mathrm{N}} \in N_{\mathbb{R}_+}(r_{\mathrm{N}}), \tag{21}$$

if $g_{\mathrm{N}} \leqslant 0$ and $r_{\mathrm{N}} = 0$ otherwise. An inclusion form of the Coulomb friction for the tangential part can also be proposed: let $D(c)$ be the disk of radius c:

$$D(c) := \{x \in \mathbb{R}^2 \mid \|x\| \leqslant c\}. \tag{22}$$

For the Coulomb friction, we get

$$-\, u_{\mathrm{T}} \in N_{D(\mu r_{\mathrm{N}})}(r_{\mathrm{T}}). \tag{23}$$

Since $D(\mu r_{\mathrm{N}})$ is not a cone, the inclusion (23) is not a complementarity problem, but a variational inequality. The formulation (23) is often related to Moreau's maximum dissipation principle of the frictional behavior:

$$r_{\mathrm{T}} \in \underset{\|z\| \leqslant \mu r_{\mathrm{N}}}{\arg\max}\; z^{\top} u_{\mathrm{T}}. \tag{24}$$

This means that the couple $(r_{\mathrm{T}}, u_{\mathrm{T}})$ maximizes the energy lost through dissipation.

SOCCP Formulation of the Signorini–Coulomb Model

In [1, 3], another formulation is proposed, inspired by the so-called bipotential [26, 27, 105]. The goal is to form a complementarity problem out of (21) and (23). To this end, we introduce the modified relative velocity $\hat{u} \in \mathbb{R}^3$ defined by

$$\hat{u} = u + [\mu \|u_{\mathrm{T}}\|, 0, 0]^{\top}. \tag{25}$$

The entire contact model (20) can be put into a Second Order Cone Complementarity Problem (SOCCP) as

$$K^{\star} \ni \hat{u} \perp r \in K, \tag{26}$$

if $g_{\mathrm{N}} \leqslant 0$ and $r = 0$ otherwise (Fig. 2).

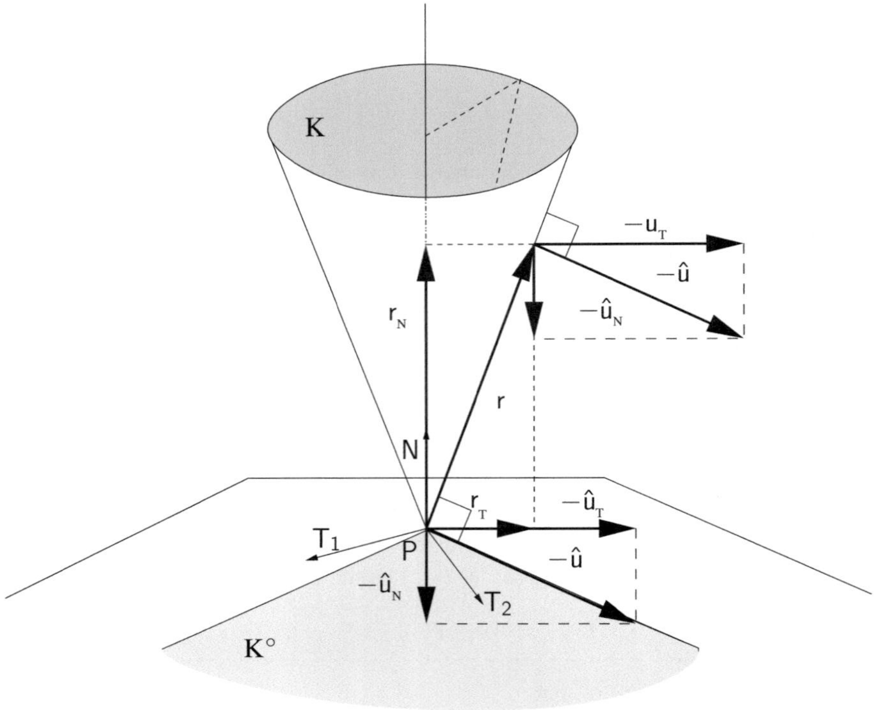

Fig. 2 Coulomb's friction law in the sliding case

2.2 Frictional Contact Discrete Problems

We assume that a finite set of n_c contact points and their associated local frames have been defined. In general, this task is not straightforward and amounts to correctly discretizing the contact surfaces. For more details, we refer to [76, 119]. For each contact $\alpha \in \{1, \ldots, n_c\}$, the local velocity is denoted by $u^\alpha \in \mathbb{R}^3$, the normal velocity by $u^\alpha_{\mathrm{N}} \in \mathbb{R}$ and the tangential velocity by $u^\alpha_{\mathrm{T}} \in \mathbb{R}^2$ with $u^\alpha = [u^\alpha_{\mathrm{N}}, (u^\alpha_{\mathrm{T}})^\top]^\top$. The vectors u, u_{N}, u_{T} respectively collect all the local velocities $u = [(u^\alpha)^\top, \alpha = 1 \ldots n_c]^\top$, all the normal velocities $u_{\mathrm{N}} = [u^\alpha_{\mathrm{N}}, \alpha = 1 \ldots n_c]^\top$, and all the tangential velocities $u_{\mathrm{T}} = [(u^\alpha_{\mathrm{T}})^\top, \alpha = 1 \ldots n_c]^\top$. For a contact α, the modified local velocity, denoted by $\hat{u}^\alpha$, is defined by

$$\hat{u}^\alpha = u^\alpha + g^\alpha(u) \qquad \text{where} \qquad g^\alpha(u^\alpha) = [\mu^\alpha \|u^\alpha_{\mathrm{T}}\|, 0, 0]^\top. \tag{27}$$

The vector $\hat{u}$ and the function g collect all the modified local velocities at each contact $\hat{u} = [\hat{u}^\alpha, \alpha = 1 \ldots n_c]^\top$ and the function $g(u) = [[\mu^\alpha \|u^\alpha_{\mathrm{T}}\|, 0, 0]^\top, \alpha = 1 \ldots n_c]^\top$.

For each contact α, the reaction vector $r^\alpha \in \mathbb{R}^3$ is also decomposed in its normal part $r^\alpha_{\mathrm{N}} \in \mathbb{R}$ and the tangential part $r^\alpha_{\mathrm{T}} \in \mathbb{R}^2$ as $r^\alpha = [r^\alpha_{\mathrm{N}}, (r^\alpha_{\mathrm{T}})^\top]^\top$. The Coulomb friction cone for a contact α is defined by $K^\alpha = \{r^\alpha \in \mathbb{R}^3 \mid \|r^\alpha_{\mathrm{T}}\| \leqslant \mu^\alpha |r^\alpha_{\mathrm{N}}|\}$ and the set $K^{\alpha,\star}$ is its dual. The set K is the Cartesian product of Coulomb's friction cone at each contact, that is,

$$K = \prod_{\alpha=1,\ldots,n_c} K^\alpha \quad \text{and} \quad K^\star \text{ is its dual.} \tag{28}$$

In this chapter, we investigate the case when the problem is given in its *reduced* form. We consider that the discretized and linearized dynamics is of the form

$$Mv = Hr + f, \tag{29}$$

with M a positive-definite matrix. The local velocities at the point of contact are given by

$$u = H^\top v + w. \tag{30}$$

More information on the term w is given later in this section. The (global) velocities v can be substituted in (30) by using a Schur-complement technique. This yields

$$u = H^\top M^{-1} H r + H^\top M^{-1} f + w. \tag{31}$$

Let us define W, often called the *Delassus matrix*, as

$$W := H^\top M^{-1} H \tag{32}$$

and the vector q as

$$q := H^\top M^{-1} f + w. \tag{33}$$

We are now ready to define the mathematical problem we want to solve.

Problem FC (Discrete frictional contact problem). Given:

- a positive semi–definite matrix $W \in \mathbb{R}^{m\times m}$ called the Delassus matrix,
- a vector $q \in \mathbb{R}^m$,
- a vector $\mu \in \mathbb{R}^{n_c}$ of coefficients of friction,

find a vector $r \in \mathbb{R}^m$ such that

$$\begin{cases} K^\star \ni \hat{u} \perp r \in K \\ u = Wr + q \\ \hat{u} = u + g(u), \end{cases} \tag{34}$$

with $g(u) = [[\mu^\alpha \|u^\alpha_{\mathrm{T}}\|, 0, 0]^\top, \alpha = 1 \ldots n_c]^\top$.
An instance of the problem is denoted by FC(W, q, μ) □

Remark 1 We do not assume that the Delassus matrix W is symmetric in the general case. In most of the applications, the Delassus matrix is symmetric, since it represents either a mass matrix or a stiffness matrix. Nevertheless, in the rigid body applications, or more generally, when large rotations are taken into account, the Delassus matrix is not symmetric. Indeed, in an implicit time-discretization, the Jacobian matrix of the gyroscopic forces brings a skew symmetric matrix into the Delassus matrix.

2.3 Existence of Solutions

The question of the existence of a solution to the Problem FC has been studied in [3, 68] with different analysis techniques, under the assumption that the Delassus matrix is symmetric. The key assumption for the existence of solutions in both articles is as follows:

$$\exists v \in \mathbb{R}^m \,:\, H^\top v + w \in \operatorname{int} K^\star, \tag{35}$$

or equivalently,

$$w \in \operatorname{im} H + \operatorname{int} K^\star. \tag{36}$$

Under the previous assumption, the Problem FC has a solution. Therefore, it makes sense to design a procedure to solve the problem. In the sequel, we will compare numerical methods only when this assumption is satisfied.

This assumption is easily verified in numerous applications. For applications in nonsmooth dynamics where the unknown v is a relative contact velocity, the term w vanishes if we have only scleronomic constraints. For $w \in \operatorname{im}(H^\top)$ (and especially $w = 0$), the assumption is trivially satisfied. As explained in [2], the term w has several possible sources. If the constraints are formulated at the velocity level, an input term of w is given in the dynamics by the impact laws (see Eq. (16)). In the case of the Newton impact law, it holds that $w \in \operatorname{im}(H^\top)$. For other impact laws, this is not clear. Another input in w is given by constraints that depend explicitly on time. In that case, we can have $w \notin \operatorname{im}(H^\top)$ and the non-existence of solutions. If the constraints are written at the position level, w can be given by initial terms that come from the velocity discretization. In those cases, the existence is also not ensured.

The assumption is also satisfied whenever $\operatorname{im} H = \mathbb{R}^m$ or, in other words, if $H^\top$ has full row rank. Unfortunately, in a large number of applications, $H^\top$ is rank deficient. From the mechanical point of view, the rank deficiency of H and the amount of friction seem to play a fundamental role in the question of the existence (and uniqueness) of solutions. In the numerical comparisons, we will attempt to get a deeper understanding of the role of these assumptions in the convergence of the algorithms. The rank deficiency of H is related to the number of constraints that are imposed on the system with respect to the number of degrees-of-freedom in the system. This is closely related to the concept of hyperstaticity in overconstrained mechanical systems. In the most favorable cases, it yields indeterminate Lagrange multipliers, but also unfeasible problems, and then the loss of solutions in the worst cases. The second assumption about the amount of friction is also well-known. The frictionless problem is easy to solve if it is feasible. It is clear that large friction coefficients prevent sliding, and therefore increase the degree of hyperstaticity of the system.

3 Alternative Formulations

In this section, various equivalent formulations of Problem FC are given. Our goal is to show that such problems can be recast into several well-known problems in the mathematical programming and optimization community. These formulations will serve as a basis for numerical solution procedures that we develop in later sections.

3.1 *Variational Inequality (VI) Formulations*

Let us recall the definition of a finite-dimensional VI(X, F): find $z \in X$ such that

$$F^\top(z)(y - z) \geqslant 0 \quad \text{for all } y \in X, \tag{37}$$

with X a nonempty subset of $\mathbb{R}^n$ and F a mapping from $\mathbb{R}^n$ into itself. We refer to [39, 47] for the standard theory of finite–dimensional variational inequalities. The easiest way to state equivalent VI formulations of Problem FC is to use the following equivalences:

$$K^\star \ni \hat{u} \perp r \in K \iff -\hat{u} \in N_K(r) \iff \hat{u}^\top(s - r) \geqslant 0, \text{ for all } s \in K. \tag{38}$$

For Problem FC, the equivalent formulation in VI is directly obtained from

$$-(Wr + q + g(Wr + q)) \in N_K(r). \tag{39}$$

The resulting VI is denoted by VI$(F_{\text{vi}}, X_{\text{vi}})$ with

$$F_{\text{vi}}(r) := Wr + q + g(Wr + q) \quad \text{and} \quad X_{\text{vi}} := K. \tag{40}$$

Uniqueness Properties

In the general case, it is difficult to prove uniqueness of solutions to (40). If the matrix H has full rank and the friction coefficients are "small", a classical argument for the uniqueness of the solution to VIs can be satisfied. Note that the full rank hypothesis on H implies that W is positive-definite. Therefore, we have $(x - y)^\top W(x - y) \geqslant C_W\|x - y\|^2$ with $C_W > 0$. Using this relation (40), it yields

$$
\begin{aligned}
(F_{\rm vi}(x) - F_{\rm vi}(y))^\top (x-y) = (x-y)^\top W(x-y) \\
+\textstyle\sum_{\alpha=1}^{n_c} \mu^\alpha (x^\alpha_{\rm N} - y^\alpha_{\rm N})[\|[Wx+q]^\alpha_{\rm T}\| - \|[Wy+q]^\alpha_{\rm T}\|] \\
\geqslant C_W\|x-y\|^2 + \textstyle\sum_{\alpha=1}^{n_c} \mu^\alpha (x^\alpha_{\rm N} - y^\alpha_{\rm N})[\|[Wx+q]^\alpha_{\rm T}\| - \|[Wy+q]^\alpha_{\rm T}\|].
\end{aligned} \tag{41}
$$

Note that for small values of the coefficients of friction, the first term on the right-hand side dominates the second one. Hence, the mapping $F_{\rm vi}$ is strictly monotone, and this ensures that the VI has, at most, one solution [39, Theorem 2.3.3]. The fact that H is full rank also implies that the Assumption (35) for the existence of solutions is trivially satisfied. Hence, there exists a unique solution to the VI($F_{\rm vi}$, $X_{\rm vi}$).

3.2 Quasi-Variational Inequalities (QVI)

Let us recast Problem FC into the QVI framework. A QVI is a generalization of the VI, where the feasible set is allowed to depend on the solution. Let us define this precisely: let X be a multi-valued mapping $\mathbb{R}^n \rightrightarrows \mathbb{R}^n$ and let F be a mapping from $\mathbb{R}^n$ into itself. The quasi-variational inequality problem, denoted by QVI(X, F), is to find a vector $z \in X(z)$ such that

$$
F^\top(z)(y-z) \geqslant 0, \ \forall y \in X(z). \tag{42}
$$

The QVI formulation of the frictional contact problems is obtained by considering the inclusions (21) and (23). We get

$$
u^\top(s-r) \geqslant 0, \ \text{ for all } s \in C(r_{\rm N}), \tag{43}
$$

where $C(r_{\rm N})$ is the Cartesian product of the semi–cylinders of radius $\mu^\alpha r^\alpha_{\rm N}$ defined as

$$
C(r_{\rm N}) := \prod_{\alpha=1}^{n_c} \left\{ s \in \mathbb{R}^3 \mid s_{\rm N} \geqslant 0, \|s_{\rm T}\| \leqslant \mu^\alpha r^\alpha_{\rm N} \right\}. \tag{44}
$$

Note that the QVI (43) involves only u and not $\hat{u}$: this is the main interest of this formulation. The price to pay is the dependence on r of the set $C(r_{\rm N})$. Problem FC can be expressed as a QVI by substituting the expression of u, which yields

$$
(Wr+q)^\top(s-r) \geqslant 0, \ \text{ for all } s \in C(r_{\rm N}). \tag{45}
$$

This expression is compactly rewritten as QVI($F_{\rm qvi}$, $X_{\rm qvi}$), with

$$
F_{\rm qvi}(r) := Wr + q \quad \text{and} \quad X_{\rm qvi}(r) := C(r_{\rm N}). \tag{46}
$$

Since W is assumed to be a positive semi-definite matrix, $F_{\rm qvi}$ is monotone. Thus, we get an affine monotone QVI($F_{\rm qvi}$, $X_{\rm qvi}$) for Problem FC.

3.3 Nonsmooth Equations

In this section, we expose a classical approach to solving a VI or a QVI, based on a reformulation of the inclusion as a nonsmooth equation. The term nonsmooth equation highlights that the mapping we consider fails to be differentiable. This is the price to pay for this reformulation. We can apply fixed-point and Newton-like algorithms to solve the resulting equation. Given the nonsmooth nature of the problem, applying Newton's method appears challenging, but it can still be done for some reformulations. More precisely for Problem FC, we search for an equation of the type

$$G(r) = 0, \tag{47}$$

where G is generally only locally Lipschitz continuous. The mapping G is such that the zeroes of (47) are the solutions to (34).

Natural and Normal Maps for the VI Formulations

A general-purpose reformulation of VI is obtained by using the normal and natural maps (see [39] for details). The natural map $F^{\mathsf{nat}}\colon \mathbb{R}^n \to \mathbb{R}^n$ associated with the VI (37) is defined by

$$F^{\mathsf{nat}}(z) := z - P_X(z - F(z)), \tag{48}$$

where P_X is the Euclidean projector on the set X. A well-known result (see [39]) states that the solutions to a VI are related to the zeroes of the natural map:

$$z \text{ solves } \mathrm{VI}(X, F) \iff F^{\mathsf{nat}}(z) = 0. \tag{49}$$

Using (37), it is easy to see that if z solves VI(X, F), then it is also a solution to VI$(X, \rho F)$ for any $\rho > 0$. Therefore, we can define a parametric variant of the natural map by

$$F_\rho^{\mathsf{nat}}(z) = z - P_X(z - \rho F(z)). \tag{50}$$

The relations given in (49) continue to hold for the parametric mapping. Using those equivalences, the frictional contact problem can be restated as zeroes of nonsmooth functions. With the natural map, Problem FC under the VI form (40) can be reformulated as

$$F_{\mathsf{vi}}^{\mathsf{nat}}(r) := \left[r - P_K\left(r - \rho(Wr + q + g(Wr + q))\right)\right] = 0. \tag{51}$$

Following the same lines, the normal map may also be used to derive algorithms. The normal map $F^{\mathsf{nor}}\colon \mathbb{R}^n \to \mathbb{R}^n$ is defined by

$$F^{\mathsf{nor}}(x) := F(P_X(x)) + x - P_X(x), \tag{52}$$

and its parametric variant

$$F_{\rho}^{\mathsf{nor}}(x) = \rho F(P_X(x)) + x - P_X(x). \quad (53)$$

An equivalent result holds:

$$z \text{ solves VI}(X, F) \iff z = P_X(x) \text{ for some } x \text{ such that } F^{\mathsf{nor}}(x) = 0. \quad (54)$$

The normal map based formulation of VI is also obtained in the same way.

In the seminal work of [106], iterative methods for solving monotone VIs are based on the natural map and fixed point iterations. The role of ρ is recognized to be very important for the rate of convergence. To improve the methods, [106] proposes using a "skewed" projector based on a non-Euclidean metric. Given a positive definite matrix $R \in R^{n\times n}$, a skewed projector $P_{X,R}$ onto X is defined as follows: $z = P_{X,R}(x)$ is the unique solution to the convex program

$$\begin{cases} \min \dfrac{1}{2}(y-x)^\top R(y-x), \\ s.t.\ y \in X. \end{cases} \quad (55)$$

The skew natural map can also be defined and yields the following nonsmooth equation:

$$F_R^{\mathsf{nat}}(z) = z - P_{X,R}(z - R^{-1}F(z)). \quad (56)$$

The zeros of $F_R^{\mathsf{nat}}(z)$ are also the solution to the VI(X, F). Considering the skew natural map, we obtain, for Problem FC under the VI form (40),

$$F_{\mathrm{vi},R}^{\mathsf{nat}}(r) := \left[r - P_{K,R}\left(r - R^{-1}(Wr + q + g(Wr + q))\right)\right]. \quad (57)$$

The previous case is retrieved by choosing $R = \rho^{-1} I_{n\times n}$.

Jean–Moreau's and Alart–Curnier's Functions

Using the alternative inclusion formulations (21)–(23) with a given set of parameters $\rho_{\mathrm{N}}, \rho_{\mathrm{T}}$ such that

$$\begin{cases} -\rho_{\mathrm{N}} u_{\mathrm{N}} \in N_{\mathbb{R}_+^{n_c}}(r_{\mathrm{N}}), & \rho_{\mathrm{N}} > 0, \\ -\rho_{\mathrm{T}} u_{\mathrm{T}} \in N_{D(\mu(r_n)_+)}(r_{\mathrm{T}}), & \rho_{\mathrm{T}} > 0, \end{cases} \quad (58)$$

we can substitute P_K into $P_{\mathbb{R}_+^{n_c}}$ and $P_{D(\mu(r_n)_+)}$, where

$$D(\mu(r_n)_+) = \prod_{\alpha=1\ldots n_c} D(\mu^\alpha (r_{\mathrm{N}}^\alpha)_+). \quad (59)$$

defines the Cartesian product of the Coulomb disks for each contact. The notation x_+ stands for $x_+ = \max(0, x)$. Using this procedure, [23, 61] propose the following nonsmooth equation formulation of the frictional contact condition:

$$\begin{cases} r_{\rm N} - P_{\mathbb{R}^{n_c}_+}(r_{\rm N} - \rho_{\rm N} u_{\rm N}) = 0, \\ r_{\rm T} - P_{D(\mu(r_{\rm N})_+)}(r_{\rm T} - \rho_{\rm T} u_{\rm T}) = 0. \end{cases} \tag{60}$$

The parameters $\rho_{\rm N}$, $\rho_{\rm T}$ may also be chosen contact by contact. Problem FC is then reformulated as

$$F_{\rm mj}(r) := \begin{bmatrix} r_{\rm N} - P_{\mathbb{R}^{n_c}_+}(r_{\rm N} - \rho_{\rm N}(Wr+q)_{\rm N}) \\ r_{\rm T} - P_{D(\mu(r_{\rm N})_+)}(r_{\rm T} - \rho_{\rm T}(Wr+q)_{\rm T}) \end{bmatrix} = 0. \tag{61}$$

In the seminal work of Alart and Curnier [10, 24], the augmented Lagrangian approach is invoked (see Remark 3) to obtain a similar formulation motivated by the development of nonsmooth (or generalized) Newton methods (see Sect. 5.2). To be accurate, the original Alart–Curnier function is given by

$$\begin{cases} r_{\rm N} - P_{\mathbb{R}^{n_c}_+}(r_{\rm N} - \rho_{\rm N} u_{\rm N}) = 0, \\ r_{\rm T} - P_{D(\mu(r_{\rm N} - \rho u_{\rm N})_+)}(r_{\rm T} - \rho_{\rm T} u_{\rm T}) = 0. \end{cases} \tag{62}$$

The difference between (60) and (62) is in the radius of the disk: $D(\mu(r_{\rm N} - \rho u_{\rm N})_+)$ rather than $D(\mu(r_{\rm N})_+)$. Problem FC can also be reformulated as in (61) using (62). This yields

$$F_{\rm ac}(r) := \begin{bmatrix} r_{\rm N} - P_{\mathbb{R}^{n_c}_+}(r_{\rm N} - \rho_{\rm N}(Wr+q)_{\rm N}) \\ r_{\rm T} - P_{D(\mu(r_{\rm N} - \rho_{\rm N} u_{\rm N})_+)}(r_{\rm T} - \rho_{\rm N}(Wr+q)_{\rm T}) \end{bmatrix} = 0. \tag{63}$$

Remark 2 From the QVI formulation (43), the following nonsmooth equation can also be written:

$$r = P_{C(r_{\rm N})}(r - \rho u), \tag{64}$$

which corresponds to (60).

Remark 3 In the literature of computational mechanics [10, 24, 107], very similar expressions are obtained using the concept of augmented Lagrangian functions. This concept, introduced in the general framework of Optimization by [57] and developed and popularized by [101, 102], is a strong theoretical tool for analyzing the existence and regularity of solutions to constrained optimization problems. Its numerical interest is still a subject of intense debate in the mathematical programming community. In the nonconvex nonsmooth context of frictional contact problems, its invocation is not so clear, but it has enabled the design of robust numerical techniques. Nevertheless, it is worth noting that some of these methods appear as variants of the methods developed to solve variational inequalities in other contexts. The method developed by [107] is a dedicated version of fixed point with projection for VI (see Algorithm 1) and the method of [10] is a tailored version of semi–smooth Newton methods (see Sect. 5). Nevertheless, the concept of an augmented Lagrangian has never been used in the optimization literature for this purpose.

Xuewen–Soh–Wanji Functions

Following the earlier work of [77, 96], the following function is proposed in [120]:

$$F_{\text{xsw}}(r) := \begin{bmatrix} \min(u_{\text{N}}, r_n) \\ \min(\|u_{\text{T}}\|, \mu r_{\text{N}} - \|r_{\text{T}}\|) = 0 \\ |u_{\text{T}_1} r_{\text{T}_2} - u_{\text{T}_2} r_{\text{T}_1} + \max(0, u_{\text{T}_1} r_{\text{T}_1}) = 0 \end{bmatrix} = 0. \tag{65}$$

In [120], the system is solved by a generalized Newton method with a line-search procedure.

Hüeber–Stadler–Wolhmuth Functions

In [60, 109], and subsequently in [71], another function is used to reformulate the problem FC:

$$F_{\text{hsw}}(r) := \begin{bmatrix} r_{\text{N}} - P_{\mathbb{R}_+^{n_c}}(r_{\text{N}} - \rho_{\text{N}}(Wr + q)_{\text{N}}) \\ \max(\mu(r_{\text{N}} - \rho_{\text{N}} u_{\text{N}}), \|r_t - \rho_t u_{\text{T}}\|) r_{\text{T}} - \mu \max(0, r_n - \rho_{\text{N}} u_{\text{N}})(r_t - \rho_{\text{T}} u_{\text{T}}) \end{bmatrix} = 0. \tag{66}$$

In [60], this function is used considering the constraints at the position level, as opposed to in [71], in which the formulation is at the velocity level.

General SOCC-Functions

More generally, a large family of reformulations of the SOCCP (26) in terms of equations can be obtained by using a so-called Second-Order Cone Complementarity (SOCC) function. Let us consider the following SOCCP over a symmetric cone $K^\star = K$. A SOCC-function ϕ is defined by

$$K \ni x \perp y \in K \quad \Longleftrightarrow \quad \phi(x, y) = 0. \tag{67}$$

The frictional contact problem can be written as a SOCCP over symmetric cones by applying the following transformations:

$$x = T_x \hat{u} = \begin{bmatrix} \hat{u}_{\text{N}} \\ \mu \hat{u}_{\text{T}} \end{bmatrix} \text{ and } y = T_y r = \begin{bmatrix} \mu r_{\text{N}} \\ r_{\text{T}} \end{bmatrix}. \tag{68}$$

Clearly, the nonsmooth equations of the previous sections provide several examples of SOCC-functions and the natural map offers the simplest one. In [43], the standard complementarity functions for Nonlinear Complementarity Problems (NCP) such as the celebrated Fischer–Burmeister function are extended to the SOCCP by means of Jordan algebra. Smoothing functions are also given with their Jacobians, and they studied their properties in view of the application of Newton's method. For the second-order cone, the Jordan algebra can be defined with the following non-associative Jordan product:

$$x \cdot y = \begin{bmatrix} x^\top y \\ y_{\text{N}} x_{\text{T}} + x_{\text{N}} y_{\text{T}} \end{bmatrix} \tag{69}$$

and the usual componentwise addition $x + y$. The vector x^2 denotes $x \cdot x$, and there exists a unique vector $x^{1/2} \in K$, the square root of $x \in K$, defined as

$$(x^{1/2})^2 = x^{1/2} \cdot x^{1/2} = x. \tag{70}$$

A direct calculation for the SOC in $\mathbb{R}^3$ yields

$$x^{1/2} = \begin{bmatrix} s \\ \dfrac{x_{\mathrm{T}}}{2s} \end{bmatrix}, \quad \text{where } s = \sqrt{(x_{\mathrm{N}} + \sqrt{x_{\mathrm{N}}^2 - \|x_{\mathrm{T}}\|^2})/2}. \tag{71}$$

We adopt the convention that $0^{1/2} = 0$. The vector $|x| \in K$ denotes $(x^2)^{1/2}$. Thanks to this algebra and its associated operator, the projection onto K can be written as

$$P_K(x) = \frac{x + |x|}{2}. \tag{72}$$

This formula provides a new expression for the natural map and its associated non-smooth equations. This is exactly what is done in [55], where the natural map (48) is used, together with an expression of the projection operator based on the Jordan algebra calculus. The resulting SOCCP is then solved with a semi–smooth Newton method, and a smoothing parameter can be added.

Most of the calculus in Jordan algebra is based on the spectral decomposition, a basic concept in Jordan algebra (see [43] for more details). For $x = (x_{\mathrm{N}}, x_{\mathrm{T}}) \in \mathbb{R} \times \mathbb{R}^2$, the spectral decomposition is defined by

$$x = \lambda_1 u_1 + \lambda_2 u_2, \tag{73}$$

where $\lambda_1, \lambda_2 \in \mathbb{R}$ and $u_1, u_2 \in \mathbb{R}^3$ are the spectral values and the spectral vectors of x given by

$$\lambda_i = x_{\mathrm{N}} + (-1)^i \|x_{\mathrm{T}}\|, \quad u_i = \begin{cases} \frac{1}{2}\begin{bmatrix} 1 \\ (-1)^i \dfrac{x_{\mathrm{T}}}{\|x_{\mathrm{T}}\|} \end{bmatrix}, & \text{if } x_{\mathrm{T}} \neq 0 \\ \frac{1}{2}\begin{bmatrix} 1 \\ (-1)^i w \end{bmatrix}, & \text{if } x_{\mathrm{T}} = 0 \end{cases} \quad i = 1, 2, \tag{74}$$

with $w \in \mathbb{R}^2$ any unit vector. Note that the decomposition is unique whenever $x_{\mathrm{T}} \neq 0$. The spectral decomposition enjoys very nice properties that simplify the computation of basic functions such that

$$\begin{aligned} x^{1/2} &= \sqrt{\lambda_1} u_1 + \sqrt{\lambda_2} u_2, \text{ for any } x \in K, \\ P_K(x) &= \max(0, \lambda_1) u_1 + \max(0, \lambda_2) u_2. \end{aligned} \tag{75}$$

More interestingly, general SOCC-functions can also be extended, and a smoothed version of this function can also be developed (see [43]). Let us start with the Fischer–Burmeister function

$$\phi_{\mathsf{FB}}(x, y) = x + y - (x^2 + y^2)^{1/2}. \tag{76}$$

It can be shown that the zeroes of ϕ_{FB} are solutions of the SOCCP (67) using the Jordan algebra associated with K. Using the spectral decomposition, the Fischer–Burmeister function can be easily computed as

$$\phi_{\mathsf{FB}}(x, y) = x + y - (\sqrt{\bar{\lambda}_1}\bar{u}_1 + \sqrt{\bar{\lambda}_2}\bar{u}_2), \tag{77}$$

where $\bar{\lambda}_1, \bar{\lambda}_2 \in \mathbb{R}$ and $\bar{u}_1, \bar{u}_2 \in \mathbb{R}^3$ are the spectral values and the spectral vectors of $x^2 + y^2$, that is,

$$\begin{aligned} \bar{\lambda}_i &= \|x\|^2 + \|y\|^2 + 2(-1)^i \|x_{\mathrm{N}} x_{\mathrm{T}} + y_{\mathrm{N}} y_{\mathrm{T}}\| \\ \bar{u}_i &= \begin{cases} \frac{1}{2}\begin{bmatrix} 1 \\ (-1)^i \dfrac{x_{\mathrm{N}} x_{\mathrm{T}} + y_{\mathrm{N}} y_{\mathrm{T}}}{\|x_{\mathrm{N}} x_{\mathrm{T}} + y_{\mathrm{N}} y_{\mathrm{T}}\|} \end{bmatrix}, \text{ if } x_{\mathrm{N}} x_{\mathrm{T}} + y_{\mathrm{N}} y_{\mathrm{T}} \neq 0 \\ \frac{1}{2}\begin{bmatrix} 1 \\ (-1)^i w \end{bmatrix}, \text{ if } x_{\mathrm{N}} x_{\mathrm{T}} + y_{\mathrm{N}} y_{\mathrm{T}} = 0 \end{cases} \end{aligned} \quad , i = 1, 2. \tag{78}$$

Finally, Problem FC is then reformulated as

$$F_{\mathsf{FB}}(u, r) := \begin{bmatrix} u - Wr - q \\ \Phi_{\mathsf{FB}}\left(\begin{bmatrix} \mu r_{\mathrm{N}} \\ r_{\mathrm{T}} \end{bmatrix}, \begin{bmatrix} \frac{1}{\mu}(u_{\mathrm{N}} + \mu\|u_{\mathrm{T}}\|) \\ u_{\mathrm{T}} \end{bmatrix}\right) \end{bmatrix} = 0, \tag{79}$$

where the mapping $\Phi_{\mathsf{FB}} : \mathbb{R}^{3n_c} \times \mathbb{R}^{3n_c} \to \mathbb{R}^{3n_c}$ is defined as

$$\Phi_{\mathsf{FB}}(x, y) = \left[(\phi(x^\alpha, y^\alpha), \alpha = 1 \ldots n_c)^\top\right]. \tag{80}$$

3.4 Optimization Problems

In this section, several optimization-based formulations are proposed. The quest for an efficient optimization formulation of the frictional problem is a hard task. Since the problem is nonsmooth and nonconvex, the use of an associated optimization problem is interesting from the numerical point of view if we want to improve the robustness and the stability of the numerical methods.

A straightforward optimization problem can be written whose cost function is the scalar product $r^\top \hat{u}$. Indeed, this product is always positive and vanishes at the solution. Let us consider this first optimization formulation:

$$\begin{cases} \min\, r^\top \hat{u} = r^\top u + \sum_{\alpha=1}^{n_c} \mu^\alpha r_{\mathrm{N}}^\alpha \|u_{\mathrm{T}}^\alpha\| \\ s.t.\ \hat{u} \in K^\star, \\ \quad r \in K, \end{cases} \tag{81}$$

which amounts to minimizing the DeSaxcé's bipotential function [26] over $K^\star \times K$. A first simplification can be made by noting that

$$\hat{u} \in K^\star \Longleftrightarrow u_{\mathrm{N}} \geqslant 0, \tag{82}$$

which leads to

$$\begin{cases} \min\, r^\top u + \sum_{\alpha=1}^{n_c} \mu^\alpha r_{\mathrm{N}}^\alpha \|u_{\mathrm{T}}^\alpha\| \\ s.t.\ u_{\mathrm{N}} \geqslant 0 \\ \quad r \in K. \end{cases} \tag{83}$$

Starting from Problem FC, a direct substitution of $u = Wr + q$ yields

$$\begin{cases} \min\, r^\top (Wr + q) + \sum_{\alpha=1}^{n_c} \mu^\alpha r_{\mathrm{N}}^\alpha \|(Wr + q)_{\mathrm{T}}^\alpha\| \\ s.t.\ (Wr + q)\mathsf{N} \geqslant 0, \\ \quad r \in K, \end{cases} \tag{84}$$

which is a nonlinear optimization problem with a nonsmooth and nonconvex cost function. From the numerical point of view, this problem may be very difficult, and we have to ensure that the cost function has to be zero at the solution, which is not guaranteed if some local minima are reached in the minimization process.

Other optimization-based formulations have been proposed in the literature. They are not direct optimization formulation, but they try to identify an optimization subproblem that is well-posed and for which efficient numerical methods are available. Three approaches can be listed in three categories: (a) the *alternating optimization* problems, (b) the *successive approximation* method, and (c) the *convex SOCP* approach.

The Panagiotopoulos Alternating Optimization Approach

The Panagiotopoulos alternating optimization approach aims at solving the frictional contact problem by alternatively solving the Signorini condition for a fixed value of the tangential reaction r_{T}, and solving the Coulomb friction model for a fixed value of the normal reaction r_{N}. Let us split the matrix W and the vector q in the following way:

$$u = Wr + q \quad \Longleftrightarrow \quad \begin{bmatrix} u_{\mathrm{N}} \\ u_{\mathrm{T}} \end{bmatrix} = \begin{bmatrix} W_{\mathrm{NN}} & W_{\mathrm{NT}} \\ W_{\mathrm{TN}} & W_{\mathrm{TT}} \end{bmatrix} \begin{bmatrix} r_{\mathrm{N}} \\ r_{\mathrm{T}} \end{bmatrix} + \begin{bmatrix} q_{\mathrm{N}} \\ q_{\mathrm{T}} \end{bmatrix}. \tag{85}$$

Two sub-problems can therefore be identified: the first one is to find u_{N} and r_{N} such that

$$\begin{cases} u_{\mathrm{N}} = W_{\mathrm{NN}} r_{\mathrm{N}} + \tilde{q}_{\mathrm{N}}, \\ 0 \leqslant u_{\mathrm{N}} \perp r_{\mathrm{N}} \geqslant 0, \end{cases} \tag{86}$$

where $\tilde{q}_{\mathrm{N}} = q_{\mathrm{N}} + W_{\mathrm{NT}} r_{\mathrm{T}}$. The second problem is to find u_{T} and r_{T} such that

$$\begin{cases} u_{\mathrm{T}} = W_{\mathrm{TT}} r_{\mathrm{T}} + \tilde{q}_{\mathrm{T}}, \\ -u_{\mathrm{T}} \in N_{D(\mu \tilde{r}_{\mathrm{N}})}(r_{\mathrm{T}}), \end{cases} \tag{87}$$

where $\tilde{r}_{\mathrm{N}}$ is fixed and $\tilde{q}_{\mathrm{T}} = q_{\mathrm{T}} + W_{\mathrm{TN}} r_{\mathrm{N}}$.

If we assume for a while that the Delassus W is a symmetric positive semi–definite matrix, W_{NN} and W_{TT} are also symmetric semi–definite positive matrices. Therefore, two convex optimization problems can be formulated:

$$\begin{cases} \min \ \frac{1}{2} r_{\mathrm{N}}^{\top} W_{\mathrm{NN}} r_{\mathrm{N}} + r_{\mathrm{N}}^{\top} \tilde{q}_{\mathrm{N}} \\ \text{s.t. } r_{\mathrm{N}} \geqslant 0 \end{cases} \tag{88}$$

and

$$\begin{cases} \min \ \frac{1}{2} r_{\mathrm{T}}^{\top} W_{\mathrm{TT}} r_{\mathrm{T}} + r_{\mathrm{T}}^{\top} \tilde{q}_{\mathrm{T}} \\ \text{s.t. } r_{\mathrm{T}} \in D(\mu \tilde{r}_{\mathrm{N}}). \end{cases} \tag{89}$$

This approach has been proposed by [93] for two–dimensional applications in soil foundation computing. It has also been used in other finite element applications in [13, 116] and studied from the mathematical point of view in [50, 51].

Remark 4 If the Delassus matrix is an unsymmetric matrix but semi-definite positive, the following quadratic programming problem is equivalent to (86):

$$\begin{cases} \min \ r_{\mathrm{N}}^{\top} W_{\mathrm{NN}} r_{\mathrm{N}} + r_{\mathrm{N}}^{\top} \tilde{q}_{\mathrm{N}} \\ \text{s.t. } r_{\mathrm{N}} \geqslant 0 \\ \qquad W_{\mathrm{NN}} r_{\mathrm{N}} + \tilde{q}_{\mathrm{N}} \geqslant 0. \end{cases} \tag{90}$$

The Successive Approximation

The successive approximation method identifies a single optimization problem by introducing a function that maps the normal reaction to itself (or the friction threshold) such that

$$h(r_{\mathrm{N}}) = r_{\mathrm{N}}. \tag{91}$$

Using this artifact, we can define a new problem from Problem FC such that

$$\begin{cases} \theta = h(r_{\mathrm{N}}) \\ u = Wr + q \\ -u_{\mathrm{N}} \in N_{\mathbb{R}_+^{nc}}(r_{\mathrm{N}}) \\ -u_{\mathrm{T}} \in N_{D(\mu\theta)}(r_{\mathrm{T}}). \end{cases} \tag{92}$$

If we assume for a while that the Delassus W is a symmetric positive semi–definite matrix, the last three lines are equivalent to a convex optimization problem over the product of semi–cylinders $C(\mu, \theta)$, that is,

$$\begin{cases} \theta = h(r_{\mathrm{N}}) \\ \begin{cases} \min \ \frac{1}{2} r^{\top} W r + r^{\top} q \\ \text{s.t.} \ \ r \in C(\mu, \theta). \end{cases} \end{cases} \tag{93}$$

The method of successive approximation has been extensively used for proving the existence and uniqueness of solutions to the discrete frictional contact problems. We refer to [51], which summarizes the seminal work of the Czech school [48, 49, 91]. We will see in the sequel that this approach also provides us with very efficient numerical solvers in Sect. 7.2.

The Convex SOCP

The convex SOCP approach is in the same vein as the previous one, with the difference that a SOCQP sub-problem is identified. To this aim, we augment the problem by introduction of an auxiliary variable s, the image of $g(u)$ introduced in (27). We then obtain

$$\begin{cases} s = g(u) \\ \hat{u} = Wr + q + s \\ K^{\star} \ni \hat{u} \perp r \in K. \end{cases} \tag{94}$$

Since W is a positive semi-definite matrix, a new convex optimization sub-problem can be defined:

$$\begin{cases} s = g(u) \\ \begin{cases} \min \ \frac{1}{2} r^{\top} W r + r^{\top} (q + s) \\ \text{s.t.} \ \ r \in K. \end{cases} \end{cases} \tag{95}$$

This formulation introduced in [18] and developed in [2, 3] has been used to give an existence criteria to the discrete frictional contact problems. Furthermore, this existence criteria can be numerically checked by solving a linear program of a second-order cone (SOCLP).

4 Numerical Methods for VIs

4.1 Fixed Point and Projection Methods for VI

Starting from the VI formulations (37), or more precisely, an associated nonsmooth equation through the natural map,

$$F_R^{\mathrm{nat}}(z) = z - P_{X,R}(z - R^{-1}F(z)). \tag{96}$$

The basic idea of the algorithm is to perform fixed point iterations on the mapping

$$z \mapsto P_{X,R}(z - R^{-1}F(z)), \tag{97}$$

yielding to Algorithm 1 with the specific choice of $R = \rho_k^{-1} I$. The choice of the updating rule of ρ_k is detailed in Sect. 4.2.

Algorithm 1 Fixed point iterations for the VI (37)

Require: F, X Data of VI (37)
Require: z_0 initial values
Require: tol $>$ 0 a tolerance value and $\mathrm{iter}_{\max} > 0$ the max number of iterations
Require: ρ_0 initial value for ρ
Ensure: z solution of VI (37)
  $k \leftarrow 0$
  while error $>$ tol and $k < \mathrm{iter}_{\max}$ **do**
    Update the value of ρ_k
    $z_{k+1} \leftarrow P_X(z_k - \rho_k F(z_k))$
    Evaluate error.
    $k \leftarrow k+1$
  end while
  $z \leftarrow z_k$

For the formulation (40), the following iterations are performed:

$$r_{k+1} \leftarrow P_{K,R}(r_k - R^{-1}(Wr_k + q + g(Wr_k + q))). \tag{98}$$

In the sequel, when a parameter ρ is specified, it is assumed that $R = \rho^{-1} I$.

The convergence of such methods is generally shown for strongly monotone VI. In our case, this assumption is not satisfied, but we will see in the sequel that such methods can converge in practice.

Remark 5 Algorithm 1 with the iteration rule (98) and a fixed value of ρ_k was originally proposed in [27, 28]. The algorithm is called Uzawa's algorithm as reference to the algorithm credited to Uzawa in computing the optimal values of a convex program by primal-dual techniques[42, 44]. Note that the algorithm in [107] is similar to the fixed point algorithm with projection, though based on an augmented Lagrangian concept (see Remark 3).

Extragradient Methods

The extragradient method [70] is also a well-known method for VI that improves the previous projection method. It can be described as

$$\begin{aligned} \bar{z}_k &\leftarrow P_X(z_k - \rho F(z_k)) \\ z_{k+1} &\leftarrow P_X(z_k - \rho\, F(\bar{z}_k)) \end{aligned} \tag{99}$$

and formally defined in Algorithm 2. The convergence of this method is guaranteed under the following assumptions: there exists a solution, and the function F is Lipschitz–continuous and pseudo–monotone.

Algorithm 2 Extragradient method for the VI (37)

Require: F, X Data of VI (37)
Require: z_0 initial values
Require: tol $>$ 0 a tolerance value and iter$_{\max}$ $>$ 0 the max number of iterations
Ensure: z solution of VI (37)

```
k ← 0
while error > tol and k < iter_max do
   Update the value of ρ_k
   z̄_k ← P_X(z_k − ρ_k F(z_k))
   z_{k+1} ← P_X(z_k − ρ_k F(z̄_k))
   Evaluate error.
   k ← k + 1
end while
z ← z_k
```

4.2 *Self-adaptive Step-Size Rules*

A key ingredient in this efficiency and the convergence of the numerical methods for VI presented above is the choice of the sequence $\{\rho_k\}$. A sensible work has been done in the literature, mainly motivated by some convergence proofs under specific assumptions. Besides the relaxation of the assumption for the convergence, we are interested in improving the numerical efficiency and robustness. We present in this section the most popular approach for choosing the sequence $\{\rho_k\}$.

In [65], a method is proposed for improving the extragradient method of [70] by adapting ρ_k in the following way. The goal is to find ρ_k that satisfies

$$0 < \rho_k \leqslant \min\left\{\bar{\rho}, L\frac{\|z_k - \bar{z}_k\|}{\|F(z_k) - F(\bar{z}_k)\|}\right\} \text{ with } L \in (0,1), \tag{100}$$

where $\bar{\rho}$ is the maximum value of ρ_k chosen in light of the specific problem. The objective is to find a coefficient that is bounded by the local Lipschitz constant. The standard way to do that is to use an Armijo–type procedure by successively trying some values of $\rho_k = \bar{\rho}\nu^m$ with $m \in \mathbb{N}$ and $\nu \in (0, 1)$, with a typical value of $2/3$. In the original article by [65], there is no procedure for sizing $\bar{\rho}$ or updating it. In [56] and in the context of prediction–correction, the authors propose using the rule $\rho_k = \rho_{k-1}\nu^m$, and if the criteria (100) is largely satisfied for ρ_k, the value is increased. In [46], a similar procedure is used for the extragradient method by adding an increasing step of ρ_k, which is done after the correction, as in [56]. The criteria (100) is verified by computing the ratio

$$r_k \leftarrow \frac{\rho_k \|F(z_k) - F(\bar{z}_k)\|}{\|z_k - \bar{z}_k\|}. \tag{101}$$

In [108], a similar Armijo–like technique is used, and the ratio r_k is computed as follows:

$$r_k \leftarrow \frac{\rho_k (z_k - \bar{z}_k)^\top (F(z_k) - F(\bar{z}_k))}{\|z_k - \bar{z}_k\|^2}. \tag{102}$$

The approach is summarized in Algorithm 3. The parameter L typically chosen around 0.9 is a safety coefficient in the evaluation of ρ_k. The parameter $\mathsf{L}_{\mathsf{min}}$ that triggers an increase of ρ_k is chosen around 0.3.

In [46], the update of the Armijo rule $\rho_k \leftarrow \nu\, \rho_k$ can also be replaced with $\rho_k \leftarrow \nu\, \rho_k \min\{1, 1/r_k\}$, but it appears that this trick does improve the self–adaptive procedure. Other more evolved step–length strategies that have been tried in this study can be found in [117].

Algorithm 3 Updating rule for ρ_k

Require: F, X
Require: Search and safety parameters. $\mathsf{L} \in (0, 1)$, $0 < \mathsf{L}_{\mathsf{min}} < \mathsf{L}$, $\nu \in (0, 1)$
Require: Initial values $\mathsf{z_k} \in \mathsf{X}$, $\rho_{\mathsf{k}-1} > 0$

  $\rho_\mathsf{k} \leftarrow \rho_{\mathsf{k}-1}$
  $\bar{\mathsf{z}}_\mathsf{k} \leftarrow \mathsf{P_X}(\mathsf{z_k} - \rho_\mathsf{k}\mathsf{F}(\mathsf{z_k}))$
  Evaluate r_k with (101) (or (102))
  while $\mathsf{r_k} > \mathsf{L}$ **do**
    $\rho_\mathsf{k} \leftarrow \nu\, \rho_\mathsf{k}$
    $\bar{\mathsf{z}}_\mathsf{k} \leftarrow \mathsf{P_X}(\mathsf{z_k} - \rho_\mathsf{k}\mathsf{F}(\mathsf{z_k}))$
    Evaluate r_k with (101) (or (102))
  end while
  Perform the correction step of extragradient or prediction–correction method.
  if $\mathsf{r_k} < \mathsf{L}_{\mathsf{min}}$ **then**
    $\rho_\mathsf{k} = \frac{1}{\nu}\rho_\mathsf{k}$
  end if

Table 1 Naming convention for the algorithms based on VI formulations

Name	Algorithm	Additional information
FP-DS	1	Iteration rule (98) and fixed ρ
FP-VI-UPK	1 and 3	Iteration rule (98) and updating rule (101)
FP-VI-UPTS	1 and 3	Iteration rule (98) and updating rule (102)
EG-VI-UPK	2 and 3	Iteration rule (99) and updating rule (101)
EG-VI-UPTS	2 and 3	Iteration rule (99) and updating rule (102)

4.3 Nomenclature

A nomenclature for the algorithms based on the VI formulation is given in Table 1.

5 Newton-Based Methods

5.1 Principle of the Nonsmooth Newton Methods

In Sect. 3.3, several formulations of the frictional contact problem by means of nonsmooth equations have been presented. These nonsmooth equations call for the use of nonsmooth Newton's methods. Remember that the standard Newton method consists in solving

$$G(z) = 0 \tag{103}$$

by performing the following Newton iteration:

$$z_{k+1} = z_k - J^{-1}(z_k)G(z_k). \tag{104}$$

If the mapping G is smooth, the matrix J is the Jacobian matrix of G with respect to z, that is, $J(z) = \nabla_z G(z)$. Whenever G is nonsmooth but locally Lipschitz continuous, the Jacobian matrix J is replaced with an element $\Phi(z)$ of the generalized Jacobian at z: $\Phi(z) \in \partial G(z)$. Let us recall the definition of the generalized Jacobian. By Rademacher's Theorem, if G is locally Lipschitz continuous, then G is almost everywhere differentiable, and let us define the set D_G by

$$D_G := \{z \mid G \text{ is differentiable at } z\}. \tag{105}$$

The generalized Jacobian of G at z can be defined by

$$\partial G(z) = \text{conv}\partial_B G(z), \tag{106}$$

with

$$\partial_B G(z) = \{\lim_{\bar{z}\to z, \bar{z}\in D_G} \nabla G(\bar{z})\}. \tag{107}$$

If $\Phi(z)$ is nonsingular, then an iteration of the nonsmooth Newton method is given by

$$z_{k+1} = z_k - \Phi^{-1}(z_k)(G(z_k)). \tag{108}$$

The resulting nonsmooth Newton method is detailed in Algorithm 4.

Algorithm 4 Nonsmooth Newton method for (103)

Require: G data of Problem (103)
Require: z_0 initial values
Require: tol > 0 a tolerance value and $\mathrm{iter}_{\max} > 0$ the max number of iterations
Ensure: z solution of Problem (103)

```
k ← 0
while error > tol and k < iter_max do
  compute (select) Φ(z_k) ∈ ∂G(z_k)
  z_{k+1} ← z_k − Φ^{-1}(z_k)(G(z_k))
  Evaluate error.
  k ← k + 1
end while
z ← z_k
```

The convergence of nonsmooth Newton methods is based on the assumption of the semi–smoothness of the nonsmooth function in (103). For this reason, they are often called *semi–smooth Newton methods* (see [39, Sect. 7.5] and references therein).

5.2 *Application to the Discrete Frictional Contact Problem*

We use the Alart–Curnier function $F_{\mathrm{ac}}(u, r)$ in (63), Jean–Moreau function $F_{\mathrm{mj}}(u, r)$ in (61), Fischer–Burmeister function $F_{\mathsf{FB}}(u, r)$ in (79), and the natural map $F^{\mathsf{nat}}_{\mathrm{vi}}$ in (51) to define a Newton method for the Problem FC.

Computation of an Element of ∂G

For any r_0 in the nonsmooth domain of G, we compute $\Phi(r_0) = \lim_{t\to 0} \Phi(r(t))$ with $t \to r(t)$ a parametrization such that $\lim_{t\to 0} r(t) = r_0$ with $r(t)$ in the smooth domain for all t. Similar computations can also be found in [62], where a Newton method based on the formulation (51) is used contact by contact in a Gauss–Seidel loop.

Lipschitz Continuity Properties

For the mappings $F^{\mathsf{nat}}_{\mathrm{vi}}$, F_{ac}, F_{FB}, F_{mj}, F_{xsw}, whose expressions are mostly made of the Lipschitz functions P_X, min, max and $\|.\|$, the local Lipschitz properties can be shown without difficulty. For the mapping F_{FB}, the proof of Lipschitz continuity of ϕ_{FB} can be found in [111] and references therein. This ensures the consistency of the definition of the generalized Jacobians.

5.3 Convergence and Robustness Issues

The local convergence of the nonsmooth Newton methods is based on the semi-smoothness of the mapping G and the fact that all elements of the generalized Jacobian at the solution point $z^\star$, $\Phi(z^\star) \in \partial G(z^\star)$ are non-singular (see [97] and Chap. 1 of [98] for a survey of mathematical results). For our application, the semi-smoothness of the mapping F_{ac}, F_{mj}, or F_{hsw} is proven in several papers [22, 60]. The strong semi–smoothness of ϕ_{FB} can be found in [111].

On the other hand, the regularity of all elements of the generalized Jacobians is not guaranteed. The first reason is the possible rank deficiency of the matrix W, which is usual in rigid body applications, as discussed in Sect. 2.3. Even if we consider a full rank matrix W, as in the standard one contact case for instance, the invertibility of all the elements of the generalized Jacobian at the solution point is not straightforward. For the mapping F_{ac}, F_{mj}, some results are given in [8, 9, 63]. Some of the results depend on the value of the coefficient of friction and the exact penalty ρ, ρ_{N}, ρ_{T} parameters. For the mapping F_{hsw}, some other results can be found in [60].

In the numerical practice, and even if W is full-rank, it may happen that the elements of the generalized Jacobians are not regular or very badly conditioned when we are far from the solution. This fact is reported in [8, 9, 60, 63, 71]. Some divergence of the Newton algorithm can be encountered. A few works has been done towards understanding this problem. Among them, we cite [60], in which some modifications of the elements of the generalized Jacobian are performed far from the solution to keep the Newton iteration matrix regular and well-conditioned when the function F_{hsw} is chosen. This very interesting work opens new directions of research for the other mappings. In [71], some other heuristics are developed to try to avoid divergence of the Newton loop. In the two next sections, we present two complementary ways to partly solve this problem by consistently choosing the parameters ρ, ρ_{N}, ρ_{T} and applying some line–search techniques to globalize the convergence.

5.4 Estimation of ρ, ρ_{N}, ρ_{T} Parameters

One of the key parameters in the efficiency of the nonsmooth Newton methods is the choice of the parameter ρ in the parameterized natural map (50) and the parameters ρ_{N} and ρ_{T} in the Jean–Moreau and Alart–Curnier functions (61) and (63). The default choice is to set these parameters equal to 1, but the numerical practice shows that the convergence of the nonsmooth solvers is drastically deteriorated, especially if the norm or the conditioning of the matrix W is far from this unit value. There are no theoretical rules through which to size these parameters, but some heuristics may be found in the literature for a single contact problem that we expose in the sequel.

Inverse of a Norm of W

A first simple choice is to consider the inverse of a norm of the matrix W. With these heuristics, we set the ρ parameter before the Newton loop as follows:

$$\rho = \frac{1}{\|W\|}, \quad \rho_{\mathrm{N}} = \rho_{\mathrm{T}} = \frac{1}{\|W\|}. \tag{109}$$

This choice is mainly based on a guess of the inverse of the local Lipschitz constant of the operator $Wr + q$. In the case of the natural map, it amounts to neglecting the nonlinear contribution of g. For the norm, whenever the matrix is symmetric definite positive, choosing the 2-norm based on the spectral radius $\|W\|_2 = \rho(W) = \lambda_{\max}(W)$ would yield

$$\rho = \frac{1}{\lambda_{\max}(W)}, \quad \rho_{\mathrm{N}} = \rho_{\mathrm{T}} = \frac{1}{\lambda_{\max}(W)}. \tag{110}$$

Estimation Based on the Splitting W_{NN} and W_{TT}

A second possible choice for the map (61) and (63) is to use the fact that the problem is split with respect to the normal and the tangent directions. In that case, we compute a value of ρ_{N} that is based on the eigenvalues of W_{NN} and a value of ρ_{T} based on the eigenvalue of W_{TT}. For a single contact, we set

$$\rho_{\mathrm{N}} = \frac{1}{W_{\mathrm{NN}}}, \quad \rho_{\mathrm{T}} = \frac{1}{\lambda_{\max}(W_{\mathrm{TT}})} \tag{111}$$

A third option is also to take into account the conditioning of the matrix W_{TT} by choosing

$$\rho_{\mathrm{N}} = \frac{1}{W_{\mathrm{NN}}}, \quad \rho_{\mathrm{T}} = \frac{\lambda_{\min}(W_{\mathrm{NN}})}{\lambda^2_{\max}(W_{\mathrm{TT}})}. \tag{112}$$

Again, these heuristics implicitly assume that the Delassus matrix W is symmetric definite positive.

Adaptive Estimation of the Parameters

In [71], an adaptive way of updating ρ is proposed that has not been implemented for our experiments.

Default Choices

By default, we use the rule (111) for the mapping (61) and (63) and the rule (110) for the natural map. When other rules are chosen in the comparison, they are specified.

5.5 *Damped Newton and Line-Search Procedures*

We use mainly two types of line-search procedures: the Goldstein–Price and the Armijo line-search. Usually, strong mathematical assumptions are needed to guarantee their success, especially on the smoothness of the merit function $\mathcal{M}(x)$. For the Newton method, we use as the merit function the half of the norm of G, that is,

$$\mathscr{M}(x) = \frac{1}{2}\|G(x)\|, \tag{113}$$

with G taken accordingly to the formulation equals to $F_{\text{vi}}^{\text{nat}}$, F_{ac}, F_{FB}, F_{mj}, F_{xsw}. Clearly, the smoothness assumptions are not satisfied in our case. Even if the assumptions are fulfilled, and despite the mathematical proofs, in practice, it is recommended that some additional stopping criteria be added during extrapolation and interpolation phases to avoid infinite loops. In the sequel, we use the recommendations in Chap. 3 of [15], where the reader can find all the mathematical explanations as to why they terminate, under some assumptions about the merit function. The choice of the values for the parameters m_1, m_2 for the Goldstein–Price line-search and the parameter m_1 alone for the Armijo line-search is also discussed, and it is advised to choose $m_1 < \frac{1}{2}$ and $m_2 > \frac{1}{2}$.

Termination requires the existence of a function $q \in C^1(\mathbb{R})$ with $q'(0) < 0$, which is the value of the merit function in a given direction d. This function has to be bounded from below. In our case, this function is $q : t \to \frac{1}{2}\|G(r + td)\|$. An additional stopping criterion is implemented as a maximum number of iterations, and when the line-search fails, the Newton loop is continued with the last value of the step found by the line–search.

The Goldstein–Price (GP) line search and Armijo line search are described in Algorithms 5 and 6.

5.6 Nomenclature

A nomenclature for the algorithms based on the nonsmooth Newton methods is listed in Table 2.

6 Splitting Techniques and Proximal Point Algorithm

Splitting techniques are standard techniques for solving VI(F, X) when the function F is affine, that is, $F(z) = Mz + q$ and the set X can be decomposed into a Cartesian product of independent smaller sets $X = \Pi_i X_i$. Usually, a block splitting of the matrix M is performed and a Projected Successive Over Relaxation (PSOR) method is used to solve the VI. Since the cone K is a product of second-order cones in $\mathbb{R}^3$, a natural way to split the problem is to form sub-problems by using single contact as a building block. The sub-problems can be solved by any method for the VI that has been presented in the previous sections. In the same way, the proximal point algorithm can also be used, which amounts to solving the original VI(F, X) by solving a sequence of (easier) VIs.

Algorithm 5 Goldstein–Price (GP) line search

Require: x, the starting point of the line-search.
Require: d, the direction of search.
Require: t, an initial stepsize-value.
Require: $t \to q(t)$, for $t \geqslant 0$, with $q \in C^1$ bounded from below and $q'(0) < 0$, a merit function representing $f(x+td)$
Require: m_1, m_2, parameters with $0 < m_1 < m_2 < 1$
Require: a, with $a > 1$, parameter for extrapolation
Ensure: a finite line-search

```
t_L ← 0
t_R ← 0
Δ ← (q(t) − q(0))/t
while m_2 q'(0) > Δ or Δ > m_1 q'(0) do
  if m_1 q'(0) < Δ then
    t_R ← t
  end if
  if Δ < m_2 q'(0) then
    t_L ← t
  end if
  if t_R = 0 then
    t ← at
  else
    t ← (t_L + t_R)/2
  end if
  Δ ← (q(t) − q(0))/t
end while
```

Algorithm 6 Armijo(A) line search

Require: x, the starting point of the line-search.
Require: d, the direction of search.
Require: t, an initial stepsize-value.
Require: $t \to q(t)$, for $t \geqslant 0$, with $q \in C^1$ bounded from below and $q'(0) < 0$, a merit function representing $f(x+td)$
Require: m_1, a parameter with $0 < m_1 < 1$
Require: a, with $a > 1$, parameter for extrapolation
Ensure: a finite line-search

```
while m_1 q'(0) < (q(t) − q(0))/t do
  if t_R = 0 then
    t ← at
  else
    t_R ← t
    t ← t_R/2
  end if
end while
```

6.1 Splitting and Relaxation Techniques

The particular structure of the cone K as a product of second-order cones in $\mathbb{R}^3$ calls for a splitting of the problem contact by contact. For Problem FC, the relation

Table 2 Naming convention for the algorithms based on the nonsmooth Newton (NSN) method

Name	Algorithm	Additional information
NSN-NM	4	Natural map formulation (51)
NSN-AC	4	Alart–Curnier formulation (63)
NSN-JM	4	Jean–Moreau formulation (61)
NSN-FB	4	Fischer–Burmeister formulation (79)
NSN-NM-GP	4 and 5	Natural map formulation (51) and the Goldstein–Price (GP) line search
NSN-AC-GP	4 and 5	Alart–Curnier formulation (63) and the Goldstein–Price (GP) line search
NSN-JM-GP	4 and 5	Jean–Moreau formulation (61) and the Goldstein–Price (GP) line search
NSN-FB-GP	4 and 5	Fischer–Burmeister formulation (79) and the Goldstein–Price (GP) line search
NSN-NM-A	4 and 6	Natural map formulation (51) and the Armijo(A) line search
NSN-AC-A	4 and 6	Alart–Curnier formulation (63) and the Armijo(A) line search
NSN-JM-A	4 and 6	Jean–Moreau formulation (61) and the Armijo(A) line search
NSN-FB-A	4 and 6	Fischer–Burmeister formulation (79) and the Armijo(A) line search
NSN-AC-HYBRID	4 and 2	Alart–Curnier formulation (63) with a pre computation of the initial guess with 100 iterations of EG-VI-UPK algorithm

$$u = Wr + q \tag{114}$$

is split along each contact as follows:

$$u^\alpha = W^{\alpha\alpha} r^\alpha + \sum_{\beta\neq\alpha} W^{\alpha\beta} r^\beta + q^\alpha, \qquad \text{for all } \alpha \in 1 \dots n_c, \tag{115}$$

where the matrices α and β are used to label the variable for each contact. The matrices $W^{\alpha\beta}$ with $\alpha \in 1, \dots, n_c$ and $\beta \in 1, \dots, n_c$ are easily identified from (114). From (115), a projected Gauss–Seidel (PGS) method is obtained by using the following update rule at the kth iterate:

$$u^\alpha_{k+1} = W^{\alpha\alpha} r^\alpha_{k+1} + \sum_{\beta<\alpha} W^{\alpha\beta} r^\beta_{k+1} + \sum_{\beta>\alpha} W^{\alpha\beta} r^\beta_k + q^\alpha, \text{ for all } \alpha \in 1 \dots n_c. \tag{116}$$

A Projected Successive Over Relaxation (PSOR) scheme is derived by introducing a relaxation parameter $\omega > 0$ such that

$$u^\alpha_{k+1} = \frac{1}{\omega} W^{\alpha\alpha} r^\alpha_{k+1} - \frac{1}{\omega} W^{\alpha\alpha} r^\alpha_k + \sum_{\beta<\alpha} W^{\alpha\beta} r^\beta_{k+1} + \sum_{\beta\geqslant\alpha} W^{\alpha\beta} r^\beta_k + q^\alpha, \text{ for all } \alpha \in 1 \dots n_c. \tag{117}$$

At the kth iteration, the following problem is solved for each contact α:

$$\begin{cases} u^\alpha_{k+1} = \bar{W}^{\alpha\alpha} r^\alpha_{k+1} + \bar{q}^\alpha_{k+1}, \\ \hat{u}^\alpha_{k+1} = u^\alpha_{k+1} + g(u^\alpha_{k+1}), \\ K^{\alpha,\star} \ni \hat{u}^\alpha_{k+1} \perp r^\alpha_{k+1} \in K^\alpha, \end{cases} \tag{118}$$

where

$$\begin{cases} \bar{W}^{\alpha\alpha} = \frac{1}{\omega} W^{\alpha\alpha} \\ \bar{q}^\alpha_{k+1} = -\frac{1}{\omega} W^{\alpha\alpha} r^\alpha_k + \sum_{\beta<\alpha} W^{\alpha\beta} r^\beta_{k+1} + \sum_{\beta\geqslant\alpha} W^{\alpha\beta} r^\beta_k + q^\alpha \end{cases}, \text{ for all } \alpha \in 1 \dots n_c. \tag{119}$$

The problem (118) has exactly the same structure as Problem FC, but is of lower size, since it is only for one contact. It is solved by a *local solver*, which can be any of the algorithms presented in this chapter or even an analytical method (enumerating all the possible cases, as in [16]).

The PSOR algorithm is summarized in Algorithm 7 and the NSGS correspond to the case $\omega = 1$.

Algorithm 7 PSOR algorithm for Problem FC

Require: W, q, μ
Require: r_0 initial values
Require: tol > 0, $\text{tol}_{\text{local}}$ tolerance values and $\text{iter}_{\max} > 0$, $\text{iter}_{\text{local}\max} > 0$ the max number of local iterations
Require: ω a relaxation parameter
Ensure: r, u solution of Problem FC
  while error $>$ tol and k $<$ $\text{iter}_{\max}$ **do**
    for $\alpha = 1 \dots n_c$ **do**
      $\bar{W}^{\alpha\alpha}_{k+1} \leftarrow \frac{1}{\omega} W^{\alpha\alpha}$
      $\bar{q}^\alpha_{k+1} \leftarrow -\frac{1}{\omega} W^{\alpha\alpha} r^\alpha_k + \sum_{\beta<\alpha} W^{\alpha\beta} r^\beta_{k+1} + \sum_{\beta\geqslant\alpha} W^{\alpha\beta} r^\beta_k + q^\alpha$
      Solve the single contact problem $\text{FC}(\bar{W}^{\alpha\alpha}, \bar{q}^\alpha_{k+1}, \mu)$ at accuracy $\text{tol}_{\text{local}}$ with a maximum of iteration $\text{iter}_{\text{local}\max}$
    end for
    Evaluate error.
    $k \leftarrow k+1$
  end while
  $r \leftarrow r_k$
  $u \leftarrow u_k$

Applications in frictional contact date back to the work of [82, 83] for two-dimensional friction. In [63], this method is developed in the Gauss–Seidel configuration ($\omega = 1$) with a local Newton solver based on the Alart–Curnier formulation. If the local solver performs only one iteration of the VI solver based on projection, we get a standard splitting technique for VI. In Table 3, the methods based on PSOR used in the comparison are summarized.

6.2 *Proximal Points Techniques*

The first use of the proximal idea dates back to the early days of convex analysis [88]. The proximity operator of a proper, lower semi-continuous function f is defined as

$$\text{prox}_f(x) = \min_z f(z) + \tfrac{\alpha}{2}\|z - x\|^2, \quad \alpha > 0 \tag{120}$$

and the point $\text{prox}_f(x)$ is called the *proximal point*. The latter is unique whenever f is convex. Recently, there has been a surge in the use of the proximity operator in optimization. There have been applications to non-differentiable, large-scale optimization, mainly because the proximity operator enjoys nicer property: it is differentiable and it may be easier to compute in some cases. There is a wealth of literature on the use of proximal mapping in optimization [95]. The basic idea is to replace (part of) the objective function with its proximal operator. Starting from an initial x_0, a proximal algorithm produces a sequence $\{x_k\}$ by the relation $x_{k+1} = \text{prox}_f(x_k)$. The sequence is guaranteed to converge whenever f is convex. This basic algorithm can be enhanced by a proper choice of the parameter α: some acceleration techniques ensure the convergence of the sequence $\{x_k\}$ with a different α at each iteration. In the non-convex case, the mapping prox_f is still well-behaved whenever f is said to be *prox-regular* and α is small enough.

The proximal mapping can also be defined for set-valued mappings. Then, it corresponds to a regularization of the (sub-) differential of f. More precisely, it correspond to the *resolvent* of ∇f, defined as $R_\alpha := \alpha(\alpha I + T)^{-1}$. Starting from an initial guess x_0, a sequence is computed as $x_{k+1} = R_\alpha(x_k)$ (assuming the single-valuedness of R_α). Much less attention has been given to this kind of algorithm, in particular, few numerical studies have been conducted. From the theoretical point of view, the convergence is shown in [104], when the mapping T is maximal monotone, an extension of the convex case previously mentioned. With the same hypothesis, the mapping R_α is single-valued for all $\alpha > 0$. For concreteness, consider the variational inequality

$$0 \in F(x) + N_X(x) \qquad \text{also written as} \quad 0 \in T(x). \tag{121}$$

Then, the proximal point algorithm applied to this VI consists in solving, at each step, the VI

$$0 \in F(x) + \alpha I - \alpha x_k + N_X(x) \tag{122}$$

that can be compactly written as

$$0 \in F_{\alpha,x_k}(x) + N_X(x). \tag{123}$$

The parameter α can be changed for each sub-VI.

Other variants of the basic algorithm can be derived, such as adding a relaxation parameter ω:

$$x_{k+1} = (1 - \omega)x_k + \omega z_{k+1}, \tag{124}$$

where z_{k+1}. The algorithm is described in Algorithm 8.

Algorithm 8 Proximal point algorithm for the VI (37)

Require: F, X Data of VI (37)
Require: ω relaxation parameter
Require: α_0 the initial value of the proximal point parameter
Require: x_0 initial value
Require: tol > 0, tol_{in} tolerance values and $iter_{max} > 0$ the max number of iterations
Ensure: x solution of VI (37)

```
k ← 0
while error > tol and k < iter_max do
    Solve VI(F_{α_k,x_k}, X) for z_{k+1} at accuracy tol_int
    x_{k+1} ← (1 − ω)x_k + ωz_{k+1}
    Evaluate error.
    Compute α_{k+1}
    k ← k + 1
end while
x ← x_{k+1}
```

For solving the sub-problems $\mathrm{VI}(F_{\alpha,x_k}, X)$, any of the previous algorithms for VI can be used. The main interest of the proximal point algorithm is that the mapping F_{α,x_k} is nicer than F. For instance, if $F(x) = Mx + q$, then the matrix in $F(\alpha, x_k)$ is $M + \alpha I$. It is easy to see that for large enough α, $M + \alpha I$ is positive-definite, with no assumption about M. With a nonlinear operator, choosing large enough α ensures that F_{α,x_k} is monotone (with some condition on F). In practice, this implies that a greater number of algorithms are capable of solving the VI. This is a good indicator of an easier problem to solve, and we observe that this approach is able to provide some robustness to the VI-based approaches. The introduction of two additional parameters (α and tol_{int}) is the main drawback of this approach. Indeed, instead of solving just one VI, this approach calls for solving multiple sub-VIs. This additional computational effort can be reduced in two ways: the first one is to drive the proximal parameter α_k as quickly as possible to zero, in order to reduce the number of sub-VIs to solve. The other option is to set the tolerance tol_{int} to a higher value when α_k is large, so as to reduce the computational effort for the sub-VIs. The choice of tol_{int} is discussed in Sect. 6.3.

6.3 *Control of the Tolerance of Internal Solvers* tol_{int} *and* tol_{local} *in the Splitting and Proximal Approaches*

In Algorithms 7 and 8, an internal tolerance is used to control the required accuracy of the internal solver. It is generally not useful to solve the internal problem at the accuracy of the global one. For Algorithm 7, the local tolerance tol_{local} is set by default to a very low value of 10^{-14}. An adaptive local tolerance strategy has also been tested that sets the local tolerance to a fraction of the current error as, for instance, $\mathsf{tol}_{\mathsf{local}} = \mathsf{error}/10$. For the proximal point algorithm in Algorithm 8, the internal tolerance $\mathsf{tol}_{\mathsf{int}}$ is set to a fraction of the error $\mathsf{tol}_{\mathsf{int}} = \mathsf{error}/10$.

6.4 Control of the Proximal Point Parameter α_k

In Algorithm 8, the proximal point parameter α_k is updated for each sub-VI. We choose to implement two rules for its computation. The first one is inspired by the work in [45], which is based on the current error or residual of the algorithm. The parameter is computed thanks to the following rule:

$$\alpha_k = \sigma(\mathsf{error})^\nu, \tag{125}$$

where $\sigma > 0$, $\nu > 0$ are two additional parameters that influence the rate of driving α_k to zero. The other rule is an heuristic rule that starts from a given value of α_0. If the internal solver for the sub-VI succeeds in reaching the required accuracy, then α_{k+1} is decreased and set to $\alpha_{k+1} = \alpha_k/10$. If the internal solver does not succeed, then we increase α_{k+1} as $\alpha_{k+1} = 5\alpha_k$.

6.5 Nomenclature

A nomenclature for the algorithms based on the projection/splitting approach is given in Table 3.

Table 3 Naming convention for the algorithms based on splitting and proximal algorithms

Name	Algorithm	Additional information
NSGS-AC	7 with $\omega = 1$	Local solver: NSN-AC with tolerance $\mathsf{tol}_{\mathsf{local}}$
NSGS-JM	7 with $\omega = 1$	Local solver: NSN-JM with tolerance $\mathsf{tol}_{\mathsf{local}}$
NSGS-AC-GP	7 with $\omega = 1$	Local solver: NSN-AC-GP with tolerance $\mathsf{tol}_{\mathsf{local}}$
NSGS-JM-GP	7 with $\omega = 1$	Local solver: NSN-JM=GP with tolerance $\mathsf{tol}_{\mathsf{local}}$
NSGS-FP-DS-One	7 with $\omega = 1$	Local solver: one iteration of FP-DS
NSGS-FP-VI-UPK	7 with $\omega = 1$	Local solver: FP-VI-UPK with tolerance $\mathsf{tol}_{\mathsf{local}}$
NSGS-EXACT	7 with $\omega = 1$	Exact local solver
PSOR-AC	7	Local solver: NSN-AC with tolerance $\mathsf{tol}_{\mathsf{local}}$
PPA-NSN-AC	8	Internal solver: NSN-AC solver
PPA-NSGS-AC	8	Internal solver: NSGS-AC

7 Optimization-Based Methods

In this section, the Delassus matrix is assumed to be symmetric in order to be able to state simple convex optimization problems.

7.1 Alternating Optimization Problem

The Panagiotopoulos approach described in Sect. 3.4 generates a family of solvers by choosing two specific solvers for the normal contact problem (88) and the tangential contact problem (89), respectively. This method may be viewed as a two-block Gauss–Seidel method (as pointed out by [116]). More precisely, the following choices may be made for the normal and tangent problems.

The normal contact problem

$$\begin{cases} \min \ \frac{1}{2} r_{\mathrm{N}}^\top W_{\mathrm{NN}} r_{\mathrm{N}} + r_{\mathrm{N}}^\top \tilde{q}_{\mathrm{N}} \\ \text{s.t. } r_{\mathrm{N}} \geqslant 0 \end{cases} \quad \text{with } \tilde{q}_{\mathrm{N}} = q_{\mathrm{N}} + W_{\mathrm{NT}} r_{\mathrm{T},k} \tag{126}$$

is a convex quadratic program with simple bound constraints. In the literature, a large number of solvers has been developed to solve such problems. Among others, we might cite the active set strategy solvers [40, 92], which are mainly dedicated to small–scale systems, the projected gradient [19] and projected conjugate gradient methods [86, 87], which are more dedicated to large–scale systems. Note that there also exists a wealth of methods in the literature that improves the methods of [86] for large–scale systems. For the reader interested in those details, we refer to the book by [33](see especially Sect. 8.7 for a review of the different approaches). It is clear that we might also use semi-smooth Newton methods or interior point methods, but our experience has shown that such methods are not efficient when $\ker(W_{\mathrm{NN}}) \neq \{0\}$. The optimality conditions of this quadratic problem reduced to a linear complementarity problem with a semi-definite matrix. In that case, it is also possible to solve the problem with PSOR techniques with line-searches. Due to space constraints, we decided in this work to use the projected Gauss–Seidel (PGS) algorithm and the projected gradient algorithm of [19] to solve the normal problem described. The projected gradient algorithm solved the following QP for a convex set C:

$$\begin{cases} \min \ q(r) := \frac{1}{2} r^\top W r + r^\top b \\ \text{s.t. } r \in C, \end{cases} \tag{127}$$

with the algorithm described in Algorithm 9.

Algorithm 9 Projected gradient algorithm for QP (127)

Require: W, b that defines $q(r)$
Require: C a convex set
Require: r_0 initial values
Require: tol > 0 a tolerance value and $iter_{max} > 0$ the maximum number of iterations
Require: $\rho_0 > 0, l, \sigma \in (0, 1)$
Require: i_{lsmax} maximum number of line-search iterations
Ensure: r solution of Problem (127)

```
r_k ← r_0 ; θ_0 ← q(r_0) ; k ← 0
while error > tol and k < iter_max do
  Armijo like–search procedure
  i_ls ← 0
  while criterion > 0 and i_ls < i_lsmax do
    ρ ← ρ_0 l^{i_ls}
    r ← P_C(r_k − ρ(M r_k + b))
    θ ← q(r)
    criterion ← θ − θ_k − σ(M r_k + b)^⊤ (r − r_k)
    i_ls ← i_ls + 1
  end while
  r_k ← r ; θ_k ← q(r);
  evaluate error.
end while
```

The tangential problem

$$\begin{cases} \min \dfrac{1}{2} r_{\mathrm{T}}^\top W_{\mathrm{TT}} r_{\mathrm{T}} + r_{\mathrm{T}}^\top \tilde{q}_{\mathrm{T}} \\ \text{s.t. } r_{\mathrm{T}} \in D(\mu \tilde{r}_{\mathrm{N}}) \end{cases} \quad \text{with } \tilde{q}_{\mathrm{T}} = q_{\mathrm{T}} + W_{\mathrm{TN}} r_{\mathrm{N},k+1}, \tag{128}$$

is also a convex program but with a more complex structure, since the constraints are quadratic. There exists a dedicated algorithm, as in [34], for QP with convex constraints. Earlier application of projected gradient and projected gradient techniques for the frictionless problem can also be found in [14], including a comparison with PSOR techniques.

In this section, we will use either (a) a reformulation of the optimality conditions of this problem as a variational inequality, applying the fixed point algorithm and the extra gradient algorithm of Sect. 4, or (b) an adaptation of one of the splitting techniques detailed in Sect. 6. The algorithm is described in Algorithm 10. In Table 4, we detailed the algorithms used in the present study.

Algorithm 10 Panagiotopoulos decomposition algorithm for Problem FC

Require: W, q, μ
Require: r_0 initial values
Require: tol > 0, tol_{int} tolerance values and $iter_{max} > 0$ the max number of iterations
Ensure: r, u solution of Problem FC
  $r_k \leftarrow r_0$; $k \leftarrow 0$
  while error $>$ tol and $k < iter_{max}$ **do**
    $\tilde{q}_N \leftarrow q_N + W_{NT} r_{T,k}$
    solve (126) for $r_{N,k+1}$ at accuracy tol_{int}
    $\tilde{q}_T \leftarrow q_T + W_{TN} r_{N,k+1}$
    solve (128) for $r_{T,k+1}$ at accuracy tol_{int}
    $k \leftarrow k+1$
    evaluate error.
  end while
  $r \leftarrow r_k$
  $u \leftarrow Wr + q$

7.2 Successive Approximation Method

The method of successive approximation is a natural tool for the numerical realization of Problem FC. It is based on the Tresca approximation of the Coulomb cone, as described in Sect. 3.4, and the work of the celebrated Czech school [48, 49, 51, 91]. Each iterative step is represented by an auxiliary contact problem with a given friction threshold described by a quadratic program over a cylinder (93), which we recall there:

$$\begin{cases} \theta = h(r_{\rm N}) \\ \min \ \frac{1}{2} r^\top W r + r^\top q \\ \text{s.t.} \ \ r \in C(\mu, \theta). \end{cases} \tag{129}$$

The radius of the cylinder is then updated in an iterative procedure. The algorithm is described in Algorithm 11.

Algorithm 11 Tresca approximation algorithm for Problem FC

Require: W, q, μ
Require: r_0 initial values
Require: tol > 0, tol_{int} tolerance values and $iter_{max} > 0$ the max number of iterations
Ensure: r, u solution of Problem FC
  $r_k \leftarrow r_0$; $k \leftarrow 0$
  while error $>$ tol and $k < iter_{max}$ **do**
    $\theta \leftarrow h(r_{N,k})$
    solve (129) for $r_{N,k+1}, r_{T,k+1}$ at accuracy tol_{int}
    $k \leftarrow k+1$
    evaluate error.
  end while
  $r \leftarrow r_k$
  $u \leftarrow Wr + q$

In the literature, the successive approximation technique has been used in the bidimensional case in [37, 52] with improved and dedicated QP solvers over box-constraints. Two strategies are implemented: (a) the classical Tresca iteration (called FPMI) and (b) the Panagiotopoulos decomposition plus a Fixed point (called FPMII). They use a specific QP solver for box constraint [32], which is an improvement of the Moré–Toraldo method [86]. This technique has been directly extended in the three-dimensional case with a faceting of the cone in [53]. In the latter case, the problem is still a box-constrained QP, since it contains only polyhedral constraints. In [54], the authors propose a successive approximation technique in 3D with the special solver of [74, 75], which is itself an extension to disk constraints of the Polyak method (conjugate gradient with an active set on the bounds constraint) and its improvements [32, 36]. Other improvements of the method may be found in [35] with a last improvement of the method in [34]. All this work is summarized and detail in [33].

7.3 ACLM Approach

In the convex SOCCP approach described in Sect. 3.4, we have to solve, for a given value, the following problem:

$$\begin{cases} \min \dfrac{1}{2} r^\top W r + r^\top (q+s) \\ \text{s.t.} \quad r \in K \end{cases} \tag{130}$$

which is again a convex quadratic program over a second–order cone. The approach listed above could again be used to solve this problem. In this work, we solve it by three different ways: (a) an adaptation of one of the splitting techniques detailed in Sect. 6, (b) using the projected gradient algorithm dedicated to convex QP described in Algorithm 9 or (c) the fixed point algorithm and the extra gradient algorithm of Sect. 4 . The algorithm is described in Algorithm 12 and we detailed the algorithms we use in the present study in Table 4.

Algorithm 12 ACLM approximation algorithm for Problem FC

Require: W, q, μ
Require: r_0 initial values
Require: tol > 0, tol_{int} tolerance values and $\text{iter}_{\max} > 0$ the max number of iterations
Ensure: r, u solution of Problem FC

```
u_0 ← W r_0 + q ; k ← 0
while error > tol and k < iter_max do
    s ← g(u_k)
    solve (130) for r_{k+1} at accuracy tol_int
    u_{k+1} ← W r_{k+1} + q
    k ← k + 1
    evaluate error.
end while
r ← r_k
u ← u_k
```

Table 4 Naming convention for optimization based algorithms

Name	Algorithm	Additional information
PANA-PGS-FP-VI-UPK	10	The normal problem is solved with a PGS algorithm and the tangent problem is solved with the FP-VI-UPK algorithm
PANA-PGS-EG-VI-UPK	10	The normal problem is solved with a PGS algorithm and the tangent problem is solved with the EG-VI-UPK algorithm
PANA-PGS-CONVEXQP-PG	10 and 9	The normal problem is solved with a PGS algorithm and the tangent problem is solved with Algorithm 9
PANA-CONVEXQP-PG	10 and 9	Both normal and tangent problems are solved with Algorithm 9
TRESCA-NSGS-FP-VI-UPK	11	The problem (129) is solved with the FP-VI-UPK algorithm
TRESCA-FP-VI-UPK	11 and 1	The problem (129) is solved with the FP-VI-UPK algorithm
TRESCA-EG-VI-UPK	11 and 2	The problem (129) is solved with the EG-VI-UPK algorithm
TRESCA-CONVEXQP-PG	11 and 9	The problem (129) is solved with Algorithm 9
ACLM-NSGS-FP-VI-UPK	12	The problem (130) is solved with the NSGS-FP-VI-UPK algorithm
ACLM-FP-VI-UPK	12 and 1	The problem (130) is solved with the FP-VI-UPK algorithm
ACLM-EG-VI-UPK and 2	12	The problem (130) is solved with the EG-VI-UPK algorithm
ACLM-CONVEXQP-PG	12	The problem (130) is solved with the Algorithm 9

A nomenclature for the algorithms based on the optimisation approach is given in Table 4.

7.4 Convex Relaxation and the SOCCP Approach

Finally, we propose comparing the optimization-based algorithm to a complete convex relaxation of the problem by solving the convex SOCCP (130) with $s = 0$. This procedure is very similar to the approach in [12, 112, 113], in which only the convex problem is solved.

7.5 *Control of the Tolerance of Internal Solvers* $\mathsf{tol}_{\mathsf{int}}$ *in the Optimization Approach*

In Algorithms 10, 11 and 12, an internal tolerance is used to control the accuracy of the internal solver. It is generally not useful to solve the internal problem at the accuracy of the global one. In the comparison study, we set the internal tolerance $\mathsf{tol}_{\mathsf{int}}$ to $\mathsf{error}/10$.

8 Comparison Framework

In this section, we present our comparison framework. In particular, we specify how the performance is measured and how the performance profiles are built.

8.1 *Measuring Errors*

A key parameter in the measurement of performance of the solver is the definition of the error. The absolute error is given by the norm of the natural map. A relative error is computed with respect to the norm of the vector q. More precisely, the error is given by

$$\mathrm{error} = \frac{\|F^{\mathsf{nat}}_{\mathrm{vi}}(r)\|}{\|q\|}, \tag{131}$$

assuming that $\|q\|$ is larger than the machine accuracy. If not, we may assume that $q = 0$, and a trivial solution can be computed. For all solvers, the error in (131) is compared to the required tolerance tol given by the user.

For some iterative solvers such as VI-FP, VI-EG, NSGS and PSOR, the computation of the error (131) at each iteration penalizes the performance of the solver: it amounts to computing a matrix-vector product, an operation that is more computationally expensive than one iteration of the solver. Hence, a cheaper error measurement is used inside the main loop in Algorithms 1, 2 and 7. This cheaper error measurement is given by

$$\mathrm{error}_{\mathrm{cheap}} = \frac{\|r_{k+1} - r_k\|}{\|r_k\|}. \tag{132}$$

The tolerance of the solver is then self-adapted in the loop to meet the required tolerance based on the error given by (131).

8.2 Performance Profiles

The concept of performance profiles was introduced in [31] for bench-marking optimization solvers over a large set of problems. For a set P of n_p problems, and a set S of n_s solvers, we define a performance criterion for a solver s, a problem p and a required precision tol by

$$t_{p,s} = \text{computing time required for } s \text{ to solve } p \text{ at precision } \mathsf{tol}. \tag{133}$$

A performance ratio over all the solvers is defined by

$$r_{p,s} = \frac{t_{p,s}}{\min\{t_{p,s},\ s \in S\}} \geqslant 1\,. \tag{134}$$

For $\tau \geqslant 1$, we define a distribution function ρ_s for the performance ratio for a solver s as

$$\rho_s(\tau) = \frac{1}{n_p}\text{card}\{p \in P,\ r_{p,s} \leqslant \tau\} \leqslant 1. \tag{135}$$

This distribution computes the number of problems p that are solved with a performance ratio below a given threshold τ. In other words, $\rho_s(\tau)$ represents the probability that the solver s has a performance ratio no larger than a factor τ of the best solver. It is worth noting that $\rho_s(1)$ represents the probability that the solver s beats the other solvers, and $\rho_s(\tau)$ characterizes the robustness of the method for large values of τ. To summarize: the higher ρ_s is, the better the method is. In the sequel, the term *performance profile* denotes a graph of the functions $\rho_s(\tau)$, $\tau \geqslant 1$.

The computational time is used to measure performance in (133). Other criteria can be used, like the number of floating point operations (flops). It is a better measure of performance, since it is independent of the computer. Unfortunately, it is usually difficult to measure in an automatic and robust way over various platforms. Whence, we stick with the computational time.

In our experiments, we decided to fix the required accuracy with the tolerance of each solver. Another performance criteria could also be used: for instance, a timeout could be defined and the metric would be the error at that time. This is a way to measure the ability of a solver to give an approximate solution within a prescribed time limit that may be interesting for real-time applications. Another way to measure performance may also be to divide the computational time by the number of contacts in order to judge the ability of the solver to be scalable. For the sake of conciseness, this has not been done in this chapter.

8.3 Benchmarks Presentation

To perform the comparison of the solvers on a fair basis, we use a large set of problems that comes from various applications. This collection is FCLib (Frictional

Contact library), which is an open collection of problems in a hdf5 format described in [4].[3] In this work, we used the version `v1.0` for the comparisons that contains 2368 problems.[4]

The test sets are illustrated in Fig. 3 and details on each test are given in Table 5. All the problems have been generated thanks to the software codes LMGC90[5] and Siconos. In Table 5, the number of degrees-of-freedom n corresponds to the degrees-of-freedom of the system before its condensation (or reduction) to local variables. In other words, the number of rows of the matrix M and H in (1). The contact density c is the ratio of the number of contact unknowns over the number of degrees-of-freedom:

$$c = \frac{3n_c}{n} = \frac{m}{n}. \tag{136}$$

The coefficient c also corresponds also to the ratio between the number of rows of H over its number of columns. If this number is larger than 1, the matrix H can not be full row rank, and thus the matrix W is also rank deficient. Whenever $m > n$, we can observe, in Table 5, that this number c is a good approximation of the rank ratio of the matrix W in our applications. The estimation of the rank of matrix W shows that it is very close to the number of degrees-of-freedom of the system when $c > 1$. For $c \ggg 1$, the contact density is really high and the system suffers from hyperstaticity as we discussed in Sect. 2.3. In Table 5, we also give an estimation of the conditioning of the matrix W. When it was possible from a computational point of view, we performed a singular value decomposition (SVD) of the matrix W to estimate the spectral radius, and then the conditioning, by cutting the small eigenvalues. This process has two drawbacks. Firstly, the computation of the SVD decomposition can be really expensive for large dense matrices. Secondly, the value of the condition number of the matrix is very sensitive to the threshold for cutting off the small eigenvalues. This is the reason why we also use the LSMR [41] algorithm to give a better approximation of the condition number of the rank deficient matrix.

The four first tests in Table 5, Cubes_H8_2, Cubes_H8_5, Cubes_H8_20 and LowWall_FEM, are examples that involve flexible elastic bodies meshed by finite element methods. Due to a consistent choice of the space-discretization of the contact surfaces, the Delassus matrix W in that case is full rank. In the sequel, we will call these sets of examples the *flexible test sets*.

8.4 Software and Implementation Details

All the solvers that are used in this chapter are implemented in standard C99 in the component of the open source software Siconos called numerics. The aim of Siconos is to provide a common platform for the modeling, simulation, analysis and

[3]More information can be found at https://frictionalcontactlibrary.github.io.

[4]The whole collection of problems can be found at https://github.com/FrictionalContactLibrary/fclib-library.

[5]https://git-xen.lmgc.univ-montp2.fr/lmgc90/lmgc90_user/wikis/home.

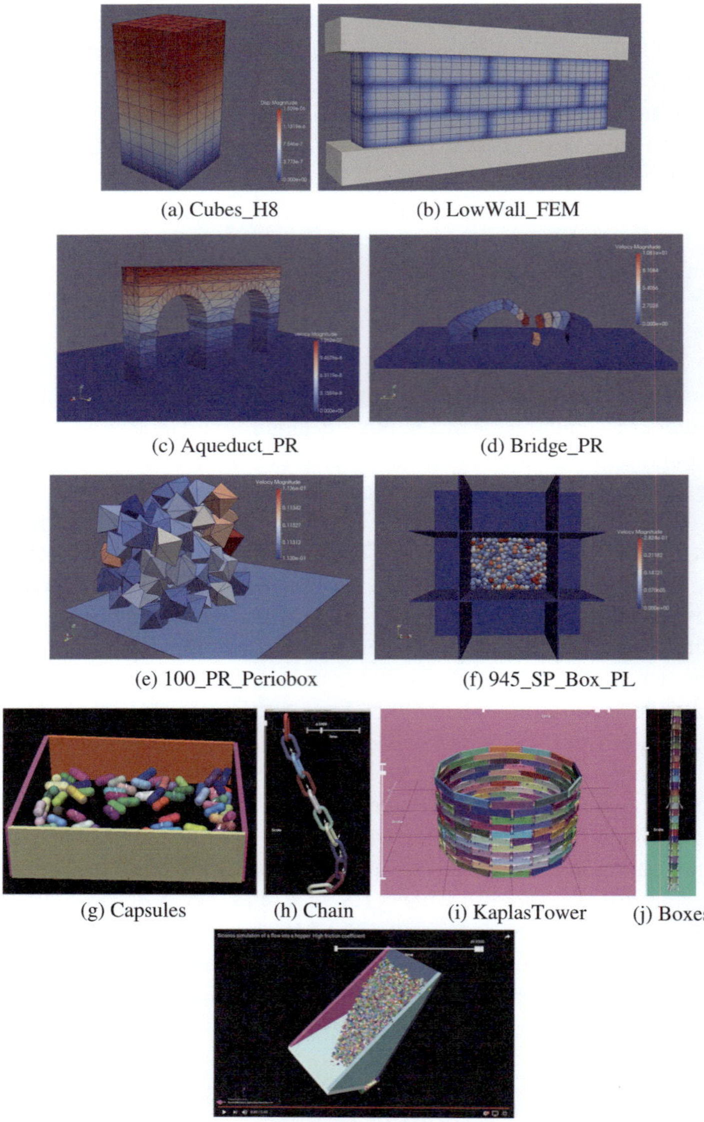

(a) Cubes_H8 (b) LowWall_FEM

(c) Aqueduct_PR (d) Bridge_PR

(e) 100_PR_Periobox (f) 945_SP_Box_PL

(g) Capsules (h) Chain (i) KaplasTower (j) BoxesStack

(k) Chute_1000,Chute_4000,Chute_local_problems

Fig. 3 Illustrations of the FClib test problems

Table 5 Description of the test sets of FCLib library (v1.0)

Test set	Code	Friction coefficient μ	# of problems	# of d.o.f.	# of contacts	Contact density c	Rank ratio(W)	cond(W)	cond(W) LSMR	$\mu(\|W - W^T\|)$ symmetry of W
Cubes_H8_2	LMGC90	0.3	15	162	[3 : 5]	[0.02 : 0.09]	1	[$2.2.10^1$: $1.3.10^3$]	[$8.1.10^5$: $1.5.10^6$]	$3.2.10^{-4}$
Cubes_H8_5	LMGC90	0.3	50	1296	[17 : 36]	[0.02 : 0.09]	1	[$3.3.10^4$: $7.2.10^4$]	[$1.3.10^6$: $3.1.10^6$]	$4.2.10^{-4}$
Cubes_H8_20	LMGC90	0.3	50	55566	[361 : 388]	[0.019 : 0.021]	1	[$2.4.10^5$: $2.5.10^5$]	[$1.3.10^6$: $5.2.10^6$]	$5.2.10^{-5}$
LowWall_FEM	LMGC90	0.83	50	{7212}	[624 : 688]	[0.28 : 0.29]	1	–	[$9.3.10^2$: $5.0.10^5$]	$5.2.10^{-2}$
Aqueduct_PR	LMGC90	0.8	10	{1932}	[4337 : 4811]	[6.81 : 7.47]	[6.80 : 7.46]	[$4.7.10^7$: $3.4.10^8$]	[$6.7.10^1$: $1.5.10^2$]	$1.1.10^{-15}$
Bridge_PR	LMGC90	0.9	50	{138}	[70 : 108]	[1.5 : 2.3]	[2.27 : 2.45]	[$8.3.10^4$: $1.1.10^5$]	[$1.9.10^3$: $2.6.10^4$]	$5.8.10^{-18}$
100_PR_Periobox	LMGC90	0.8	106	{606}	[14 : 578]	[0.2 : 3]	[1.76 : 3.215]	[$4.3.10^2$: $1.0.10^6$]	[$6.3.10^5$: $3.5.10^6$]	$8.8.10^{-20}$
945_SP_Box_PL	LMGC90	0.8	60	{5700}	[2322 : 5037]	[1.22 : 2.65]	[1.0 : 2.66]	[$2.2.10^4$: $4.4.10^5$]	[$2.9.10^1$: $9.2.10^2$]	$1.3.10^{-10}$
Capsules	Siconos	0.7	249	[96:600]	[17 : 304]	[0.53 : 1.52]	[1.08 : 1.55]	–	[4.8 : $1.6.10^2$]	$3.3.10^{-02}$
Chain	Siconos	0.3	242	{60}	[8 : 28]	[0.5 : 1.3]	[1.05 : 1.6]	[$7.4.10^4$: $4.0.10^9$]	[$1.5.10^1$: $4.7.10^5$]	$3.7.10^{-02}$
KaplasTower	Siconos	0.7	201	[72 : 792]	[48 : 933]	[3.0 : 3.6]	[2.0 : 3.53]	[67 : 2174]	[8 : 67]	$5.4.10^{-08}$
BoxesStack	Siconos	0.7	255	[6 : 300]	[1 : 200]	[1.86 : 2.00]	[1.875 : 2.0]	[$3.8.10^4$: $2.5.10^7$]	[9.0 : $5.4.10^3$]	$2.23.10^{-14}$
Chute_1000	Siconos	1.0	156	[276 : 5508]	[74 : 5056]	[0.69 : 2.95]	[1.0 : 2.95]	[$2.1.10^1$: $1.9.10^3$]	$6.6.10^{-02}$	
Chute_4000	Siconos	1.0	40	[17280 : 20034]	[15965 : 19795]	[2.51 : 3.06]	–	–	[$5.5.10^1$: $9.0.10^3$]	$8.9.10^{-14}$
Chute_local_problems	Siconos	1.0	834	3	1	1	1	[1.04 : 4.66]	[2.6 : $2.6.10^1$]	$1.76.10^{-09}$

Table 6 Parameters of the simulation campaign

Test set	Precision	Prescribed time limit (s)	Mean performance of the fastest solver $\mu\{\min\{t_{p,s}, s \in S\}\}$	Std. deviation performance of the fastest solver $\sigma(\min\{t_{p,s}, s \in S\})$	Mean performance of the fastest solver by contact $\mu\{\min\{t_{p,s} / n_{c,p}, s \in S\}\}$	Std. deviation performance of the fastest solver by contact $\sigma(\min\{t_{p,s}/n_{c,p}, s \in S\})$	# of unsolved problems
Cubes_H8_⋆	10^{-08}	100	1.73	2.13	4.83^{-03}	5.78^{-03}	0
Cubes_H8_⋆ II	10^{-04}	100	0.92	1.06	2.66^{-03}	2.83^{-03}	0
LowWall_FEM	10^{-08}	400	13.1	3.50	1.91^{-02}	5.09^{-03}	0
LowWall_FEM II	10^{-04}	400	14.8	2.85	2.16^{-02}	4.54^{-03}	0
Aqueduct_PR	10^{-04}	200	5.80	6.36	4.90^{-04}	3.03^{-04}	0
Bridge_PR	10^{-08}	400	10.3	12.9	1.23^{-01}	2.88^{-01}	0
Bridge_PR II	10^{-04}	100	0.048	0.038	1.30^{-03}	1.42^{-03}	0
100_PR_Periobox	10^{-04}	100	0.064	0.062	1.56^{-04}	1.22^{-04}	0
945_SP_Box_PL	10^{-04}	100	3.20	1.71	6.45^{-04}	3.36^{-04}	0
Capsules	10^{-08}	50	$1.46.10^{-02}$	$1.74.10^{-02}$	5.67^{-05}	6.26^{-05}	0
Chain	10^{-08}	50	$6.19.10^{-04}$	$3.68.10^{-04}$	$3.15.10^{-05}$	$1.46.10^{-05}$	0
KaplasTower	10^{-08}	200	$1.27.10^{-01}$	$3.75.10^{-01}$	$1.84.10^{-04}$	$4.57.10^{-04}$	0
KaplasTower II	10^{-04}	100	$2.84.10^{-02}$	$1.51.10^{-01}$	$3.39.10^{-05}$	$1.84.10^{-04}$	0
BoxesStack	10^{-08}	100	$3.42.10^{-02}$	$8.87.10^{-02}$	$3.24.10^{-04}$	$9.77.10^{-04}$	0
Chute_1000	10^{-04}	200	2.62	3.06	6.76^{-04}	6.58^{-04}	0
Chute_4000	10^{-04}	200	10.52	7.88	5.71^{-04}	4.07^{-04}	0
Chute_local_problems	10^{-08}	10	$1.80.10^{-04}$	$1.57.10^{-05}$	$1.80.10^{-04}$	$1.57.10^{-05}$	0

control of general nonsmooth dynamical systems.[6] The linear algebra operations are based on BLAS/LAPACK. The algorithms VI-FP, VI-EG, NSGS and PSOR use the sparse block structure of the Delassus matrix W. The NSN solvers relies on a standard sparse implementation given by csparse.[7] We solve linear systems with the LU factorization method embedded in csparse. The simulations are performed on the University of Grenoble-Alpes cluster CIMENT.[8]

8.5 Simulation Campaign

The simulation campaign is described in Table 6. For some test sets, two simulation runs have been performed with different precisions and prescribed time limits. A trade-off between the time limit and the precision has been chosen such that all the problems of the test sets are solved by at least one solver. In Sects. 9 and 10, we report the results for the simulation campaign, which includes more that 27000 runs. Given this wealth of data, we have chosen not to report profiles in which a family of solvers failed to solve the instances in this chapter.[9]

9 Comparison of Methods by Family

In this section, we perform a comparison of the solvers by family. The goal is to study the influence of the various parameters and possible strategies on the performance of the solvers.

9.1 *Numerical Methods for VI:* FP-DS, FP-VI-$\star$ *and* FP-EG-$\star$

In Fig. 4, we compare the different VI numerical solvers described in Sect. 4. With the exception of the FP-DS solver, the solvers FP-VI-$\star$ and FP-EG-$\star$ are very robust. Nevertheless, they are quite slow to converge in practice for large problems and/or with tight tolerances. Only the test sets for which the solvers have reached the precision before the prescribed time limit are presented. For that reason, the results for the test sets LowWall_FEM, LowWall_FEM II, Cubes_H8, Bridge_PR,

---

[6]More information on the software is available at http://siconos.gforge.inria.fr and the software can be downloaded at https://github.com/siconos/siconos.

[7]http://people.sc.fsu.edu/~jburkardt/c_src/csparse/csparse.html.

[8]https://ciment-grid.ujf-grenoble.fr/.

[9]Nevertheless, the reader can have access to the complete list of performance profiles at https://github.com/siconos/faf/blob/master/TeX/Full-test/full-test_current.pdf.

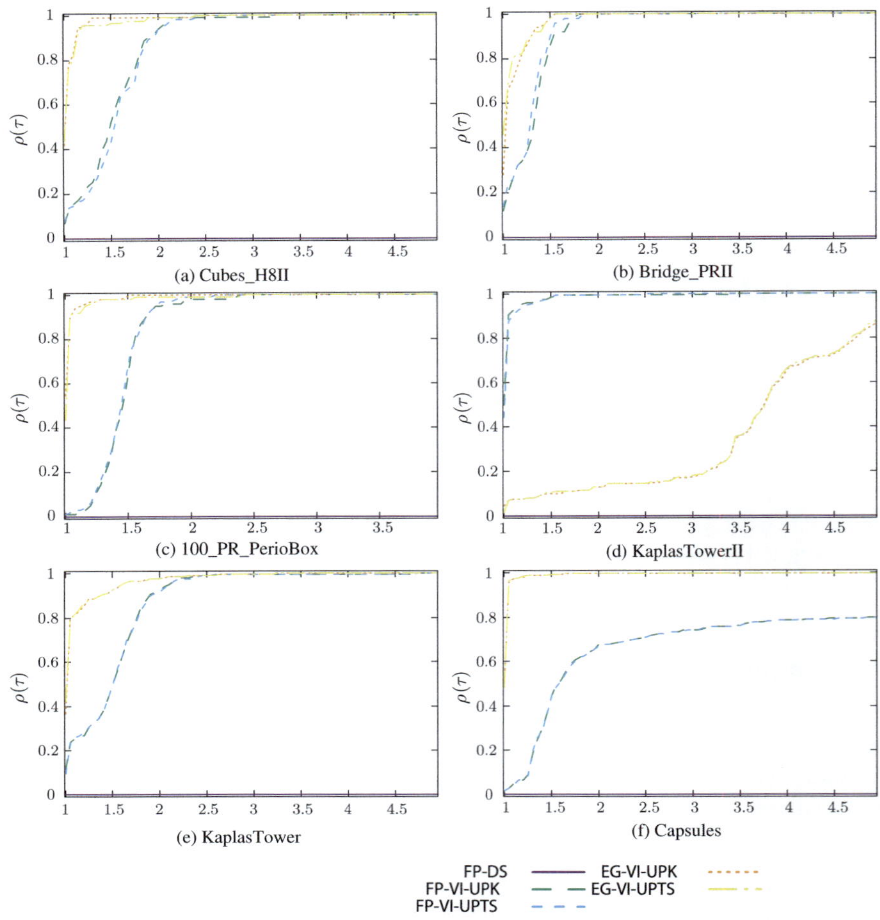

Fig. 4 Comparison of numerical methods FP-DS, FP-VI-⋆ and FP-EG-⋆

AqueducPR, 945_SP_Box_PL, BoxesStack, Chute_4000 and Chute_1000 are not depicted. The main conclusions are as follows:

1. The solver FP-DS suffers from robustness problems and a lot of divergence has been observed. This is mainly due to the fact that we set a priori the ρ parameter in Algorithm 1 to a fixed value equal to 1, independently of the problem.
2. The solvers FP-VI-⋆ and FP-EG-⋆ are really robust but slow. They are able to solve all the problems, but they require a lot of time. We did not observe divergence issues on all the test sets for these solvers. Compared with FP-DS, the self-adaptive rule for sizing the parameter ρ_k is of utmost importance for the robustness and the convergence rate.

3. Except for the test set KaplasTower II, the FP-EG-$\star$ performs better than FP-VI-$\star$. Otherwise, the performances are quite similar, since we plot the performance for a quite narrow range of values of $\tau \in [1, 5]$
4. The difference between the adaptive strategies for sizing ρ_k, UPK and UPTS, is negligible in all the test sets. Therefore, the choice of the update rule is not really important.

9.2 *Splitting-Based Algorithms:* NSGS-$\star$ *and* PSOR-$\star$

In this section, we compare the family of solvers based on splitting and relaxation techniques described in Sect. 6.1. Firstly, we start by comparing the choice of the local solvers in NSGS-$\star$, and then the effect of the local tolerance $\mathsf{tol}_{\mathsf{local}}$. Secondly, we study the influence of the order of the contact list. Finally, we study the effect of the relaxation parameter ω in PSOR-$\star$ solvers.

Influence of the Local Solver in NSGS-$\star$ Algorithms

In Fig. 5, we report the performance profiles of the NSGS-$\star$ for the different local solvers. The main conclusions are:

1. When the prescribed time limit is sufficiently large and the tolerance is low (10^{-4}), we observe that the NSGS-$\star$ solvers are robust. Indeed, we are able to find a local solver for each test set that is able to give a solution at the required accuracy. Nevertheless, there is no universal efficient local solver that outperforms the other ones.
2. When the tolerance is equal to 10^{-8}, the NSGS-$\star$ solvers have some difficulty in reaching convergence for all the problems within the prescribed time limit. This is the case for the test sets LowWall_FEM, Cubes_H8, Bridge_PR, Chain, Capsules and BoxesStack. Generally, the convergence is so slow that it is difficult to reach tight tolerance within a reasonable time limit.
3. With the exception of the test sets KaplasTower II and BoxesStack, the solver NSGS-EXACT behaves poorly. This is mainly due to the fact that the local solver is not robust enough to find a solution when the unknowns are far from the global solution for all the other contacts. This behavior was already reported in [25], in which another solver based on a nonsmooth Newton technique is used when the exact solution is not satisfactory.
4. The NSGS-FP-DS-One solver is most efficient on the test sets Bridge_PR II, KaplasTower II, Chain and BoxesStack. In these tests sets, a portion of the problems seems easier to solve and the NSGS-FP-DS-One solver seems sufficient to get a global convergence. Nevertheless, this local solver seems slow or suffers from robustness issues for other test sets.
5. On the flexible test sets, Cubes_H8_$\star$, LowWall_FEM and the rigid test sets 945_SP_Box_PL and Chute_4000, the best solver is NSGS-FP-VI-UPK for a relatively low required tolerance ($\mathsf{tol}_{\mathsf{local}} = 10^{-06}$). For these test sets, an

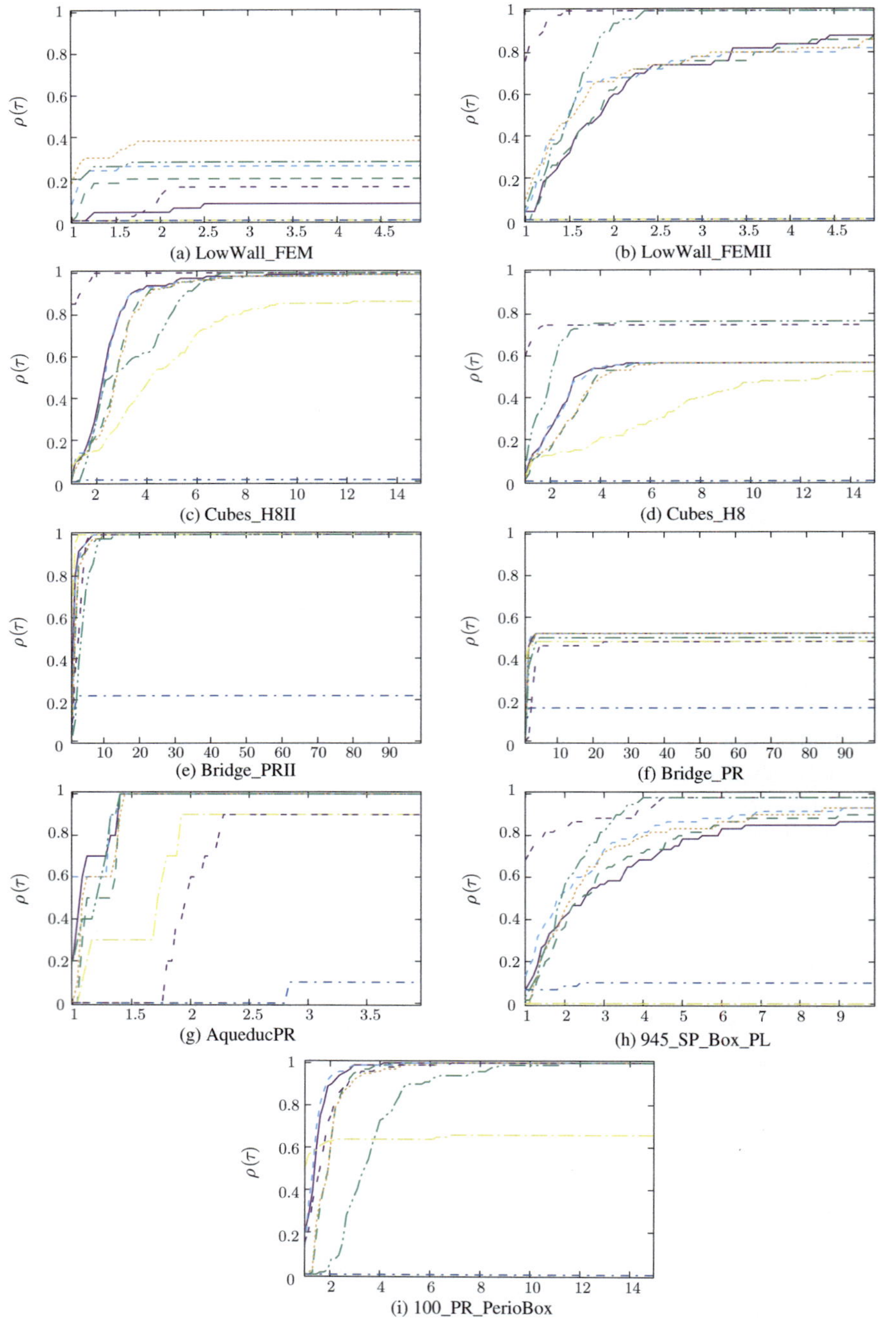

Fig. 5 Influence of the local solver in NSGS-$\star$ algorithms

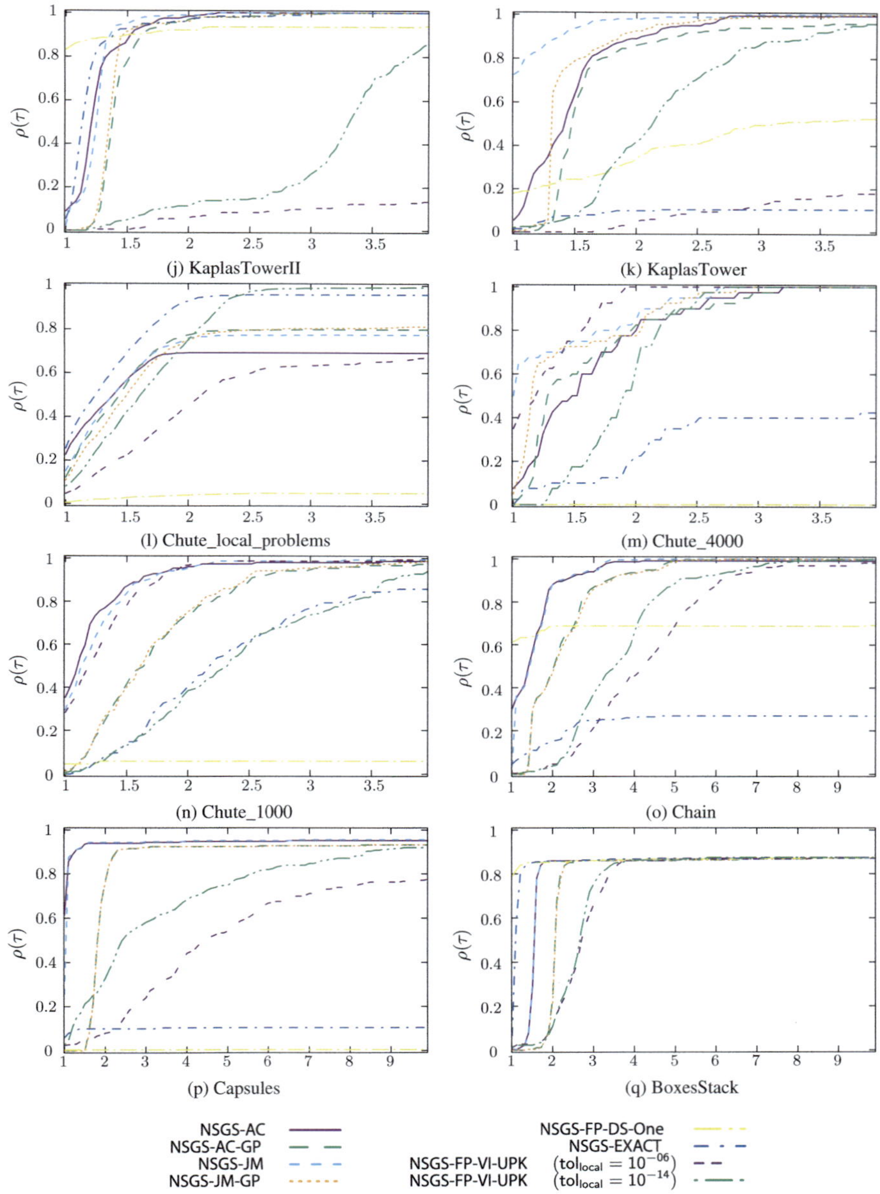

Fig. 5 (continued)

approximate solution of the single contact problems seems sufficient to ensure an efficient convergence towards the solution without entailing robustness.
6. On the test sets 100_PR_PerioBox, KaplasTower, Chain, and Capsules, the solvers NSGS-NSN-⋆ are the best solvers and behave very well on Bridge_PR II. It seems that when a tight accuracy is required, the solvers NSGS-NSN-⋆ are useful and help, with a tight local tolerance, to speed-up the convergence.
7. For the Chute_1000, Chute_4000 test sets, we observe large differences between the local formulations of the nonsmooth equations for the Newton solvers (NSGS-NSN-AC and NSGS-NSN-JM). The solver NSGS-NSN-JM is the best solver, genuinely better than NSGS-NSN-AC, although their theoretical formulations are very close. These two test sets are characterized by difficult local problems in which the Delassus matrix W is unsymmetric with large extra-diagonal terms.
8. For almost all the tests, the line–search procedures slow down the solvers without increasing the robustness. The only test sets in which it has a positive outcome is Chute_4000, for which the NSGS-AC solver fails to get a solution and the line–search seems to stabilize the algorithm.

Influence of the Tolerance of the Local Solver $\mathsf{tol}_{\mathsf{local}}$ in NSGS-FP-VI-UPK Algorithms

In this paragraph, the tolerance of the local solver $\mathsf{tol}_{\mathsf{local}}$ is varied and its effect on the global convergence of the solver is reported. In Fig. 6, we report the performance profiles of NSGS-FP-VI-UPK algorithms for the $\mathsf{tol}_{\mathsf{local}}$ within the range $[10^{-04}, 10^{-16}]$. We also report the efficiency of the adaptive strategy for sizing the value of the local tolerance (see Sect. 6.3). The main observations are:

1. For the test sets that are quickly solved (see Table 6), such as Capsules, a tight tolerance on the local solver 10^{-16} improves the efficiency of the NSGS-FP-VI-UPK solver. Similar results are obtained for BoxesStack, Chain, KaplasTower and KaplasTower II; they are not depicted.
2. For the other problems that are harder to solve, that is, when we expect more iterations of the NSGS-FP-VI-UPK solver, the adaptive rule, or a tight local tolerance, is better.

From these results, it is quite difficult to guess in advance the internal dynamics of the solver. By internal dynamics, we mean the propagation in the algorithm of the error and the values of the unknowns, between the local problem solvers and the global loop over contacts. Note that the range of τ that we used in the graph is quite small, so the difference in performance between the solvers is not crucial.

Influence of the Tolerance of the Local Solver $\mathsf{tol}_{\mathsf{local}}$ in NSGS-AC-GP Algorithms

In Fig. 7, we report the performance profiles of NSGS-AC-GP algorithms for the $\mathsf{tol}_{\mathsf{local}}$ in the range $[10^{-04}, 10^{-16}]$. We also test the adaptive strategy for the local tolerance. Except for the test set Chute_local_problems, the main observation is that the local tolerance does not noticeably change the convergence of the solver. For the test set Chute_local_problems, there is no internal dynamics of the main loop of

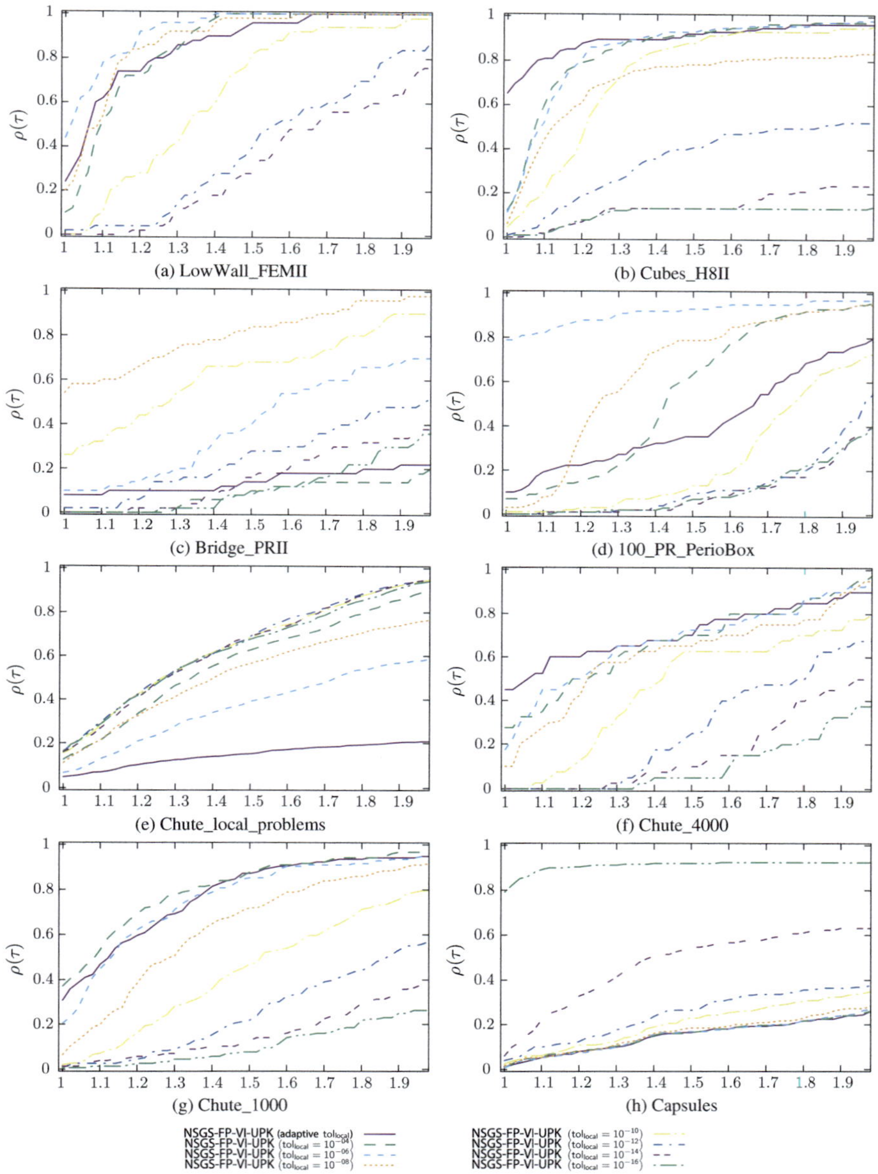

Fig. 6 Influence of the tolerance of the local solver $\mathsf{tol}_{\mathsf{local}}$ in NSGS-FP-VI-UPK algorithms

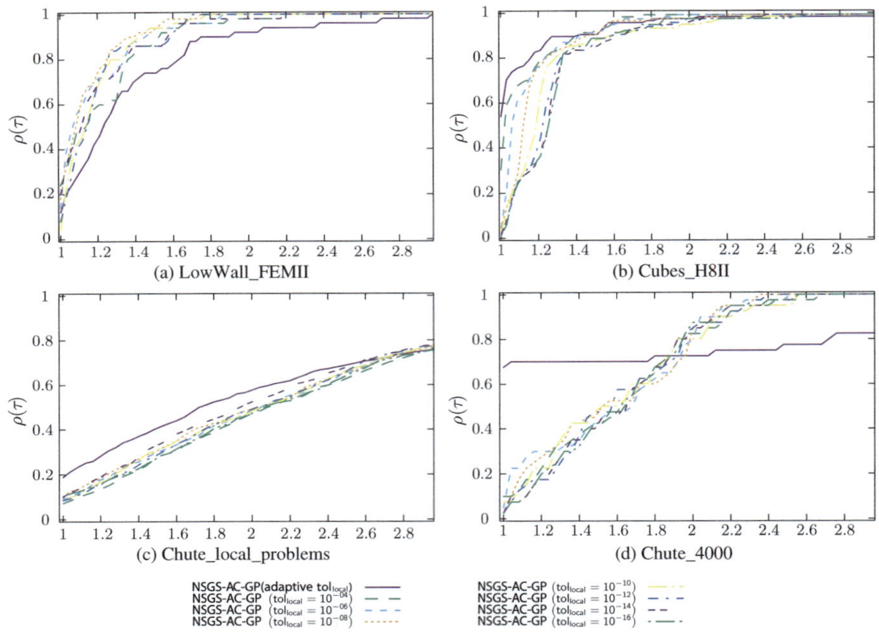

Fig. 7 Influence of the tolerance of the local solver $\mathsf{tol}_{\mathsf{local}}$ in NSGS-FP-NSN-AC-GP algorithms

the NSGS, since there is only one contact. It is therefore reasonable to see that the adaptive strategy performs better than the other.

Influence of the Choice of the Parameters ρ_{N}, ρ_{T} in the Local Solver of the NSGS-AC Algorithms

In Fig. 8, we evaluate the influence of the choice of the parameters ρ_{N}, ρ_{T} on the convergence of the solver. The main conclusions are:

1. For the test sets 945_SP_Box_PL, 100_PR_PerioBox, KaplasTower II, KaplasTower, Chute_local_problems, Chute_4000, Chute_1000, and Capsules, a fixed value of $\rho_{\mathrm{N}} = \rho_{\mathrm{T}} = 1$ has a dramatic effect on the convergence of the algorithm. The scaling of ρ is of utmost importance for the efficiency and robustness of the solver. Note that the rule (112) that takes into account the condition number of the local Delassus matrix W deteriorates the performance for Chute_4000, Chute_1000. In these problems, the local matrix is unsymmetric with large extra-diagonal terms due to large gyroscopic effects.
2. For the other tests, such as LowWall_FEM II, the choice of ρ_{N}, ρ_{T} does not really change the results, mainly due the fact that the order of magnitude of the chosen ρ with the rules (110), (111) or (112) is in $[10^{-01}, 1]$. Cubes_H8 II, Cubes_H8, Bridge_PR II, Bridge_PR, 100_PR_PerioBox, Chain, BoxesStack and AqueducPR are not displayed, since the results are similar.

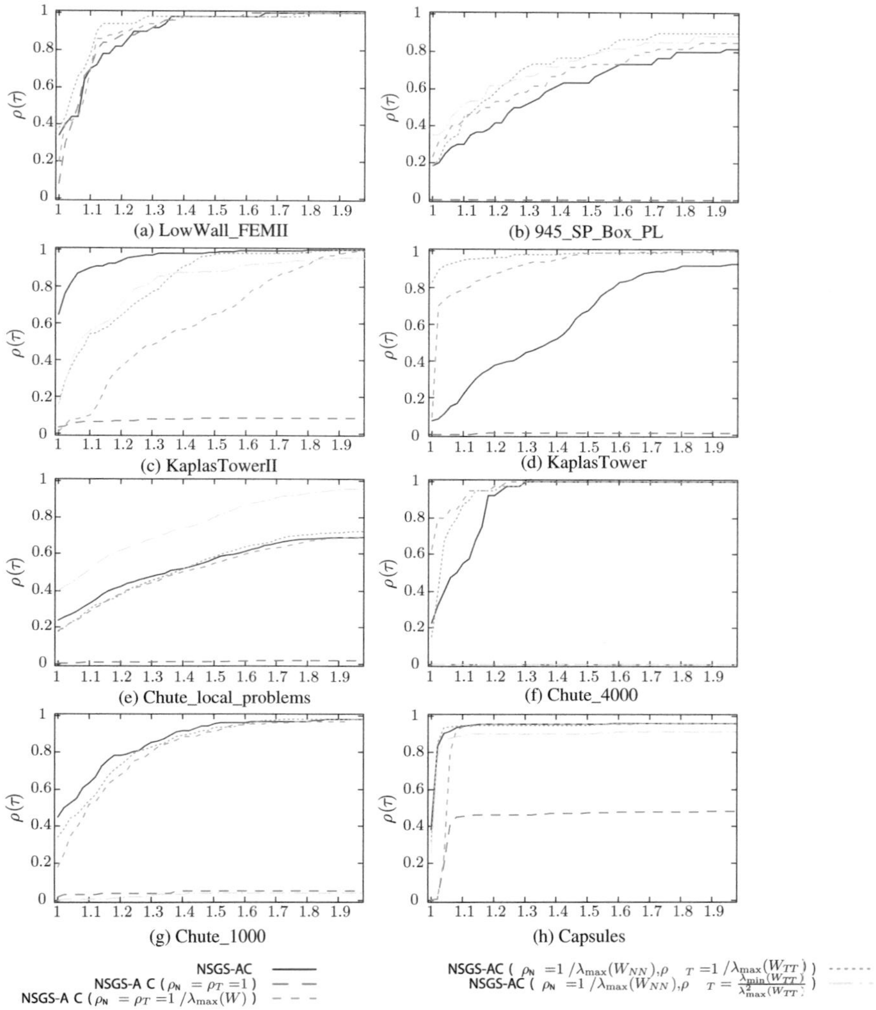

Fig. 8 Influence of the choice of the parameters $\rho_{\mathrm{N}}, \rho_{\mathrm{T}}$ in the local solver of the NSGS-AC algorithms

One of the conclusions of this study is as follows: the rules (110), (111) improve some simulations a lot without increasing the computational cost for the others. Therefore, it is strongly advised that they be used. Some further theoretical studies are needed to understand the effect of ρ on the convergence. In particular, the rule (112) is usually better, but sometimes completely destroys the convergence.

Influence of the Order of Contacts in NSGS Algorithms

In this section, we study the influence of the contact order within the loop of the NSGS-AC-GP solver. We reproduce in Fig. 9 the result of the solvers with the

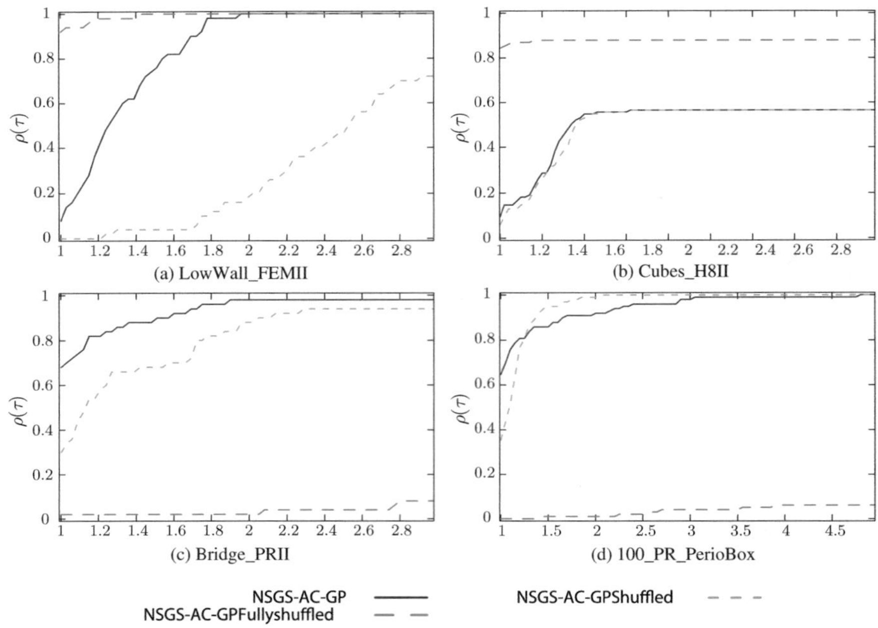

Fig. 9 Influence of the contacts order in NSGS algorithms

original contact list of the problem (NSGS-AC-GP) and with two other ways of iterating over the contacts. The solver NSGS-AC-GP Shuffled corresponds to a single randomization of the list of contacts at the beginning of the algorithm. In the solver NSGS-AC-GP Fully shuffled, the list is shuffled at each iteration. The following observations can be made:

1. The solver NSGS-AC-GP Fully shuffled performs significantly better on the flexible test sets (Cubes_H8_⋆, LowWall_FEM).
2. For the rigid test sets, we reproduce here only the test set 100_PR_PerioBox, because the other test sets behave similarly. The NSGS-AC-GP Fully shuffled has a really bad influence on the convergence of the solver. It seems that it modifies the internal dynamics of the solver in such a way that the rate of convergence is significantly decreased.

Comparison of the PSOR Algorithm with Respect to the Relaxation Parameter ω

In Fig. 10, the relaxation parameter ω is varied ranging in [0.5, 1.8]. Two conclusions can be drawn:

1. For the flexible tests Cubes_H8_⋆ and LowWall_FEM, the efficiency of the solver improved significantly as we decreased the value of ω. Moreover, this is done without destroying the robustness of the solver.
2. For the rigid tests, the effect of the relaxation is not so clear. For values of ω greater than 1.0, the efficiency is improved but the robustness deteriorates. We

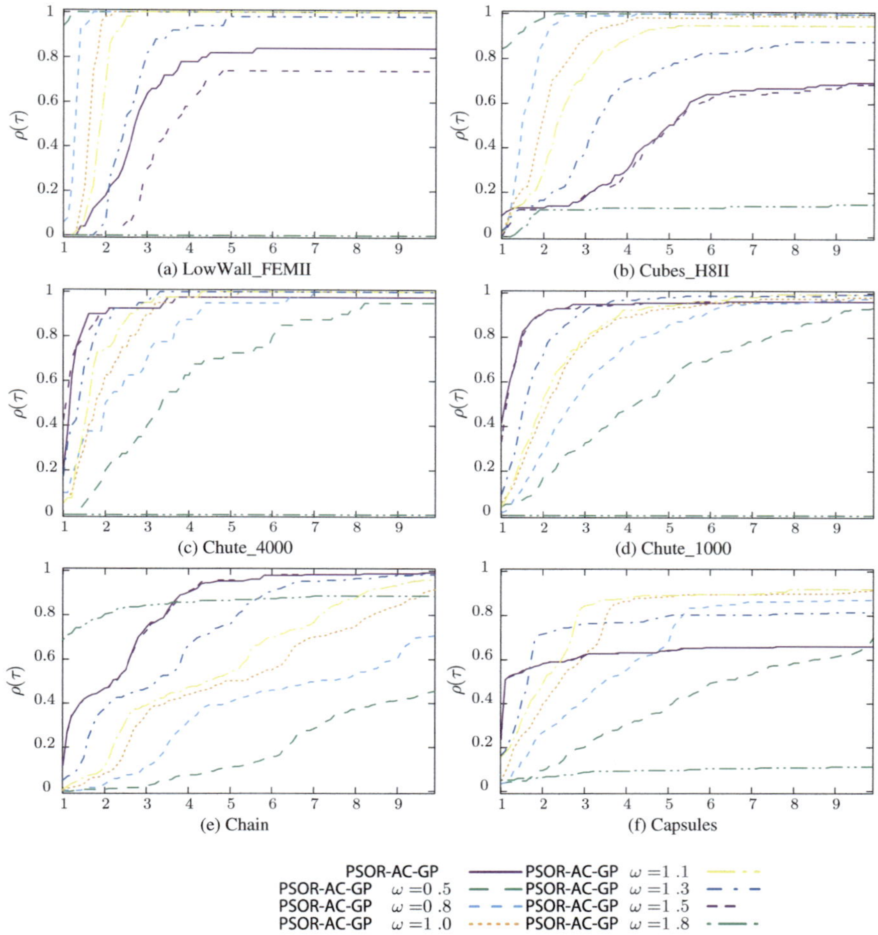

Fig. 10 Effect of relation coefficient ω in PSOR-AC-GP algorithm

observe the contrary for the ω less than 1.0. Note, in particular, that, for the test sets Chute_1000 and Chute_4000, the convergence is completely destroyed for $\omega = 1.8$.

To conclude, it is difficult to advise use of the PSOR algorithm with $\omega \neq 1$. It drastically accelerates the rate of convergence of the algorithm for some problems, but deteriorates the convergence for others. Further studies would be needed to design self–adaptive schemes for sizing ω.

9.3 Comparison of NSN-⋆ Algorithms

In this section, the nonsmooth Newton methods are compared. The performance profiles are depicted in Fig. 11 for the test sets for which the NSN-⋆ are able to solve at least 10% of the problems. The main conclusions are as follows:

1. For the flexible tests Cubes_H8_⋆ and LowWall_FEM, most of the Newton methods succeed in solving the problems within the prescribed time limit. The solver NSN-AC-HYBRID appears to be the best solver. The effect of computing an initial guess with a robust method such as EG-VI-UPK improves the convergence. In practice, we observe that the computation allows one to roughly determine the set of closed and sliding contacts and it helps a lot with the convergence of the Newton solvers. The solvers without a line–search procedure also perform better than those with a line–search procedure, which seems to slow down the convergence without improving the robustness. For the different formulations, the NSN-AC and NSN-JM give equivalent results and are better than the NSN-NM solver, which is, in turn, better than the NSN-FB solver. Note that the Goldstein–Price line search is usually better than the Armijo, despite the fact that the merit function is not necessarily smooth. Finally, we note that NSN-FB and NSN-FB-A are really the slowest solvers within these flexible examples.
2. For the rigid test sets with a high value of the rank ratio or the contact density c (see Table 5), the Newton methods fail to converge, and a lot of divergence issues have been noted in practice. This is the case for the test sets Bridge_PR II, Bridge_PR, AqueducPR, 945_SP_Box_PL, and 100_PR_PerioBox, which are not depicted in Fig. 11.
3. For the rigid test sets with a low value of the rank ratio or contact density c less than 1, such as Chute_1000 and Chain, we observe that the Newton methods are able to solve some problems. We note also that in the Chain test set, the use of a fixed value of ρ significantly penalizes the convergence of the solver. Contrary to flexible test sets, the use of a line–search procedure helps in obtaining a better robustness of the solver. This is particularly true for NSN-NM-GP.
4. Finally, for the test sets KaplasTower and Capsules, the NSN-FB-GP is able to solve more than 80% of the tests in a very efficient way. Some further studies would be needed to understand why this specific solver performs so much better than the others.

As a general conclusion, the success of the NSN-⋆ algorithms is conditioned by the rank of the Delassus matrix W, and then by the contact density value c. For full rank matrix W, the solvers are robust and efficient. For values of c no larger than 1, the methods are able to find a solution with tight accuracy. For larger values of c and larger rank ratio, the nonsmooth Newton methods are not robust and generally diverge.

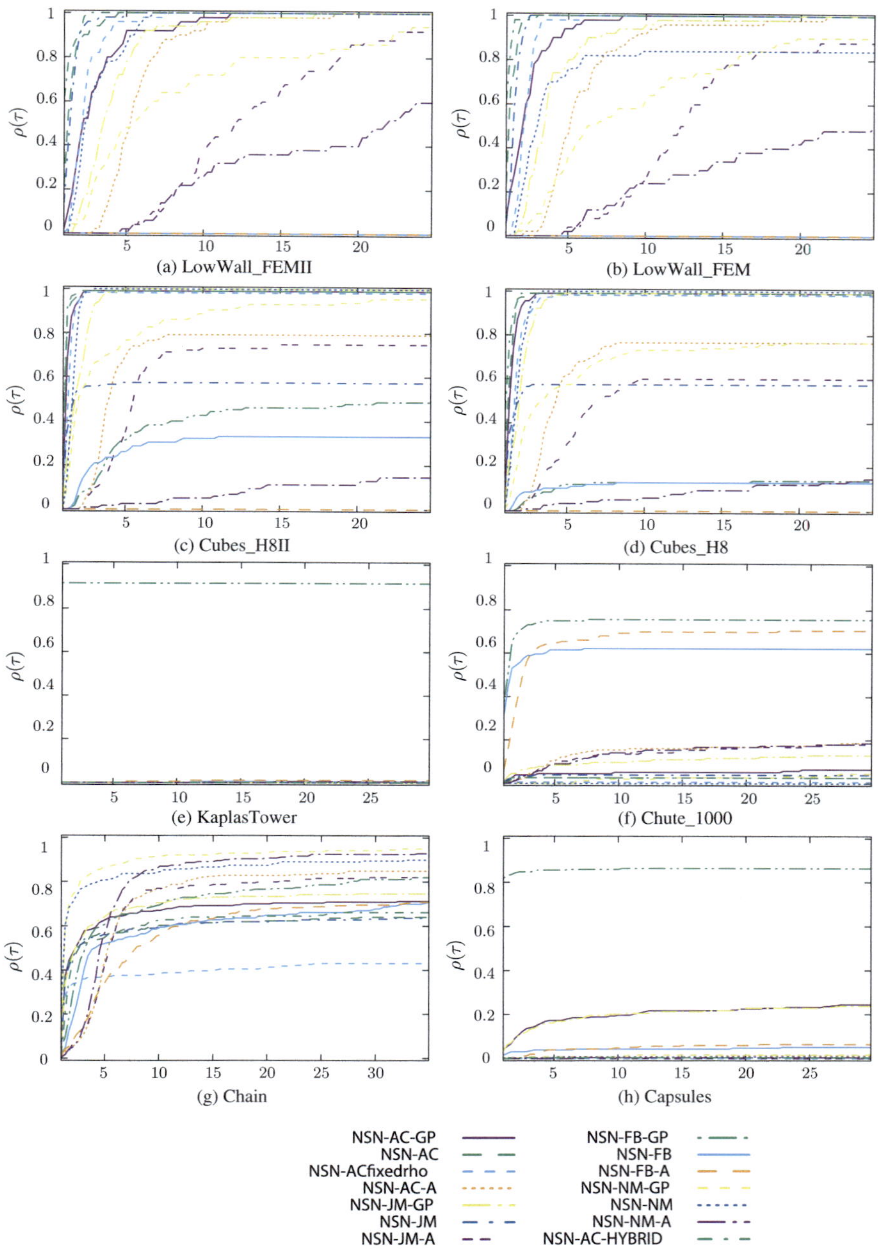

Fig. 11 Comparison of NSN-⋆ algorithms

9.4 Comparison of the Proximal Point Algorithm PPA-NSN-⋆ *and* PPA-NSGS-⋆ *Algorithms*

In Fig. 12, we compare the proximal point approach with various internal solvers based on nonsmooth Newton methods NSN-⋆. The main observations are:

1. For the flexible test sets (see, for an illustration, the test set LowWall_FEM II), for which the nonsmooth Newton solvers work pretty well, the use of a proximal point algorithm has no interest, since it slows down the convergence of the algorithm by performing a first iteration with a given, and possibly large, value of the parameter α.
2. For the test sets KaplasTower, Chute_1000, Chain, Capsules and BoxesStack, the proximal point approach greatly improves the efficiency of the NSN-AC-GP solver and often also improves its reliability (see, for comparison, Fig. 11). Clearly, the regularization introduced in the proximal point algorithm increases the rank of the matrix W and has a strong effect on the convergence of the nonsmooth Newton methods.
3. The efficiency of the proximal point algorithm strongly depends on the internal solver.
4. The strategy for updating the regularization parameter α also plays an important role. Quite surprisingly, for the Bridge_PR test set, the adaptive rule that does not take into account the current error is very efficient and allows us to get a robust and efficient solver with respect to the others. Unfortunately, there is no updating rule for the parameter α that works for all test sets.

In Fig. 13, we compare the NSGS-AC solver when it is used directly or inside the proximal point algorithm. On most of the test sets, such as KaplasTower, a direct application of the NSGS-AC solver is already efficient, and its embedding into a proximal point algorithm does not bring any improvements. Nevertheless, we can see, in Fig. 13, that the proximal point algorithm improves the robustness and the efficiency for the test sets 945_SP_Box_PL and Capsules.

9.5 Comparison of Optimization-Based Algorithms PANA-⋆, TRESCA-⋆ and ACLM-⋆

In Fig. 14, we compare the algorithms based on the optimization approach presented in Sect. 7. The pure convex relaxation SOCLCP-NSGS-PLI method has been added so that we might be able to understand the effect of the nonconvexity of the problems on the efficiency and robustness of the solvers. The main conclusions are:

1. The pure convex relaxation in SOCLCP-NSGS-PLI drastically simplifies the problems in the test sets LowWall_FEM II, AqueducPR, KaplasTower, and BoxesStack and is slightly better in the Bridge_PR II, 100_PR_PerioBox, and KaplasTower II test sets. In particular, we note that if we want to reach better

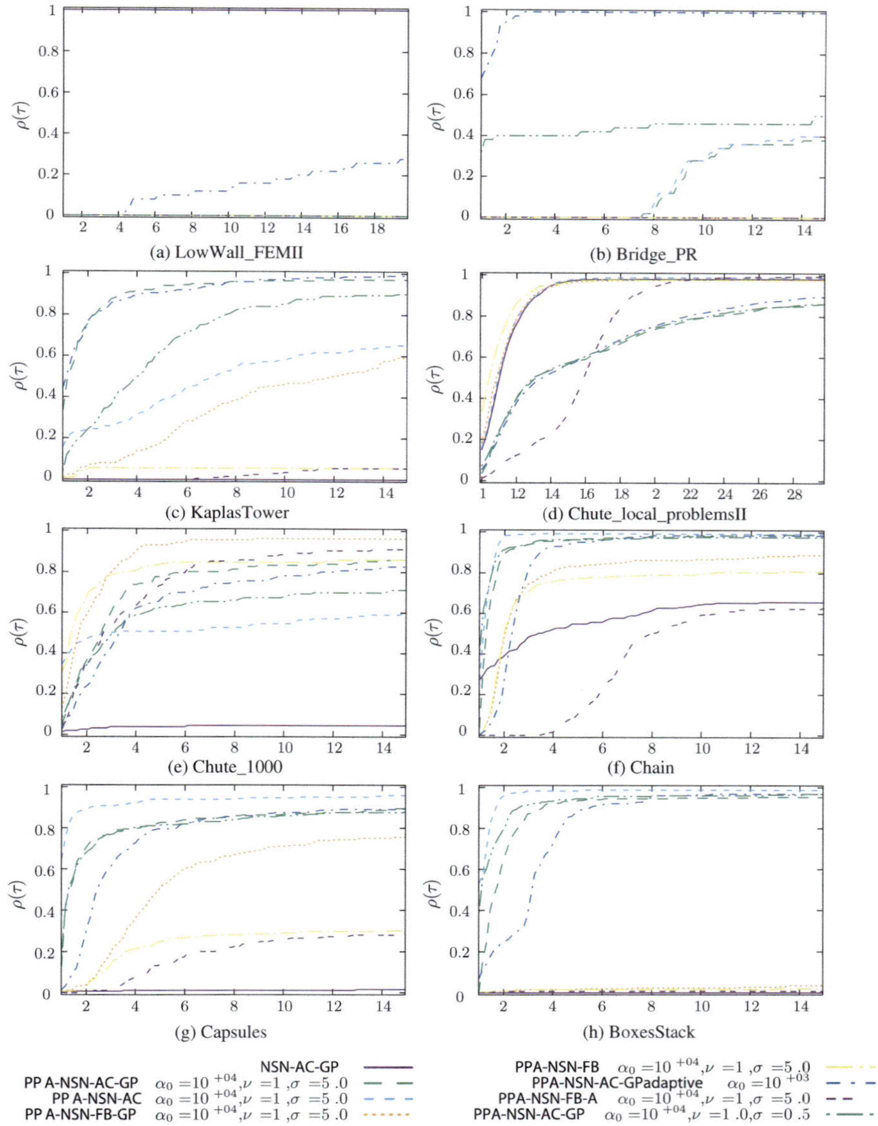

Fig. 12 Comparison of internal solvers in PPA-NSN-⋆ algorithms

accuracy, as in the KaplasTower test set, the convex relaxation helps much, but this conclusion cannot be made for the Bridge_PR test set. Let us also note that the convex relaxation does not help a lot in the test sets Cubes_H8, Bridge_PR, Chute_1000, Chute_4000 and Capsules. One of the conclusions we can draw may be that the nonconvexity of the problem is not the only difficulty in such problems. Using a convex relaxation is not sufficient to solve all the problems.

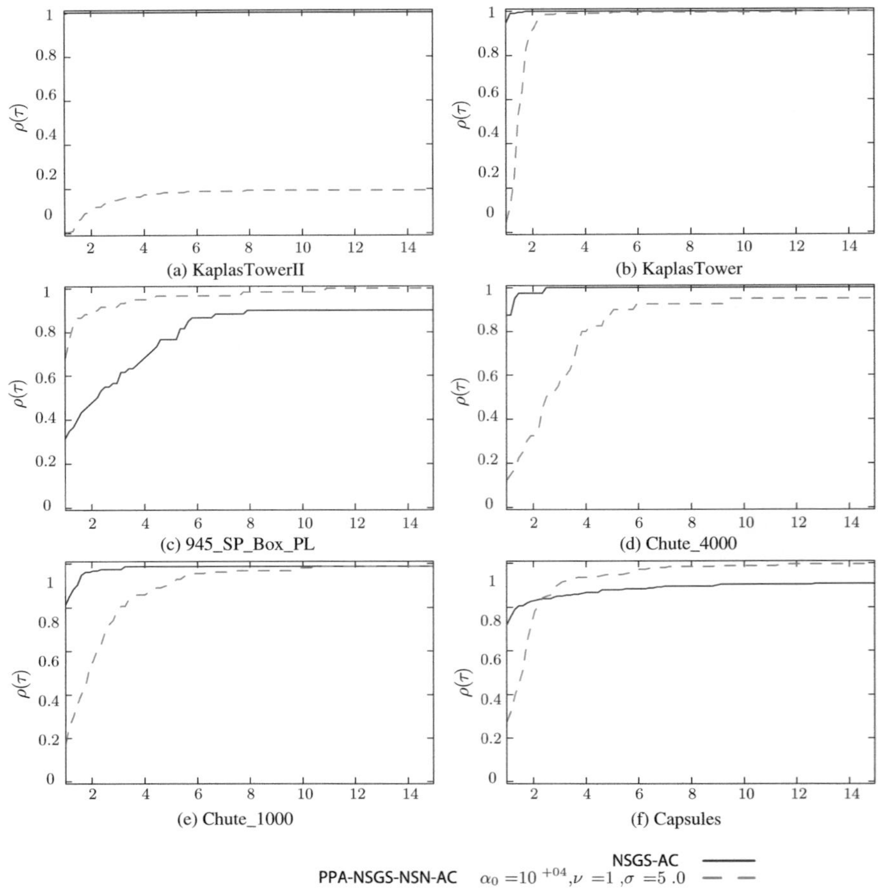

Fig. 13 Comparison of internal solvers in PPA-NSGS-⋆ algorithms

2. The solvers based on the optimization approach are generally robust but slow. There are two primary reasons for this. Firstly, we use iterative first order solvers as an internal solver with a slow convergence rate. The fact that the Delassus matrix does not have full rank in the rigid tests prevents the use of second-order methods as nonsmooth Newton methods. For the flexible test, it could be of interest to implement new dedicated solvers of the internal convex problems based on nonsmooth Newton methods. Furthermore, the tests with off-the-shelf implementations of optimization methods were not really conclusive. The general convex solvers are not capable of exploiting the particular structure of the constraints given by a Cartesian product of a large number of second-order cones in $\mathbb{R}^3$. Secondly, the fixed point iteration that drives the convergence is generally slow. Once again, it would be valuable to implement a second-order method to drive the external loop.

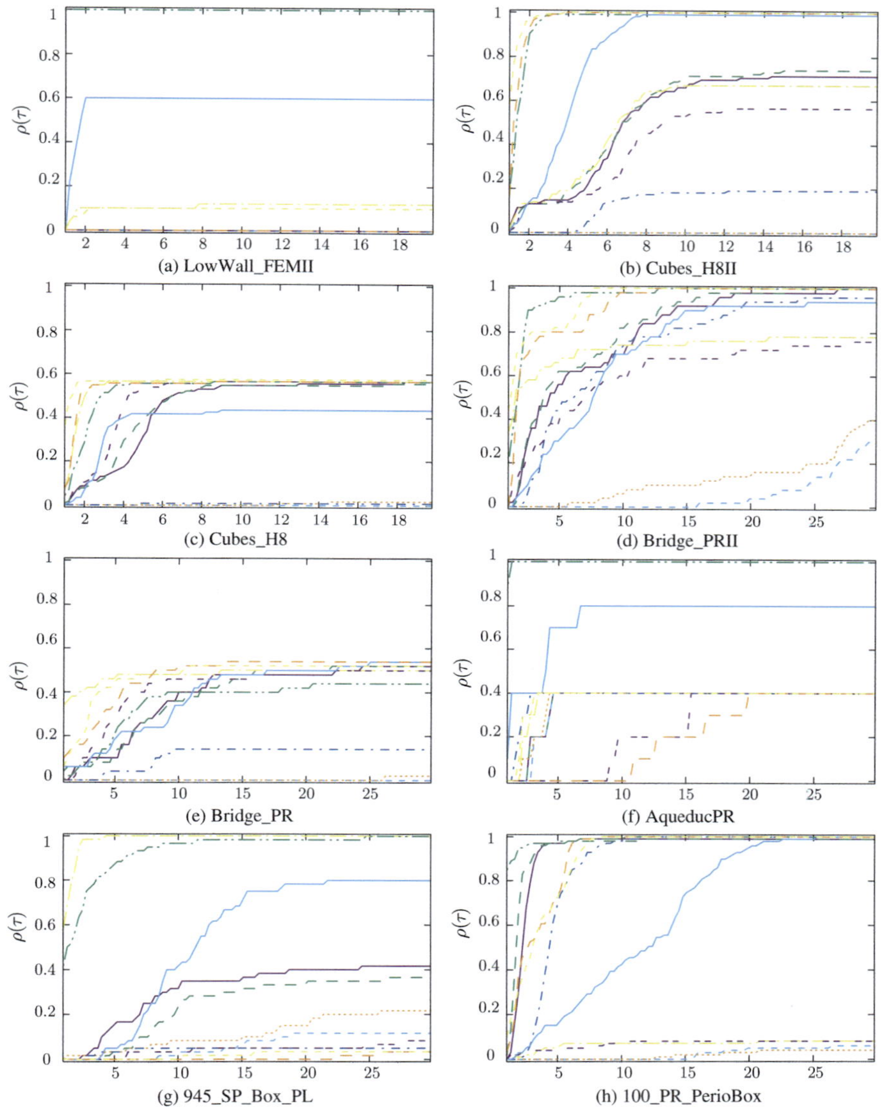

Fig. 14 Comparison of the optimization based solvers

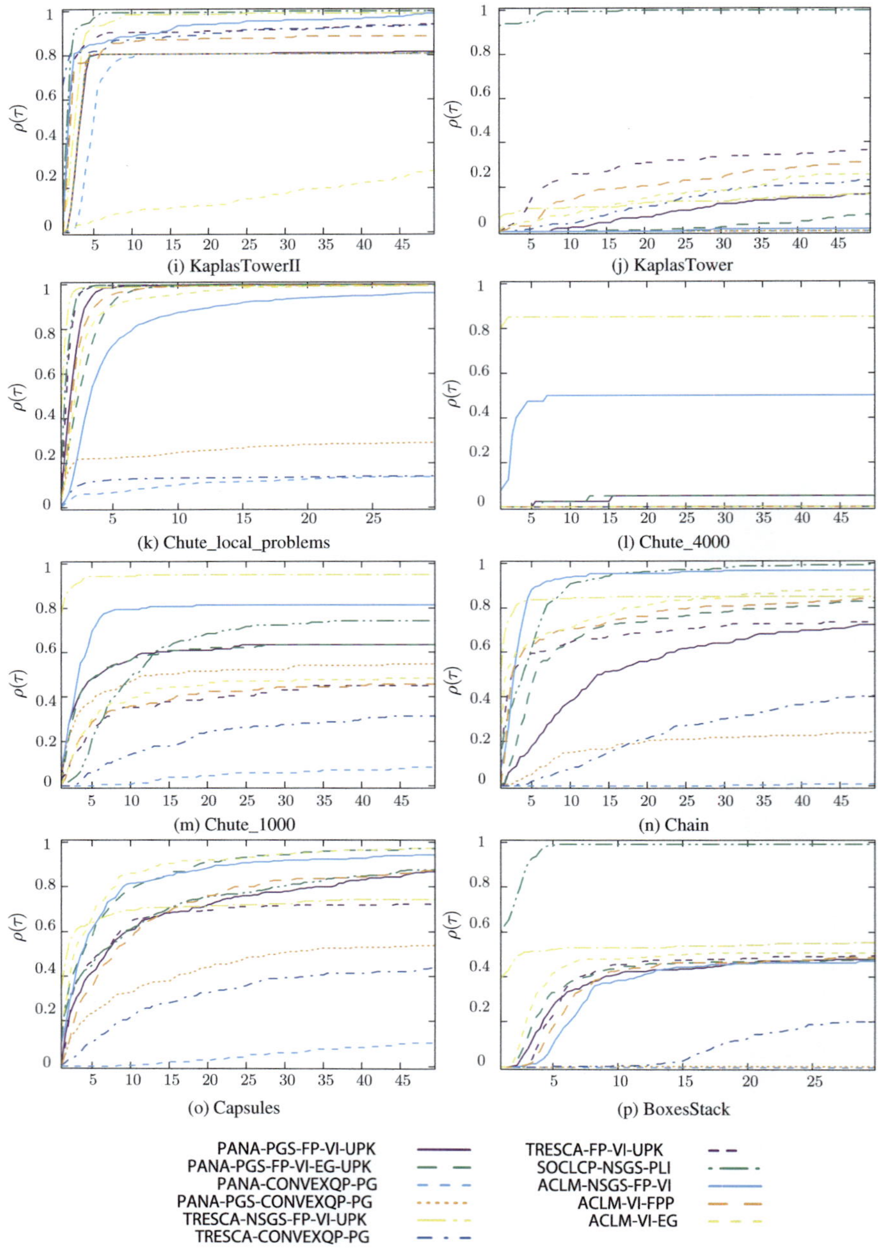

Fig. 14 (continued)

3. On the choice of a specific optimization-based strategy with respect to the others, we can observe that the comparison is really problem–dependent. For the test sets Cubes_H8, Bridge_PR, Bridge_PR II, LowWall_FEM and AqueducPR, the ACLM-⋆ solvers are the best. For the test problems KaplasTower, 945_SP_Box_PL, Chute_4000, Chute_1000 and BoxesStack, the TRESCA-⋆ solvers are better. Finally, the PANA-⋆ solvers are better on the 100_PR_PerioBox test set. Since the convex relaxation of the internal problem is effected in different manners, it is expected that the different families of solvers will behave differently. In particular, if the coefficient of friction is large or if the number of sliding contacts is low, we expect the ACLM-⋆ solvers to behave better, because the s variable in the fixed point iteration will not drastically influence the convergence. On the contrary, when the coefficient of friction is low, we may expect the splitting introduced in the PANA-⋆ to be better. An analysis of the contact status (closed, sliding, sticking) in the problems would be a next step in understanding the performance of each family.

10 Comparison of Different Families of Solvers

In this last section, we compare the most efficient solvers for each family. The performance profiles are reported in Fig. 15. The main conclusions are as follows:

1. First of all, we can observe that, for all the test sets, at least one solver is capable of solving all the problems within the prescribed time. Unfortunately, there is no universal solver that outperforms all the other solvers for all the test sets.
2. For the flexible test sets, the nonsmooth Newton solvers NSN-⋆ are the best solvers. In the test set LowWall_FEM II, the NSN-⋆ are followed NSGS-FP-VI-UPK and NSGS-AC solvers. On this test set, the required accuracy is limited to 10^{-04}, and the NSGS-⋆ are still able to reach the tolerance in a competitive time. Between the test sets Cubes_H8 II and Cubes_H8, and between LowWall_FEM II and LowWall_FEM, the required accuracy is decreased to 10^{-08}. With a tighter tolerance, we observe that the relative efficiency of the NSN-⋆ solvers increases. This was already noted in [6]. In other words, on the flexible tests, we are able to use nonsmooth Newton methods efficiently, since the Delassus matrix W has full rank. In that case, the quadratic convergence rate helps in reaching tighter tolerances. Note that in the flexible test sets, the proximal point algorithms PPA-NSN-⋆ are not really interesting, but as the required accuracy decreased, they began to compete with the NSGS-⋆ algorithms.
3. For most of the rigid test sets with a low required accuracy of 10^{-04}, such as AqueducPR, 945_SP_Box_PL, 100_PR_PerioBox, KaplasTower, Chute_4000 and Chute_1000, the NSGS-⋆ are the most efficient and robust solvers. In the case of the test sets Chute_4000 and Chute_1000, the NSGS-FP-VI-UPK solvers are better than the NSGS-AC-⋆, due to some robustness issues in the local solvers based on nonsmooth Newton methods. These solvers are generally followed by

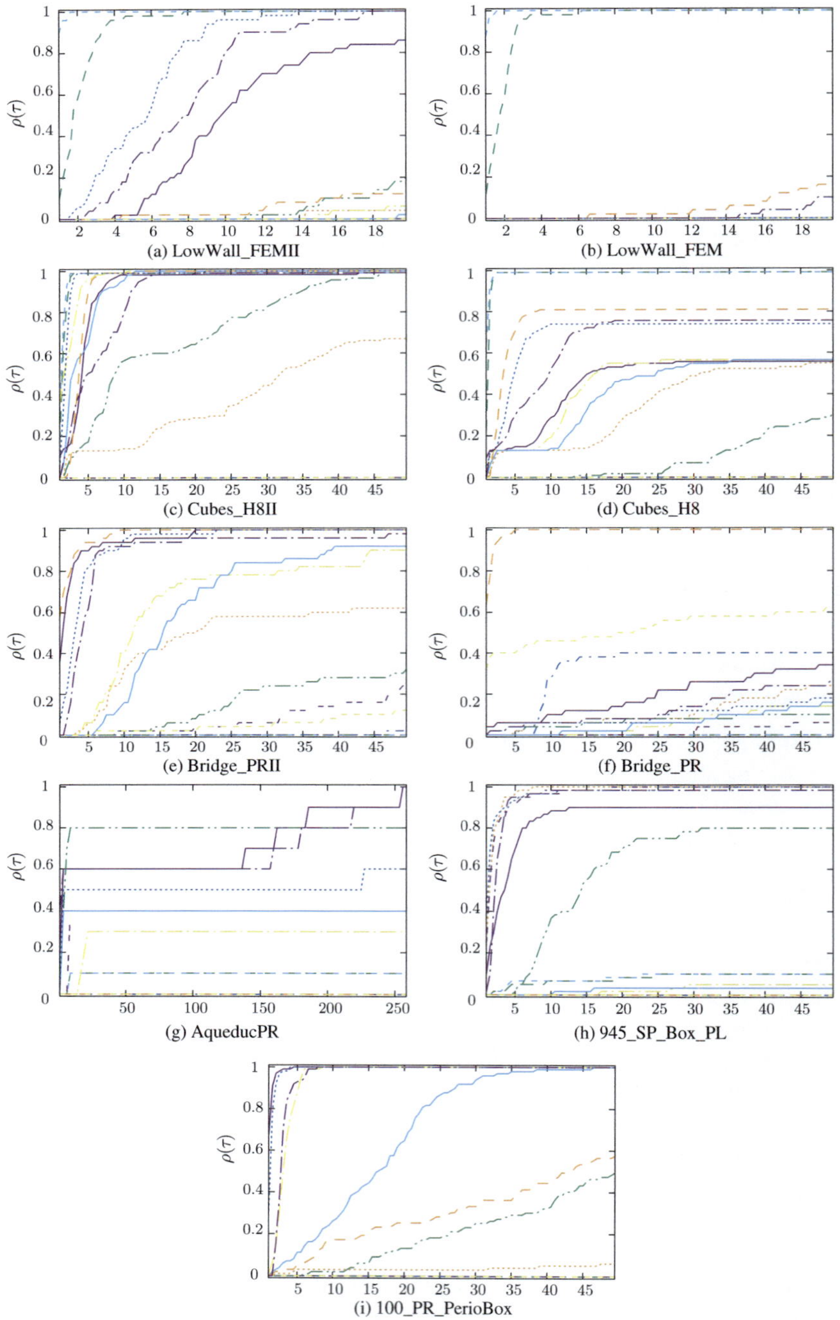

Fig. 15 Comparison of the solvers between families

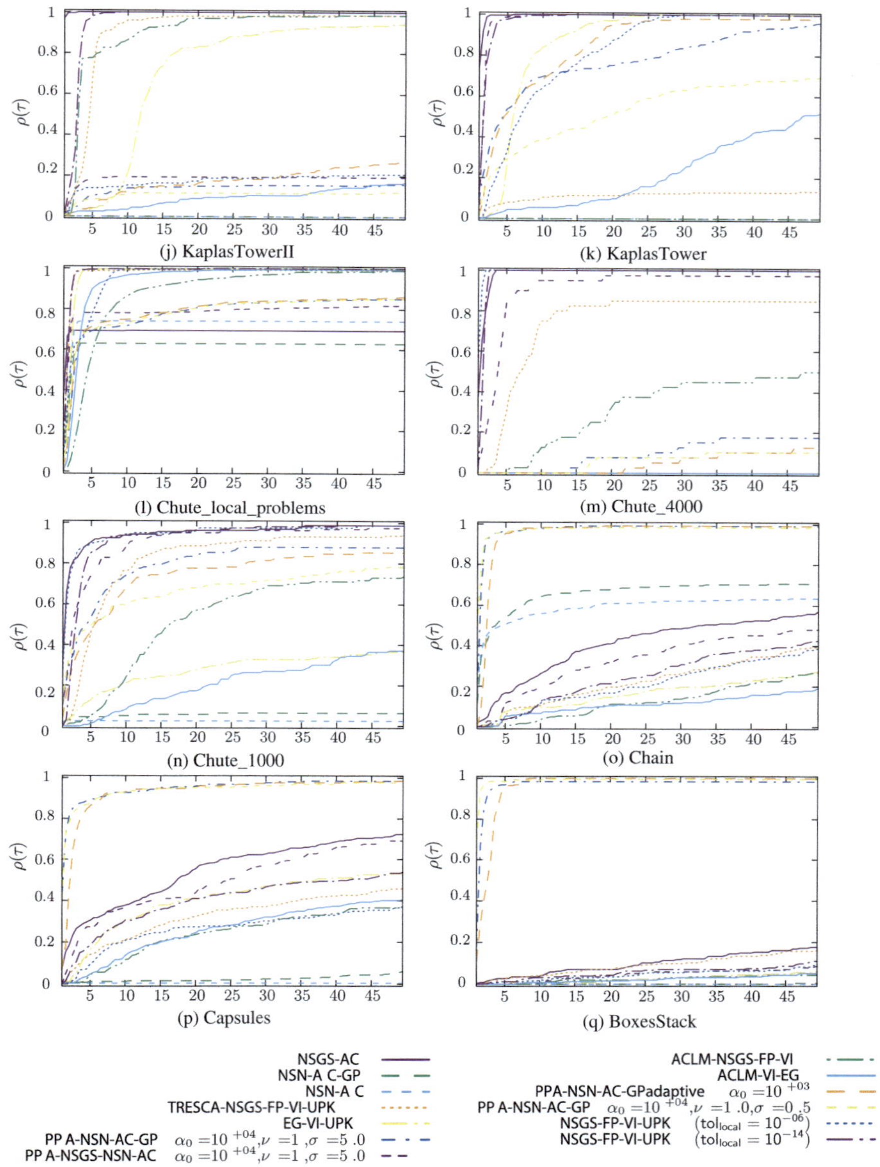

Fig. 15 (continued)

optimization-based solvers such as ACLM-$\star$ and TRESCA-$\star$, except for the test set 945_SP_Box_PL, for which the more robust solver is TRESCA-NSGS-FP-VI-UPK.

4. For the rigid test sets with a required accuracy of 10^{-08}, such as Bridge_PR, Chain, Capsules and BoxesStack, the solvers PPA-$\star$ are the most efficient and robust solvers. The regularization of the Delassus matrix introduced by the proximal point algorithm has a very positive effect. In particular, it enables the use of nonsmooth Newton techniques that help in reaching a tighter accuracy thanks to their quadratic convergence rates. The PPA-$\star$ algorithms are generally followed by NSGS-$\star$, except in the case of the Chain test set, for which the NSN-$\star$ are able to solve 60% of the problems quite efficiently. In the case of the Bridge_PR test set, the proximal point technique PPA-NSN-AC-GP $\alpha_0 = 10^{+03}$ is the only one capable of solving all the problems at the tolerance of 10^{-08}. As discussed in Sect. 9.4, the rule for updating the proximal point parameter α plays an important role and deserves further study.
5. In the case of the Chute_local_problems test set, we observe that the optimization-based solvers are the best and allow one to circumvent the issues of robustness of NSGS-AC-$\star$ solvers, which are reduced, in that case, to the NSN-$\star$ solvers. We recall that these local problems are extracted from Chute_4000 and selected as the most difficult local problems. These problems are characterized by strongly unsymmetric matrices with large extra-diagonal terms compared to the diagonal ones. In that case, the optimization solvers based on a convexification help in solving the problems, although the local Delassus matrix is not necessarily symmetric. We can also note, as in Chute_4000 and Chute_1000, that the NSGS-FP-VI-UPK solvers are less sensitive to this asymmetry of the Delassus matrix.

11 General Conclusions

In this chapter, we have reviewed several formulations of the discrete contact problem with Coulomb friction. These formulations open the way to various solving procedures that have been detailed. Some are already well-known: (a) the splitting and relaxation techniques (NSGS-$\star$ and PSOR-$\star$ solvers), (b) the nonsmooth Newton methods (NSN-$\star$ solvers) and (c) the optimization-based solvers (PANA-$\star$, TRESCA-$\star$ and ACLM-$\star$ solvers). For the first time, we present general solvers based on the variational inequality formulation (FP-VI-$\star$ and FP-EG-$\star$). These methods extend the standard fixed point iteration (FP-DS, also known as Uzawa's algorithm) in various directions and provide some self-adaptive rules for updating the ρ parameter that appears to be crucial in practice for the efficiency of the methods. As far as we know, it is also the first application of the proximal point algorithms (PPA-$\star$) to the discrete frictional contact problem. This new family of solvers appears to be a promising alternative when we want to reach tight accuracy for collections of rigid bodies such as granular materials.

Then, we presented a thorough comparison of solvers over a large set of test problems. Using performance profiles, the solvers have been compared family by family, and then altogether. The main conclusions and perspectives of this study are as follows:

- The methods based on variational inequality formulations (FP-VI-$\star$) are robust, if a consistent self-adaptive rule for the parameter ρ is used. We presented two rules that yield very satisfactory results. Thanks to their robustness, these methods provide reliable solvers for the local problem in splitting techniques. Nevertheless, the convergence is slow: these methods have difficulty obtaining a solution within the prescribed time for tight tolerances, or if the problem size is large. The main perspectives for these methods are (a) to adapt the values of ρ contact by contact to try to improve the convergence speed and (b) to perform computation in parallel for large-scale systems. Indeed, each iteration of the FP-VI-$\star$ solvers may be straightforwardly implemented on distributed computer architectures.
- The methods based on splitting techniques, the NSGS-$\star$ solvers, provide us with robust and efficient solvers, provided the local solver is robust. They are generally more efficient than the FP-VI-$\star$ methods, since they exploit the particular structure of the problem (sparse block sparsity and local solver routines). However, they suffer from the same problems as the FP-VI-$\star$ solvers: the convergence rate is low and high accuracy is difficult to reach within the prescribed time. The main perspective for this solver is to improve the robustness and the efficiency of the local solver, for instance, by using proximal point techniques or optimization-based solvers. Regarding the PSOR-$\star$ solvers, for some values of the relaxation parameter ω, the convergence rate is greatly improved with respect to the NSGS-$\star$ solvers. However, guessing the correct value of the parameter ω is challenging, as some values may increase the computational effort or make the algorithm diverge. Clearly, a self-adaptive rule for sizing the relaxation parameter ω would be a notable improvement.
- The nonsmooth Newton solvers NSN-$\star$ appear to be a very efficient family of solvers for problems that have a full-rank Delassus matrix or a very low contact density. For instance, in the case of the flexible tests, they are the best solvers among others and they are capable of reaching tight tolerances that are not reachable with the FP-VI-$\star$ and NSGS-$\star$ solvers. For the other test sets, they suffer from robustness issues. To overcome this, we work on several options: (a) the choice of the ρ parameters in the equation-based formulation, (b) using line-search procedures to stabilize the convergence (at the expense of the convergence speed) and (c) improving the initial starting point of the solver with a FP-VI-$\star$. All these improvements appear to increase the robustness. Unfortunately, it was not sufficient to circumvent all the divergence problems. Some pointers in the literature try to modify the iteration matrix in the Newton loop to improve robustness when the iterates are far from the solution. This solution has not been tested. The main perspectives for these solvers are to improve their robustness by testing modifications of the iteration matrix or the self-adapting rule for sizing ρ. The question of the scaling and the preconditioning must be more deeply studied. When these

solvers are robust, they are also highly parallelizable for large systems, since we can rely on massively parallel solvers for linear systems such as MUMPS.

- As we discussed before, the PPA-⋆ solvers are a possible solution for improving the robustness of the NSN-⋆ methods while keeping their convergence rates. This solution proves its efficiency on a lot of test sets. Nevertheless, we were not able to find a universal rule for updating the parameter α such that it works for all test sets. Clearly, this aspect deserves further study.
- The optimization-based solvers (PANA-⋆, TRESCA-⋆ and ACLM-⋆ might also exhibit good robustness properties. Unfortunately, they suffer from the slow convergence of the external loop based on a fixed point updating, which is not compensated for by the efficiency of the convex problem solver. As we have seen, the nonconvexity of the problems is not the only difficulty: most of the time, the rank deficiency of the Delassus matrix is the main cause of the slow convergence or divergence. Finally, it would be worthwhile investigating why an optimization formulation is better than another for certain test sets. One of the reasons might be that the contact status (closed, sticking, sliding) is not distributed in the same way along the test sets. A study based on the contact status would be complementary to the measure of the rank ratio and the contact density for guessing the cause of the issues.

Acknowledgements The authors are grateful to Pierre Alart, Paul Armand, Florent Cadoux, Frédéric Dubois, Claude Lémaréchal, Jérôme Malick and Mathieu Renouf for all of the stimulating discussion.

Appendix 1. Basics in Convex Analysis

Definition 1 ([103]) Let $X \subseteq \mathbb{R}^n$. A multivalued (or point-to-set) mapping $T: X \rightrightarrows X$ is said to be (strictly) monotone if there exists $c(>) \geqslant 0$ such that, for all $\hat{x}, \widetilde{x} \in X$,

$$(\hat{v} - \widetilde{v})^\top(\hat{x} - \widetilde{x}) \geqslant c\|\hat{x} - \widetilde{x}\| \qquad \text{with } \hat{v} \in T(\hat{x}), \widetilde{v} \in T(\widetilde{x}). \tag{137}$$

Moreover, T is said to be maximal when it is not possible to add a pair (x, v) to the graph of T without destroying the monotonicity.

The Euclidean projector P_X onto a closed convex set X: for a vector $x \in \mathbb{R}^n$, the projected vector $z = P_X(x)$ is the unique solution to the convex quadratic program

$$\begin{cases} \min \ \dfrac{1}{2}(y - x)^\top(y - x), \\ \text{s.t. } y \in X. \end{cases} \tag{138}$$

The following equivalences are classical:

$$y = P_K(x) \iff \begin{array}{l} \min \frac{1}{2}(y-x)^\top(y-x) \\ \text{s.t.} \quad y \in K \end{array} \tag{139}$$

$$\iff -(y-x) \in N_K(y) \tag{140}$$

$$\iff (x-y)^\top(y-z) \geqslant 0, \forall z \in K \tag{141}$$

$$-F(x) \in N_K(x) \iff -\rho F(x)^\top(y-x) \geqslant 0, \forall y \in K \tag{142}$$

$$\iff (x-(x-\rho F(x))^\top(y-x) \geqslant 0, \forall y \in K \tag{143}$$

$$\iff x = P_K(x-\rho F(x)) \text{ thanks to (141).} \tag{144}$$

Sub-differential of the Euclidean Norm

The sub-differential of the Euclidean norm in $\mathbb{R}^n$ is given by

$$\partial\|z\| = \begin{cases} \dfrac{z}{\|z\|}, & z \neq 0 \\ \{x, \|x\| \leqslant 1\}, & z = 0. \end{cases} \tag{145}$$

Euclidean Projection on the Unit Ball

Let $B = \{x \in \mathbb{R}^n, \|x\| \leqslant 1\}$. The Euclidean projection on the unit ball is given by

$$P_B(z) = \begin{cases} z & \text{if } z \in B \\ \dfrac{z}{\|z\|} & \text{if } z \notin B. \end{cases} \tag{146}$$

Its subdifferential can be computed as

$$\partial P_B(z) = \begin{cases} I & \text{if } z \in B \setminus \partial B \\ I + (s-1)zz^\top, s \in [0,1] & \text{if } z \in \partial B \\ \dfrac{I}{\|z\|} - \dfrac{zz^\top}{\|z\|^3} & \text{if } z \notin B. \end{cases} \tag{147}$$

Euclidean Projection on the Second-Order Cone of $\mathbb{R}^3$

Let $K = \{x = [x_{\text{N}} x_{\text{T}}]^T \in \mathbb{R}^3, x_{\text{N}} \in \mathbb{R}, \|x_{\text{T}}\| \leqslant \mu x_{\text{N}}\}$ be the second-order cone in $\mathbb{R}^3$. The Euclidean projection on K is

$$P_K(z) = \begin{cases} z & \text{if } z \in K \\ 0 & \text{if } -z \in K^* \\ \dfrac{1}{1+\mu^2}(z_{\text{N}} + \mu\|z_{\text{T}}\|) \begin{bmatrix} 1 \\ \mu \dfrac{z_{\text{T}}}{\|z_{\text{T}}\|} \end{bmatrix} & \text{if } z \notin K \text{ and } -z \notin K^*. \end{cases} \tag{148}$$

Direct Computation of an Element of the Subdifferential
The computation of the subdifferential of P_K is given as follows:

- if $z \in K \setminus \partial K$, $\partial_z P_K(z) = I$,
- if $-z \in K^* \setminus \partial K^*$, $\partial_z P_K(z) = 0$,
- if $z \notin K$ and $-z \notin K^*$ and, $\partial_z P_K(z) = 0$, we get

$$\partial_{z_{\mathrm{N}}} P_K(z) = \frac{1}{1+\mu^2}\begin{bmatrix} 1 \\ \mu z_{\mathrm{T}} \end{bmatrix} \tag{149}$$

and

$$\partial_{z_{\mathrm{T}}} [P_K(z)]_{\mathrm{N}} = \frac{\mu}{1+\mu^2}\frac{z_{\mathrm{T}}}{\|z_{\mathrm{T}}\|} \tag{150}$$

$$\partial_{z_{\mathrm{T}}} [P_K(z)]_{\mathrm{T}} = \frac{\mu}{(1+\mu^2)}\left[\mu\frac{z_{\mathrm{T}}}{\|z_{\mathrm{T}}\|}\frac{z_{\mathrm{T}}^\top}{\|z_{\mathrm{T}}\|} + (z_{\mathrm{N}} + \mu\|z_{\mathrm{T}}\|)\left(\frac{I_2}{\|z_{\mathrm{T}}\|} - \frac{z_{\mathrm{T}} z_{\mathrm{T}}^\top}{\|z_{\mathrm{T}}\|^3}\right)\right], \tag{151}$$

that is,

$$\partial_{z_{\mathrm{T}}} [P_K(z)]_{\mathrm{T}} = \frac{\mu}{(1+\mu^2)\|z_{\mathrm{T}}\|}\left[(z_{\mathrm{N}} + \mu\|z_{\mathrm{T}}\|)\, I_2 + z_{\mathrm{N}}\frac{z_{\mathrm{T}} z_{\mathrm{T}}^\top}{\|z_{\mathrm{T}}\|^2}\right]. \tag{152}$$

Computation of the Subdifferential Using the Spectral Decomposition

In [55], the computation of the Clarke subdifferential of the projection operator is also done by inspecting the different cases using the spectral decomposition

$$\partial P_K(x) = \begin{cases} I & (\lambda_1 > 0, \lambda_2 > 0) \\ \dfrac{\lambda_2}{\lambda_1+\lambda_2} I + Z & (\lambda_1 < 0, \lambda_2 > 0) \\ 0 & (\lambda_1 < 0, \lambda_2 < 0) \\ \mathrm{co}\{I, I+Z\} & (\lambda_1 = 0, \lambda_2 > 0) \\ \mathrm{co}\{0, Z\} & (\lambda_1 < 0, \lambda_2 = 0) \\ \mathrm{co}\{0 \cup I \cup S\} & (\lambda_1 = 0, \lambda_2 = 0), \end{cases} \tag{153}$$

where

$$\begin{aligned} Z &= \tfrac{1}{2}\begin{bmatrix} -y_{\mathrm{N}} & y_{\mathrm{T}}^\top \\ y_{\mathrm{T}} & -y_{\mathrm{N}} y_{\mathrm{T}} y_{\mathrm{T}}^\top \end{bmatrix}, \\ S &= \left\{ \tfrac{1}{2}(1+\beta) I + \tfrac{1}{2}\begin{bmatrix} -\beta & w^\top \\ w & -\beta w w^\top \end{bmatrix} \,\middle|\, -1 \leqslant \beta \leqslant 1, \|w\| = 1 \right\}, \end{aligned} \tag{154}$$

with $y = x/\|x_{\mathrm{T}}\|$. A simple verification shows that the previous computation is an element of the subdifferential.

Appendix 2. Computation of Generalized Jacobians for Nonsmooth Newton Methods

Computation of Components of a Subgradient of $F_{\text{vi}}^{\text{nat}}$

Let us introduce the following notation for an element of the subdifferential:

$$\Phi(u,r) = \begin{bmatrix} \rho I & -\rho W \\ \Phi_{ru}(u,r) & \Phi_{rr}(u,r) \end{bmatrix} \in \partial F_{\text{vi}}^{\text{nat}}(u,r), \tag{155}$$

where $\Phi_{xy}(u,r) \in \partial_x [F_{\text{vi}}^{\text{nat}}]_y(u,r)$. Since $\Phi_{uu}(u,r) = I$, a reduction of the system is performed in practice and Algorithm 4 is applied or $z = r$ with

$$\begin{cases} G(z) = [F_{\text{vi}}^{\text{nat}}]_r(Wr+q,r) \\ \Phi(z) = \Phi_{rr}(r, Wr+q) + \Phi_{ru}(r, Wr+q)W. \end{cases} \tag{156}$$

Let us introduce the following notation for an element of the sub–differential with an obvious simplification:

$$\Phi(v,r) = \begin{bmatrix} \rho M & -\rho H & \\ -\rho H^\top & \rho I & 0 \\ 0 & \Phi_{ru}(v,u,r) & \Phi_{rr}(v,u,r) \end{bmatrix} \in \partial F_{\text{vi}}^{\text{nat}}(u,r), \tag{157}$$

where $\Phi_{xy}(v,u,r) \in \partial_x [F_{\text{vi-1}}^{\text{nat}}]_y(v,u,r)$. A possible computation of $\Phi_{ru}(v,u,r)$ and $\Phi_{rr}(v,u,r)$ is directly given by (159) and (158). In this case, the variable u can also be substituted.

For one contact, a possible computation of the remaining parts in $\Phi(u,r)$ is given by

$$\Phi_{ru}(u,r) = \begin{cases} 0 & \text{if } r - \rho(u+g(u)) \in K \\ I - \partial_r[P_K(r-\rho(u+g(u)))] & \text{if } r - \rho(u+g(u)) \notin K \end{cases} \tag{158}$$

$$\Phi_{ru}(u,r) = \begin{cases} \rho\left(I + \begin{bmatrix} 0 & 0\,0 \\ \frac{u_{\text{T}}}{\|u_{\text{T}}\|} & 0\,0 \end{bmatrix}\right) & \text{if } \begin{cases} r - \rho(u+g(u)) \in K \\ u_{\text{T}} \neq 0 \end{cases} \\ \rho\left(I + \begin{bmatrix} 0\,0\,0 \\ s\,0\,0 \end{bmatrix}\right), s \in \mathbb{R}^2, \|s\| = 1 & \text{if } \begin{cases} r - \rho(u+g(u)) \in K \\ u_{\text{T}} = 0 \end{cases} \\ I + \rho\left(I + \begin{bmatrix} 0 & 0\,0 \\ \frac{u_{\text{T}}}{\|u_{\text{T}}\|} & 0\,0 \end{bmatrix}\right) \partial_u[P_K(r-\rho(u+g(u)))] & \text{if } r - \rho(u+g(u)) \notin K. \end{cases} \tag{159}$$

The computation of an element of ∂P_K is given in Appendix 11.

Alart–Curnier Function and Its Variants

For one contact, a possible computation of the remaining parts in $\Phi(u, r)$ is given by

$$\Phi_{r_{\mathrm{N}}u_{\mathrm{N}}}(u, r) = \begin{cases} \rho_{\mathrm{N}} & \text{if } r_{\mathrm{N}} - \rho_{\mathrm{N}}u_{\mathrm{N}} > 0 \\ 0 & \text{otherwise} \end{cases} \tag{160}$$

$$\Phi_{r_{\mathrm{N}}r_{\mathrm{N}}}(u, r) = \begin{cases} 0 & \text{if } r_{\mathrm{N}} - \rho_{\mathrm{N}}u_{\mathrm{N}} > 0 \\ 1 & \text{otherwise} \end{cases} \tag{161}$$

$$\Phi_{r_{\mathrm{T}}u_{\mathrm{N}}}(u, r) = \begin{cases} 0 & \text{if } \|r_{\mathrm{T}} - \rho_{\mathrm{T}}u_{\mathrm{T}}\| \leqslant \mu \max(0, r_{\mathrm{N}} - \rho_{\mathrm{N}}u_{\mathrm{N}}) \\ 0 & \text{if } \begin{cases} \|r_{\mathrm{T}} - \rho_{\mathrm{T}}u_{\mathrm{T}}\| > \mu \max(0, r_{\mathrm{N}} - \rho_{\mathrm{N}}u_{\mathrm{N}}) \\ r_{\mathrm{N}} - \rho_{\mathrm{N}}u_n \leqslant 0 \end{cases} \\ \mu\rho_{\mathrm{N}} \dfrac{r_{\mathrm{T}} - \rho_{\mathrm{T}}u_{\mathrm{T}}}{\|r_{\mathrm{T}} - \rho_{\mathrm{T}}u_{\mathrm{T}}\|} & \text{if } \begin{cases} \|r_{\mathrm{T}} - \rho_{\mathrm{T}}u_{\mathrm{T}}\| > \mu \max(0, r_{\mathrm{N}} - \rho_{\mathrm{N}}u_{\mathrm{N}}) \\ r_{\mathrm{N}} - \rho_{\mathrm{N}}u_n > 0 \end{cases} \end{cases} \tag{162}$$

$$\Phi_{r_{\mathrm{T}}u_{\mathrm{T}}}(u, r) = \begin{cases} \rho_{\mathrm{T}} & \text{if } \|r_{\mathrm{T}} - \rho_{\mathrm{T}}u_{\mathrm{T}}\| \leqslant \mu \max(0, r_{\mathrm{N}} - \rho_{\mathrm{N}}u_{\mathrm{N}}) \\ \mu\rho_{\mathrm{T}}(r_{\mathrm{N}} - \rho_{\mathrm{N}}u_{\mathrm{N}})_+ \Gamma(r_{\mathrm{T}} - \rho_{\mathrm{T}}u_{\mathrm{T}}) & \text{if } \begin{cases} \|r_{\mathrm{T}} - \rho_{\mathrm{T}}u_{\mathrm{T}}\| > \mu \max(0, r_{\mathrm{N}} - \rho_{\mathrm{N}}u_{\mathrm{N}}) \\ r_{\mathrm{N}} - \rho_{\mathrm{N}}u_n > 0 \end{cases} \end{cases} \tag{163}$$

$$\Phi_{r_{\mathrm{T}}r_{\mathrm{N}}}(u, r) = \begin{cases} 0 & \text{if } \|r_{\mathrm{T}} - \rho_{\mathrm{T}}u_{\mathrm{T}}\| \leqslant \mu \max(0, r_{\mathrm{N}} - \rho_{\mathrm{N}}u_{\mathrm{N}}) \\ 0 & \text{if } \begin{cases} \|r_{\mathrm{T}} - \rho_{\mathrm{T}}u_{\mathrm{T}}\| > \mu \max(0, r_{\mathrm{N}} - \rho_{\mathrm{N}}u_{\mathrm{N}}) \\ r_{\mathrm{N}} - \rho_{\mathrm{N}}u_n \leqslant 0 \end{cases} \\ -\mu \dfrac{r_{\mathrm{T}} - \rho_{\mathrm{T}}u_{\mathrm{T}}}{\|r_{\mathrm{T}} - \rho_{\mathrm{T}}u_{\mathrm{T}}\|} & \text{if } \begin{cases} \|r_{\mathrm{T}} - \rho_{\mathrm{T}}u_{\mathrm{T}}\| > \mu \max(0, r_{\mathrm{N}} - \rho_{\mathrm{N}}u_{\mathrm{N}}) \\ r_{\mathrm{N}} - \rho_{\mathrm{N}}u_n > 0 \end{cases} \end{cases} \tag{164}$$

$$\Phi_{r_{\mathrm{T}}r_{\mathrm{T}}}(u, r) = \begin{cases} 0 & \text{if } \|r_{\mathrm{T}} - \rho_{\mathrm{T}}u_{\mathrm{T}}\| \leqslant \mu \max(0, r_{\mathrm{N}} - \rho_{\mathrm{N}}u_{\mathrm{N}}) \\ I_2 - \mu(r_{\mathrm{N}} - \rho_{\mathrm{N}}u_{\mathrm{N}})_+ \Gamma(r_{\mathrm{T}} - \rho_{\mathrm{T}}u_{\mathrm{T}}) & \text{if } \begin{cases} \|r_{\mathrm{T}} - \rho_{\mathrm{T}}u_{\mathrm{T}}\| > \mu \max(0, r_{\mathrm{N}} - \rho_{\mathrm{N}}u_{\mathrm{N}}) \\ r_{\mathrm{N}} - \rho_{\mathrm{N}}u_n > 0, \end{cases} \end{cases} \tag{165}$$

with the function $\Gamma(\cdot)$ defined by

$$\Gamma(x) = \frac{I_{2\times 2}}{\|x\|} - \frac{x\,x^\top}{\|x\|^3}. \tag{166}$$

If the variant (60) is chosen, the computation of $\Phi_{r_{\mathrm{T}}\bullet}$ simplifies to

$$\Phi_{r_{\mathrm{T}}u_{\mathrm{N}}}(u, r) = 0 \tag{167}$$

$$\Phi_{r_{\mathrm{T}}u_{\mathrm{T}}}(u, r) = \begin{cases} \rho_{\mathrm{T}} & \text{if } \|r_{\mathrm{T}} - \rho_{\mathrm{T}}u_{\mathrm{T}}\| \leqslant \mu r_{\mathrm{N}} \\ -\mu\rho_{\mathrm{T}}r_{n,+}\Gamma(r_{\mathrm{T}} - \rho_{\mathrm{T}}u_{\mathrm{T}}) & \text{if } \|r_{\mathrm{T}} - \rho_{\mathrm{T}}u_{\mathrm{T}}\| > \mu r_{\mathrm{N}} \end{cases} \tag{168}$$

$$\Phi_{r_{\mathrm{T}}r_{\mathrm{N}}}(u, r) = \begin{cases} 0 & \text{if } \|r_{\mathrm{T}} - \rho_{\mathrm{T}}u_{\mathrm{T}}\| \leqslant \mu r_{\mathrm{N}} \\ 0 & \text{if } \begin{cases} \|r_{\mathrm{T}} - \rho_{\mathrm{T}}u_{\mathrm{T}}\| > \mu r_n \\ r_{\mathrm{N}} \leqslant 0 \end{cases} \\ -\mu\dfrac{r_{\mathrm{T}} - \rho_{\mathrm{T}}u_{\mathrm{T}}}{\|r_{\mathrm{T}} - \rho_{\mathrm{T}}u_{\mathrm{T}}\|} & \text{if } \begin{cases} \|r_{\mathrm{T}} - \rho_{\mathrm{T}}u_{\mathrm{T}}\| > \mu r_n \\ r_{\mathrm{N}} > 0 \end{cases} \end{cases} \tag{169}$$

$$\Phi_{r_{\mathrm{T}}r_{\mathrm{T}}}(u, r) = \begin{cases} 0 & \text{if } \|r_{\mathrm{T}} - \rho_{\mathrm{T}}u_{\mathrm{T}}\| \leqslant \mu r_{\mathrm{N}} \\ I_2 - \mu(r_{\mathrm{N}})_+\Gamma(r_{\mathrm{T}} - \rho_{\mathrm{T}}u_{\mathrm{T}}) & \text{if } \|r_{\mathrm{T}} - \rho_{\mathrm{T}}u_{\mathrm{T}}\| > \mu r_{\mathrm{N}}. \end{cases} \tag{170}$$

References

1. Acary V, Brogliato B (2008) Numerical methods for nonsmooth dynamical systems. Applications in mechanics and electronics. Lecture notes in applied and computational mechanics, vol 35. Springer, Berlin, p xxi, 525 pp
2. Acary V, Cadoux F (2013) Applications of an existence result for the Coulomb friction problem. In: Stavroulakis GE (ed) Recent advances in contact mechanics. Lecture notes in applied and computational mechanics, vol 56. Springer, Berlin
3. Acary V, Cadoux F, Lemaréchal C, Malick J (2011) A formulation of the linear discrete coulomb friction problem via convex optimization. ZAMM - J Appl Math Mech/Zeitschrift für Angewandte Mathematik und Mechanik 91(2):155–175. https://doi.org/10.1002/zamm.201000073
4. Acary V, Brémond M, Koziara T, Pérignon F (2014) FCLIB: a collection of discrete 3D frictional contact problems. Technical Report RT-0444, INRIA. https://hal.inria.fr/hal-00945820
5. Acary V, Brémond M, Huber O, Pérignon F (2015) An introduction to Siconos. Technical Report TR-0340, second version, INRIA. http://hal.inria.fr/inria-00162911/en/
6. Acary V, Brémond M, Dubois F (2017) Méthodes de Newton non–lisses pour les problèmes de contact frottant dans les systèmes de multi–corps flexibles. In: CSMA 2017 - 13ème Colloque National en Calcul des Structures, Presqu'île de Giens (Var), France, p 8. https://hal.inria.fr/hal-01562706
7. Al-Fahed A, Stavroulakis G, Panagiotopulos P (1991) Hard and soft fingered robot grippers. The linear complementarity approach. Zeitschrift für Angewandte Mathematik und Mechanik 71:257–265
8. Alart P (1993) Injectivity and surjectivity criteria for certain mappings of $\mathbb{R}^n$ into itself; application to contact mechanics. (Critères d'injectivité et de surjectivité pour certaines applications de $\mathbb{R}^n$ dans lui-même; application à la mécanique du contact.). RAIRO, Modélisation Math Anal Numér 27(2):203–222
9. Alart P (1995) Méthode de newton généralisée en mécanique du contact. Journal de Mathématiques Pures et Appliquées

10. Alart P, Curnier A (1991) A mixed formulation for frictional contact problems prone to Newton like solution method. Comput Methods Appl Mech Eng 92(3):353–375
11. Anitescu M, Potra F (1997) Formulating dynamic multi-rigid-body contact problems with friction as solvable linear complementarity problems. Nonlinear Dyn Trans ASME 14:231–247
12. Anitescu M, Tasora A (2010) An iterative approach for cone complementarity problems for nonsmooth dynamics. Comput Optim Appl 47(2):207–235. https://doi.org/10.1007/s10589-008-9223-4
13. Barbosa H, Feijóo R (1985) A numerical algorithm for signorini problem with Coulomb friction. In: [26]
14. Barbosa H, Raupp F, Borges C (1997) Numerical experiments with algorithms for bound constrained quadratic programming in mechanics. Comput Struct 64(1–4):579–594
15. Bonnans J, Gilbert J, Lemaréchal C, Sagastizábal C (2003) Numerical optimization: theoretical and practical aspects. Springer, Berlin
16. Bonnefon O, Daviet G (2011) Quartic formulation of Coulomb 3D frictional contact. Technical Report RT-0400, INRIA. https://hal.inria.fr/inria-00553859
17. Breitkopf P, Jean M (1999) Modélisation parallèèle des matéériaux granulaires. In: Actes du 4ème Colloque National en Calcul des Structures, Giens(Var), pp 387–392
18. Cadoux F (2009) Analyse convexe et optimisation pour la dynamique non–regulière. PhD thesis, Université Joseph Fourier, Grenoble I
19. Calamai P, More J (1987) Projected gradient methods for linearly constrained problems. Math Program 39(1):93–116. https://doi.org/10.1007/BF02592073
20. Chabrand P, Dubois F, Raous M (1998) Various numerical methods for solving unilateral contact problems with friction. Math Comput Model 28(4–8):97–108. https://doi.org/10.1016/S0895-7177(98)00111-3, http://www.sciencedirect.com/science/article/pii/S0895717798001113, Recent advances in contact mechanics
21. Chaudhary A, Bathe K (1986) A solution method for static and dynamic analysis of three-dimensional contact problems with friction. Comput Struct 24(6):855–873
22. Christensen P, Pang J (1998) Frictional contact algorithms based on semismooth newton methods. In: Qi MFL (ed) Reformulation - nonsmooth, piecewise smooth, semismooth and smoothing methods. Kluwer Academic Publishers, Dordrecht, pp 81–116
23. Christensen P, Klarbring A, Pang J, Stromberg N (1998) Formulation and comparison of algorithms for frictional contact problems. Int J Numer Meth Eng 42:145–172
24. Curnier A, Alart P (1988) A generalized Newton method for contact problems with friction. Journal de Mécanique Théorique et Appliquée supplément no 1 to 7:67–82
25. Daviet G, Bertails-Descoubes F, Boissieux L (2011) A hybrid iterative solver for robustly capturing coulomb friction in hair dynamics. ACM Trans Graph 30(6):139:1–139:12. https://doi.org/10.1145/2070781.2024173, https://hal.inria.fr/hal-00647497
26. De Saxcé G (1992) Une généralisation de l'inégalité de Fenchel et ses applications aux lois constitutives. Comptes Rendus de l'Académie des Sciences t 314,série II:125–129
27. De Saxcé G, Feng ZQ (1991) New inequality and functional for contact with friction: the implicit standard material approach. Mech Struct Mach 19(3):301–325
28. De Saxcé G, Feng ZQ (1998) The bipotential method: a constructive approach to design the complete contact law with friction and improved numerical algorithms. Math Comput Model 28(4):225–245
29. Del Piero G, Maceri F (eds) (1983) Unilateral problems in structural analysis. CISM courses and lectures, vol 288. Springer, Ravello
30. Del Piero G, Maceri F (eds) (1985) Unilateral problems in structural analysis – II. CISM courses and lectures, vol 304. Springer, Prescudin
31. Dolan E, Moré J (2002) Benchmarking optimization software with performance profiles. Math Program 91(2):201–213
32. Dostál Z (1997) Box constrained quadratic programming with proportioning and projections. SIAM J Optim 7(3):871–887. https://doi.org/10.1137/S1052623494266250

33. Dostál Z (2016) Scalable algorithms for contact problems. Springer, New York. https://doi.org/10.1007/978-1-4939-6834-3_8
34. Dostál Z, Kozubek T (2012) An optimal algorithm and superrelaxation for minimization of a quadratic function subject to separable convex constraints with applications. Math Program 135(1-2, Ser. A):195–220. https://doi.org/10.1007/s10107-011-0454-2
35. Dostál Z, Kučera R (2010) An optimal algorithm for minimization of quadratic functions with bounded spectrum subject to separable convex inequality and linear equality constraints. SIAM J Optim 20(6):2913–2938. https://doi.org/10.1137/090751414
36. Dostál Z, Schöberl J (2005) Minimizing quadratic functions subject to bound constraints with the rate of convergence and finite termination. Comput Optim Appl 30(1):23–43. https://doi.org/10.1007/s10589-005-4557-7
37. Dostál Z, Haslinger J, Kučera R (2002) Implementation of the fixed point method in contact problems with Coulomb friction based on a dual splitting type technique. J Comput Appl Math 140(1–2):245–256. https://doi.org/10.1016/S0377-0427(01)00405-8
38. Dostál Z, Kozubek T, Horyl P, Brzobohatý T, Markopoulos A (2010) A scalable tfeti algorithm for two-dimensional multibody contact problems with friction. J Computat Appl Math 235(2):403 – 418. https://doi.org/10.1016/j.cam.2010.05.042, http://www.sciencedirect.com/science/article/pii/S0377042710003328, special Issue on Advanced Computational Algorithms
39. Facchinei F, Pang JS (2003) Finite-dimensional variational inequalities and complementarity problems. Springer series in operations research, vol I, II. Springer, Berlin
40. Fletcher R (1987) Practical methods of optimization. Wiley, Chichester
41. Fong DCL, Saunders M (2011) Lsmr: an iterative algorithm for sparse least-squares problems. SIAM J Sci Comput 33(5):2950–2971. https://doi.org/10.1137/10079687X
42. Fortin M, Glowinski R (1983) Augmented Lagrangian methods. Studies in mathematics and its applications, vol 15. North-Holland Publishing Co., Amsterdam, applications to the numerical solution of boundary value problems, Translated from the French by Hunt B, Spicer DC
43. Fukushima M, Luo Z, Tseng P (2001) Smoothing functions for second-order-cone complementarity problems. SIAM J Optim 12(2):436–460. https://doi.org/10.1137/S1052623400380365
44. Glowinski R, JL L, Trémoliéres R, (1976) Approximations des Inéquations Variationnelles. Dunod, Paris
45. Hager WW, Zhang H (2008) Self-adaptive inexact proximal point methods. Comput Optim Appl 39(2):161–181. https://doi.org/10.1007/s10589-007-9067-3
46. Han D, Lo HK (2002) Two new self-adaptive projection methods for variational inequality problems. Comput Math Appl 43(12):1529–1537. https://doi.org/10.1016/S0898-1221(02)00116-5, http://www.sciencedirect.com/science/article/pii/S0898122102001165
47. Harker P, Pang JS (1990) Finite-dimensional variational inequality and complementarity problems: a survey of theory, algorithms and applications. Math Program 48:160–220
48. Haslinger J (1983) Approximation of the signorini problem with friction, obeying the coulomb law. Math Methods Appl Sci 5:422–437
49. Haslinger J (1984) Least square method for solving contact problems with friction obeying coulomb's law. Appl Math 29(3):212–224. http://dml.cz/dmlcz/104086
50. Haslinger J, Panagiotopoulos PD (1984) The reciprocal variational approach to the signorini problem with friction. approximation results. Proc R Soc Edinb Section A Math 98:365–383. http://journals.cambridge.org/article_S0308210500013536
51. Haslinger J, Hlaváček I, Nečas J (1996) Numerical methods for unilateral problems in solid mechanics. In: Ciarlet P, Lions J (eds) Handbook of numerical analysis, vol IV, Part 2. North-Holland, Amsterdam, pp 313–485
52. Haslinger J, Dostál Z, Kučera R (2002) On a splitting type algorithm for the numerical realization of contact problems with Coulomb friction. Comput Methods Appl Mech Engrg 191(21-22):2261–2281. https://doi.org/10.1016/S0045-7825(01)00378-4

53. Haslinger J, Kučera R, D Z (2004) An algorithm for the numerical realization of 3D contact problems with Coulomb friction. In: Proceedings of the 10th international congress on computational and applied mathematics (ICCAM-2002), vol 164/165, pp 387–408. https://doi.org/10.1016/j.cam.2003.06.002
54. Haslinger J, Kučera R, Vlach O, Baniotopoulos C (2012) Approximation and numerical realization of 3d quasistatic contact problems with coulomb friction. Math Comput Simul 82(10):1936 – 1951. https://doi.org/10.1016/j.matcom.2011.01.004, http://www.sciencedirect.com/science/article/pii/S0378475411000310, the fourth IMACS conference: Mathematical modelling and computational methods in applied sciences and engineering" Devoted to Owe Axelsson in ocassion of his 75th birthday
55. Hayashi S, Yamashita N, Fukushima M (2005) A combined smoothing and regularization method for monotone second-order cone complementarity problems. SIAM J Optim 15(2):593–615. https://doi.org/10.1137/S1052623403421516
56. He B, Liao L (2002) Improvements of some projection methods for monotone nonlinear variational inequalities. J Optim Theory Appl 112(1):111–128. https://doi.org/10.1023/A:1013096613105
57. Hestenes M (1969) Multiplier and gradient methods. J Optim Theory Appl 4:303–320
58. Heyn T (2013) On the modeling, simulation, and visualization of many–body dynamics problems with friction and contact. PhD thesis, University of Wisconsin–Madison
59. Heyn T, Anitescu M, Tasora A, Negrut D (2013) Using Krylov subspace and spectral methods for solving complementarity problems in many-body contact dynamics simulation. Int J Numer Methods Engrg 95(7):541–561. https://doi.org/10.1002/nme.4513
60. Hüeber S, Stadler G, Wohlmuth BI (2008) A primal-dual active set algorithm for three-dimensional contact problems with coulomb friction. SIAM J Sci Comput 30(2):572–596. https://doi.org/10.1137/060671061
61. Jean M, Moreau J (1987) Dynamics in the presence of unilateral contacts and dry friction: a numerical approach. In: Del Pietro G, Maceri F (eds) Unilateral problems in structural analysis. II, CISM 304. Spinger, pp 151–196
62. Joli P, Feng ZQ (2008) Uzawa and newton algorithms to solve frictional contact problems within the bi-potential framework. Int J Numer Methods Eng 73(3):317–330. https://doi.org/10.1002/nme.2073
63. Jourdan F, Alart P, Jean M (1998) A Gauss Seidel like algorithm to solve frictional contact problems. Comput Methods Appl Mech Eng 155(1):31–47
64. Katona MG (1983) A simple contact–friction interface element with applications to buried culverts. Int J Numer Anal Methods Geomech 7(3):371–384. https://doi.org/10.1002/nag.1610070308
65. Khobotov E (1987) Modification of the extra-gradient method for solving variational inequalities and certain optimization problems. USSR Comput Math Math Phys 27(5):120–127. https://doi.org/10.1016/0041-5553(87)90058-9, http://www.sciencedirect.com/science/article/pii/0041555387900589
66. Kikuchi N, Oden JT (1988) Contact problems in elasticity: a study of variational inequalities and finite element methods. SIAM studies in applied mathematics, vol 8. Society for Industrial and Applied Mathematics (SIAM), Philadelphia. https://doi.org/10.1137/1.9781611970845
67. Klarbring A (1986) A mathematical programming approach to three-dimensional contact problem with friction. Compt Methods Appl Math Engrg 58:175–200
68. Klarbring A, Pang JS (1998) Existence of solutions to discrete semicoercive frictional contact problems. SIAM J Optim 8(2):414–442
69. Kleinert J, Simeon B, Obermayr M (2014) An inexact interior point method for the large-scale simulation of granular material. Comput Methods Appl Mech Eng 278(0):567–598. https://doi.org/10.1016/j.cma.2014.06.009, http://www.sciencedirect.com/science/article/pii/S0045782514001959
70. Korpelevich G (1976) The extragradient method for finding saddle points and other problems. Matecon 12(747–756)

71. Koziara T, Bićanić N (2008) Semismooth newton method for frictional contact between pseudo-rigid bodies. Comput Methods Appl Mech Eng 197(33–40):2763–2777. https://doi.org/10.1016/j.cma.2008.01.006, http://www.sciencedirect.com/science/article/pii/S0045782508000194
72. Koziara T, Bićanić N (2011) A distributed memory parallel multibody contact dynamics code. Int J Numer Methods Eng 87(1–5):437–456. https://doi.org/10.1002/nme.3158
73. Krabbenhoft K, Lyamin A, Huang J, da Silva MV (2012) Granular contact dynamics using mathematical programming methods. Comput Geotech 43:165–176. https://doi.org/10.1016/j.compgeo.2012.02.006, http://www.sciencedirect.com/science/article/pii/S0266352X12000262
74. Kučera R (2007) Minimizing quadratic functions with separable quadratic constraints. Optim Methods Softw 22(3):453–467. https://doi.org/10.1080/10556780600609246
75. Kučera R (2008) Convergence rate of an optimization algorithm for minimizing quadratic functions with separable convex constraints. SIAM J Optim 19(2):846–862. https://doi.org/10.1137/060670456
76. Laursen T (2003) Computational contact and impact mechanics – fundamentals of modeling interfacial phenomena in nonlinear finite element analysis. Springer, Berlin, 1st ed 2002. Corr 2nd printing
77. Leung A, Guoqing C, Wanji C (1998) Smoothing Newton method for solving two- and thre-dimensional frictional contact problems. Int J Numer Meth Eng 41:1001–1027
78. Mijar A, Arora J (2000) Review of formulations for elastostatic frictional contact problems. Struct Multidiscip Optim 20(3):167–189. https://doi.org/10.1007/s001580050147
79. Mijar A, Arora J (2000) Study of variational inequality and equality formulations for elastostatic frictional contact problems. Arch Comput Methods Eng 7(4):387–449. https://doi.org/10.1007/BF02736213
80. Mijar A, Arora J (2004) An augmented Lagrangian optimization method for contact analysis problems, 1: formulation and algorithm. Struct Multidiscip Optim 28(2-3):99–112. https://doi.org/10.1007/s00158-004-0423-y
81. Mijar A, Arora J (2004) An augmented Lagrangian optimization method for contact analysis problems, 2: numerical evaluation. Struct Multidiscip Optim 28(2-3):113–126. https://doi.org/10.1007/s00158-004-0424-x
82. Mitsopoulou E, Doudoumis I (1987) A contribution to the analysis of unilateral contact problems with friction. Solid Mech Arch 12(3):165–186
83. Mitsopoulou E, Doudoumis I (1988) On the solution of the unilateral contact frictional problem for general static loading conditions. Comput Struct 30(5):1111–1126
84. Miyamura T, Kanno Y, Ohsaki M (2010) Combined interior-point method and semismooth newton method for frictionless contact problems. Int J Numer Methods Eng 81(6):701–727. https://doi.org/10.1002/nme.2707
85. Morales JL, Nocedal J, Smelyanskiy M (2008) An algorithm for the fast solution of symmetric linear complementarity problems. Numerische Mathematik 111(2):251–266. https://doi.org/10.1007/s00211-008-0183-5
86. Moré J, Toraldo G (1991) On the solution of large quadratic convex programming problems with bound constraints. SIAM J Optim 1(1):93–113
87. Moré JJ, Toraldo G (1989) Algorithms for bound constrained quadratic programming problems. Numerische Mathematik 55(4):377–400. https://doi.org/10.1007/BF01396045
88. Moreau J (1965) Proximité et dualité dans un espace hilbertien. Bulletin de la société mathématique de France 93:273–299
89. Moreau J (1988) Unilateral contact and dry friction in finite freedom dynamics. In: Moreau J, PD P (eds) Nonsmooth mechanics and applications, no. 302 in CISM, Courses and lectures, CISM 302, Spinger, Wien- New York, pp 1–82, formulation mathematiques tire du livre Contacts mechanics
90. Mylapilli H, Jain A (2017) Complementarity techniques for minimal coordinates contact dynamics. ASME J Comput Nonlinear Dyn 12(2)

91. Nečas J, Jarušek J, Haslinger J (1980) On the solution of the variational inequality to the Signorini problem with small friction. Bollettino UMI 5(17-B):796–811
92. Nocedal J, Wright S (1999) Numerical optimization. Springer, Berlin
93. Panagiotopoulos P (1975) A nonlinear programming approach to the unilateral contact-, and friction-boundary value problem in the theory of elasticity. Ingenieur-Archiv 44(6):421–432. https://doi.org/10.1007/BF00534623
94. Pang J, Trinkle J (1996) Complementarity formulations and existence of solutions of dynamic multi-rigid-body contact problems with Coulomb friction. Math Program 73:199–226
95. Parikh N, Boyd S et al (2014) Proximal algorithms. Found Trends® Optim 1(3):127–239
96. Park J, Kwak B (1994) Three dimensional frictional contact analysis using the homotopy method. J Appl Mech Trans ASME 61:703–709
97. Qi L, Sun J (1993) A nonsmooth version of Newton's method. Math Program 58:353–367
98. Qi L, Sun D, Ulbrich M (eds) (2018) Semismooth and smoothing Newton methods. Springer, Berlin
99. Raous M, Chabrand P, Lebon F (1988) Numerical methods for frictional contact problems and applications. J Méc Théor Appl 7(1):111–18
100. Renouf M, Dubois F, Alart P (2004) A parallel version of the non smooth contact dynamics algorithm applied to the simulation of granular media. J Comput Appl Math 168:375–382
101. Rockafellar R (1974) Augmented Lagrange multiplier functions and duality in nonconvex programming. SIAM J Control 12:268–285
102. Rockafellar R (1993) Lagrange multipliers and optimality. SIAM Rev 35(2):183–238
103. Rockafellar R, Wets R (1997) Variational analysis, vol 317. Springer, New York
104. Rockafellar RT (1976) Monotone operators and the proximal point algorithm. SIAM J Control Optim 14(5):877–898
105. Saxcé GD, Feng ZQ (1998) The bipotential method: a constructive approach to design the complete contact law with friction and improved numerical algorithms. Math Comput Model 28(4–8):225–245. https://doi.org/10.1016/S0895-7177(98)00119-8, http://www.sciencedirect.com/science/article/pii/S0895717798001198, Recent advances in contact mechanics
106. Sibony M (1970) Méthodes itératives pour les équations et inéquations aux dérivées partielles non linéaires de type monotone. Calcolo 7:65–183
107. Simo J, Laursen T (1992) An augmented Lagrangian treatment of contact problems involving friction. Comput Struct 42(1):97–116
108. Solodov M, Tseng P (1996) Modified projection-type methods for monotone variational inequalities. SIAM J Control Optim 34(5):1814–1830. http://citeseer.ist.psu.edu/article/solodov95modified.html
109. Stadler G (2004) Semismooth Newton and augmented Lagrangian methods for a simplified friction problem. SIAM J Optim 15(1):39–62. https://doi.org/10.1137/S1052623403420833
110. Stewart D, Trinkle J (1996) An implicit time-stepping scheme for rigid body dynamics with inelastic collisions and Coulomb friction. Int J Numer Methods Eng 39(15), reference tiree du site WILEY
111. Sun D, Sun J (2005) Strong semismoothness of the fischer-burmeister sdc and soc complementarity functions. Math Program. 103(3):575–581. https://doi.org/10.1007/s10107-005-0577-4
112. Tasora A, Anitescu M (2009) A fast NCP solver for large rigid-body problems with contacts, friction, and joints. In: Multibody dynamics, computational methods and applications science, vol 12. Springer, Berlin, pp 45–55
113. Tasora A, Anitescu M (2011) A matrix-free cone complementarity approach for solving large-scale, nonsmooth, rigid body dynamics. Comput Methods Appl Mech Engrg 200(5-8):439–453. https://doi.org/10.1016/j.cma.2010.06.030
114. Tasora A, Anitescu M (2013) A complementarity-based rolling friction model for rigid contacts. Meccanica 48(7):1643–1659. https://doi.org/10.1007/s11012-013-9694-y
115. Temizer I, Abdalla M, Gürdal Z (2014) An interior point method for isogeometric contact. Comput Methods Appl Mech Eng 276(0):589–611. https://doi.org/10.1016/j.cma.2014.03.018, http://www.sciencedirect.com/science/article/pii/S0045782514001042

116. Tzaferopoulos M (1993) On an efficient new numerical method for the frictional contact problem of structures with convex energy density. Comput Struct 48(1):87–106
117. Wang X, He B, Liao LZ (2010) Steplengths in the extragradient type methods. J Comput Appl Math 233(11):2925–2939. https://doi.org/10.1016/j.cam.2009.11.037, http://www.sciencedirect.com/science/article/pii/S0377042709007845
118. Wohlmuth BI, Krause RH (2003) Monotone multigrid methods on nonmatching grids for nonlinear multibody contact problems. SIAM J Sci Comput 25(1):324–347. https://doi.org/10.1137/S1064827502405318
119. Wriggers P (2006) Computational contact mechanics, 2nd edn. Springer, Berlin. Originally published by Wiley, 2002
120. Xuewen L, Soh AK, Wanji C (2000) A new nonsmooth model for three-dimensional frictional contact problems. Comput Mech 26:528–535

Erratum to: Nonsmooth Modal Analysis: From the Discrete to the Continuous Settings

Anders Thorin and Mathias Legrand

Erratum to: Chapter "Nonsmooth Modal Analysis: From the Discrete to the Continuous Settings" in: R. I. Leine et al. (eds.), *Advanced Topics in Nonsmooth Dynamics*, https://doi.org/10.1007/978-3-319-75972-2_5

The original version of the book was inadvertently published with incorrect author name "Acary V" in reference [125] of chapter "Nonsmooth Modal Analysis: From the Discrete to the Continuous Settings", which has to be now changed as "Thorin A". The erratum chapter and the book have been updated with the change.

The updated online version of this chapter can be found at
https://doi.org/10.1007/978-3-319-75972-2_5

R. I. Leine et al. (eds.), *Advanced Topics in Nonsmooth Dynamics*,
https://doi.org/10.1007/978-3-319-75972-2_11

MIX
Papier aus verantwortungsvollen Quellen
Paper from responsible sources
FSC® C105338

If you have any concerns about our products,
you can contact us on
ProductSafety@springernature.com

In case Publisher is established outside the EU,
the EU authorized representative is:
Springer Nature Customer Service Center GmbH
Europaplatz 3, 69115 Heidelberg, Germany

Printed by Libri Plureos GmbH
in Hamburg, Germany